Grundlehren der mathematischen Wissenschaften 314

A Series of Comprehensive Studies in Mathematics

Editors

M. Artin S. S. Chern J. Coates J. M. Fröhlich
H. Hironaka F. Hirzebruch L. Hörmander
C. C. Moore J. K. Moser M. Nagata W. Schmidt
D. S. Scott Ya. G. Sinai J. Tits M. Waldschmidt
S. Watanabe

Managing Editors

M. Berger B. Eckmann S. R. S. Varadhan

Springer
Berlin
Heidelberg
New York
Barcelona
Hong Kong
London
Milan
Paris
Singapore
Tokyo

David R. Adams   Lars Inge Hedberg

Function Spaces and Potential Theory

Springer

[redacted]

David R. Adams
Department of Mathematics
University of Kentucky
Lexington, KY 40506-0027, USA
e-mail: dave@ms.uky.edu

Lars Inge Hedberg
Department of Mathematics
Linköping University
58183 Linköping, Sweden
e-mail: lahed@math.liu.se

Corrected Second Printing 1999

The Library of Congress has cataloged the original printing as follows:

Adams, David R., 1941– . Function spaces and potential theory /
David R. Adams, Lars Inge Hedberg. p. cm. –
(Grundlehren der mathematischen Wissenschaften; 314)
Including bibliographical references and index.
ISBN 3-540-57060-8 (hardcover: alk. paper)
1. Function spaces. 2. Potential theory (Mathematics).
I. Hedberg, Lars Inge, 1935– . II. Title. III. Series.
QA323.A33 1996 515'.73–dc20 95-23396 CIP

Mathematics Subject Classification (1991):
46-02, 46E35, 31-02, 31C45, 31B15, 31C15, 30E10

ISSN 0072-7830

ISBN 3-540-57060-8 Springer-Verlag Berlin Heidelberg New York

Cover design: MetaDesign plus GmbH, Berlin
Photocomposed from the authors' TeX files after editing and reformatting by Kurt Mattes, Heidelberg, using the MathTime fonts and a Springer TeX macro-package
Printed on acid-free paper SPIN: 10691706 41/3143Ko - 5 4 3 2 1 0

[redacted]

Preface

Function spaces, especially those spaces that have become known as Sobolev spaces, and their natural extensions, are now a central concept in analysis. In particular, they play a decisive role in the modern theory of partial differential equations (PDE).

Potential theory, which grew out of the theory of the electrostatic or gravitational potential, the Laplace equation, the Dirichlet problem, etc., had a fundamental role in the development of functional analysis and the theory of Hilbert space. Later, potential theory was strongly influenced by functional analysis. More recently, ideas from potential theory have enriched the theory of those more general function spaces that appear naturally in the study of nonlinear partial differential equations. This book is motivated by the latter development.

The connection between potential theory and the theory of Hilbert spaces can be traced back to C. F. Gauss [181], who proved (with modern rigor supplied almost a century later by O. Frostman [158]) the existence of equilibrium potentials by minimizing a quadratic integral, the energy. This theme is pervasive in the work of such mathematicians as D. Hilbert, Ch.-J. de La Vallée Poussin, M. Riesz, O. Frostman, A. Beurling, and the connection was made particularly clear in the work of H. Cartan [97] in the 1940's. In the thesis of J. Deny [119], and in the subsequent work of J. Deny and J. L. Lions [122] in the early 1950's, this point of view, blended with the L. Schwartz theory of distributions, led to a new understanding of the function spaces of Hilbert type.

According to the classical Dirichlet principle, one obtains the solution of Dirichlet's problem for the Laplace equation in a domain Ω by minimizing the Dirichlet integral, $\int_\Omega |\nabla u(x)|^2\,dx$, over a certain class of functions taking given values on the boundary $\partial\Omega$. The natural explanation is that solutions to the Laplace equation describe an equilibrium state, a state attained when the energy carried by the system is at a minimum.

Also, the classical electrostatic capacity (or capacitance, as it is known to physicists) of a compact set $K \subset \mathbf{R}^3$ differs only by a constant factor from $C(K) = \inf \int_\Omega |\nabla u(x)|^2\,dx$, where the infimum is taken over all smooth compactly supported functions u, such that $u \geq 1$ on K.

As emphasized by H. Cartan, taking these infima of the Dirichlet integral is equivalent to taking projections in a Hilbert space normed by the square root of the Dirichlet integral.

A slight modification of the norm defined by the Dirichlet integral leads to the simplest of the function spaces treated in this work, the Sobolev space denoted by H^1 or $W^{1,2}$. It is normed by the square root of $\|u\|_{H^1}^2 = \int_{\mathbf{R}^N} (|u|^2 + |\nabla u|^2)\, dx$. This Hilbert space is an indispensable tool in the study of linear PDE of the second order. The higher Sobolev spaces, H^m or $W^{m,2}$, whose definition involves derivatives of order up to m, are similarly related to linear PDE of order $2m$.

For the study of nonlinear equations it is advantageous to introduce Sobolev spaces $W^{m,p}$, modeled on L^p. The methods of classical potential theory based on a Hilbert space approach are insufficient to investigate analogous questions in these spaces at the same depth as those modeled on L^2, and this has led to the creation of a new, nonlinear potential theory. This potential theory can be viewed as a natural extension of the classical, linear theory, and when specialized to $p = 2$, it includes a surprisingly large part of the linear theory as a special case.

The present book is devoted to this interplay of potential theory and function spaces, with the purpose of studying the properties of functions belonging to Sobolev spaces, or to some of their natural extensions, such as Bessel potential spaces, Besov spaces, and Lizorkin–Triebel spaces.

Although there are earlier roots, one can date the birth of this theory to the work of V. G. Maz′ya and J. Serrin in the early 1960's. The theory took a new turn in the writings of several authors in the years around 1970: B. Fuglede, N. G. Meyers, Yu. G. Reshetnyak, V. P. Havin, and V. G. Maz′ya. Over the last decades it has continued to develop, and it has found numerous applications. It has greatly clarified the properties of elements of Sobolev spaces and their generalizations, and many problems have been given definitive answers in terms of this theory. Some of these originate with the book of L. Carleson [92], and part of the original motivation for writing this book was to make these results more easily accessible. By now the theory has reached a level of maturity and beauty that makes the time ripe for a book on the subject.

The germ of the book lies in lectures given by the authors — by the Swedish author at Indiana University, Bloomington, Indiana, during the academic year 1978–79, and by the American author at the University of Umeå, Sweden, during the months of March and April 1981. The idea of writing a joint book goes back almost to the same time. The first synopsis was in fact written during the AMS Summer Research Institute at Berkeley in the summer of 1983, although the writing did not start in earnest until some six years later.

The reader we have in mind should have a good graduate course in real analysis, but is not required to know anything about capacities and potentials. We often refer to the well-known books by W. Rudin [367, 368], and E. M. Stein [389] for basic facts needed herein.

We describe very briefly the contents of the book. More information is found in the introduction of each chapter.

Chapter 1 gives some background, and can be consulted as needed. Chapter 2 contains some of the central material of the book, especially in Sections 2.2, 2.3, and 2.5. This includes definitions of (α, p)-capacities and the associated nonlinear potentials. The contents are described in more detail in Section 2.1.

Chapter 3, which is largely independent of the theory of Chapter 2, is mainly devoted to a number of important inequalities. In Chapter 4 we show that the general theory of Chapter 2, which was first developed with Sobolev spaces in mind, can also be applied to spaces of Besov and Lizorkin–Triebel type, and that their representations by "smooth atoms" are well suited for this purpose. Since we do not want to assume any previous knowledge of these spaces, we present their theory from the beginning.

Chapter 5 is mainly devoted to comparisons of (α, p)-capacities with other, better known set functions (Hausdorff measures). The results are sharp, and extend estimates which are well known in the case $p = 2$. In Chapter 6 we apply the theory to a close study of local continuity properties of functions in Sobolev (and more general) spaces, including a non-trivial extension of the classical theory of thin sets. Chapters 7 and 8 are devoted to trace and imbedding theorems, and to inequalities of Poincaré type, respectively.

Chapters 6–8 show that many questions in analysis can be given final answers in terms of (α, p)-capacities, capacities which are shown in Chapter 5 to be very well understood.

Chapters 9 and 10 have a character which differs somewhat from the preceding chapters. They are mainly devoted to the proof of a theorem which describes the kernel of a trace operator on arbitrary sets, but they use different methods. In Chapter 9 the methods are mainly potential theoretical, and depend on the results of Chapters 6 and 8. In Chapter 10 a more general result due to Yu. V. Netrusov is proved using the powerful methods presented in Chapter 4.

In the final Chapter 11, most of which can be read immediately after Chapter 6, we give complete solutions in terms of capacities to some approximation problems for holomorphic and harmonic functions.

Netrusov communicated the results presented in Chapter 10 to the authors in December 1991, which necessitated some changes in the original plans for the book, and he presented detailed proofs during visits to Linköping in 1992 and 1993. The authors are most grateful to him for permitting the inclusion of these results and their proofs here before their full publication elsewhere, and for his most valuable help in preparing the presentation.

In each chapter most references to the literature are collected in "Notes" at the end of the chapter. These notes are sometimes quite detailed, but there is no uniformity in the depth to which the history of different ideas is traced. The discussion reflects the authors' interests and knowledge, and is not based on systematic historical research. Concerning the difficulties involved in writing such historical remarks we refer to J. L. Doob [125], p. 793.

The same apologies apply to the "Further Results" sections, where we give a number of results that complement the main body of the text.

The bibliography, although quite extensive, is limited to works mentioned in the text, and does not pretend to completeness.

There is not much overlap between this book and other books on function spaces. In fact, most of the contents have not previously been presented in book form. Also, there are many parts of the theory that are not treated by us. An important omission is the Hardy spaces H^p for $p \leq 1$, and spaces derived from them, and also the related theory of interpolation spaces. These subjects are treated in many places, e.g. in several books by H. Triebel [404, 405, 406]. A second omission is the space BV of functions of bounded variation and related spaces, treated e.g. in the book by W. P. Ziemer [438]. We also leave aside the situation in spaces defined on domains with irregular boundaries. This is an area with many open problems, and we refer to the book by V. G. Maz′ya [308] for more information on this and many other subjects.

Finally, we largely omit the nonlinear potential theory which is special to quasilinear second order PDE. A natural generalization of the Laplace equation is obtained by minimizing the integral $\int_\Omega |\nabla u(x)|^p \, dx$ over functions taking given boundary values. This leads to the so called p-Laplace equation, $\Delta_p u = 0$, with the p-Laplacian Δ_p defined by $\Delta_p u = \operatorname{div}(\nabla u |\nabla u|^{p-2})$. However, there is no potential representation of solutions to the equation $\Delta_p u = f$ for $p \neq 2$ which corresponds to the Newtonian potential in the case of the Laplacian. Nevertheless, concepts like superharmonic functions, harmonic measure, and the Perron method for solving the Dirichlet problem have been successfully extended to this setting. This theory is (at least for the time being) limited to second order equations and inequalities, and consequently it is concerned with Sobolev spaces (with and without weight) of order one. But it should be noted that, although the methods are quite different, there are many parallels and similarities with the theory presented in this book, especially in such key places as the theory of capacity and the theory of thin sets. We refer the reader to the recent book by J. Heinonen, T. Kilpeläinen, and O. Martio [221] for an exposition.

The authors owe a large dept of gratitude to Yu. V. Netrusov. In addition to his contributions described above, he has read the entire manuscript carefully, and suggested a number of significant improvements of the presentation. The more important ones of these are acknowledged in the text. We are also grateful to V. G. Maz′ya, Linköping, for many enlightening comments.

The fact that both authors have worked in this area for many years does not diminish their feelings of gratitude to those who taught them mathematics, and above all to L. Carleson, and N. G. Meyers. They have influenced this book by their teaching and example, and directly by their work, some of which was mentioned above.

It is a pleasure to acknowledge that this transatlantic cooperation has been supported by a number of travel grants from the Swedish Natural Science Research Council (NFR), and by Linköping University.

The manuscript was prepared on a Macintosh, using $\mathcal{AMS}$-LaTeX (a TeX macro system owned by the American Mathematical Society), and a macro package provided by Springer-Verlag.

Lexington, Kentucky, and Linköping *David R. Adams*
December 1994 *Lars Inge Hedberg*

Table of Contents

1. Preliminaries

We begin with a review of some of the basic results that will be needed in subsequent chapters. Not all of these results are prerequisites for understanding the book, and the reader is advised to refer to them as the need arises. They are generally stated without proof, although we attempt to provide good references where the proofs can be found. Also, we take this opportunity to introduce some of our notational conventions.

1.1 Basics

1.1.1 Convention

We mention already at this point that throughout the book we will use the letter A to denote various unspecified positive constants, whose value can change within a sequence of inequalities. Thus the estimate $|f(x)| \le A$ implies $2|f(x)| \le A$. This convention does not apply to constants denoted by other letters, or by A_1, A_2, etc.

1.1.2 Notation

The set of natural numbers is denoted by **N**, i.e., $\mathbf{N} = \{0, 1, 2, \dots\}$. The integers and the real numbers are denoted by **Z** and **R**, respectively.

$\mathbf{R}^N$ is Euclidean N-dimensional space, $x \in \mathbf{R}^N$ is denoted by $x = (x_1, \dots, x_N)$, the Euclidean scalar product is $x \cdot y = x_1 y_1 + \cdots + x_N y_N$, and the Euclidean norm is $|x| = (x \cdot x)^{1/2}$.

If $E \subset \mathbf{R}^N$, then $\overline{E}$ denotes the closure of E, $E^c = \mathbf{R}^N \setminus E$ is the complement of E in $\mathbf{R}^N$, and $E^0 = \operatorname{int} E$ is the interior of E.

The N-dimensional Lebesgue measure of E is denoted by $|E|$; also m and dx will be used. The characteristic function of E is denoted by χ_E or $\chi(E)$, i.e., $|E| = m(E) = \int_{\mathbf{R}^N} \chi_E \, dm = \int_E dx$.

$B(x_0, r)$ is the open ball centered at x_0 of radius r, in other words $B(x_0, r) = \{x : |x - x_0| < r\}$. We shall often use "dyadic balls", of radius 2^{-n}, $n \in \mathbf{Z}$, and we write $B(x_0, 2^{-n}) = B_n(x_0)$, and for balls centered at the origin, $B_n(0) = B_n$.

The positive and negative parts of a function f are denoted f_+ and f_-, i.e., $f_+(x) = \max\{f(x), 0\}$, and $f_-(x) = \max\{-f(x), 0\}$.

We write $(\partial/\partial x_j)f = \partial_j f$, the gradient of f is $\nabla f = (\partial_1 f, \ldots, \partial_N f)$, and $\Delta f = (\partial_1^2 + \cdots + \partial_N^2)f$ is the Laplacian of f. Higher partial derivatives are also denoted $D^\sigma f = \partial_1^{\sigma_1} \ldots \partial_N^{\sigma_N} f$, where $\sigma \in \mathbf{N}^N$ is a *multiindex* whose *length* (without risk of confusion) is denoted $|\sigma| = \sigma_1 + \cdots + \sigma_N$. Similarly $x^\sigma = x_1^{\sigma_1} \cdots x_N^{\sigma_N}$, and $\sigma! = \sigma_1! \cdots \sigma_N!$. Then the Taylor polynomial at a point x of a function f can be written concisely

$$P_x^m f(x+y) = \sum_{|\sigma| \leq m} \frac{1}{\sigma!} D^\sigma f(x)\, y^\sigma \quad .$$

We also write $\nabla^m f$ for the vector $(D^\sigma f)_{|\sigma|=m}$ of all derivatives of order m, and we denote by $|\nabla^m f|$ any suitable norm, for example the Euclidean norm, of this vector.

1.1.3 Spaces of Functions and Their Duals

For $0 < p < \infty$, and a measurable set $E \subset \mathbf{R}^N$, $L^p(E)$ is the Lebesgue space of functions (equivalence classes of measurable functions on the measurable set E) with norm (quasinorm for $p < 1$)

$$\|f\|_{L^p(E)} = \left(\int_E |f(x)|^p\, dx \right)^{1/p} < \infty \; .$$

For $p = \infty$, $\|f\|_{L^\infty(E)}$ is the usual essential supremum norm. We write L^p for $L^p(\mathbf{R}^N)$ and $\|f\|_p$ for $\|f\|_{L^p}$. The subset (cone) of nonnegative elements in $L^p(E)$ is $L_+^p(E)$. If ν is a positive measure on some set we write $L^p(\nu)$, $L_+^p(\nu)$, etc. The dual space of $L^p(E)$, $1 \leq p < \infty$, is $L^{p'}(E)$, $1/p + 1/p' = 1$, $1 < p' \leq \infty$.

$C(E)$ is the space of (usually real valued) continuous functions on $E \subset \mathbf{R}^N$, equipped with the topology of uniform convergence on compact subsets of E. If K is compact, $C(K)$ is usually normed with the supremum norm, $\|\cdot\|_{L^\infty(K)}$. If $\Omega \subset \mathbf{R}^N$ is an open set or a domain (connected open set), then $C_0(\Omega)$ is the subset of $C(\Omega)$ consisting of functions with compact support contained in Ω.

$C^m(\Omega)$, $m \in \mathbf{N}$, is the space of functions f in $C(\Omega)$ such that all partial derivatives $D^\sigma f$, $|\sigma| \leq m$, also belong to $C(\Omega)$, and $C_0^m(\Omega)$ is the subset of functions with compact support in Ω.

The dual of the space $C_0(\Omega)$ is denoted by $\mathcal{M}(\Omega)$, the *Radon measures* on Ω. (Note that according to a common terminology, as in W. Rudin [367, 368], these are measures only locally, i.e., when restricted to compact subsets of Ω. See, however, L. Schwartz [373], Ch. I, §1, or L. Hörmander [230], Theorem 2.1.6.) The cone of positive elements in $\mathcal{M}(\Omega)$, i.e., the set of positive measures on Ω, is denoted by $\mathcal{M}^+(\Omega)$.

The space of measures supported by a compact set K, i.e., the dual of $C(K)$, is denoted by $\mathcal{M}(K)$, and the cone of positive elements is $\mathcal{M}^+(K)$.

A sequence $\{\mu_n\}_1^\infty$ in $\mathcal{M}(\Omega)$ (or in $\mathcal{M}(K)$) is said to converge in the *weak* topology* to μ if

$$\lim_{n\to\infty} \int_\Omega \varphi\, d\mu_n = \int_\Omega \varphi\, d\mu$$

for all $\varphi \in C_0(\Omega)$ (or $\varphi \in C(K)$).

A function F on (a subset of) $\mathcal{M}(\Omega)$ is *lower semicontinuous* in the weak* topology if $F(\mu) \le \liminf_{n\to\infty} F(\mu_n)$ for all sequences $\{\mu_n\}_1^\infty$ converging weak* to μ in $\mathcal{M}(\Omega)$.

1.1.4 Maximal Functions

A tool that will often be used is the Hardy–Littlewood *maximal function*. The Hardy–Littlewood maximal function of a locally integrable function f is

$$Mf(x) = M_0 f(x) = \sup_{r>0} \frac{1}{|B(x,r)|} \int_{B(x,r)} |f(y)|\, dy \ . \tag{1.1.1}$$

We shall also use the fractional maximal function, defined for $0 < \alpha < N$ by

$$M_\alpha f(x) = \sup_{r>0} \frac{1}{|B(x,r)|^{\frac{N-\alpha}{N}}} \int_{B(x,r)} |f(y)|\, dy \ , \tag{1.1.2}$$

and, for $0 \le \alpha < N$, and $\delta > 0$, the "inhomogeneous" versions of these functions,

$$M_{\alpha,\delta} f(x) = \sup_{0<r\le\delta} \frac{1}{|B(x,r)|^{\frac{N-\alpha}{N}}} \int_{B(x,r)} |f(y)|\, dy \ . \tag{1.1.3}$$

The basic result is due to G. H. Hardy and J. E. Littlewood [195] for $N = 1$, and to N. Wiener [434] in the general case.

Theorem 1.1.1 (Hardy–Littlewood–Wiener). *Let $f \in L^p(\mathbf{R}^N)$, $1 \le p \le \infty$. There is a constant A, depending only on p and N, such that:*

(a) *if $p = 1$, then*

$$|\{x : Mf(x) > \lambda\}| \le \frac{A}{\lambda} \|f\|_1 \quad \textit{for all } \lambda > 0 \ ;$$

(b) *if $1 < p \le \infty$, then*

$$\|Mf\|_p \le A \|f\|_p \ .$$

The proof is found e.g. in the book by E. M. Stein [389], Theorem I.1, or in the original paper by Wiener.

There is a deeper maximal theorem for l^q-valued functions due to C. Fefferman and E. M. Stein [144]. An l^q-valued function on $\mathbf{R}^N$ is a sequence of functions $\{f_n\}$, $n \in \mathbf{Z}$ or $n \in \mathbf{N}$, such that $\|f_n(x)\|_{l^q}^q = \sum |f_n(x)|^q < \infty$ a.e., and we say that $\{f_n\} \in L^p(l^q)$ if the mixed norm

$$\|\{f_n\}\|_{L^p(l^q)} = \Big\| \|\{f_n(\cdot)\}\|_{l^q} \Big\|_{L^p}$$

is finite.

Theorem 1.1.2 (Fefferman–Stein). *Suppose $1 < p < \infty$ and $1 < q \le \infty$. Then there is a constant A such that*

$$\|\{Mf_n\}\|_{L^p(l^q)} \le A \, \|\{f_n\}\|_{L^p(l^q)}$$

for all sequences $\{f_n\} \in L^p(l^q)$ on $\mathbf{R}^N$.

For the proof we refer to the original source [144], or to A. Torchinsky [400], Theorem XII:1.1, or E. M. Stein [390], Section II.1 (for $q = 2$). In contrast to Theorem 1.1.1, which is trivially true for $p = \infty$, Theorem 1.1.2 is false in this case; see [390].

1.1.5 Integral Inequalities

On several occasions we will use the general Minkowski inequality: for $1 \le p \le \infty$

$$\left(\int_F \left|\int_E f(x,y)\,d\mu(x)\right|^p d\nu(y)\right)^{1/p} \le \int_E \left(\int_F |f(x,y)|^p\,d\nu(y)\right)^{1/p} d\mu(x)\ , \tag{1.1.4}$$

for any $f = f(x, y)$ measurable on the product set $E \times F$ with respect to the product measure $\mu \times \nu$. Or, more concisely,

$$\left\|\int_E f(x,\cdot)\,d\mu(x)\right\|_{L^p(\nu)} \le \int_E \|f(x,\cdot)\|_{L^p(\nu)} d\mu(x)\ . \tag{1.1.5}$$

See e.g. A. Zygmund [440], inequality (9.12) in Chapter I. A useful special case is the inequality

$$\|f * g\|_p \le \|f\|_1 \|g\|_p\ , \tag{1.1.6}$$

which is also a special case of the more general W. H. Young inequality,

$$\|f * g\|_r \le \|f\|_p\,\|g\|_q,\quad \frac{1}{r} = \frac{1}{p} + \frac{1}{q} - 1\ . \tag{1.1.7}$$

See e.g. E. M. Stein and G. Weiss [391], Section V.1. Here and throughout the book $f * g$ is the *convolution* of f and g;

$$f * g(x) = \int_{\mathbf{R}^N} f(x-y)\,g(y)\,dy = \int_{\mathbf{R}^N} f(y)\,g(x-y)\,dy\ .$$

1.1.6 Distributions

We shall also need some of the most basic facts about distributions. The class of infinitely differentiable functions on Ω is denoted $C^\infty(\Omega)$, and $C_0^\infty(\Omega)$ is the subset of functions with compact support in Ω. The Schwartz class of rapidly decreasing C^∞-functions on $\mathbf{R}^N$ is denoted by $\mathcal{S}$. Elements in $C_0^\infty(\Omega)$ or $\mathcal{S}$ are also called *test functions*.

We let the spaces of test functions be equipped with their usual topologies in distribution theory. Then the space of continuous linear functionals on $C_0^\infty(\Omega)$ is the space of Schwartz distributions on Ω, denoted by $\mathcal{D}'(\Omega)$. The continuous linear functionals on $\mathcal{S}$ are called *temperate*, or *tempered distributions*, and the space is denoted $\mathcal{S}'$. The pairing between distributions and test functions is denoted $\langle\cdot,\cdot\rangle$. A distribution T is *positive*, if $\langle T, \varphi\rangle \ge 0$ for all nonnegative test functions φ.

From the theory of distributions we require the following basic theorem of L. Schwartz [373]. See [373], Ch. I, Théorème V, or L. Hörmander [230], Theorem 2.1.7, for the proof.

Theorem 1.1.3. *The positive distributions are positive measures.*

1.1.7 The Fourier Transform

If $f \in L^1$, its Fourier transform is the bounded and continuous function

$$\mathcal{F}f(\xi) = \widehat{f}(\xi) = \int_{\mathbf{R}^N} f(x)\, e^{-ix\cdot\xi}\, dx \ .$$

If also $\widehat{f}$ is integrable, Fourier's inversion formula gives

$$f(x) = \frac{1}{(2\pi)^N} \int_{\mathbf{R}^N} \widehat{f}(\xi)\, e^{ix\cdot\xi}\, d\xi \ ,$$

or in other words, $\mathcal{F}^2 f(x) = (2\pi)^N f(-x)$. If f and g are functions in L^1, then $f * g \in L^1$, and $\mathcal{F}(f * g) = \mathcal{F}(f)\mathcal{F}(g)$. If f is sufficiently regular, we have $\mathcal{F}\partial_j f(\xi) = i\xi_j \widehat{f}(\xi)$, and $\mathcal{F}\Delta f(\xi) = -|\xi|^2 \widehat{f}(\xi)$. We also define positive powers of the Laplace operator by

$$(-\Delta)^{\alpha/2} f(x) = \mathcal{F}^{-1}\big(|\xi|^\alpha \widehat{f}(\xi)\big) \ , \quad \alpha > 0, \quad f \in \mathcal{S} \ . \tag{1.1.8}$$

The Fourier transformation $\mathcal{F}$ is a bijection on $\mathcal{S}$. By Parseval's formula $\|f\|_2 = (2\pi)^{-N/2}\|\widehat{f}\|_2$, and it follows that $\mathcal{F}$ can be extended by continuity to a bijection on L^2 (Plancherel's theorem). The definition of $\mathcal{F}$ is extended to $S \in \mathcal{S}'$ by the relation $\langle \widehat{S}, \varphi\rangle = \langle S, \widehat{\varphi}\rangle$ for all $\varphi \in \mathcal{S}$, and in this way $\mathcal{F}$ becomes a bijection on $\mathcal{S}'$. References for these ideas include the books L. Schwartz [373], W. Rudin [368] and L. Hörmander [230].

1.1.8 The Riesz Transform and Singular Integrals

Another operator that is easily defined by means of the Fourier transform is the *Hilbert transform*, and its higher dimensional generalization, the *Riesz transforms* (named for M. Riesz). The Riesz transform $\mathcal{R}_j$, $j = 1, \ldots, N$, is defined, if $f \in L^2$, by

$$\mathcal{R}_j f = \mathcal{F}^{-1}\big(-i\xi_j |\xi|^{-1} \widehat{f}(\xi)\big) \ .$$

By the Plancherel theorem, $\mathcal{R}_j$ is a bounded mapping on L^2. We see that $\partial_j\partial_k f = \mathcal{R}_j\mathcal{R}_k(-\Delta) f$, and $\partial_j f = -\mathcal{R}_j(-\Delta)^{1/2} f$. See Stein [389], Section III.1. (Observe that Stein defines the Fourier transform as $\int f(x)\, e^{+2\pi i x\cdot\xi}\, dx$.)

The Riesz transform can also be defined by the principal value integral

$$\mathcal{R}_j f = \lim_{\delta\to 0} \int_{|y|\geq\delta} R_j(x-y)\, f(y)\, dy \ , \tag{1.1.9}$$

where R_j is the singular Riesz kernel,

$$R_j(x) = A_N \frac{x_j}{|x|^{N+1}} \ , \tag{1.1.10}$$

whose Fourier transform (in the sense of distributions) is $\widehat{R}_j(\xi) = -i\xi_j/|\xi|$, which makes $A_N = \Gamma(\frac{N+1}{2})\pi^{-(N+1)/2}$. This integral is well defined if f is smooth, for example if $f \in \mathcal{S}$. The definition can again be extended to L^2 by means of the Plancherel theorem.

That the Riesz transform can be extended to a bounded operator on L^p, $1 < p < \infty$, is a fundamental result from the theory of singular integral operators of A. P. Calderón and A. Zygmund [89]. More precisely, the result is the following.

Theorem 1.1.4. *Let $f \in L^p$, $1 \le p < \infty$. Then the integral* (1.1.9) *converges for a.e. x, and there is a constant A, depending only on p and N, such that*:

(a) *if $p = 1$, then*

$$|\{x : |\mathcal{R}_j f(x)| > \lambda\}| \le \frac{A}{\lambda} \|f\|_1 \quad \textit{for all } \lambda > 0 \ ;$$

(b) *if $1 < p < \infty$, then*

$$\|\mathcal{R}_j f\|_p \le A \, \|f\|_p \ .$$

We refer to Stein [389], Section II.2, for the proof. In the same chapter the following more general result is also proved (Theorem 1, p. 29, and Corollary, p. 34).

Theorem 1.1.5. *Let $k \in L^1_{\text{loc}}(\mathbf{R}^N)$. Suppose that $\widehat{k}$ is an essentially bounded function, and that there is a constant B such that*

$$\|\widehat{k}\|_\infty \le B \ ,$$

and

$$\int_{|x|\ge 2t} |k(x-y) - k(x)|\, dx \le B \ , \quad |y| \le t \ , \tag{1.1.11}$$

for all $t > 0$. Then the transformation $f \mapsto Kf = \mathcal{F}^{-1}\big(\widehat{k}\mathcal{F}(f)\big)$ extends to a bounded transformation $K : L^p \to L^p$ for $1 < p < \infty$, and there is A, only depending on p and N, such that

$$\|Kf\|_p \le A \, B \, \|f\|_p \ .$$

In this form the theorem is due to L. Hörmander, who introduced the condition (1.1.11). See Hörmander [227], Theorem 2.1, p. 114.

1.2 Sobolev Spaces and Bessel Potentials

1.2.1 Sobolev Spaces

We now introduce the main object of study of this book, the Sobolev spaces on $\mathbf{R}^N$. We give the definition for general open sets. The Sobolev space $W^{\alpha,p}(\Omega)$ for an open set $\Omega \subset \mathbf{R}^N$ is the collection of L^p functions f, all of whose distribution (or weak) derivatives $D^\sigma f$ with $|\sigma| \le \alpha \in \mathbf{N}$ belong to $L^p(\Omega)$. This means that there are functions $g_\sigma \in L^p(\Omega)$, $|\sigma| \le \alpha$, such that f satisfies

$$\int_\Omega f\, D^\sigma \varphi\, dx = (-1)^{|\sigma|} \int_\Omega g_\sigma\, \varphi\, dx$$

for all $\varphi \in C_0^\infty(\mathbf{R}^N)$; then $D^\sigma f$ is just g_σ. The norm on $W^{\alpha,p}(\Omega)$ is

$$\|f\|_{W^{\alpha,p}(\Omega)} = \left(\sum_{|\sigma|\le\alpha} \int_\Omega |D^\sigma f|^p\, dx \right)^{1/p} .$$

We write $W^{\alpha,p}(\mathbf{R}^N) = W^{\alpha,p}$.

The Sobolev spaces are named for S. L. Sobolev, who used these spaces systematically from the mid 1930's. (See e.g. [383, 384, 385]). However, the history of these spaces goes back at least to the work of Beppo Levi in the beginning of the century. References to the early work are given in e.g. the book by C. B. Morrey, Jr., [333], Chapter 1.8.

The following theorem gives an alternative characterization of $W^{\alpha,p}(\Omega)$.

Theorem 1.2.1. *Let $f \in L^p(\Omega)$. Then $f \in W^{\alpha,p}(\Omega)$ if and only if there exists a sequence $\{\varphi_n\}_1^\infty$ of functions in $C^\infty(\Omega)$ that converges to f in $L^p(\Omega)$, and for which the derivatives $\{D^\sigma\varphi_n\}_1^\infty$, $|\sigma| \le \alpha$, form Cauchy sequences in $L^p(\Omega)$. The assumption implies that $\{D^\sigma\varphi_n\}_1^\infty$ converges to $D^\sigma f$ in $L^p(\Omega)$.*

This theorem was first proved by J. Deny and J. L. Lions [122], Théorème 2.3, p. 312 (for $\alpha = 1$), and by N. G. Meyers and J. Serrin [323]. In the case $\Omega = \mathbf{R}^N$ it is Proposition V.1 in Stein [389]. The general result is also found in e.g. V. G. Maz′ya [308], Theorem 1.1.5/1, or in W. P. Ziemer [438], Theorem 3.16.

The Sobolev space $W_0^{\alpha,p}(\Omega)$ is defined as the closure of $C_0^\infty(\Omega)$ in $W^{\alpha,p}(\Omega)$. It is an easy consequence of the Leibniz formula that $W_0^{\alpha,p}(\mathbf{R}^N) = W^{\alpha,p}(\mathbf{R}^N)$.

A domain $\Omega \subset \mathbf{R}^N$ is called an *extension domain* if for any domain Ω' such that $\overline{\Omega} \subset \Omega'$ there is for every positive integer m a bounded linear extension operator $E_m : W^{m,p}(\Omega) \to W_0^{m,p}(\Omega')$. Thus, for any $f \in W^{m,p}(\Omega)$ there is $E_m f \in W_0^{m,p}(\Omega')$ such that $E_m f|_\Omega = f$.

Theorem 1.2.2. *Every bounded domain in $\mathbf{R}^N$ with C^∞ boundary is an extension domain.*

This theorem is proved by a generalized reflection method, which is due to M. R. Hestenes [224] in the case of the space C^m. See e.g. R. A. Adams [26], Theorem 4.26, G. B. Folland [150], Proposition 6.40, or V. G. Maz′ya [308], Theorem 1.1.16. According to a deeper theorem of A. P. Calderón [88] (see also Stein [389], Theorem VI.5, p. 181) all Lipschitz domains have the extension property, and for still more general domains the property was proved by P. W. Jones [237]. Further references and results are found in Maz′ya [308], Section 1.5.

We shall also have occasion to use a homogeneous seminorm,

$$\|f\|_{\dot{W}^{\alpha,p}(\Omega)} = \left(\sum_{|\sigma|=\alpha} \int_\Omega |D^\sigma f|^p \, dx \right)^{1/p} ,$$

and a "homogeneous version" of $W_0^{\alpha,p}(\Omega)$, denoted $\dot{W}_0^{\alpha,p}(\Omega)$, which is defined as the completion of $C_0^\infty(\Omega)$ in $\|f\|_{\dot{W}^{\alpha,p}(\Omega)}$, which is a norm on $C_0^\infty(\Omega)$. If Ω is bounded, then by the elementary Poincaré inequality we have $\|f\|_{L^p(\Omega)} \le A\|\nabla f\|_{L^p(\Omega)}$ for $f \in C_0^\infty(\Omega)$; thus $\|f\|_{W^{\alpha,p}(\Omega)} \le A\|f\|_{\dot{W}^{\alpha,p}(\Omega)}$, and it follows that $W_0^{\alpha,p}(\Omega) = \dot{W}_0^{\alpha,p}(\Omega)$ in this case. In general the elements in $\dot{W}_0^{\alpha,p}(\Omega)$ belong to L^p only locally. If $\alpha p \ge N$, $\dot{W}_0^{\alpha,p}(\Omega)$ is not always a space of distributions. This question will be discussed in Section 11.1.

Our vision of the Sobolev spaces changes, when we view them on $\mathbf{R}^N$ and look at pointwise representations. Let $f \in \mathcal{S}$, and $N \ge 3$. Then it is well known, and easy to prove by means of Green's formula, that $f(x)$ can be represented as a Newton potential,

$$f(x) = -\frac{1}{(N-2)\omega_{N-1}} \int_{\mathbf{R}^N} \frac{\Delta f(y)}{|x-y|^{N-2}} \, dy , \tag{1.2.1}$$

where

$$\omega_{N-1} = 2\pi^{N/2} / \Gamma(N/2) \tag{1.2.2}$$

is the $(N-1)$-dimensional area of the unit sphere in $\mathbf{R}^N$. See e.g. F. John [235], p. 9. Thus, writing

$$I_2(x) = \frac{1}{(N-2)\omega_{N-1}} \frac{1}{|x|^{N-2}} , \tag{1.2.3}$$

the Newton kernel, we have the representation $f = I_2 * (-\Delta f)$. In the language of distributions this can also be expressed as $-\Delta I_2 = \delta_0$, the Dirac distribution at 0. In other words, $-I_2$ is a fundamental solution of the Laplace operator. See e.g. L. Hörmander [230], Theorem 3.3.2.

A related formula is

$$f(x) = \frac{1}{\omega_{N-1}} \int_{\mathbf{R}^N} \frac{\nabla f(y) \cdot (x-y)}{|x-y|^N} \, dy . \tag{1.2.4}$$

See e.g. Stein [389], Section V.2.3.

There are extensions of these representations to higher orders of differentiation, but we prefer to take a different approach, using the Fourier transform; see however T. Bagby [44], Lemma 2.

1.2.2 Riesz Potentials

Let $0 < \alpha < N$. Then the function $|\xi|^{-\alpha}$ is locally integrable, and it can be viewed as an element in $\mathcal{S}'$. It follows that we can extend the definition (1.1.8) to certain negative powers of $-\Delta$, $(-\Delta)^{-\alpha/2}$ for $0 < \alpha < N$, and define an operator $\mathcal{I}_\alpha : \mathcal{S} \to \mathcal{S}'$ by

$$\mathcal{I}_\alpha f = (-\Delta)^{-\alpha/2} f = \mathcal{F}^{-1}\big(|\xi|^{-\alpha} \widehat{f}\big) \quad \text{for } 0 < \alpha < N \ . \tag{1.2.5}$$

If I_α is defined as the inverse Fourier transform of $|\xi|^{-\alpha}$ (in the sense of distributions), one can show that

$$I_\alpha(x) = \frac{\gamma_\alpha}{|x|^{N-\alpha}} \ , \tag{1.2.6}$$

where γ_α is a certain constant, whose exact value is

$$\gamma_\alpha = \Gamma(\tfrac{N-\alpha}{2}) \big/ \big(\pi^{N/2} 2^\alpha \Gamma(\tfrac{\alpha}{2})\big) \ .$$

See e.g. Stein [389], Section V.1, or N. S. Landkof [266], Chapter 1. Note that $\gamma_2^{-1} = (N-2)\omega_{N-1}$, which agrees with (1.2.3). The function I_α is known as the *Riesz kernel*. It follows immediately from the rules for manipulating Fourier transforms that any $f \in \mathcal{S}$ can be written as a *Riesz potential*,

$$f(x) = \mathcal{I}_\alpha g(x) = (I_\alpha * g)(x) = \gamma_\alpha \int_{\mathbf{R}^N} \frac{g(y)}{|x-y|^{N-\alpha}} \, dy \ , \quad 0 < \alpha < N \ , \tag{1.2.7}$$

where $g = (-\Delta)^{\alpha/2} f$. In particular,

$$f = I_1 * \left(\sum_{j=1}^{N} \mathcal{R}_j \partial_j f \right) \ . \tag{1.2.8}$$

Thus we see that at least on a dense subset of $W^{\alpha,p}(\mathbf{R}^N)$, we can write elements as Riesz potentials. In view of Theorem 1.1.4, which implies that the operator $(-\Delta)^{\alpha/2}$ extends to a bounded operator from $W^{\alpha,p}$ to L^p, we would expect to be able to pass to the limit in (1.2.7), and extend the representation to functions in $W^{\alpha,p}$, at least for "most values" of $x \in \mathbf{R}^N$.

1.2.3 Bessel Potentials

The above motivates our desire to study more closely the Riesz potentials $f = I_\alpha * g$ with $g \in L^p(\mathbf{R}^N)$, $1 < p < \infty$. However, there are problems with this representation. It soon becomes evident that because of the fact that $I_\alpha \notin L^1(\mathbf{R}^N)$, the operator $g \to I_\alpha * g$ is not bounded on L^p, and thus not a bounded operator from L^p to $W^{\alpha,p}$. A remedy would be to replace I_α with a function that has the same singularity at 0 but a more rapid decay at ∞, such as $I_\alpha(x)e^{-|x|}$. There are certain advantages in proceeding via the Fourier transform and replacing the operator $(-\Delta)^{\alpha/2}$ by $(I-\Delta)^{\alpha/2}$, where I as usual denotes the identity operator. In other words, we want to replace the fundamental solution of the operator $-\Delta$ by that of $I - \Delta$. Thus, we define the *Bessel kernel*

$$G_\alpha = \mathcal{F}^{-1}\big((1+|\xi|^2)^{-\alpha/2}\big) \ .$$

Here we can allow any exponent $\alpha \in \mathbf{R}$ (or even $\alpha \in \mathbf{C}$), since the function $\widehat{G}_\alpha(\xi) = (1+|\xi|^2)^{-\alpha/2}$ can be identified with a distribution in $\mathcal{S}'$ for all α. Moreover, if $f \in \mathcal{S}$, then clearly $\widehat{G}_\alpha \widehat{f} \in \mathcal{S}$. Denoting

$$\mathcal{G}_\alpha = (I - \Delta)^{-\alpha/2} \ , \quad \alpha \in \mathbf{R} \ ,$$

we immediately obtain the *Bessel potential* representation

$$f = \mathcal{G}_\alpha g = G_\alpha * g \ , \quad \alpha \in \mathbf{R} \ ,$$

with

$$g = \mathcal{G}_{-\alpha} f = (I - \Delta)^{\alpha/2} f \ ,$$

where $g \in \mathcal{S}$ for all $f \in \mathcal{S}$, and $\mathcal{G}_\alpha : \mathcal{S} \to \mathcal{S}$ is a bijection.

It is clear from the definition that $\{\mathcal{G}_\alpha\}_{\alpha \in \mathbf{R}}$ is a group of operators, i.e., it satisfies

$$\mathcal{G}_\alpha \mathcal{G}_\beta = \mathcal{G}_{\alpha+\beta} \ , \quad \text{and} \quad \mathcal{G}_\alpha^{-1} = \mathcal{G}_{-a} \ , \quad \alpha, \beta \in \mathbf{R} \ . \tag{1.2.9}$$

The idea of replacing Riesz kernels by Bessel kernels appears in J. Deny [119], p. 129. Bessel potentials were introduced systematically by N. Aronszajn and K. T. Smith [40], and by A. P. Calderón (see [88]).

1.2.4 Bessel Kernels

In order to be able to use the representation we need to know some of the properties of G_α, at least for $\alpha > 0$. Because of the fact that $\widehat{G}_\alpha(\xi)$ decays at ∞ like $\widehat{I}_\alpha(\xi) = |\xi|^{-\alpha}$, we would expect $G_\alpha(x)$ to be a function with the same singularity at 0 as $I_\alpha(x)$. The important difference between $\widehat{G}_\alpha(\xi)$ and $\widehat{I}_\alpha(\xi)$ is that the former is real analytic on $\mathbf{R}$, which leads us to expect $G_\alpha(x)$ to have exponential decay at ∞.

That this is in fact so can be seen from the following integral formulas for I_α and G_α.

$$I_\alpha(x) = a_\alpha \int_0^\infty t^{\frac{\alpha-N}{2}} e^{-\frac{\pi|x|^2}{t}} \frac{dt}{t} \ , \quad 0 < \alpha < N \ ; \tag{1.2.10}$$

$$G_\alpha(x) = a_\alpha \int_0^\infty t^{\frac{\alpha-N}{2}} e^{-\frac{\pi|x|^2}{t} - \frac{t}{4\pi}} \frac{dt}{t} \ , \quad \alpha > 0 \ , \tag{1.2.11}$$

where $1/a_\alpha = (4\pi)^{\alpha/2}\Gamma(\alpha/2)$. The proofs are easy, and given in e.g. Stein [389], Section V.3.1.

It follows immediately that $G_\alpha(x)$ is positive, and that

$$0 < G_\alpha(x) < I_\alpha(x) \quad \text{for } 0 < \alpha < N \ . \tag{1.2.12}$$

An application of Fubini's theorem gives that $G_\alpha \in L^1$, and thus

$$\|G_\alpha\|_1 = \int_{\mathbf{R}^N} G_\alpha \, dx = \widehat{G}_\alpha(0) = 1 \ . \tag{1.2.13}$$

Moreover, G_α is radial, i.e. $G_\alpha(x)$ depends only on $|x|$, and it is decreasing as a function of $|x|$. Comparing the formulas (1.2.10) and (1.2.11), it follows easily that

$$G_\alpha(x) \sim I_\alpha(x) \ , \quad \text{as } |x| \to 0 \ , \quad 0 < \alpha < N \ , \tag{1.2.14}$$

(in the sense that the limit of the ratio of the two sides is 1), and an examination of (1.2.11) shows without much effort that for any $c < 1$,

$$G_\alpha(x) = O(e^{-c|x|}) \ , \quad \text{as } |x| \to \infty \ , \quad \alpha > 0 \ . \tag{1.2.15}$$

See Stein [389], Section V.3.1.

A further easy consequence of (1.2.11) that we will sometimes use is the estimate

$$G_\alpha(x) \le A\, G_\alpha(x+y) \ , \quad |x| \ge 2 \ , \quad |y| \le 1 \ . \tag{1.2.16}$$

1.2.5 Some Classical Formulas for Bessel Functions

Another approach to the Bessel kernel is by means of the Fourier inversion formula. If $\alpha > N$, then $\widehat{G} \in L^1(\mathbf{R}^N)$. Consequently, G_α is continuous, and

$$G_\alpha(x) = \frac{1}{(2\pi)^N} \int_{\mathbf{R}^N} \frac{e^{ix\cdot\xi}}{(1+|\xi|^2)^{\alpha/2}} \, d\xi \ .$$

By a formula which goes back to Poisson and Cauchy (see S. Bochner [65], Satz 56, p. 187) this gives

$$G_\alpha(x) = (2\pi)^{-N/2} |x|^{-(N-2)/2} \int_0^\infty \frac{t^{N/2}}{(1+t^2)^{\alpha/2}} J_{(N-2)/2}(|x|t) \, dt \ , \tag{1.2.17}$$

where J_ν denotes the Bessel function of order ν. By means of another classical formula (see the standard treatise of G. N. Watson [426], 13.6(2), p. 434, or E. C. Titchmarsh [399], (7.11.6), p. 201) one obtains

$$G_\alpha(x) = c_\alpha \, |x|^{-(N-\alpha)/2} \, K_{(N-\alpha)/2}(|x|) \ , \tag{1.2.18}$$

where $1/c_\alpha = 2^{(\alpha-2)/2}(2\pi)^{N/2}\Gamma(\alpha/2)$, and K_ν is known as a *modified Bessel function of the third kind*, or a *Macdonald function*. (For this reason N. Aronszajn and K. T. Smith introduced the name "Bessel potentials" in [40], p. 386.)

As we shall now briefly explain, this formula can be extended to all α. The function $K_\nu(z)$ is an analytic function of the complex variable z for $z \neq 0$, and it is an even, entire analytic function of ν for each $z \neq 0$.

On the other hand $\widehat{G}_\alpha(\xi) = (1+|\xi|^2)^{-\alpha/2}$ is an entire analytic function of α for each ξ. This implies that the distribution G_α is an analytic function of α in the sense that $\langle G_\alpha, \varphi \rangle$ is analytic for each $\varphi \in \mathcal{S}$. It follows by analytic continuation that (1.2.18) remains valid for all $\alpha \neq 0, -2, -4, \ldots$; the restriction appears because of the poles of $\Gamma(\alpha/2)$, which make $c_\alpha = 0$. Here the right hand side of (1.2.18) should be understood as a finite part in the sense of distributions, unless $\operatorname{Re}\alpha > 0$, when it is an integrable function. For the exceptional values the fact that $\mathcal{F}1 = \delta_0$ immediately gives

$$G_\alpha = (1 - \Delta)^2 \delta_0 \ , \quad \alpha = 0, -2, -4, \ldots \ . \tag{1.2.19}$$

See N. Aronszajn and K. T. Smith [40], and L. Schwartz [373], formulas (II,3;20), p. 47, and (VII,7;23), p. 260.

The functions K_ν and other Bessel functions have been studied in great detail (see G. N. Watson [426]), and the following asymptotic formulas for real ν and $r > 0$ are classical:

$$K_\nu(r) \sim 2^{\nu-1}\Gamma(\nu)\, r^{-\nu}, \quad \text{as } r \to 0 \text{ for } \nu > 0 \ ; \tag{1.2.20}$$

$$K_0(r) \sim \log(1/r), \quad \text{as } r \to 0 \ ; \tag{1.2.21}$$

$$K_\nu(r) \sim (\pi/2r)^{1/2}\, e^{-r}, \quad \text{as } r \to \infty \text{ for all } \nu. \tag{1.2.22}$$

This again implies (1.2.14). In addition we see that $G_N(x)$ behaves at 0 like the logarithmic kernel,

$$G_N(x) \sim c_N \log \frac{1}{|x|}, \quad \text{as } |x| \to 0 \ , \tag{1.2.23}$$

with $c_N^{-1} = 2^{N-1}\pi^{N/2}\Gamma(N/2)$ (in particular, $c_2 = 1/(2\pi)$ for $N = 2$), and that (1.2.15) can be improved to

$$G_\alpha(x) \sim c_\alpha (\pi/2)^{1/2} |x|^{(\alpha-N-1)/2} e^{-|x|} \quad \text{as } |x| \to \infty \text{ for all } \alpha > 0 \ . \tag{1.2.24}$$

In particular, (1.2.15) is true with $c = 1$ for $\alpha \leq N + 1$. See N. Aronszajn and K. T. Smith [40] and N. Aronszajn, F. Mulla and P. Szeptycki [38]. (It could be noted that an easy estimate in (1.2.11) shows that $G_N(x) \leq A \log(2/|x|)$ for $|x| \leq 1$, which is enough for most purposes.)

The formula (1.2.22) follows easily from the Laplace transform type representation,

$$K_\nu(r) = b_\nu r^{-\frac{1}{2}} e^{-r} \int_0^\infty e^{-t} t^{\nu-\frac{1}{2}} \left(1 + \frac{t}{2r}\right)^{\nu-\frac{1}{2}} dt, \tag{1.2.25}$$

with $b_\nu = (\pi/2)^{\frac{1}{2}} / \Gamma(\nu + \frac{1}{2})$, which is valid for $r > 0$ and $\nu > -\frac{1}{2}$. See Watson [426], 7.3., p. 206.

It also might be noted that (1.2.25) shows that K_ν is an elementary function, when $\nu - \frac{1}{2}$ is integer. In particular, $K_{\pm\frac{1}{2}} = (\pi/2)^{\frac{1}{2}} r^{-\frac{1}{2}} e^{-r}$, and thus

$$G_{N-1}(x) = c_{N-1}(\pi/2)^{1/2} |x|^{-1} e^{-|x|} \ ,$$

$$G_{N-3}(x) = (I - \Delta) G_{N-1}(x) = c_{N-3} (\pi/2)^{1/2} \left(|x|^{-3} + |x|^{-2}\right) e^{-|x|} \ ,$$

etc.

Occasionally we shall also need estimates for the derivatives of G_α. Writing $G_\alpha(x) = G_\alpha(r)$, $r = |x|$, and differentiating (1.2.11), we find by (1.2.18)

$$G'_\alpha(r) = -a_\alpha 2\pi r \int_0^\infty t^{\frac{\alpha-2-N}{2}} e^{-\frac{\pi r^2}{t}-\frac{t}{4\pi}} \frac{dt}{t}$$

$$= -a_\alpha (2/\pi)^{1/2} r^{(\alpha-N)/2} K_{(N-\alpha+2)/2}(r) \ . \tag{1.2.26}$$

This implies for $\alpha > 1$,

$$G'_\alpha(r) \sim -(N-\alpha) G_{\alpha-1}(r) \ , \quad \text{as } r \to 0 \ , \tag{1.2.27}$$

and

$$G'_\alpha(r) \sim -a_\alpha r^{(\alpha-N-1)/2} e^{-r} \sim -c\, G_\alpha(r) \ , \quad \text{as } r \to \infty \tag{1.2.28}$$

with $c = 2^{(N+\alpha-3)/2} \pi^{(N-\alpha-1)/2}$.

Remark. We would like to emphasize that the finer properties of the Bessel kernel G_α are not really important to the theory we are going to develop. What is important is the nature of the singularity at the origin, the monotonicity, and the fact that the decay at infinity is sufficiently rapid to make it integrable. The group property, and the property that G_2 is a fundamental solution of $I - \Delta$, are also useful.

1.2.6 Bessel Potential Spaces

After this long preamble we now define the *Bessel potential spaces* $L^{\alpha,p} = L^{\alpha,p}(\mathbf{R}^N)$ by

$$L^{\alpha,p}(\mathbf{R}^N) = \{ f : f = G_\alpha * g,\ g \in L^p(\mathbf{R}^N) \} \ , \quad \alpha \in \mathbf{R} \ . \tag{1.2.29}$$

The norm on $L^{\alpha,p}$ is $\|f\|_{\alpha,p} = \|g\|_p$. Occasionally, we shall also use the notation

$$\dot{L}^{\alpha,p}(\mathbf{R}^N) = \{ f : f = I_\alpha * g,\ g \in L^p(\mathbf{R}^N) \} \ , \quad 0 < \alpha < N \tag{1.2.30}$$

for the spaces of Riesz potentials. The following fundamental theorem of A. P. Calderón [88] is one of main motivations for the theory that will be exposed in this book.

Theorem 1.2.3 (A. P. Calderón). *For $\alpha \in \mathbf{N}$, $W^{\alpha,p}(\mathbf{R}^N) = L^{\alpha,p}(\mathbf{R}^N)$, $1 < p < \infty$, with equivalence of norms, i.e., there is a constant A such that for all f*

$$A^{-1} \|f\|_{\alpha,p} \le \|f\|_{W^{\alpha,p}} \le A \|f\|_{\alpha,p} \ .$$

The theorem is a consequence of Theorem 1.1.4. Different proofs are found in [88], and in Stein [389], Theorem V.3.

It is easy to prove that C_0^∞ and $\mathcal{S}$ are dense in $L^{\alpha,p}$, and that the dual of $L^{\alpha,p}$ is $L^{-\alpha,p'}$ for for all $\alpha \in \mathbf{R}$, $1 < p < \infty$, and $pp' = p + p'$.

The Bessel potential spaces were systematically studied by N. Aronszajn and coauthors. See N. Aronszajn and K. T. Smith [40], N. Aronszajn, F. Mulla, and P. Szeptycki [38], and N. Aronszajn [37].

1.2.7 The Sobolev Imbedding Theorem

We also want to state the following imbedding theorem of S. L. Sobolev [384] at this point, although we will return to it in Chapter 3. (See Theorem 3.1.4 and Corollary 3.1.5.)

Theorem 1.2.4 (S. L. Sobolev). *Let* $\alpha > 0$, *and* $1 < p < \infty$. *Then*

(a) $L^{\alpha,p}(\mathbf{R}^N) \subset L^q(\mathbf{R}^N)$ *for all* q, $p \le q \le p^*$, $\frac{1}{p^*} = \frac{1}{p} - \frac{\alpha}{N}$, *when* $\alpha p < N$;
(b) $L^{\alpha,p}(\mathbf{R}^N) \subset L^q_{\text{loc}}(\mathbf{R}^N)$ *for all* q, $p \le q < \infty$, *when* $\alpha p = N$;
(c) $L^{\alpha,p}(\mathbf{R}^N) \subset C(\mathbf{R}^N)$, *when* $\alpha p > N$.

Each of these statements is a continuous embedding, meaning that there is a corresponding inequality between the norms involved; here $C(\mathbf{R}^N)$ is equipped with the supremum norm. When $\alpha p > N$ this last fact is an immediate consequence of the Hölder inequality, $|G_\alpha * g(x)| \le \|G_\alpha\|_{p'} \|g\|_p$, since $G_\alpha \in L^{p'}(\mathbf{R}^N)$ precisely when $\alpha p > N$. The continuity of $G_\alpha * g$, i.e. (c), then follows from the continuity of translation in the L^p-norm.

See also e.g. the books by Robert A. Adams [26], Chapter V, V. G. Maz'ya [308], Section 1.4, E. M. Stein [389], Sections V.1.2, and V.2.2, and W. P. Ziemer [438], Chapter 2, for other approaches to Sobolev's theorem, including the important case $p = 1$, and for many other related results that will not be treated here.

1.3 Banach Spaces

We will have the opportunity to work with several (usually real) Banach spaces. Most of the basic results needed are the standard ones found in e.g. W. Rudin [367]; the Hahn–Banach theorem, the uniform boundedness principle, weak* compactness of the unit ball. A somewhat more special property that will be used in Chapter 2 is that of *uniform convexity.*

Definition 1.3.1. *A Banach space is uniformly convex if for every* $\varepsilon > 0$ *there is a* $\delta > 0$ *such that if* $\|f\| < 1 + \delta$, $\|g\| < 1 + \delta$, *and* $\|\frac{1}{2}(f + g)\| \ge 1$, *then* $\|f - g\| < \varepsilon$.

The concept of a uniformly convex Banach space is due to J. A. Clarkson [107], who also proved the following theorem.

Theorem 1.3.2. L^p *is uniformly convex for* $1 < p < \infty$.

Proof. In order to prove the theorem, Clarkson proved the following beautiful inequalities, whose proof we omit. See Clarkson [107], G. Köthe [259], pp. 355–359, or E. Hewitt and K. Stromberg [225], §15.

(a) For $2 \le p < \infty$

$$2^{1-p}\left(\|f\|_p^p + \|g\|_p^p\right) \le \left\|\tfrac{1}{2}(f - g)\right\|_p^p + \left\|\tfrac{1}{2}(f + g)\right\|_p^p \le \tfrac{1}{2}\left(\|f\|_p^p + \|g\|_p^p\right) .$$

(b) For $1 < p \le 2$

$$2^{1-p}\left(\|f\|_p^p + \|g\|_p^p\right) \ge \left\|\tfrac{1}{2}(f-g)\right\|_p^p + \left\|\tfrac{1}{2}(f+g)\right\|_p^p \ge \tfrac{1}{2}\left(\|f\|_p^p + \|g\|_p^p\right) .$$

(c) For $1 < p \le 2$

$$\left\|\tfrac{1}{2}(f-g)\right\|_p^{p'} + \left\|\tfrac{1}{2}(f+g)\right\|_p^{p'} \le \left(\tfrac{1}{2}\left(\|f\|_p^p + \|g\|_p^p\right)\right)^{p'-1} .$$

(d) For $2 \le p < \infty$

$$\left\|\tfrac{1}{2}(f-g)\right\|_p^{p'} + \left\|\tfrac{1}{2}(f+g)\right\|_p^{p'} \ge \left(\tfrac{1}{2}\left(\|f\|_p^p + \|g\|_p^p\right)\right)^{p'-1} .$$

The theorem now follows easily from the second inequality in (a) and from (c). Other proofs can be found in O. Hanner [190], C. Morawetz [332], and J. Diestel [123].

The following corollaries will play an important role for us in Chapter 2.

Corollary 1.3.3. *Let B be a uniformly convex Banach space, and let $\{f_n\}_1^\infty$ be a sequence in B such that $\lim_{n\to\infty}\|f_n\| = 1$ and*

$$\liminf_{n,m\to\infty}\left\|\tfrac{1}{2}(f_n + f_m)\right\| \ge 1 .$$

Then $\{f_n\}_1^\infty$ is strongly convergent in B.

Proof. Let $\varepsilon > 0$. For any $\delta > 0$ we can find n_0 so that $\|f_n\| < 1 + \delta$ and $\|\frac{1}{2}(f_n + f_m)\| > 1 - \delta$ if $n, m \ge n_0$. By the definition of uniform convexity, applied to $f_n/(1-\delta)$, $\|f_n - f_m\| < \varepsilon$ if δ is small enough.

Corollary 1.3.4. *If Ω is a convex subset of a uniformly convex Banach space, then there is a unique element in $\overline{\Omega}$ (the closure of Ω in B) with least norm; in fact any sequence $\{f_n\}_{n=1}^\infty$ in Ω minimizing the norm is Cauchy.*

Proof. If $0 \in \overline{\Omega}$ there is nothing to prove, so we can assume that $\inf_\Omega \|f\| = 1$. Let $\{f_n\}_{n=1}^\infty$ be a sequence in Ω such that $\lim_{n\to\infty}\|f_n\| = 1$. Then the convexity of Ω implies that $\|\frac{1}{2}(f_n + f_m)\| \ge 1$ for all n and m, and the result follows from the previous corollary.

The following important consequence of the Hahn–Banach theorem is known as Mazur's lemma.

Theorem 1.3.5. *The weak closure and the strong closure of a convex set in a Banach space coincide.*

See e.g. Rudin [368], Theorem 3.12.

1.4 Two Covering Lemmas

The following "simple Vitali lemma" plays an essential role in the proof of Theorem 1.1.1. This method of proof is due to N. Wiener; see [434], where the lemma appears as Lemma C′. It will be used in a similar way in Chapter 6. The lemma is also found in Stein [389], Section I.1.6, and in Rudin [367], Lemma 7.3.

Theorem 1.4.1. *Let E be a measurable subset of $\mathbf{R}^N$ which is covered by the union of a family of balls of bounded diameter. Then from this family we can select a subsequence of disjoint balls $B(x_1, r_1)$, $B(x_2, r_2)$, ..., $B(x_j, r_j)$, ..., such that $E \subset \bigcup_1^\infty B(x_j, 5r_j)$, and thus $|E| \leq 5^N \sum_1^\infty |B(x_j, r_j)|$.*

The other covering lemma is the Whitney decomposition, due to H. Whitney [430].

By *dyadic cubes* in $\mathbf{R}^N$ we mean cubes with side 2^{-n}, $n \in \mathbf{Z}$, whose vertices belong to the lattice of points $\{ 2^{-n}k : k \in \mathbf{Z}^N \}$.

Theorem 1.4.2. *Let F be a non-empty closed set in $\mathbf{R}^N$, set $\Omega = F^c$. Then there is a collection of closed dyadic cubes, $\{Q_n\}_1^\infty$, such that*

(a) $\bigcup_1^\infty Q_n = \Omega$;
(b) *The interiors $(Q_n)^0$ are mutually disjoint*;
(c) $\operatorname{diam} Q_n \leq \operatorname{dist}(Q_n, F) \leq 4 \operatorname{diam} Q_n$ *for all n.*

Here diam Q *is the diameter of Q and* dist(Q, F) *is the distance of Q from F.*

We refer to Stein [389], Section VI.1, for the proof.

2. L^p-Capacities and Nonlinear Potentials

2.1 Introduction

As a motivation for what follows we here give a rather heuristic definition of classical Newton capacity. We begin with a positive charge distribution (a Radon measure) μ on a body K, by which we simply mean a compact subset of $\mathbf{R}^3$. The total charge on K is $\mu(K)$. A unit test charge placed at a point x in $\mathbf{R}^3 \setminus K$ then, according to the laws of physics, experiences a force on it which is equal to $-\nabla U^\mu(x)$, where (with properly chosen units)

$$U^\mu(x) = \int_K \frac{d\mu(y)}{|x-y|} \ ,$$

the Newton potential of μ at x. The potential $U^\mu(x)$ is interpreted as the amount of work necessary to move a unit charge from ∞ to x, and the total energy stored in the system is given by one half of the *energy integral*:

$$E(\mu) = \int_K U^\mu(x)d\mu(x) \ .$$

Now, if K is a conductor, this means that the initial charge distribution μ is allowed to distribute itself freely on K. It will do so in a manner that minimizes the energy, and as C. F. Gauss [181] realized, the new charge distribution ν will satisfy:

(a) $\nu(K) = \mu(K)$;
(b) $E(\nu) \leq E(\mu)$;
(c) $U^\nu(x) = \text{constant} = V$ for $x \in K$;
(d) $U^\nu(x) \leq V$ for $x \in \mathbf{R}^3 \setminus K$.

This new distribution ν is called the *equilibrium measure* and U^ν the *equilibrium potential*. In terms of these the *capacity* of K is then defined as

$$C(K) = \frac{\nu(K)}{V} \ ,$$

so that the capacity of K is the charge that makes the equilibrium potential equal to one. If one normalizes ν in this way, i.e., so that $\nu(K) = C(K)$, it follows that $E(\nu) = C(K)$. Moreover, one can prove that

$$C(K) = \inf\{ E(\mu) : \ \mu \in \mathcal{M}^+(K),\ U^\mu(x) \ge 1 \ \text{ on } K \} .$$

It is an important observation that the energy integral is a quadratic form on the space of measures. This leads to a connection with Hilbert space which becomes more explicit if one notices that $-\Delta U^\mu = 4\pi\mu$ (in the sense of distributions), and that by Gauss's theorem

$$\int_{\mathbf{R}^3} |\nabla U^\mu|^2 \, dx = -\int_{\mathbf{R}^3} U^\mu \cdot \Delta U^\mu \, dx = 4\pi E(\mu) \ .$$

Thus the Hilbert space defined with the square root of the energy integral as norm is closely related to the space we have denoted by $W^{1,2}$. Furthermore, one can show that the capacity $C(K)$ has the following definition:

$$C(K) = \inf \left\{ \frac{1}{4\pi} \int_{\mathbf{R}^3} |\nabla\varphi|^2 \, dx : \varphi \in C_0^\infty(\mathbf{R}^3) \text{ and } \varphi(x) \ge 1 \text{ on } K \right\} .$$

In fact, the extremal φ of this last formulation turns out to be precisely U^ν with ν equal to normalized equilibrium measure.

We are going to take this last definition as our starting point for the development that follows, and in this chapter we shall prove the statements made above in a more general context.

Later we shall see that $C(\cdot)$ is a natural measure for the class $W^{1,2}$ in the sense that such functions can be defined up to sets of such capacity zero, in the same way that L^2 functions can only be defined up to sets of Lebesgue measure zero. And since a $W^{1,2}$ function is in some sense more regular than an arbitrary L^2 function, we would expect the null sets of C to be smaller in general than the null sets of Lebesgue measure. Thus we might think of C as a "refined Lebesgue measure" on $\mathbf{R}^3$. However, it is not a measure; in fact it is not even finitely additive on disjoint sets. So in a sense, our "measure" loses one of the more useful properties of such devices—additivity—when we try to construct a set function that distinguishes between the negligible sets of $W^{1,2}$ and those of L^2. This loss is compensated by the fact that many problems can be given a complete solution in terms of capacity.

In this chapter, our aim is to extend the notion of capacity in several directions, mainly with the purpose of developing a theory of capacities and potentials that is related to the spaces $W^{\alpha,p}$ and $L^{\alpha,p}$ in much the same way as the classical theory is related to $W^{1,2}$. Here the most important changes in the theory come about because we allow the exponent p to be different from 2. Indeed, this introduces an element of nonlinearity into the theory, which has motivated the introduction of the term *nonlinear potential theory*.

It is quite a surprising fact that so many of the notions and results of classical, linear potential theory can be generalized to this new situation, even if the generalization sometimes is far from obvious and requires new tools.

In Section 2.2 we define one of the central objects of this book, the (α, p)-capacity, and prove a number of its basic properties, including the existence of extremal functions, and a dual definition of capacity. In Section 2.3 we make a new start, and define capacities in a more general situation. The advantages of this approach will be seen clearly in Chapter 4, where we deal with Besov and Lizorkin–Triebel spaces.

The basic duality result depends on the minimax theorem, which is the subject of Section 2.4. The following section, Section 2.5 is central to the whole theory. In Section 2.6 the results are specialized to capacities associated to radially decreasing convolution kernels, and a number of classical results are extended.

Finally, in Section 2.7 we study an alternative definition of capacity, which appears naturally in connection with the study of removable singularities of solutions of partial differential equations.

2.2 A First Version of (α, p)-Capacity

As discussed above it is natural to define capacities related to general function spaces, such as the Sobolev spaces $W^{\alpha,p}$, $\alpha \in \mathbf{N}$, $1 \leq p < \infty$. In particular it is natural to try to measure the lack of continuity of functions in $W^{\alpha,p}(\mathbf{R}^N)$ when $\alpha p \leq N$ by some sort of (α, p)-capacity. In this section we present a first version of these ideas, which will later in the chapter be subsumed under a more far-reaching, general theory. We first make a preliminary definition.

Definition 2.2.1. Let $K \subset \mathbf{R}^N$ be compact. Then

$$C'_{\alpha,p}(K) = \inf\{ \|\varphi\|^p_{W^{\alpha,p}} : \varphi \in C_0^\infty,\ \varphi \geq 1 \text{ on } K \} .$$

We extend this definition to open sets in the following way.

Definition 2.2.2. Let $G \subset \mathbf{R}^N$ be open. Then

$$C'_{\alpha,p}(G) = \sup\{ C'_{\alpha,p}(K) : K \subset G,\ K \text{ compact} \} .$$

Proposition 2.2.3. *Let $K \subset \mathbf{R}^N$ be compact. Then*

$$C'_{\alpha,p}(K) = \inf\{ C'_{\alpha,p}(G) : G \supset K,\ G \textit{ open} \} .$$

Proof. Let K be compact, and let $\varepsilon > 0$. In Definition 2.2.1 it clearly makes no difference if the condition $\varphi \geq 1$ is replaced by $\varphi > 1$. Thus, there is a $\varphi \in C_0^\infty$ such that $\|\varphi\|^p_{W^{\alpha,p}} < C'_{\alpha,p}(K) + \varepsilon$ and $\varphi(x) > 1$ on K. Set $K_1 = \{x : \varphi(x) \geq 1\}$. Then K_1 is compact, $C'_{\alpha,p}(K_1) \leq \|\varphi\|^p_{W^{\alpha,p}}$, and $K \subset (K_1)^0 \subset K_1$. It follows easily from the definitions that $C'_{\alpha,p}((K_1)^0) \leq C'_{\alpha,p}(K_1) < C'_{\alpha,p}(K) + \varepsilon$. The proposition follows.

Thanks to Proposition 2.2.3 we can now extend the definitions to arbitrary sets.

Definition 2.2.4. Let $E \subset \mathbf{R}^N$ be arbitrary. Then

$$C'_{\alpha,p}(E) = \inf\{ C'_{\alpha,p}(G) : G \supset E,\ G \text{ open} \} .$$

A capacity that has this property is called an *outer capacity*.

Definition 2.2.5. A property that holds true for all x except those belonging to a set E with $C'_{\alpha,p}(E) = 0$ is said to be true *(α, p)-quasieverywhere*, abbreviated *(α, p)-q.e.*

For $\alpha = 1$ and $p = 2$ it is classical, and easily seen (see e.g. G. B. Folland [150] or F. John [236]), that an extremal function in Definition 2.2.1 is a weak solution of the linear second order partial differential equation

$$-\Delta u + u = 0$$

on the complement of K, and one is lead to classical potential theory (with the Newton kernel I_2 replaced by the Bessel kernel G_2). For $p \neq 2$, however, the corresponding equations are nonlinear, and much more difficult to handle. For example, for $\alpha = 1$ one obtains

$$-\operatorname{div}(\nabla u|\nabla u|^{p-2}) + u|u|^{p-2} = 0 \ .$$

(The operator $\operatorname{div}(\nabla u|\nabla u|^{p-2})$, which coincides with the Laplace operator Δu for $p = 2$, is often denoted $\Delta_p u$, and is known as the p-Laplace operator.)

However, it turns out that by redefining (α, p)-capacity slightly, one is lead to extremal functions that have a simple representation and give a rich potential theory.

The key to this observation is A. P. Calderón's theorem, Theorem 1.2.3, about the representation of elements in $W^{\alpha,p}(\mathbf{R}^N)$, $1 < p < \infty$, as Bessel potentials, or in other words, the fact that $W^{\alpha,p}(\mathbf{R}^N) = L^{\alpha,p}(\mathbf{R}^N)$ for $1 < p < \infty$.

We redefine (α, p)-capacity in the following way. Notice that we now allow α to be any non-negative real number.

Definition 2.2.6. Let $K \subset \mathbf{R}^N$ be compact, and set

$$\omega_K = \{\varphi \in \mathcal{S} : \varphi \geq 1 \text{ on } K\} \ ,$$

so that ω_K is a convex subset of the Schwartz class $\mathcal{S}$. Let $\alpha > 0$, and $1 < p < \infty$. Then

$$C_{\alpha,p}(K) = \inf\{ \|\varphi\|^p_{\alpha,p} : \varphi \in \omega_K \} \ .$$

The definition is extended to arbitrary sets as in Definitions 2.2.2 and 2.2.4.

Clearly there are constants A_1 and A_2, depending only on α, p, and N, such that

$$A_1 C'_{\alpha,p}(E) \leq C_{\alpha,p}(E) \leq A_2 C'_{\alpha,p}(E)$$

for all sets E. We say that capacities satisfying such inequalities are *equivalent*. Thus, in Definition 2.2.5 it makes no difference if we use $C'_{\alpha,p}$, $C_{\alpha,p}$, or some other equivalent capacity.

The apparently slight change made in the definition of capacity has very important consequences for the extremal functions. In fact, it is now quite easy to prove the following fundamental theorem. ($\overline{\omega}_K$ denotes the closure of ω_K in $L^{\alpha,p}$.)

Theorem 2.2.7. *Let $K \subset \mathbf{R}^N$ be compact, let $\alpha > 0$, and $1 < p < \infty$. Then there is a unique extremal element $F^K = G_\alpha * f^K$ in $\overline{\omega}_K$ such that*

$$\|f^K\|_p^p = C_{\alpha,p}(K) \ .$$

There is a $\mu^K \in \mathcal{M}^+(K)$, called a capacitary measure for K, such that

$$f^K = (G_\alpha * \mu^K)^{p'-1} \ ,$$

and thus,

$$F^K = G_\alpha * (G_\alpha * \mu^K)^{p'-1} \ . \tag{2.2.1}$$

Moreover, the extremals satisfy

$$C_{\alpha,p}(K) = \int_{\mathbf{R}^N} (G_\alpha * \mu^K)^{p'}\, dx = \int_K F^K\, d\mu^K \ , \tag{2.2.2}$$

$$F^K(x) \leq 1 \quad \textit{everywhere on } \operatorname{supp} \mu^K \ , \tag{2.2.3}$$

$$\mu^K(K) = C_{\alpha,p}(K) \ , \tag{2.2.4}$$

and (α, p)-capacity has the dual definition

$$C_{\alpha,p}(K) = \sup_{\mu \in \mathcal{M}^+(K)} \left(\frac{\mu(K)}{\|G_\alpha * \mu\|_{p'}} \right)^p \ . \tag{2.2.5}$$

Remark. We cannot assert that $F^K(x) \geq 1$ everywhere on K. It is true, however, that $F^K(x) \geq 1$ (α, p)-q.e. on K, but in order to avoid repetition we prefer to prove this in the more general context of the next section. It follows, in fact, immediately from Proposition 2.3.9. See also Proposition 2.6.7, and Chapter 6. Theorem 2.2.7 is generalized in Theorems 2.5.3 and 2.5.5.

Functions of the type F^K will play an important role in what follows, so we give them a special name.

Definition 2.2.8. For any $\mu \in \mathcal{M}^+(\mathbf{R}^N)$ the function $G_\alpha * (G_\alpha * \mu)^{p'-1}$ is denoted $V_{\alpha,p}^\mu$ and called a *nonlinear potential* of μ.

The nonlinear potential F^K is then called the *capacitary potential* for K, and f^K is the corresponding *capacitary function*

We observe that for $p = 2$, by the group property (1.2.9) of the Bessel kernels,

$$V_{\alpha,p}^\mu = G_\alpha * (G_\alpha * \mu) = G_{2\alpha} * \mu \ ,$$

so that we have a classical, linear potential. Then Theorem 2.2.7 is a well-known result in classical potential theory, although it is more often formulated for the Newton kernel I_2 or the Riesz kernels I_α, $0 < \alpha < N$, than for G_α.

Clearly

$$\int V^{\mu}_{\alpha,p}\,d\mu = \int_{\mathbf{R}^N} (G_\alpha * \mu)^{p'}\,dx \;, \tag{2.2.6}$$

and by analogy with the case $p = 2$ this can be considered as a generalized energy integral. Cf. Section 2.1.

Proof of Theorem 2.2.7. By the uniform convexity of L^p for $1 < p < \infty$ (see Theorem 1.3.2, and Corollary 1.3.4) there is a unique element, $F^K = G_\alpha * f^K$, in the $L^{\alpha,p}$-closure $\overline{\omega}_K$ of ω_K, such that

$$\|F^K\|^p_{\alpha,p} = \inf_{\varphi\in\omega_K} \|\varphi\|^p_{\alpha,p} = C_{\alpha,p}(K) \;.$$

In order to prove the existence of a measure μ^K satisfying (2.2.1) we let $\varphi = G_\alpha * \psi$ be a non-negative function in $\mathcal{S}$. Then $F^K + t\varphi \in \overline{\omega}_K$ for all $t \geq 0$, so that

$$\int_{\mathbf{R}^N} |f^K + t\psi|^p\,dx \geq \int_{\mathbf{R}^N} |f^K|^p\,dx \;, \quad t \geq 0 \;.$$

Taking the derivative of $\int |f^K + t\psi|^p\,dx$ at $t = 0$, we obtain

$$\int_{\mathbf{R}^N} |f^K|^{p-2} f^K \psi\,dx \geq 0$$

for all $\psi \in \mathcal{S}$ such that $G_\alpha * \psi \geq 0$.

Set $|f^K|^{p-2} f^K = h$. Then $h \in L^{p'}$, and $|h|^{p'} = |f^K|^p$. Consequently, there is a distribution $\mu^K = G_{-\alpha} * h$, belonging to $L^{-\alpha,p'}(\mathbf{R}^N)$, such that $h = G_\alpha * \mu^K$, and thus

$$\int_{\mathbf{R}^N} (G_\alpha * \mu^K)\,\psi\,dx \geq 0 \;.$$

But, by the properties of convolutions of distributions, this is the same thing as saying that

$$\langle \mu^K, G_\alpha * \psi\rangle = \langle \mu^K, \varphi\rangle \geq 0 \;,$$

$\langle\cdot,\cdot\rangle$ denoting the action of a distribution on a test function.

This being true for all positive test functions, μ^K must be a positive Radon measure by L. Schwartz's theorem (Theorem 1.1.3). Thus

$$f^K = h^{p'-1} = (G_\alpha * \mu^K)^{p'-1} \;.$$

To prove that $\mu^K \in \mathcal{M}^+(K)$ we repeat the reasoning with a $\varphi \in \mathcal{S}$ with arbitrary sign such that $\operatorname{supp}\varphi \subset K^c$. Then $F^K + t\varphi \in \overline{\omega}_K$ for all $t \in \mathbf{R}$. It follows that $\langle \mu^K, \varphi\rangle = 0$ for all such φ, and thus $\operatorname{supp}\mu^K \subset K$ as required.

For (2.2.2) we observe that by Fubini's theorem

$$\begin{aligned}\int_K F^K\,d\mu^K &= \int_K (G_\alpha * f^K)\,d\mu^K = \int_{\mathbf{R}^N} f^K (G_\alpha * \mu^K)\,dx \\ &= \int_{\mathbf{R}^N} (G_\alpha * \mu^K)^{p'}\,dx = \int_{\mathbf{R}^N} (f^K)^p\,dx = C_{\alpha,p}(K) \;.\end{aligned}$$

For the proof of (2.2.3) we notice that $F^K = G_\alpha * f^K$ is a lower semicontinuous function, so that the set $\{x : F^K(x) > 1\}$ is open. It follows that for all test functions φ with $\operatorname{supp}\varphi \subset \{x : F^K(x) > 1\}$ we have $F^K + t\varphi \in \overline{\omega}_K$ for every t with $|t|$ sufficiently small. Again we find that $\langle \mu^K, \varphi\rangle = 0$ for all such φ, so that $\operatorname{supp}\mu^K \subset \{x : F^K(x) \leq 1\}$, which is (2.2.3).

Now let $\mu \in \mathcal{M}^+(K)$, and let $F = G_\alpha * f \in \omega_K$. Then

$$\mu(K) \leq \int_K F\,d\mu = \int_K (G_\alpha * f)\,d\mu = \int_{\mathbf{R}^N} (G_\alpha * \mu) f\,dx \leq \|G_\alpha * \mu\|_{p'} \|f\|_p \ .$$

It follows from a passage to the limit, using Hölder's inequality, that the same inequality holds true for $F \in \overline{\omega}_K$. In particular, $F = F^K$ gives

$$\mu(K) \leq \|G_\alpha * \mu\|_{p'}\, C_{\alpha,p}(K)^{1/p} \ ,$$

and thus

$$\sup_{\mu \in \mathcal{M}^+(K)} \left(\frac{\mu(K)}{\|G_\alpha * \mu\|_{p'}}\right)^p \leq C_{\alpha,p}(K) \ .$$

Choosing $\mu = \mu^K$, this gives by (2.2.2) that

$$\mu^K(K) \leq \|G_\alpha * \mu^K\|_{p'}\, C_{\alpha,p}(K)^{1/p} = C_{\alpha,p}(K) \ .$$

On the other hand, (2.2.3) gives

$$\mu^K(K) \geq \int_K F^K\,d\mu^K \ ,$$

and from (2.2.2)

$$\int_K F^K\,d\mu^K = C_{\alpha,p}(K) = \|G_\alpha * \mu^K\|_{p'}\, C_{\alpha,p}(K)^{1/p} \ .$$

Thus

$$\mu^K(K) = C_{\alpha,p}(K) \ ,$$

and

$$\left(\frac{\mu^K(K)}{\|G_\alpha * \mu^K\|_{p'}}\right)^p = C_{\alpha,p}(K) \ ,$$

which proves (2.2.4) and (2.2.5).

2.3 A General Theory for L^p-Capacities

Before embarking on a more detailed study of nonlinear potentials, we want to move to a more general situation, which will contain the theory of (α, p)-capacities as a special case.

Definition 2.3.1. Let $\mathbf{M}$ be a space, equipped with a positive measure ν and a family of measurable sets. By a *kernel* on $\mathbf{R}^N \times \mathbf{M}$ we shall mean any nonnegative function g on $\mathbf{R}^N \times \mathbf{M}$, such that $g(\cdot, y)$ is lower semicontinuous on $\mathbf{R}^N$ for each $y \in \mathbf{M}$, and $g(x, \cdot)$ is measurable on $\mathbf{M}$ for each $x \in \mathbf{R}^N$.

Let $\mu \in \mathcal{M}^+(\mathbf{R}^N)$ and let f be a nonnegative, ν-measurable function. We define potentials (cf. B. Fuglede [165]) $\mathcal{G}f$, and $\check{\mathcal{G}}\mu$ by

$$\mathcal{G}f(x) = \int_{\mathbf{M}} g(x, y)\, f(y)\, d\nu(y), \quad x \in \mathbf{R}^N \ ; \tag{2.3.1}$$

$$\check{\mathcal{G}}\mu(y) = \int_{\mathbf{R}^N} g(x, y)\, d\mu(x), \quad y \in \mathbf{M} \ . \tag{2.3.2}$$

Then $\mathcal{G}f$ and $\check{\mathcal{G}}\mu$ are well defined everywhere if we allow them to take the value $+\infty$.

We also define a "mutual energy" which, by Fubini's theorem, can be written

$$\mathcal{E}_g(\mu, f) = \int_{\mathbf{R}^N} \mathcal{G}f\, d\mu = \int_{\mathbf{M}} \check{\mathcal{G}}\mu\, f\, d\nu \ . \tag{2.3.3}$$

One advantage of formulating the theory for general kernels is that we will be able to make small modifications of the kernels we are interested in, without changing the theory. But the full advantages of this setup will only become clear later, when we discuss capacities associated to Besov and other spaces. See e.g. Sections 4.4, 4.5, 4.7, and 6.3 below.

Proposition 2.3.2. *Let g be a kernel, let f be a fixed nonnegative ν-measurable function, and let y be fixed in $\mathbf{M}$. Then*

(a) *$x \mapsto \mathcal{G}f(x)$ is lower semicontinuous on $\mathbf{R}^N$;*
(b) *$\mu \mapsto \check{\mathcal{G}}\mu(y)$ is lower semicontinuous on $\mathcal{M}^+(\mathbf{R}^N)$ in the weak* topology;*
(c) *$\mu \mapsto \mathcal{E}_g(\mu, f)$ is lower semicontinuous on $\mathcal{M}^+(\mathbf{R}^N)$ in the weak* topology.*

Proof. (a) Let $x_0 \in \mathbf{R}^N$, and let $\{x_i\}_1^\infty$ be a sequence converging to x_0, such that $\lim_{i\to\infty} \mathcal{G}f(x_i) = \liminf_{x\to x_0} \mathcal{G}f(x)$. By the lower semicontinuity of $g(\cdot, y)$ and by Fatou's lemma,

$$\begin{aligned} \mathcal{G}f(x_0) = \int_{\mathbf{M}} g(x_0, y)\, f(y)\, d\nu(y) &\le \int_{\mathbf{M}} \liminf_{i\to\infty} g(x_i, y)\, f(y)\, d\nu(y) \\ &\le \liminf_{i\to\infty} \int_{\mathbf{M}} g(x_i, y)\, f(y)\, d\nu(y) = \liminf_{x\to x_0} \mathcal{G}f(x) \ , \end{aligned}$$

so $\mathcal{G}f(\cdot)$ is semicontinuous.

(b) Let $\mu \in \mathcal{M}^+(\mathbf{R}^N)$ and let $\{\mu_i\}_1^\infty$ converge to μ weak*. Let $\{h_n\}_1^\infty$ be an increasing sequence of continuous functions with compact support on $\mathbf{R}^N$, such that $h_n(x)$ converges to $g(x, y)$ for all x. Such a sequence exists by the assumption on semicontinuity (see e.g. Rudin [367], Chapter 2, Exercise 22). Then

$$\int_{\mathbf{R}^N} h_n\, d\mu = \lim_{i\to\infty} \int_{\mathbf{R}^N} h_n\, d\mu_i \le \liminf_{i\to\infty} \int_{\mathbf{R}^N} g(\cdot\,, y)\, d\mu_i = \liminf_{i\to\infty} \check{\mathcal{G}}\mu_i(y)\ .$$

By monotone convergence

$$\check{\mathcal{G}}\mu(y) = \lim_{n\to\infty} \int_{\mathbf{R}^N} h_n\, d\mu \le \liminf_{i\to\infty} \check{\mathcal{G}}\mu_i(y)\ .$$

(c) The proof is similar to (b). Let μ and $\{\mu_i\}_1^\infty$ be as in (b), and let $\{h_n\}_1^\infty$ be as in (b) but such that $h_n(x)$ converges to $\mathcal{G}f(x)$ everywhere. Such a sequence exists by (a). Then

$$\int_{\mathbf{R}^N} h_n\, d\mu = \lim_{i\to\infty} \int_{\mathbf{R}^N} h_n\, d\mu_i \le \liminf_{i\to\infty} \int_{\mathbf{R}^N} \mathcal{G}f\, d\mu_i\ .$$

By monotone convergence

$$\mathcal{E}_g(\mu, f) = \lim_{i\to\infty} \int_{\mathbf{R}^N} h_n\, d\mu \le \liminf_{i\to\infty} \int_{\mathbf{R}^N} \mathcal{G}f\, d\mu_i = \liminf_{i\to\infty} \mathcal{E}_g(\mu_i, f)\ .$$

For a given kernel on $\mathbf{R}^N \times \mathbf{M}$ we now define the L^p-capacity of an arbitrary $E \subset \mathbf{R}^N$.

Definition 2.3.3. Let $1 \le p < \infty$, and let $E \subset \mathbf{R}^N$. Denote

$$\Omega_E = \{\, f : f \in L^p_+(\nu),\ \mathcal{G}f(x) \ge 1 \text{ for all } x \in E\,\}\ ,$$

so that Ω_E is a convex subset of $L^p(\nu)$. Then

$$C_{g,p}(E) = \inf\left\{ \int_{\mathbf{M}} f^p\, d\nu : f \in \Omega_E \right\}\ .$$

If $\Omega_E = \emptyset$ we set $C_{g,p}(E) = \infty$.

If $\mathbf{M} = \mathbf{R}^N$ with $\nu = m$, i.e. ν is Lebesgue measure, and $g(x, y) = G_\alpha(x - y)$, then $C_{G_\alpha,p}(\cdot) = C_{\alpha,p}(\cdot)$. The proof of this fact, although not deep, is not entirely evident, so we postpone it for the moment. See Proposition 2.3.13 below.

The following proposition is obvious.

Proposition 2.3.4.

$$E_1 \subset E_2 \quad\Longrightarrow\quad C_{g,p}(E_1) \le C_{g,p}(E_2).$$

We then prove that the capacity we have defined is an outer capacity (cf. Definition 2.2.4).

Proposition 2.3.5. *For any* $E \subset \mathbf{R}^N$

$$C_{g,p}(E) = \inf\{\, C_{g,p}(G) : G \supset E,\ G \text{ open} \,\} .$$

Proof. We assume that $C(E) < \infty$. If $0 < \varepsilon < 1$ there is a measurable, nonnegative function f on $\mathbf{M}$ such that $\mathcal{G}f(x) \geq 1$ on E, and $\int_{\mathbf{M}} f^p \, d\nu < C(E) + \varepsilon$. But $\mathcal{G}f$ is lower semicontinuous on $\mathbf{R}^N$, so $\{x : \mathcal{G}f(x) > 1 - \varepsilon\}$ is an open set, G. Clearly,

$$C_{g,p}(G) \leq (1-\varepsilon)^{-p} \int_{\mathbf{M}} f^p \, d\nu < (1-\varepsilon)^{-p} \big(C_{g,p}(E) + \varepsilon \big) ,$$

which proves the proposition.

The above definition gives a simple proof of the subadditivity of capacity.

Proposition 2.3.6. *Let* $E_i \subset \mathbf{R}^N$, $i = 1, 2, \ldots$, *and* $E = \bigcup_1^\infty E_i$. *Then*

$$C_{g,p}(E) \leq \sum_{i=1}^{\infty} C_{g,p}(E_i) .$$

Proof. Let $\varepsilon > 0$ and let $f_i \geq 0$ be such that $\mathcal{G}f_i(x) \geq 1$ on E_i, and $\int_{\mathbf{M}} f_i^p \, d\nu < C(E_i) + \varepsilon 2^{-i}$. Define $f(x) = \sup_i f_i(x)$. Then $\mathcal{G}f(x) \geq 1$ on E, and

$$\int_{\mathbf{M}} f^p \, d\nu \leq \sum_i^{\infty} \int_{\mathbf{M}} f_i^p \, d\nu < \sum_i^{\infty} C(E_i) + \varepsilon .$$

The following result characterizes the sets of capacity zero.

Proposition 2.3.7. *Let* $E \subset \mathbf{R}^N$. *Then* $C_{g,p}(E) = 0$ *if and only if there is an* $f \in L^p_+(\nu)$ *such that* $E \subset \{x : \mathcal{G}f(x) = +\infty\}$.

Proof. We first note that by definition

$$C_{g,p}(\{x : \mathcal{G}f(x) \geq \lambda\}) \leq \lambda^{-p} \int_{\mathbf{M}} f^p \, d\nu$$

for any $\lambda > 0$ and $f \in L^p_+(\nu)$, and thus $C_{g,p}(\{x : \mathcal{G}f(x) = +\infty\}) = 0$.

For the converse, assume that $C_{g,p}(E) = 0$, and choose $f_i \geq 0$, $i = 1, 2, \ldots$, such that $\mathcal{G}f_i(x) \geq 1$ on E and $\int_{\mathbf{M}} f_i^p \, d\nu < 2^{-ip}$. Then $f = \sum_i f_i$ has the required properties.

As in Definition 2.2.5 we say that a property holds (g, p)-quasieverywhere, or (g, p)-q.e., if it holds except on a set E with $C_{g,p}(E) = 0$.

We extend the definition of the potentials $\mathcal{G}f$ to arbitrary ν-measurable functions f by setting

$$\mathcal{G}f(x) = \mathcal{G}f^+(x) - \mathcal{G}f^-(x) ,$$

whenever at least one of the terms on the right is finite. By Proposition 2.3.7 $\mathcal{G}f(x)$ is well defined and finite (g, p)-q.e.

The following proposition extends Egorov's theorem.

Proposition 2.3.8. *Suppose that $\{f_i\}_1^\infty$ is a Cauchy sequence in $L^p(\nu)$ with limit f. Then there is a subsequence $\{f_{i_n}\}_{n=1}^\infty$ such that $\lim_{n\to\infty} \mathcal{G} f_{i_n}(x) = \mathcal{G} f(x)$ (g, p)-q.e., uniformly outside an open set of arbitrarily small (g, p)-capacity.*

Proof. By Propositions 2.3.7 and 2.3.6 all the $\mathcal{G} f_n(x)$ and $\mathcal{G} f(x)$ are well defined, and finite outside a set F with $C_{g,p}(F) = 0$. Choose $\{i_n\}_{n=1}^\infty$ so that

$$\int_{\mathbf{M}} |f_{i_n} - f|^p \, d\nu < 4^{-np} \ .$$

Set $E_n = \{ x : \mathcal{G}|f_{i_n} - f|(x) > 2^{-n} \}$ and $G_m = \bigcup_{n=m}^\infty E_n$. Then

$$C_{g,p}(E_n) \le 2^{np} \int_{\mathbf{M}} |f_{i_n} - f|^p \, d\nu \le 2^{-np} \ ,$$

and

$$C_{g,p}(G_m) \le \sum_{n=m}^{\infty} 2^{-np} \ ,$$

so that

$$C_{g,p}\Big(\bigcap_{m=1}^\infty G_m\Big) = 0 \ .$$

Note that if $x \notin G_m \cup F$, then

$$|\mathcal{G} f_{i_n}(x) - \mathcal{G} f(x)| \le \mathcal{G}|f_{i_n} - f|(x) \le 2^{-n}$$

for all $n \ge m$. Thus $\lim_{n\to\infty} \mathcal{G} f_{i_n}(x) = \mathcal{G} f(x)$ uniformly outside $G_m \cup F$ for any m. This proves the proposition, since F is contained in an open set of arbitrarily small capacity.

We denote by $\overline{\Omega}_E$ the closure of Ω_E in $L^p(\nu)$ (see Definition 2.3.3). Note that by Mazur's lemma, Theorem 1.3.5, the weak closure of Ω_E equals the strong closure. The following proposition gives an explicit description of $\overline{\Omega}_E$.

Proposition 2.3.9. *Let $1 \le p < \infty$, and let $E \subset \mathbf{R}^N$. Then*

$$\overline{\Omega}_E = \{ f : f \in L_+^p(\nu),\ \mathcal{G} f(x) \ge 1 \ (g, p)\text{-}q.e. \ on \ E \} \ .$$

Proof. The set defined on the right side contains Ω_E. We first prove that it is closed. Let $\{f_n\}_1^\infty$ be a sequence of elements in $L_+^p(\nu)$ such that $\mathcal{G} f_n(x) \ge 1$ on $E \setminus F_n$, where $\{F_n\}_1^\infty$ are sets such that $C_{g,p}(F_n) = 0$, and assume that the sequence converges strongly in $L^p(\nu)$ to a function f. By Proposition 2.3.8 there is a subsequence $\{f_{i_n}\}_{n=1}^\infty$ such that $\mathcal{G} f_{i_n}(x)$ converges (g, p)-q.e. to $\mathcal{G} f(x)$. It follows that $\mathcal{G} f(x) \ge 1$ (g, p)-q.e. on $E \setminus (\cup F_{i_n})$, i.e. (g, p)-q.e. on E by Proposition 2.3.6. Thus

$$\{ f : f \in L_+^p(\nu),\ \mathcal{G} f(x) \ge 1 \ (g, p)\text{-q.e. on } E \} \ ,$$

is closed, and thus it contains $\overline{\Omega}_E$.

On the other hand, if $f \in L_+^p(\nu)$, $\mathcal{G} f(x) \ge 1$ on $E \setminus F$, and $C_{g,p}(F) = 0$, it follows from Proposition 2.3.7 that there exist nonnegative h with arbitrarily small norm, such that $\mathcal{G} h(x) = +\infty$ on F. But then $f + h \in \Omega_E$, and $f + h$ is arbitrarily close to f, so $f \in \overline{\Omega}_E$.

Theorem 2.3.10. *Let $1 < p < \infty$, let $E \subset \mathbf{R}^N$, and assume $C_{g,p}(E) < \infty$. Then there is a unique f^E such that $f \in L^p_+(\nu)$ and $\mathcal{G}f^E(x) \geq 1$ (g, p)-q.e. on E, and*

$$\int_{\mathbf{M}} (f^E)^p \, d\nu = C_{g,p}(E) \ .$$

Proof. By Corollary 1.3.4 there is a unique $f^E \in \overline{\Omega}_E$ such that

$$\int_{\mathbf{M}} (f^E)^p \, d\nu = \inf_{f \in \Omega_E} \int_{\mathbf{M}} f^p \, d\nu = C_{g,p}(E) \ .$$

We call the function f^E the *capacitary function* of E, and we call $\mathcal{G}f^E$ the *capacitary potential* of E.

Remark. It is easily seen that

$$C_{g,p}(E) = \min\left\{ \|f\|^p_{L^p(\nu)} : f \in L^p(\nu),\ \mathcal{G}f(x) \geq 1 \ (g, p)\text{-q.e. on } E \right\} \ ,$$

i.e., the restriction $f \geq 0$ can be removed. In fact, if $\mathcal{G}f(x) \geq 1$, or is undefined, then $\mathcal{G}f^+(x) \geq 1$, and $\int_{\mathbf{M}} (f^+)^p \, d\nu \leq \int_{\mathbf{M}} |f|^p \, d\nu$.

A set is called *capacitable* for a capacity C if

$$C(E) = \sup\{ C(K) : K \subset E, K \text{ compact} \} = \inf\{ C(G) : G \supset E, G \text{ open} \}.$$

The following very general theorem is due to G. Choquet [102].

Theorem 2.3.11 (Capacitability Theorem). *Let $C(\cdot)$ be a function, defined for all $E \subset \mathbf{R}^N$, taking values on the extended real line, and satisfying:*

(a) $C(\emptyset) = 0$;
(b) $E_1 \subset E_2 \implies C(E_1) \leq C(E_2)$;
(c) *If K_i is a decreasing sequence of compact sets, then*

$$C\left(\bigcap_{i=1}^{\infty} K_i\right) = \lim_{i \to \infty} C(K_i) \ ;$$

(d) *If E_i is an increasing sequence of arbitrary sets, then*

$$C\left(\bigcup_{i=1}^{\infty} E_i\right) = \lim_{i \to \infty} C(E_i) \ .$$

Then all Suslin sets, and in particular all Borel sets, are capacitable for C.

We shall not prove this theorem (see the notes at the end of the chapter), but we can now easily show that $C_{g,p}$ satisfies the conditions of the theorem.

We already know that (a) and (b) are satisfied. It is easily seen that (c) is satisfied for any outer capacity. In fact, if G is open and $K = \bigcap_{i=1}^{\infty} K_i \subset G$, then $K_i \subset G$ for some i, and it follows that $C(K) \leq \lim_{i \to \infty} C(K_i) \leq \inf_{G \supset K} C(G) = C(K)$. Thus all that remains is to show (d), which we do in the next proposition.

Proposition 2.3.12. *Let $1 < p < \infty$. If $\{E_i\}_1^\infty$ is an increasing sequence of arbitrary subsets of $\mathbf{R}^N$ with union E, then*

$$C_{g,p}(E) = \lim_{i\to\infty} C_{g,p}(E_i).$$

Thus all Suslin sets are capacitable for $C_{g,p}$.

If, moreover, $C_{g,p}(E) < \infty$, then the capacitary functions f^{E_i} converge strongly in $L^p(\nu)$ to f^E.

Proof. Trivially $\lim_{i\to\infty} C_{g,p}(E_i) \le C_{g,p}(E)$. Without loss of generality we can assume that $\lim_{i\to\infty} C_{g,p}(E_i)$ is finite. Consider the sequence of capacitary functions. If $i < j$, then clearly $f^{E_j} \in \overline{\Omega}_{E_i}$, so that

$$\int_{\mathbf{M}} \left(\tfrac{1}{2}(f^{E_i} + f^{E_j})\right)^p d\nu \ge C_{g,p}(E_i) \ .$$

It follows that the assumptions of Corollary 1.3.3 are satisfied, so that $\{f^{E_i}\}_1^\infty$ converges strongly to a function f with $\int_{\mathbf{M}} f^p \, d\nu = \lim_{i\to\infty} C_{g,p}(E_i)$. Proposition 2.3.9 gives that $\mathcal{G}f(x) \ge 1$ (g, p)-q.e. on each E_i. Thus $\mathcal{G}f(x) \ge 1$ on E, except possibly on a countable union of sets of $C_{g,p}$-capacity zero. By the subadditivity of capacity, Proposition 2.3.6, $f \in \overline{\Omega}_E$, and thus $\int_{\mathbf{M}} f^p \, d\nu \ge C_{g,p}(E)$, which proves the proposition.

We shall now prove the identity of $C_{G_\alpha,p}$ and $C_{\alpha,p}$, as promised after Definition 2.3.3.

Proposition 2.3.13. *Let $1 < p < \infty$. Then $C_{G_\alpha,p}(E) = C_{\alpha,p}(E)$ for any $E \subset \mathbf{R}^N$.*

Proof. Both capacities are outer, so it is enough to prove the equality for open sets. But any open set in $\mathbf{R}^N$ is the union of an increasing sequence of compact sets, so by Proposition 2.3.12 and Definition 2.2.2 it is enough to consider compact sets.

Let K be compact. Then any function competing in Definition 2.2.6 is also competing in Definition 2.3.3 (see the remark following Theorem 2.3.10). Thus $C_{G_\alpha,p}(K) \le C_{\alpha,p}(K)$.

Now let $f \in L^p(\mathbf{R}^N)$ be a nonnegative function satisfying $G_\alpha * f(x) > 1$ on K, and $\|f\|_p^p < C_{G_\alpha,p}(K) + \varepsilon$ for some $\varepsilon > 0$. We claim that $G_\alpha * f \in \overline{\omega}_K$. Define f_n by $f_n(x) = \min\{f(x), n\}$ for $|x| \le n$ and $f_n(x) = 0$ for $|x| > n$, so that $G_\alpha * f_n$ is continuous, and $G_\alpha * f_n \nearrow G_\alpha * f$, as $n \to \infty$. By lower semicontinuity there are $\delta > 0$ and n such that $G_\alpha * f_n(x) \ge 1 + \delta$ on K, and $\|f_n\|_p^p < C_{G_\alpha,p}(K) + \varepsilon$. For any $q < \infty$ we can approximate f_n in L^q by functions $h \in C_0^\infty$. Then $G_\alpha * h \in \mathcal{S}$, and by the Sobolev imbedding theorem, Theorem 1.2.4, $G_\alpha * h$ approximates $G_\alpha * f_n$ uniformly if $\alpha q > N$. This implies that we can choose h so that $G_\alpha * h \ge 1$ on K and $C_{\alpha,p}(K) \le \|h\|_p^p < C_{G_\alpha,p}(K) + \varepsilon$, which proves the claim and the proposition.

2.4 The Minimax Theorem

Our next goal is to extend Theorem 2.2.7 to the general situation treated in the previous section. In particular, we will prove that the capacitary potential $\mathcal{G}^E$ can be represented by a positive measure as in (2.2.1), and extend the dual definition of capacity given in (2.2.5). The proof presented there depends strongly on the mapping properties of the Bessel kernel, and does not apply in this general setting. Instead, we shall invoke a general version of the von Neumann minimax theorem, the statement and proof of which are the subject of the present section.

Let X and Y be any two sets and let f be a real-valued function ($\leq +\infty$) on $X \times Y$. We shall say that f is *convex* on X if for any two elements $x_1, x_2 \in X$ and two numbers $\xi_1 \geq 0$, $\xi_2 \geq 0$, with $\xi_1 + \xi_2 = 1$, there is an element $x_0 \in X$ such that

$$f(x_0, y) \leq \xi_1 f(x_1, y) + \xi_2 f(x_2, y) \quad \text{for all } y \in Y \ .$$

Similarly, f is *concave* on Y if for any two elements $y_1, y_2 \in Y$ and two numbers $\eta_1 \geq 0$, $\eta_2 \geq 0$, with $\eta_1 + \eta_2 = 1$, there is an element $y_0 \in Y$ such that

$$f(x, y_0) \geq \eta_1 f(x, y_1) + \eta_2 f(x, y_2) \quad \text{for all } x \in X \ .$$

Clearly f is convex or concave in this sense if X or Y is a convex subset of a linear space and f is convex or concave in the ordinary sense.

Theorem 2.4.1 (Minimax Theorem). *Suppose that X is a compact Hausdorff space, Y an arbitrary set, and f a real-valued function ($\leq +\infty$) on $X \times Y$, which is lower semicontinuous in x for each fixed y, convex on X, and concave on Y. Then*

$$\min_{x \in X} \sup_{y \in Y} f(x, y) = \sup_{y \in Y} \min_{x \in X} f(x, y).$$

Proof. Note that we always have

$$\min_{x \in X} \sup_{y \in Y} f(x, y) \geq \sup_{y \in Y} \min_{x \in X} f(x, y)$$

since clearly $\sup_{y \in Y} f(x, y) \geq f(x, y)$ for all $y \in Y$ and all $x \in X$. Thus

$$\min_{x \in X} \sup_{y \in Y} f(x, y) \geq \min_{x \in X} f(x, y) \quad \text{for all } y \in Y \ .$$

Also note that the quantity $\sup_{y \in Y} \min_{x \in X} f(x, y)$ is either a real number or $+\infty$, hence so is $\min_{x \in X} \sup_{y \in Y} f(x, y)$.

The remainder of the argument is divided into 5 steps.

Step 1. Let $y_0 \in Y$ be such that $X_0 = \{x \in X : f(x, y_0) \leq 0\} \neq \emptyset$. If we replace X by X_0 and restrict f to $X_0 \times Y$, then the hypotheses of the theorem are satisfied. In fact, X_0 is a closed subset of a compact space, hence it is compact. We have to verify that f restricted to $X_0 \times Y$ is convex on X_0. Let $x_1, x_2 \in X_0$, ξ_1 and $\xi_2 \geq 0$, $\xi_1 + \xi_2 = 1$. Since f is convex on X, there is an $x_0 \in X$ such that $f(x_0, y) \leq \xi_1 f(x_1, y) + \xi_2 f(x_2, y)$, for all $y \in Y$. Hence $f(x_0, y_0) \leq 0$, so $x_0 \in X_0$.

Step 2. We claim that if $y_1, y_2 \in Y$ are such that

$$\max_{k=1,2} f(x, y_k) > 0 \quad \text{for all } x \in X \ ,$$

then there exists a $y_0 \in Y$ such that $f(x, y_0) > 0$ for all $x \in X$.

Let $X_k = \{x \in X : f(x, y_k) \le 0\}$. By the assumption X_k are disjoint compact sets. We may assume without loss of generality that each $X_k \ne \emptyset$. Now, if $x \in X_1$, then $f(x, y_1) \le 0$, and $f(x, y_2) > 0$. Then $-f(x, y_1)/f(x, y_2)$ is upper semicontinuous and nonnegative on X_1. So we can find $x_1 \in X_1$ such that

$$\max_{x \in X_1} \frac{-f(x, y_1)}{f(x, y_2)} = \frac{-f(x_1, y_1)}{f(x_1, y_2)} = \mu_1 \ge 0 \ .$$

Similarly, we can find $x_2 \in X_2$ such that

$$\max_{x \in X_2} \frac{-f(x, y_2)}{f(x, y_1)} = \frac{-f(x_2, y_2)}{f(x_2, y_1)} = \mu_2 \ge 0 \ .$$

Now $\mu_1 \mu_2 < 1$. In fact, assume that $\mu_1 \mu_2 \ne 0$. Since $f(x_1, y_1) \le 0$ and $f(x_2, y_1) > 0$, there exist numbers $\xi_1, \xi_2 \ge 0$ with $\xi_1 + \xi_2 = 1$ such that

$$\xi_1 f(x_1, y_1) + \xi_2 f(x_2, y_1) = 0.$$

Since f is convex on X, there is an $x_0 \in X$ with

$$f(x_0, y) \le \xi_1 f(x_1, y) + \xi_2 f(x_2, y) \quad \text{for all } y \in Y \ .$$

So $f(x_0, y_1) \le 0$, i.e. $x_0 \in X_1$, and hence $x_0 \notin X_2$. Thus $f(x_0, y_2) > 0$, and

$$\xi_1 f(x_1, y_2) + \xi_2 f(x_2, y_2) \ge f(x_0, y_2) > 0 \ .$$

So

$$-\xi_1 \frac{f(x_1, y_1)}{\mu_1} - \xi_2 \mu_2 f(x_2, y_1) > 0 \ ,$$

or

$$\xi_1 f(x_1, y_1) + \xi_2 \mu_1 \mu_2 f(x_2, y_1) < 0 \ ,$$

and hence

$$-\xi_2 f(x_2, y_1) + \xi_2 \mu_1 \mu_2 f(x_2, y_1) < 0 \ ,$$

which is just

$$\xi_2 f(x_2, y_1)(\mu_1 \mu_2 - 1) < 0 \ .$$

(Note that $\xi_2 > 0$, since if not, then $f(x_1, y_1) = 0$ and hence $\mu_1 = 0$.)

Now take numbers $\nu_1 > \mu_1$, $\nu_2 > \mu_2$, $\nu_1 \nu_2 = 1$, and let

$$\eta_1 = \frac{1}{1 + \nu_1} = \frac{\nu_2}{1 + \nu_2} \ ,$$

$$\eta_2 = \frac{1}{1 + \nu_2} = \frac{\nu_1}{1 + \nu_1} \ .$$

We claim that

$$\eta_1 f(x, y_1) + \eta_2 f(x, y_2) > 0 \quad \text{for all } x \in X \ .$$

If $x \notin X_1 \cup X_2$, then the claim follows trivially. If $x \in X_1$, then

$$\begin{aligned} 0 \le f(x, y_1) + \mu_1 f(x, y_2) &< f(x, y_1) + \nu_1 f(x, y_2) \\ &= (1 + \nu_1)\big(\eta_1 f(x, y_1) + \eta_2 f(x, y_2)\big) \ . \end{aligned}$$

The claim holds for $x \in X_2$ by a similar argument.

The conclusion of Step 2 now follows by the concavity of f on Y.

Step 3. We claim that if a finite set $\{y_1, \ldots, y_m\} \subset Y$ is such that

$$\max_{1 \le k \le m} f(x, y_k) > 0 \quad \text{for all } x \in X \ ,$$

then there exists $y_0 \in Y$ such that $f(x, y_0) > 0$ for all $x \in X$.

We use induction on m. For $m = 1$ there is nothing to prove. Assume that the claim is true for $m = p - 1$, and suppose that there is $\{y_1, \ldots, y_p\} \subset Y$ such that

$$\max_{1 \le k \le p} f(x, y_k) > 0 \quad \text{for all } x \in X \ .$$

Let $X_p = \{x \in X : f(x, y_p) \le 0\}$. We can assume that $X_p \neq \emptyset$, since otherwise there is nothing to prove. Then

$$\max_{1 \le k \le p-1} f(x, y_k) > 0 \quad \text{for all } x \in X_p \ .$$

By Step 1 we can apply the induction hypothesis to f restricted to the space $X_p \times Y$. It follows that there exists a $y_0' \in Y$ such that

$$f(x, y_0') > 0 \quad \text{for all } x \in X_p \ .$$

But now note that

$$\max\{f(x, y_p), f(x, y_0')\} > 0 \quad \text{for all } x \in X \ ,$$

due to the fact that $f(x, y_p) > 0$ for all $x \notin X_p$, and apply Step 2. This gives that there is a point $y_0 \in Y$ such that $f(x, y_0) > 0$ for all $x \in X$, which completes Step 3.

Step 4. We shall prove that for any real number α, either there is an $x_0 \in X$ such that

$$f(x_0, y) \leq \alpha \quad \text{for all } y \in Y \ ,$$

or there is a $y_0 \in Y$ such that

$$f(x, y_0) > \alpha \quad \text{for all } x \in X \ .$$

Suppose the first alternative is not true, and set

$$L(y; \alpha) = \{x \in X : f(x, y) \leq \alpha\} \ .$$

Then

$$\bigcap_{y \in Y} L(y; \alpha) = \emptyset \ .$$

But X is compact, so there must be a finite subfamily, $L(y_k; \alpha)$, $k = 1, \ldots, m$, with empty intersection. Thus the union of their complements is all of X, i.e.

$$\max_{k=1,\ldots,m} f(x, y_k) > \alpha \quad \text{for all } x \in X \ .$$

The result now follows by applying Step 3 to $f(x, y) - \alpha$.

Step 5. Let $\alpha = \sup_Y \min_X f(x, y)$. Then by Step 4 there is either an $x_0 \in X$ such that $f(x_0, y) \leq \alpha$ for all $y \in Y$, or there is a $y_0 \in Y$ such that $f(x, y_0) > \alpha$ for all $x \in X$. If the first alternative holds, then clearly

$$\sup_Y f(x_0, y) \leq \alpha \ ,$$

so

$$\min_X \sup_Y f((x, y) \leq \alpha = \sup_Y \min_X f(x, y) \ .$$

If the second alternative holds, then

$$\min_X f(x, y_0) > \alpha \ ,$$

since X is compact, and f is lower semicontinuous in x. Thus

$$\alpha = \sup_Y \min_X f(x, y) > \alpha \ ,$$

which is impossible.

2.5 The Dual Definition of Capacity

We shall apply the minimax theorem to the bilinear functional $\mathcal{E}_g(\mu, f)$ (see (2.3.3)) on $X \times Y$, where $X = \{\mu : \mu \in \mathcal{M}^+(K), \mu(K) = 1\}$ for a compact $K \subset \mathbf{R}^N$, and $Y = \{f : f \in L_+^p(\nu), \|f\|_{L^p(\nu)} \leq 1\}$. Thus X and Y are convex, X is compact in the weak* topology, and $\mu \mapsto \mathcal{E}_g(\mu, f)$ is lower semicontinuous on X for each fixed f by Proposition 2.3.2.

The key application is the following dual definition of capacity (cf. (2.2.5) in Theorem 2.2.7).

Theorem 2.5.1. *Let $K \subset \mathbf{R}^N$ be compact, and $1 < p < \infty$. Then*

$$C_{g,p}(K)^{1/p} = \sup\{\mu(K) : \mu \in \mathcal{M}^+(K), \|\check{\mathcal{G}}\mu\|_{L^{p'}(\nu)} \leq 1\} .$$

Proof. Notice that for any $f \in \Omega_K$ (see Definition 2.3.3)

$$\mu(K) \leq \int_{\mathbf{R}^N} \mathcal{G}f \, d\mu = \int_{\mathbf{M}} \check{\mathcal{G}}\mu \, f \, d\nu \leq \|\check{\mathcal{G}}\mu\|_{L^{p'}(\nu)} \|f\|_{L^p(\nu)} ,$$

whence

$$\mu(K) \leq \|\check{\mathcal{G}}\mu\|_{L^{p'}(\nu)} C_{g,p}(K)^{1/p} , \tag{2.5.1}$$

which implies that

$$\sup\{\mu(K) : \mu \in \mathcal{M}^+(K), \|\check{\mathcal{G}}\mu\|_{L^{p'}(\nu)} \leq 1\} \leq C_{g,p}(K)^{1/p} .$$

The proof of equality consists in identifying $\min_{\mu \in X} \sup_{f \in Y} \mathcal{E}_g(\mu, f)$ and $\sup_{f \in Y} \min_{\mu \in X} \mathcal{E}_g(\mu, f)$. We find on one hand

$$\sup_{f \in Y} \mathcal{E}_g(\mu, f) = \sup_{f \in Y} \int_{\mathbf{M}} \check{\mathcal{G}}\mu \, f \, d\nu = \|\check{\mathcal{G}}\mu\|_{L^{p'}(\nu)} ,$$

so that

$$\min_{\mu \in X} \sup_{f \in Y} \mathcal{E}_g(\mu, f) = \min_{\mu \in \mathcal{M}^+(K)} \frac{\|\check{\mathcal{G}}\mu\|_{L^{p'}(\nu)}}{\mu(K)} .$$

On the other hand

$$\min_{\mu \in X} \mathcal{E}_g(\mu, f) = \min_{x \in K} \mathcal{G}f(x) ,$$

so that, recalling Definition 2.3.3,

$$\sup_{f \in Y} \min_{\mu \in X} \mathcal{E}_g(\mu, f) = \sup_{f \in L_+^p(\nu)} \frac{\min_{x \in K} \mathcal{G}f(x)}{\|f\|_{L^p(\nu)}} = \sup_{f \in \Omega_K} \|f\|_{L^p(\nu)}^{-1} = C_{g,p}(K)^{-1/p} .$$

The conclusion follows from Theorem 2.4.1.

The dual definition of capacity can be extended to Suslin sets.

Corollary 2.5.2. *If $E \subset \mathbf{R}^N$ is a Suslin set, then*

$$C_{g,p}(E)^{1/p} = \sup\left\{ \mu(E) : \mu \in \mathcal{M}^+(\mathbf{R}^N),\ \operatorname{supp}\mu \in E,\ \|\check{\mathcal{G}}\mu\|_{L^{p'}(\nu)} \leq 1 \right\} .$$

Proof. This follows immediately from the capacitability of E.

Thanks to Theorem 2.5.1 we can now show that the capacitary function f^K (see Theorem 2.3.10) can be represented by means of a positive measure.

Theorem 2.5.3. *Let $K \subset \mathbf{R}^N$ be compact, and $1 < p < \infty$. Then there is a $\mu^K \in \mathcal{M}^+(K)$ such that $f^K = (\check{\mathcal{G}}\mu^K)^{p'-1}$, and*

$$\mu^K(K) = \int_{\mathbf{M}} (\check{\mathcal{G}}\mu^K)^{p'}\, d\nu = \int_{\mathbf{R}^N} \mathcal{G}f^K\, d\mu^K = C_{g,p}(K) .$$

Proof. Let $\{\mu_n\}_1^\infty$ be a sequence in $\mathcal{M}^+(K)$ such that $\|\check{\mathcal{G}}\mu_n\|_{L^{p'}(\nu)} = 1$ and

$$\lim_{n\to\infty} \mu_n(K) = C_{g,p}(K)^{1/p} .$$

We can assume that the sequence has a weak* limit $\mu \in \mathcal{M}^+(K)$, and then $\mu(K) = C_{g,p}(K)^{1/p}$. We know from Proposition 2.3.2(b) that $\check{\mathcal{G}}\mu(y)$ is lower semicontinuous on $\mathcal{M}^+(K)$ for each y, which ensures that $\|\check{\mathcal{G}}\mu\|_{L^{p'}(\nu)} \leq 1$, and thus $\|\check{\mathcal{G}}\mu\|_{L^{p'}(\nu)} = 1$ by Theorem 2.5.1.

Now normalize the extremal measure by defining $\mu^K = C_{g,p}(K)^{1/p'}\mu$, so that

$$\mu^K(K) = \|\check{\mathcal{G}}\mu^K\|_{L^{p'}(\nu)}^{p'} = C_{g,p}(K) .$$

The existence of this *capacitary measure* for K was proved in the case of the Bessel kernels in Theorem 2.2.7.

Let f^K be the capacitary function in Theorem 2.3.10, so that $\mathcal{G}f^K(x) \geq 1$ $C_{g,p}$-q.e. on K. Let $S = \{x \in K : \mathcal{G}f^K(x) < 1\}$. The estimate (2.5.1) implies that $\mu^K(F) = 0$ for every compact $F \subset S$, and thus $\mu^K(S) = 0$, since S is a Borel set. It follows that $\mathcal{G}f^K(x) \geq 1$ a.e.(μ^K) on K. Then, by Fubini's theorem and Hölder's inequality,

$$\begin{aligned} C_{g,p}(K) = \mu^K(K) &\leq \int_{\mathbf{R}^N} \mathcal{G}f^K\, d\mu^K = \int_{\mathbf{M}} \check{\mathcal{G}}\mu^K\, f^K\, d\nu \\ &\leq \|\check{\mathcal{G}}\mu^K\|_{L^{p'}(\nu)} \|f^K\|_{L^p(\nu)} = C_{g,p}(K)^{1/p'} C_{g,p}(K)^{1/p} = C_{g,p}(K) . \end{aligned}$$

It follows that we have equality in Hölder's inequality, and thus because of the normalization chosen,

$$(f^K)^p = (\check{\mathcal{G}}\mu^K)^{p'} ,$$

which proves the theorem.

We make a definition analogous to Definition 2.2.8.

Definition 2.5.4. For any $\mu \in \mathcal{M}^+(\mathbf{R}^N)$ the nonlinear potential $V^\mu_{g,p}$ is defined by

$$V^\mu_{g,p}(x) = \mathcal{G}(\check{\mathcal{G}}\mu)^{p'-1}(x) = \int_{\mathbf{M}} g(x,y)\left(\int_{\mathbf{R}^N} g(z,y)\,d\mu(z)\right)^{p'-1} d\nu(y)\ .$$

Again, the generalized energy is

$$\int_{\mathbf{R}^N} V^\mu_{g,p}\,d\mu = \int_{\mathbf{M}} (\check{\mathcal{G}}\mu)^{p'}\,d\nu\ .$$

The first part of the following theorem generalizes (2.2.3).

Theorem 2.5.5. *Let $K \subset \mathbf{R}^N$ be compact and $1 < p < \infty$. Then*

$$\mathcal{G}f^K(x) = V^{\mu^K}_{g,p}(x) \leq 1 \quad \textit{for all } x \in \operatorname{supp}\mu^K\ .$$

Moreover

$$C_{g,p}(K) = \max\{\,\mu(K) : \mu \in \mathcal{M}^+(K),\ V^\mu_{g,p}(x) \leq 1 \textit{ for all } x \in \operatorname{supp}\mu\,\}\ .$$

Proof. Suppose that $\mathcal{G}f^K(x_0) > 1$. Because of lower semicontinuity $\mathcal{G}f^K(x) \geq 1+\delta > 1$ on some neighborhood G of x_0. By Theorem 2.3.10 we have $\mathcal{G}f^K(x) \geq 1$ a.e.(μ^K), so

$$C_{g,p}(K) = \int_{\mathbf{R}^N} V^{\mu^K}_{g,p}\,d\mu^K \geq (1+\delta)\mu^K(G) + \mu^K(K \setminus G)$$
$$= \delta\mu^K(G) + \mu^K(K) = \delta\mu^K(G) + C_{g,p}(K)\ .$$

Thus $\mu^K(G) = 0$, which proves that $x_0 \notin \operatorname{supp}\mu^K$.

To prove the second part of the theorem we let $\mu \in \mathcal{M}^+(K)$, and assume that $V^\mu_{g,p}(x) \leq 1$ on $\operatorname{supp}\mu$. Then

$$\mu(K) \geq \int_K V^\mu_{g,p}\,d\mu = \|\check{\mathcal{G}}\mu\|^{p'}_{L^{p'}(\nu)}\ .$$

On the other hand, by (2.5.1),

$$\mu(K) \leq \|\check{\mathcal{G}}\mu\|_{L^{p'}(\nu)} C_{g,p}(K)^{1/p}\ ,$$

whence

$$\mu(K) \leq \mu(K)^{1/p'} C_{g,p}(K)^{1/p}\ ,$$

and $\mu(K) \leq C_{g,p}(K)$. But the measure μ^K gives equality, and this proves the theorem.

By Theorem 2.3.10 there is a unique capacitary function f^E corresponding to any set $E \subset \mathbf{R}^N$ with $C_{\alpha,p}(E) < \infty$. We want to extend the potential representation of the capacitary function in Theorem 2.5.3 to general sets. For this purpose we assume that the set $\mathbf{M}$ is a locally compact topological space, and we make a mild assumption on the kernel g. It is easily seen that this hypothesis is satisfied if g is e.g. a Bessel kernel, $g(x,y) = G_\alpha(x-y)$.

Theorem 2.5.6. *Let $E \subset \mathbf{R}^N$, let $1 < p < \infty$, and suppose $C_{g,p}(E) < \infty$. Suppose further that $\mathbf{M}$ has a locally compact topology, and that the kernel g is such that $\mathcal{G}\varphi$ is continuous on $\mathbf{R}^N$ and*

$$\lim_{|x|\to\infty} \mathcal{G}\varphi(x) = 0$$

for any $\varphi \in C_0(\mathbf{M})$. Then there is a $\mu^E \in \mathcal{M}^+(\overline{E})$, again called a capacitary measure (*or* outer capacitary measure) *for E, such that*

$$f^E = (\check{\mathcal{G}}\mu^E)^{p'-1} .$$

Moreover,

$$\mathcal{G}f^E(x) \geq 1 \quad C_{g,p}\text{-q.e. on } E ,$$
$$\mathcal{G}f^E(x) \leq 1 \quad \text{on } \operatorname{supp} \mu^E ,$$

and

$$\mu^E(\overline{E}) = \int_{\mathbf{M}} (\check{\mathcal{G}}\mu^E)^{p'} \, d\nu = \int_{\mathbf{R^N}} \mathcal{G}f^E \, d\mu^E = C_{g,p}(E) .$$

Proof. Since $C_{g,p}$ is an outer capacity, there is a G_δ set H such that $E \subset H \subset \overline{E}$ and $C_{g,p}(E) = C_{g,p}(H)$. Since H is $C_{g,p}$-capacitable, there exists an increasing sequence of compact sets, $\{K_i\}_1^\infty$, such that $\cup_1^\infty K_i = S \subset H$ and $C_{g,p}(S) = C_{g,p}(H)$.

By Theorem 2.5.3 we have $f^{K_i} = (\check{\mathcal{G}}\mu^{K_i})^{p'-1}$, where $\mu^{K_i}(K_i) = C_{g,p}(K_i)$. Proposition 2.3.12 shows that $\{f^{K_i}\}$ converges strongly in $L^p(\nu)$ to f^S. The uniqueness of the extremal function, Theorem 2.3.10, then gives that $f^S = f^H = f^E$. In fact, f^S, f^H, and f^E all have the same norm, and f^H belongs both to $\overline{\Omega}_E$ and to $\overline{\Omega}_S$.

Furthermore, we can assume, after extraction of a subsequence, that $\{\mu^{K_i}\}$ converges weak* to some $\mu^E \in \mathcal{M}^+(\overline{E})$ with $\mu^E(\overline{E}) \leq C_{g,p}(S) = C_{g,p}(E)$.

We claim that $f^E = (\check{\mathcal{G}}\mu^E)^{p'-1}$, and this is where we need the additional assumption on g. Let $\varphi \in C_0(\mathbf{M})$. Then, by assumption,

$$\lim_{i\to\infty} \mathcal{E}_g(\mu^{K_i}, \varphi) = \lim_{i\to\infty} \int_{\mathbf{R}^N} \mathcal{G}\varphi \, d\mu^{K_i} = \int_{\mathbf{R}^N} \mathcal{G}\varphi \, d\mu^E = \mathcal{E}_g(\mu^E, \varphi) .$$

Also, $\{(f^{K_i})^{p-1}\}_1^\infty$ is bounded in $L^{p'}(\nu)$, so we can assume that the sequence is weakly convergent in $L^{p'}(\nu)$ to $(f^E)^{p-1}$. Thus

$$\lim_{i\to\infty} \mathcal{E}_g(\mu^{K_i}, \varphi) = \lim_{i\to\infty} \int_{\mathbf{M}} (f^{K_i})^{p-1} \varphi \, d\nu = \int_{\mathbf{M}} (f^E)^{p-1} \varphi \, d\nu .$$

Hence

$$\int_{\mathbf{M}} (f^E)^{p-1} \varphi \, d\nu = \mathcal{E}_g(\mu^E, \varphi) = \int_{\mathbf{M}} \check{\mathcal{G}}\mu^E \varphi \, d\nu$$

for all $\varphi \in C_0(\mathbf{M})$. Thus $(f^E)^{p-1} = \check{\mathcal{G}}\mu^E$, which proves the claim.

We now prove that $\mathcal{G}f^E(x) \le 1$ on $\operatorname{supp}\mu^E$. Let $x \in \operatorname{supp}\mu^E$ and choose $x_i \in \operatorname{supp}\mu^{K_i}$ so that $x_i \to x$. By Theorem 2.5.5 we have $\mathcal{G}f^{K_i}(x_i) \le 1$. By choosing a subsequence we can assume that $f^{K_i}(y) \to f^E(y)$ ν-a.e. It follows from the lower semicontinuity of $g(\cdot, y)$ for each y, and Fatou's lemma that

$$\mathcal{G}f^E(x) = \int_{\mathbf{M}} g(x,y)\, f^E(y)\, d\nu(y) \le \int_{\mathbf{M}} \liminf_{i\to\infty}\big(g(x_i,y)\, f^{K_i}(y)\big)\, d\,\nu(y)$$

$$\le \liminf_{i\to\infty} \int_{\mathbf{M}} g(x_i,y)\, f^{K_i}(y)\, d\,\nu(y) \le 1\ .$$

It then follows by Fubini's theorem that

$$C_{g,p}(E) = \int_{\mathbf{M}} (f^E)^p\, d\nu = \int_{\mathbf{R}^N} \mathcal{G}f^E\, d\mu^E \le \mu^E(\overline{E})\ .$$

This proves the theorem.

2.6 Radially Decreasing Convolution Kernels

We shall say that a function g on $\mathbf{R}^N \times \mathbf{R}^N$ is a *radially decreasing convolution kernel* if $g(x,y) = g_0(|x-y|)$, where g_0 is a non-negative, lower semi-continuous, non-increasing function on $\mathbf{R}^+$ for which $\int_0^1 g_0(t)t^{N-1}\,dt < \infty$. We shall also write $g(x) = g_0(|x|)$ for $x \in \mathbf{R}^N$, as well as $\mathcal{G}f = g * f$, and $\check{\mathcal{G}}\mu = g * \mu$. In most applications the kernel g will be either a Riesz kernel I_α or a Bessel kernel G_α. In this case we shall write $\dot{C}_{\alpha,p}$ and $C_{\alpha,p}$ for the capacity $C_{g,p}$, and talk about Riesz and Bessel capacity, respectively. Similarly, we denote Riesz and Bessel nonlinear potentials by $\dot{V}^\mu_{\alpha,p}$ and $V^\mu_{\alpha,p}$.

We shall deduce some important properties of the nonlinear potentials associated with such kernels. We start by the following observation.

Proposition 2.6.1. *Let* $1 < p < \infty$. *If* g *is a radially decreasing convolution kernel, then*

(a) $\int_{\mathbf{R}^N} g(x)^{p'}\,dx < \infty$ *implies that* $C_{g,p}(\{a\}) > 0$ *for any* $a \in \mathbf{R}^N$;
(b) $\int_{|x|>1} g(x)^{p'}\,dx = \infty$ *implies that* $C_{g,p}(E) = 0$ *for all* E;
(c) $\int_{|x|>1} g(x)^{p'}\,dx < \infty$ *implies that* $|E| = 0$ *whenever* $C_{g,p}(E) = 0$.

Remark. The statements (a) and (b) mean that in order to have a capacity that is useful for measuring small sets, we should assume that $g \notin L^{p'}$, but that $\int_{|x|>1} g(x)^{p'}\,dx < \infty$. Then (c) says that in this case capacity is a more sensitive measure than N-dimensional Lebesgue measure.

Notice that the Riesz kernel I_α satisfies $I_\alpha \notin L^{p'}$ for all α, $0 < \alpha < N$, but that $\int_{|x|>1} I_\alpha^{p'}\,dx < \infty$ only for $0 < \alpha p < N$. On the other hand, the Bessel kernel G_α satisfies $G_\alpha \notin L^{p'}$ only when $\alpha p \le N$, but $\int_{|x|>1} G_\alpha^{p'}\,dx < \infty$ for all $\alpha > 0$. Thus $\dot{C}_{\alpha,p}(E) = 0$ for all nonempty E if $\alpha p \ge N$, and $C_{\alpha,p}(\{0\}) > 0$ if $\alpha p > N$.

Notice also that it follows immediately from the Sobolev imbedding theorem, Theorem 1.2.4, that there is a constant A depending only on α, p, and N, such that for all measurable sets E

$$|E|^{1-\alpha p/N} \leq A\,\dot{C}_{\alpha,p}(E) \ , \tag{2.6.1}$$

when $0 < \alpha < N$, $1 < p < N/\alpha$.

Proof. (a) Suppose that $\|g\|_{p'} < \infty$. Set $a = 0$ and let δ be the Dirac measure at 0. Then, by Theorem 2.5.1,

$$C_{g,p}(\{0\})^{1/p} = \sup\left\{\frac{\mu(\{a\})}{\|g * \mu\|_{p'}} : \mu \in \mathcal{M}^+(\{a\})\right\} = \|g * \delta\|_{p'}^{-1} = \|g\|_{p'}^{-1} > 0 \ .$$

(b) Suppose that $\int_{|x|>1} g(x)^{p'}\,dx = \infty$. It suffices to show that $C_{g,p}(B) = 0$ for the unit ball $B = B(0,1)$, since any set E can be covered by a countable number of balls with radius 1, all of which have the same capacity by the translation invariance of capacity for a convolution kernel.

But if the positive measure μ is supported on B, then

$$g * \mu(x) \geq g_0(|x| + 1)\mu(B) \ .$$

Hence $\|g * \mu\|_{p'} = \infty$, and by Theorem 2.5.1, $C_{g,p}(B) = 0$.

(c) Suppose that $\int_{|x|>1} g(x)^{p'}\,dx < \infty$. It suffices to consider measurable sets E and show that $|E \cap B| = 0$ for the unit ball B. Let $F = E \cap B$ and take $f \in L^p_+$ such that $g * f \geq 1$ on F. Then

$$|F| \leq \int_F g * f\,dx = \int_{\mathbf{R}^N} f(g * \chi_F)\,dx \leq \|f\|_p \, \|g * \chi_F\|_{p'} \ ,$$

where χ_F is the characteristic function of F. But $g * \chi_F(x) \leq A\,g(|x|/2)$ when $|x| \geq 2$, and $g * \chi_F(x) \leq A$ when $|x| < 2$, where A is a constant independent of E. Thus $|E \cap B| \leq A\,C_{g,p}(E)^{1/p}$.

We now show that a classical result, called the "boundedness principle", can be extended to nonlinear potentials. We begin by recalling the linear case.

Theorem 2.6.2. *Let g be a radially decreasing convolution kernel, and let $\mu \in \mathcal{M}^+(\mathbf{R}^N)$. Then there is a constant Q, depending only on N, such that for all $x \in \mathbf{R}^N$*

$$g * \mu(x) \leq Q \sup_{y \in \operatorname{supp}\mu} g * \mu(y) \ .$$

Proof. We assume, as we may, that $\sup_{\operatorname{supp}\mu} g * \mu(y) = 1$. Let $x \notin \operatorname{supp}\mu$, and let $\Gamma_1, \ldots, \Gamma_Q$ be closed circular cones with vertices at x and total angular opening at the vertex of $\pi/3$, such that $\bigcup_{i=1}^{Q} \Gamma_i = \mathbf{R}^N$. Thus, if $N = 2$, Q can be taken equal to 6.

Denote the restriction of μ to Γ_i by μ_i. Let x_i be a point of $\operatorname{supp}\mu_i$ such that $|x - x_i| = \operatorname{dist}(x, \operatorname{supp}\mu_i)$, let Π_i be the (hyperplane) perpendicular bisector of the line segment from x to x_i, and let Π_i^+ and Π_i^- denote the halfspaces determined by Π_i. It is seen by elementary geometry that if $x_i \in \Pi_i^-$, then $\operatorname{supp}\mu_i \subset \Pi_i^-$. Thus, if $y \in \operatorname{supp}\mu_i$, then $|y - x_i| \le |y - x|$, so that $g(x-y) \le g(x_i - y)$, and thus $g * \mu_i(x) \le g * \mu_i(x_i) \le 1$. Consequently, $g * \mu(x) \le \sum g * \mu_i(x) \le Q$.

Remark. This boundedness principle should be distinguished from the maximum principle of A. J. Maria and O. Frostman, where $Q = 1$. The latter holds when g is continuous and subharmonic on $\mathbf{R}^N \setminus \{0\}$, in particular when g is a Riesz kernel, I_α, or a Bessel kernel, G_α, with $0 < \alpha \le 2$. It is false when $\alpha > 2$. See e.g. L. Carleson [92], or N. S. Landkof [266], Theorem 1.10. Notice, however, that if the support of μ is known to be convex, then only one hyperplane is necessary, and we can take $Q = 1$ in the theorem. The same holds true if we seek a bound on the potential in terms of its values on the closed convex hull of the support.

Theorem 2.6.3 (Boundedness Principle). *Let g, μ, and Q be as in Theorem 2.6.2. Then for all $x \in \mathbf{R}^N$*

$$V_{g,p}^{\mu}(x) \le \max\{Q^{p'-1}, Q\} \sup_{y \in \operatorname{supp}\mu} V_{g,p}^{\mu}(y) . \tag{2.6.2}$$

Proof. Let $x \notin \operatorname{supp}\mu$, and let $x_0 \in \operatorname{supp}\mu$ minimize the distance from x to $\operatorname{supp}\mu$. Suppose first of all that $\operatorname{supp}\mu \subset \Pi^+$, one of the halfspaces determined by the (hyperplane) perpendicular bisector Π of the segment from x to x_0. If a point y_- belongs to the other halfspace, Π^-, we denote its reflected point in Π by $y_+ \in \Pi^+$. Then, for all $z \in \operatorname{supp}\mu$ we have $|z - y_-| > |z - y_+|$. Thus $g * \mu(y_-) \le g * \mu(y_+)$, and also $f(y_-) \le f(y_+)$ for $f(y) = (g * \mu(y))^{p'-1}$.

We now claim that $g * f(x_-) \le g * f(x_+)$ for all $x_- \in \Pi^-$. (Notice that x_0 is a possible x_+.) To see this we first proceed as if all terms below are finite. Then the claim holds if and only if

$$\int_{\Pi^-} \big(g(x_- - y) - g(x_+ - y)\big) f(y)\,dy \le \int_{\Pi^+} \big(g(x_+ - y) - g(x_- - y)\big) f(y)\,dy \tag{2.6.3}$$

for all $x_- \in \Pi^-$. But notice that

$$g(x_- - y_-) - g(x_+ - y_-) = g(x_+ - y_+) - g(x_- - y_+) , \tag{2.6.4}$$

and that since $|x_- - y_-| \le |x_+ - y_-|$ for $y_- \in \Pi^-$, both sides in (2.6.4) are non-negative. Thus, multiplying both sides of $f(y_-) \le f(y_+)$ by (2.6.4) and integrating over points and their reflections yields (2.6.3).

To see that our claim holds even if one or more terms of (2.6.3) are infinite, we replace g by a truncated kernel, for example by g_m defined by $g_m(x) = \max\{0, \min\{g(x) - m^{-1}, m\}\}$ for $m = 1, 2, \ldots$, and apply monotone convergence.

To handle arbitrary measures $\mu \in \mathcal{M}^+$, we again choose $x \notin \operatorname{supp}\mu$, and subdivide $\mathbf{R}^N = \bigcup_{i=1}^{Q} \Gamma_i$ as in Theorem 2.6.2. With μ_i and x_i as in that theorem we have $V_{g,p}^{\mu_i}(x) \le V_{g,p}^{\mu_i}(x_i)$. Now, for $1 < p \le 2$ Hölder's inequality gives

$$V_{g,p}^{\mu}(x) \leq Q^{p'-2} \sum_{i=1}^{Q} V_{g,p}^{\mu_i}(x) \ ,$$

whereas, when $p > 2$ and consequently $p' - 1 < 1$, the elementary inequality

$$\left(\sum a_i\right)^q \leq \sum a_i^q, \quad 0 < q \leq 1, \quad a_i \geq 0 \ , \tag{2.6.5}$$

gives

$$V_{g,p}^{\mu}(x) \leq \sum_{i=1}^{Q} V_{g,p}^{\mu_i}(x) \ .$$

The inequality (2.6.5) is trivial if $\sum a_i = 1$, and the general case is easily reduced to this if a_i is replaced by $a_i / \sum a_j$.

Remark 1. As in Theorem 2.6.2 we can take $Q = 1$ if $\operatorname{supp} \mu$ is convex.

Remark 2. The proof works equally well for nonlinear potentials of the form $g_1 * (g_2 * \mu)^{p'-1}$, where both g_1 and g_2 are radially decreasing convolution kernels.

Corollary 2.6.4. *If g is a radially decreasing convolution kernel and E is a set with $C_{g,p}(E) < \infty$, then the (g, p)-capacitary potential $g * f^E$ is a bounded function on $\mathbf{R}^N$.*

As an application of the boundedness principle we can now extend another result from classical potential theory, the so called "continuity principle".

Theorem 2.6.5 (Continuity Principle). *Let g be a radially decreasing convolution kernel, continuous on $\mathbf{R}^N \setminus \{0\}$. Let $\mu \in \mathcal{M}^+(\mathbf{R^N})$ be a measure with compact support,* $\operatorname{supp} \mu = K$*, and suppose that the restriction of $V_{g,p}^{\mu}$ to K belongs to $C(K)$. Then $V_{g,p}^{\mu}$ is continuous in $\mathbf{R}^N$.*

Proof. By Dini's theorem on monotone convergence the integral $g * f(x)$, where $f = (g * \mu)^{p'-1}$, converges uniformly on K in the sense that for any $\varepsilon > 0$ there is $\delta > 0$, such that

$$\int_{|x-y|<\delta} g(x-y) f(y) \, dy < \varepsilon \quad \text{for all } x \in K \ .$$

Now, if Theorem 2.6.3 and the second remark following it are applied to the kernels g and g_δ, defined by $g_\delta(x) = g(x)$ for $|x| < \delta$, $g_\delta(x) = 0$ otherwise, we see that

$$g_\delta * f(x) = \int_{|x-y|<\delta} g(x-y) f(y) \, dy < Q\varepsilon \quad \text{for all } x \in \mathbf{R}^N \ ,$$

where Q depends only on N and p. Let $x_0 \in K$, and take a sequence $x_n \to x_0$, $n = 1, 2, \ldots$. Then, by the continuity of g away from the origin

$$\limsup_{n\to\infty} V_{g,p}^{\mu}(x_n) \leq \big((g - g_\delta) * f\big)(x_0) + Q\varepsilon \leq V_{g,p}^{\mu}(x_0) + Q\varepsilon \ .$$

But $V^\mu_{g,p}$ is lower semicontinuous, so

$$V^\mu_{g,p}(x_0) \le \liminf_{n\to\infty} V^\mu_{g,p}(x_n) \ .$$

It follows that $V^\mu_{g,p}$ is continuous at all $x \in K$. But the continuity off K is clear from the continuity of g.

The following corollary is also an extension of a well known theorem in classical potential theory.

Corollary 2.6.6. *Let g be as in Theorem 2.6.5, and let $K \subset \mathbf{R}^N$ be a compact set such that $C_{g,p}(K) > 0$. Then there is a $\mu \in \mathcal{M}^+(K)$ such that $\mu \ne 0$ and $V^\mu_{g,p}$ is continuous in $\mathbf{R}^N$.*

Proof. It follows from the assumption that there is a nonzero $\mu \in \mathcal{M}^+(K)$ such that $V^\mu_{g,p}(x) \le 1$ everywhere on K. By Egorov's theorem there is a compact $K' \subset K$ such that $\mu(K') > \frac{1}{2}\mu(K)$ and $g * f(x) = g * (g * \mu)^{p'-1}(x)$ converges uniformly on K'. If we denote the restriction of μ to K' by μ', the integral $g * (g * \mu')^{p'-1}(x) = V^{\mu'}_{g,p}(x)$ also converges uniformly on K', and it follows as before that $V^{\mu'}_{g,p}$ is continuous.

In Chapter 6 we shall study the continuity properties of potentials in some detail. The following proposition is a first, rough result in this direction. See also Theorem 11.3.2.

Proposition 2.6.7. *Let $1 < p < \infty$, and let $g(x) = g_0(|x|)$ be a radially decreasing convolution kernel, continuous on $\mathbf{R}^N \setminus \{0\}$, and such that $\int_{|x|>1} g^{p'}\,dx < \infty$. Assume that there is an L and a $\delta > 0$ such that g_0 satisfies*

$$g_0(r) \le L\, g_0(2r) \quad \textit{for } 0 < r \le \delta \ . \tag{2.6.6}$$

Let $f \in L^p_+(\mathbf{R}^N)$, and suppose that $g * f(x) \ge 1$ a.e. on an open set U. Then $g * f(x) \ge 1$ everywhere on U.

Proof. Without loss of generality we can assume that $g * f(x) \ge 1$ a.e. on a neighborhood of 0, and prove that $g * f(0) \ge 1$. We can also assume that

$$g * f(0) = \int_{\mathbf{R}^N} g\, f\, dx < \infty \ .$$

Let $0 < a < b$ and define a weight function $\eta = \eta_{a,b}$ by

$$\eta(x) = \frac{g(x)}{\int_{|y|<|x|} g(y)\,dy} \quad \text{for } a < |x| < b \ , \tag{2.6.7}$$

and $\eta(x) = 0$ otherwise. If we set $\int_{|y|<r} g(y)\,dy = G(r)$ it follows that $\int_{\mathbf{R}^N} \eta\,dx = \log G(b) - \log G(a)$. But $\lim_{a\to 0} G(a) = 0$, so for an arbitrarily small b we can choose a so that $\int_{\mathbf{R}^N} \eta\,dx = 1$.

Then, clearly, for small enough b,

$$1 \leq \int_{\mathbf{R}^N} \eta\,(g * f)\,dx = \int_{\mathbf{R}^N} (\eta * g)\, f\, dy \ , \tag{2.6.8}$$

If a number ρ is fixed so that $0 < \rho \leq \delta$, then

$$\lim_{a,b\to 0} \eta * g(y) = \lim_{a,b\to 0} \int_{\mathbf{R}^N} \eta(x)\, g(y-x)\, dx = g(y) \ ,$$

uniformly for $|y| \geq \rho$, and also

$$\int_{\mathbf{R}^N} \eta(x)\, g(y-x)\, dx \leq g_0(|y| - b) \ .$$

Thus, for any $R < \infty$,

$$\lim_{a,b\to 0} \int_{\rho\leq|y|\leq R} (\eta * g)\, f\, dy = \int_{\rho\leq|y|\leq R} g\, f\, dy \ ,$$

and

$$\int_{|y|\geq R} (\eta * g)\, f\, dy \leq \|f\|_p \left(\int_{|y|\geq R-b} g^{p'}\, dy \right)^{1/p'} < \varepsilon \ ,$$

if R is large enough.

It is in order to estimate $\eta * g(y)$ for $|y| \leq \rho$ that we need the special choice (2.6.7) of the weight function η. We observe that if $|x-y| \leq \frac{1}{2}|y|$, then $|x| \geq \frac{1}{2}|y|$, and $|x| \geq |x-y|$. Thus, by the monotonicity of g,

$$\begin{aligned}
\int_{|x-y|\leq\frac{1}{2}|y|} \eta(x)\, g(y-x)\, dx &= \int_{|x-y|\leq\frac{1}{2}|y|} \frac{g(x)g(y-x)}{\int_{|t|<|x|} g(t)\, dt}\, dx \\
&\leq g(\tfrac{1}{2}y) \int_{|x-y|\leq\frac{1}{2}|y|} \frac{g(y-x)}{\int_{|t|<|x-y|} g(t)\, dt}\, dx \\
&\leq g(\tfrac{1}{2}y) \int_{\mathbf{R}^N} \eta(x)\, dx = g(\tfrac{1}{2}y) \ .
\end{aligned}$$

Again by monotonicity,

$$\int_{|x-y|\geq\frac{1}{2}|y|} \eta(x)\, g(y-x)\, dx \leq g(\tfrac{1}{2}y) \int_{\mathbf{R}^N} \eta(x)\, dx = g(\tfrac{1}{2}y) \ .$$

Then the assumption (2.6.6) gives that $\eta * g(y) \leq 2Lg(y)$, if $|y| \leq 2\delta$, and thus

$$\int_{|y|\leq\rho} (\eta * g)\, f\, dy \leq 2L \int_{|y|\leq\rho} g\, f\, dy < \varepsilon \ ,$$

if ρ is small enough.

Letting $b \to 0$, we obtain from (2.6.8)

$$1 \leq \int_{\rho\leq|y|\leq R} g\, f\, dy + 2\varepsilon \ .$$

The proposition follows if $\rho \to 0$ and $R \to \infty$.

Remark. If, for example, $g = I_\alpha$, or $g = G_\alpha$, $0 < \alpha < N$, the weighted averages with weight η in the above proof can be replaced with ordinary averages over balls. We leave this simplification to the reader.

As a corollary of Proposition 2.6.7 we get an equivalent formulation of the definition of capacity. We formulate it for (α, p)-capacity (cf. Definitions 2.2.6 and 2.3.3).

Corollary 2.6.8. *Let* $1 < p < \infty$, $\alpha > 0$, *and let* $E \subset \mathbf{R}^N$. *Then*

$$C_{\alpha,p}(E) = \inf_F \|F\|_{\alpha,p}^p ,$$

where the infimum is taken over all $F \in L^{\alpha,p}$ *such that* $F(x) \geq 1$ *a.e. on some neighborhood of* E.

Proof. Let $F \in L^{\alpha,p}$, and assume that $F(x) \geq 1$ a.e. on an open set U containing E. Then $F = G_\alpha * f$ for an $f \in L^p$, and also $G_\alpha * f_+(x) \geq 1$ a.e. on U. Proposition 2.6.7 gives that $G_\alpha * f_+(x) \geq 1$ everywhere on U, and thus $C_{\alpha,p}(U) \leq \|f\|_p^p$. The corollary now follows from the fact that $C_{\alpha,p}$ is an outer capacity (Proposition 2.3.5).

We finally give a lower estimate for nonlinear potentials that will be of use later.

Proposition 2.6.9. *Let* $g(x) = g_0(|x|)$ *be a radially decreasing convolution kernel satisfying* (2.6.6) *for a* $\delta > 0$. *Then there is a constant* $A > 0$ *such that for any measure* $\mu \in \mathcal{M}^+(\mathbf{R}^N)$

$$V_{g,p}^\mu(0) \geq A \int_0^\delta g_0(r)^{p'} r^{N-1} \mu(B(0,r))^{p'-1}\, dr .$$

Proof. Using the assumption we find

$$\begin{aligned}
V_{g,p}^\mu(0) &= \int_{\mathbf{R}^N} g(x) \Big(\int_{\mathbf{R}^N} g(y-x)\, d\mu(y) \Big)^{p'-1} dx \\
&\geq \sum_{n=0}^\infty \int_{\delta 2^{-n-1} < |x| \leq \delta 2^{-n}} g(x) \Big(\int_{|y| \leq \delta 2^{-n}} g(y-x)\, d\mu(y) \Big)^{p'-1} dx \\
&\geq \sum_{n=0}^\infty \int_{\delta 2^{-n-1} < |x| \leq \delta 2^{-n}} g(x)\, g\big(\delta 2^{-n+1}\big)^{p'-1} \mu\big(B(0, \delta 2^{-n})\big)^{p'-1} dx \\
&\geq L^{-2} \int_{B(0,\delta)} g(x)^{p'} \mu\big(B(0,|x|)\big)^{p'-1} dx \\
&= A \int_0^\delta g_0(r)^{p'} \mu\big(B(0,r)\big)^{p'-1} r^{N-1}\, dr .
\end{aligned}$$

Remark. If we specialize to the Riesz kernel I_α and the Bessel kernel G_α the proposition gives

$$\dot{V}^{\mu}_{\alpha,p}(0) \geq A \int_0^\infty \left(\frac{\mu(B(0,r))}{r^{N-\alpha p}}\right)^{p'-1} \frac{dr}{r}\,, \tag{2.6.9}$$

and

$$V^{\mu}_{\alpha,p}(0) \geq A \int_0^\delta \left(\frac{\mu(B(0,r))}{r^{N-\alpha p}}\right)^{p'-1} \frac{dr}{r}\,, \tag{2.6.10}$$

or, denoting $B(0,2^{-n})$ by $B_n(0)$, and choosing an $n_0 \in \mathbf{Z}$,

$$\dot{V}^{\mu}_{\alpha,p}(0) \geq A \sum_{n=-\infty}^{\infty} \left(2^{n(N-\alpha p)}\mu(B_n(0))\right)^{p'-1}\,, \tag{2.6.11}$$

and

$$V^{\mu}_{\alpha,p}(0) \geq A \sum_{n=n_0}^{\infty} \left(2^{n(N-\alpha p)}\mu(B_n(0))\right)^{p'-1}\,. \tag{2.6.12}$$

The quantities on the right, sometimes called Wolff potentials, and denoted $\dot{W}^{\mu}_{\alpha,p}$ and $W^{\mu}_{\alpha,p}$, will have an important part to play later. See e.g. Section 4.5.

2.7 An Alternative Definition of Capacity and Removability of Singularities

We shall now address a question that has perhaps already occurred to the reader. In Definitions 2.2.1, 2.2.6 and 2.3.3, we allowed the competing functions to be greater than or equal to one on the set. What would happen if we required the functions to be equal to one on the set?

We modify Definition 2.2.6 in the following way.

Definition 2.7.1. *Let $K \subset \mathbf{R}^N$ be compact, and set*

$$\widetilde{\omega}_K = \{\varphi \in \mathcal{S} : \varphi = 1 \textit{ on a neighborhood of } K\}\,.$$

Let $\alpha > 0$ and $1 < p < \infty$. Then

$$N_{\alpha,p}(K) = \inf\{\, \|\varphi\|^p_{\alpha,p} : \varphi \in \widetilde{\omega}_K \,\}\,.$$

The definition is extended to arbitrary sets as in Definitions 2.2.2 and 2.2.4.

We can show that there is a dual definition of $N_{\alpha,p}(K)$, similar to the one given in Theorems 2.2.7 and 2.5.1. In fact, we have the following theorem.

Theorem 2.7.2. *Let $K \subset \mathbf{R}^N$ be compact, let $\alpha > 0$, and $1 < p < \infty$. Then there is a unique element $F^K = G_\alpha * f^K$ in the closure of $\widetilde{\omega}_K$ such that $\|f^K\|_p^p = N_{\alpha,p}(K)$. Moreover, there is a distribution $T^K \in L^{-\alpha,p'} \cap \mathcal{D}'(K)$, such that*

$$f^K = (G_\alpha * T^K)\,|G_\alpha * T^K|^{p'-2}\ ; \tag{2.7.1}$$

$$\langle T^K, 1\rangle = N_{\alpha,p}(K)\ ; \tag{2.7.2}$$

$$N_{\alpha,p}(K)^{1/p} = \sup_{T\in\mathcal{D}'(K)} \frac{\langle T, 1\rangle}{\|G_\alpha * T\|_{p'}}\ . \tag{2.7.3}$$

Proof. By uniform convexity there is a unique element, $F^K = G_\alpha * f^K$, in the closure of $\widetilde{\omega}_K$ such that $\|F^K\|_{\alpha,p}^p = N_{\alpha,p}(K)$. Let $\varphi = G_\alpha * \psi$ be a function in $\mathcal{S}$ whose support does not intersect K. Then as in the proof of Theorem 2.2.7 it follows from the extremal property of F^K that

$$\int_{\mathbf{R}^N} f^K |f^K|^{p-2} \psi\, dx = 0\ .$$

Set $f^K|f^K|^{p-2} = h$, so that $h \in L^{p'}$, and $f^K = h|h|^{p'-2}$. Denote by T^K the distribution $G_{-\alpha} * h$, so that $T^K \in L^{-\alpha,p'}$. Thus

$$\langle T^K, \varphi\rangle = \int_{\mathbf{R}^N} (G_\alpha * T^K)\psi\, dx = \int_{\mathbf{R}^N} h\psi\, dx = 0\ .$$

Since this is true for all test functions φ with $\operatorname{supp}\varphi \subset K^c$, we have $\operatorname{supp} T^K \subset K$. Thus $f^K = (G_\alpha * T^K)|G_\alpha * T^K|^{p'-2}$. Here it should be noticed that, in distinction to what was the case in Theorem 2.2.7, we cannot conclude that $T^K \geq 0$.

Now let T be a distribution with $\operatorname{supp} T \subset K$, i.e. $T \in \mathcal{D}'(K)$, and let $F = G_\alpha * f \in \widetilde{\omega}_K$. Then clearly

$$\langle T, 1\rangle = \langle T, F\rangle = \int_{\mathbf{R}^N} (G_\alpha * T) f\, dx \leq \|G_\alpha * T\|_{p'} \|f\|_p\ .$$

A passage to the limit, using Hölder's inequality, shows that the same is true for F in the closure of $\widetilde{\omega}_K$, and in particular for $F^K = G_\alpha * f^K$. Thus

$$\begin{aligned}\langle T, 1\rangle = \langle T, F^K\rangle &= \int_{\mathbf{R}^N} (G_\alpha * T) f^K\, dx \\ &\leq \|G_\alpha * T\|_{p'} \|f^K\|_p = \|G_\alpha * T\|_{p'} N_{\alpha,p}(K)^{1/p}\ .\end{aligned}$$

Taking $T = T^K$ gives equality, and thus

$$\sup_{T\in\mathcal{D}'(K)} \frac{\langle T, 1\rangle}{\|G_\alpha * T\|_{p'}} = N_{\alpha,p}(K)^{1/p}\ .$$

This proves the theorem.

The absence of a representation of the extremal by means of positive measures is a great drawback in the study of $N_{\alpha,p}$ as compared to $C_{\alpha,p}$. On the other hand $N_{\alpha,p}$ sometimes appears naturally, as the next theorem illustrates. We first make a definition.

Definition 2.7.3. *Let $K \subset \mathbf{R}^N$ be compact, and let $\mathcal{L}$ be a partial differential operator defined in a neighborhood of K. Then K is said to be removable for $\mathcal{L}$ in L^p if any solution u of $\mathcal{L}u = 0$ in $O \setminus K$ for some bounded open neighborhood O of K, such that $u \in L^p(O \setminus K)$, can be extended to a function $\tilde{u} \in L^p(O)$ such that $\mathcal{L}\tilde{u} = 0$ in O.*

We give a result that can be proved without going deeply into the theory of partial differential equations.

Theorem 2.7.4. *Let $\mathcal{L}$ be an elliptic linear partial differential operator of order $\alpha < N$ with constant coefficients, and let $K \subset \mathbf{R}^N$ be compact. Then K is removable for $\mathcal{L}$ in L^p, $1 < p < \infty$, if $N_{\alpha,p'}(K) = 0$, and it is not removable if $C_{\alpha,p'}(K) > 0$.*

Proof. First assume that $C_{\alpha,p'}(K) > 0$. Then there is a nonzero $\mu \in \mathcal{M}^+(K)$ such that $G_\alpha * \mu \in L^p(\mathbf{R}^N)$. Let $\mathcal{E}$ be the fundamental solution of $\mathcal{L}$. By the properties of fundamental solutions of elliptic linear operators there is a constant A such that $|\mathcal{E}(x)| \le A\,|x|^{\alpha-N}$ for small $|x|$. This is well known if $\mathcal{L}$ is the Cauchy–Riemann operator $\partial/\partial\bar{z}$, or an integral power of the Laplacian (see (1.2.7), and L. Hörmander [230], Theorem 3.3.2). For general operators we refer to F. John [236], pp. 61–65. It follows that $\mathcal{E} * \mu \in L^p_{\mathrm{loc}}$. Moreover, $\mathcal{E} * \mu$ is a solution of $\mathcal{L}u = 0$ in K^c, which proves that K is not removable.

In the other direction, we assume that $N_{\alpha,p'}(K) = 0$. Then $|K| = 0$, otherwise we could use Lebesgue measure restricted to K in the first part of the proof. Thus, a given solution u in $O \setminus K$ is defined a.e. in O, so it can be considered as a distribution in O. Let $\varepsilon > 0$ and let $\chi \in \widetilde{\omega}_K$ satisfy $\|\chi\|_{\alpha,p'} < \varepsilon$. Let $\varphi \in C_0^\infty(O)$.

We claim that $\int_{\mathbf{R}^N} u\,\mathcal{L}^*\varphi\,dx = 0$, $\mathcal{L}^*$ denoting the adjoint operator. We have that $(1-\chi)\varphi \in C_0^\infty(O \setminus K)$, and thus by assumption

$$\int_{\mathbf{R}^N} u\,\mathcal{L}^*((1-\chi)\varphi)\,dx = 0$$

It follows that

$$\left|\int_{\mathbf{R}^N} u\,\mathcal{L}^*\varphi\,dx\right| = \left|\int_{\mathbf{R}^N} u\,\mathcal{L}^*(\chi\varphi)\,dx\right| \le \|u\|_{L^p(O)}\|\mathcal{L}^*(\chi\varphi)\|_{p'}.$$

By the Leibniz formula and the equivalence of norms in $L^{\alpha,p'}$ and $W^{\alpha,p'}$ we have $\|\mathcal{L}^*(\chi\varphi)\|_{p'} < A\,\|\chi\varphi\|_{\alpha,p'} \le A\,\varepsilon$ for some constant A. But ε is arbitrary, so the claim follows. Thus u is a weak solution in O, and the theorem follows from the regularity theory for elliptic equations, see e.g. Hörmander [230], Theorem 4.4.1.

Clearly we have the inequality $C_{\alpha,p}(K) \le N_{\alpha,p}(K)$. In view of the last theorem it is of considerable interest that these capacities are in fact equivalent, i.e. there is a constant A, depending only on α, p, and N, such that

$$C_{\alpha,p}(K) \leq N_{\alpha,p}(K) \leq A\, C_{\alpha,p}(K) \tag{2.7.4}$$

for all compact $K \subset \mathbf{R}^N$, and thus for all subsets of $\mathbf{R}^N$.

The proof of this inequality depends on certain estimates for potentials, which we prefer to treat in a larger context in the next chapter. See Corollary 3.3.4.

The result will become even more interesting when, in Chapter 5 below, we give very precise metric characterisations of sets of zero capacity.

2.8 Further Results

2.8.1. The extremal problems associated to the definition of $C'_{1,p}$ (Definition 2.2.1) for $p > 1$, and the related quasilinear elliptic partial differential equations have been the subject of much study; see e.g. J. Serrin [374], and V. G. Maz′ya [302]. More recently, concepts like subharmonic functions and the Perron method have been generalized to this setting, and a whole nonlinear potential theory has been developed by O. Martio and others in a series of papers. This theory parallels the theory presented here for $\alpha = 1$ and $p \neq 2$, as has been pointed out in the preface. An account of it falls outside the scope of this book, but a complete exposition is given in the recent book by J. Heinonen, T. Kilpeläinen and O. Martio [221], where references to the earlier work are also found. See also Section 6.5.5 below.

2.8.2. Classical potential theory is intimately related to the theory of stochastic processes, such as the Brownian motion process. See e.g. the book by J. L. Doob [125]. This connection has been extended to the context of (α, p)-capacities by M. Fukushima and H. Kaneko [173]. Further results and references are found in e.g. the C.I.M.E. lectures by Fukushima [172] and in P. Malliavin [287]. Another probabilistic interpretation of (α, p)-capacity is due to E. B. Dynkin; see the notes to Section 2.7 below.

2.9 Notes

2.1. There are several books in potential theory, with emphasis on different parts of the theory. A classical treatise is O. D. Kellogg [244], although it appeared a few years before the modern development started. The little book [428] by J. Wermer is a nice introduction. The books by L. L. Helms [222] and N. S. Landkof [266] are more comprehensive. Of these [222] limits itself to Newton potentials, whereas [266] treats general Riesz potentials. N. Du Plessis [126] should also be mentioned. L. Carleson's brief book [92] has had a strong influence on later developments, and many of the problems treated in this book give generalizations of results from it.

The properties of the equilibrium potential were proved by Gauss [181] in a way that was considered satisfactory at the time. A rigorous existence proof was given in 1935 by O. Frostman in his ground-breaking thesis [158].

2.2, 2.3, 2.5. The history of (α, p)-capacities is quite complicated, even when limited to the case $p \neq 2$. Capacities associated to general function spaces were defined in the early 1950's in a way similar to Definition 2.2.1 by N. Aronszajn and K. T. Smith [39] in their study of functional completion and exceptional sets. See also the papers by N. Aronszajn, F. Mulla, and P. Szeptycki [38], and N. Aronszajn [37]. General $(1, p)$-capacities were also studied by G. Choquet [102] (see p. 202), who proved that they satisfy his conditions for capacitability.

B. Fuglede [160] defined classes of exceptional sets in terms of the modulus of families of curves or surfaces. These classes coincide with the classes of sets of (α, p)-capacity zero (see in particular Theorem 6, p. 191, in [160], and compare to Proposition 2.3.7 above). His work extends the concept of extremal length introduced by A. Beurling in the 1930's, but not published until 1951 by L. V. Ahlfors and A. Beurling [28].

A capacity close to our $C'_{1,p}$ was studied by C. Loewner [278] in the case $p = N$, N the dimension of the space. (In Definition 2.2.1 the norm $\|\varphi\|_{W^{1,p}}$ should be replaced by the "homogeneous" $\|\nabla\varphi\|_{L^p}$.) This capacity is invariant under conformal mappings, and plays an important role in the theory of N-dimensional quasiconformal mappings. See e.g. J. Väisälä [410].

For general integral α and $1 \leq p < \infty$ the capacities $C'_{\alpha,p}$ were introduced explicitly by V. G. Maz′ya [294, 296, 299, 301] and applied by him in many subsequent papers. See his book [308] for a full account and for many references not included here. Other important early references are the papers by J. Serrin [374, 375]; see below.

The idea of defining an (α, p)-capacity by means of a kernel, and of identifying it with a dually defined capacity, seems to have occurred at about the same time to several different people: B. Fuglede [165], N. G. Meyers [318], Yu. G. Reshetnyak [362], V. P. Havin, and V. G. Maz′ya [202, 203].

The variational approach followed in Section 2.2 was used by Reshetnyak, and by Havin and Maz′ya. The latter two authors coined the term "nonlinear potential theory", and investigated the properties of nonlinear potentials and the corresponding capacities systematically. In [204] and other papers they gave many applications.

The more general approach of Sections 2.3 and 2.5, depending on the minimax theorem, is due to Fuglede, and to Meyers. In particular, Theorem 2.5.6 and its proof are taken from Meyers [318]. The minimax theorem had been used previously in potential theory by Fuglede [163] and M. Kishi [253].

There are extensions of the theory to more general spaces by S. K. Vodop′yanov [417]–[420], T. Sjödin [381], and N. Aïssaoui and A. Benkirane [34].

The capacitability of Borel sets was an open problem until G. Choquet [102] proved his general capacitability theorem, Theorem 2.3.11. See also e.g. L. Carleson [92], Choquet [103], M. Sion [379], and the books by C. Dellacherie [117], C. Dellacherie and P.-A. Meyer [118], and N. S. Landkof [266]. The class of Suslin sets is strictly larger than the class of Borel sets. The Suslin sets in $\mathbf{R}^N$ are most easily defined as the sets that can be obtained by perpendicular projection of

Borel sets (or G_δ sets) in $\mathbf{R}^{N+1}$, and a Suslin set is Borel if and only if its complement is Suslin. This class of sets was discovered in 1917 by M. Ya. Suslin [386], a student of N. N. Luzin, in connection with a mistake by H. Lebesgue. The story is well told by Lebesgue himself in his preface to Luzin's book [281]. Luzin used the term "analytic set", a usage which has survived to the present day, in spite of the completely different meaning the term has in the theory of functions of several complex variables.

2.4. The version of the von Neumann minimax theorem given in Section 2.4 is due to K. Fan [140]. The proof, which is also taken from [140], is a modification of the proof of a somewhat less general minimax theorem by H. Kneser [254]. Von Neumann's original paper is [348]. See also B. Fuglede [163].

2.6. The classical boundedness principle for a general kernel, Theorem 2.6.2, is due to T. Ugaheri [409]. The nonlinear extension, Theorem 2.6.3, is due to V. P. Havin and V. G. Maz′ya [203] in the case of Riesz kernels, and to D. R. Adams and N. G. Meyers [23] in the general case. The classical continuity principle is due to G. C. Evans [134] and F. Vasilesco [411]. Kernels satisfying this principle are called regular by B. Fuglede in [161]. For extensions to obstacle type problems, see L. Caffarelli and D. Kinderlehrer [87], and also D. R. Adams [9]. The nonlinear continuity principle, Theorem 2.6.5, is again due to V. P. Havin and V. G. Maz′ya [203]. Proposition 2.6.7 is adapted from L. Carleson [92], Theorem III:3, and Proposition 2.6.9 comes from L. I. Hedberg [207].

2.7. The history of the capacity $N_{\alpha,p}$ and Theorem 2.7.2 goes back to A. Beurling's 1947 conference report [59]. He introduced a predecessor of the Bessel kernel on $\mathbf{R}$ as the inverse Fourier transform of $(1+|\xi|)^{-\alpha}$, $0<\alpha\leq 1$. By means of (2.7.3), which he formulated without using the language of distributions, he defined a set function corresponding to $N_{\alpha/2,2}$, called the spectral measure, and proved in a rather complicated way that for all closed sets this spectral measure is equal to the capacity corresponding to $C_{\alpha/2,2}$ (Théorème 1, p. 20).

Introducing L. Schwartz's newly created theory of distributions into potential theory, J. Deny [119] studied the problem in much greater generality, extended the definition of spectral measure ($N_{g,2}$ in our notation) to very general convolution kernels g, and proved Theorem 2.7.2 in the case $p=2$ (see p. 127 in [119]). He also proved the equality of $N_{g,2}$ and $C_{g,2}$ for kernels such that $g*g$ satisfies the maximum principle, in particular for the Riesz and Bessel kernels I_α and G_α, $0<\alpha\leq 1$, (Théorème II:3, p. 144).

Compact sets K with $N_{\alpha,p}(K)=0$ were called α-p polar by W. Littman [273, 274]. (The concept of polarity is classical in potential theory. The definition of α-2 polar sets is due to L. Hörmander and J. L. Lions [231]; see also J. Deny and J. L. Lions [122], p. 368.) For $p\neq 2$ Theorem 2.7.2 is due to R. Harvey and J. C. Polking [197].

The important result (2.7.4), i.e., the fact that $C_{\alpha,p}$ is equivalent to $N_{\alpha,p}$ for all positive integers α and all $p>1$ was announced by V. G. Maz′ya in [301], and proved in [304] (Theorem 3.3), using results from nonlinear potential theory of V. P. Havin and Maz'ya [202, 203]. See also the book Maz'ya [308], Section 9.3. This theorem was extended to non-integer values of α by D. R. Adams and J. C. Polking [25]. See also J.-P. Kahane [239, 240] for the case $p = 2$. See Corollary 3.3.4 below, and also the notes to Section 3.3 at the end of Chapter 3.

The problem of characterizing removable singularities for classes of solutions to partial differential equations has been one of the motivations behind the development of nonlinear potential theory. Such theorems have been proved for many function spaces and many classes of equations, beginning with B. Riemann's theorem on holomorpic functions in a punctured disk.

A classical result is that sets of zero $(1, 2)$-capacity are removable for square integrable holomorphic functions, and for bounded harmonic functions. In the book [92] L. Carleson extended this, and proved removability theorems for holomorphic and harmonic functions in L^p (and other spaces) in terms of classical capacities and Hausdorff measures (see in particular Theorems VI:1 and VII:3). Later (but independently of [92]) J. Serrin [374] made a deep study of general second order quasilinear elliptic equations, and proved removability of sets of $(1, p)$-capacity zero for solutions belonging to suitable L^q-classes. In [375] Serrin extended Carleson's results to very general second order linear equations. Extensions to higher order equations are due to R. Harvey and J. C. Polking [196]. Theorem 2.7.4 was proved by W. Littman [273, 274]. See also V. P. Havin and V. G. Maz'ya [204].

Extending one of the results of J. Serrin [374], V. G. Maz'ya in [301, 304] (see Theorem 1.4 and Remark 1.4 in [304]) applied the equivalence of $N_{\alpha,p}$ and $C_{\alpha,p}$ to characterize the removable singularities for bounded solutions of a class of quasilinear elliptic equations of any order.

Many papers have been written on the question for nonlinear equations. A result of some interest to the authors is the necessary and sufficient condition for the removability of a compact subset K for the non-negative solutions to the equation $-\Delta u = u^p$ in domains of $\mathbf{R}^N$, $N \geq 3$, in terms of the condition $C_{2,p'}(K) = 0$; see P. Baras and M. Pierre [47], and D. R. Adams and M. Pierre [24]. For $1 < p \leq 2$ this equation has a probabilistic interpretation in terms of measure-valued, branching Markov processes, and removability is equivalent to a hitting probability being zero. See E. B. Dynkin [127, 128, 129], and E. B. Dynkin and S. E. Kuznetsov [130, 131].

A valuable survey of pre-1984 results on removable singularities for both linear and nonlinear elliptic equations is given by J. C. Polking [361]. See also the book by N. N. Tarkhanov [396]. Some further information on removability for analytic and harmonic functions is found in the notes to Chapter 11.

3. Estimates for Bessel and Riesz Potentials

Here we interrupt the development of the general theory in order to gain a deeper understanding of some of the aspects of the spaces $L^{\alpha,p}$. In Section 3.1 we give some simple pointwise estimates of potentials in terms of maximal functions. These are going to be used in several of the following chapters. We apply them here to obtain elementary proofs of certain integral inequalities, among which are the Sobolev inequalities of Theorem 1.2.4. In Section 3.2 we pursue a more special subject; we give a sharp exponential integral estimate in the "borderline case" $\alpha p = N$. Sections 3.3 and 3.5 are devoted to the question under which circumstances a function T "operates" on functions f in $L^{\alpha,p}$ in the sense that the composite function $T \circ f$ also belongs to $L^{\alpha,p}$. This is in part motivated by the desire to prove the equivalence of capacities formulated in Section 2.7. Another consequence is a one-sided approximation theorem, given in Section 3.4, which has turned out to be useful in the theory of nonlinear partial differential equations. Finally, in Section 3.6 we prove an important inequality of B. Muckenhoupt and R. L. Wheeden, comparing Riesz and Bessel potentials with the fractional maximal function $M_\alpha f$, which will have a role to play later.

3.1 Pointwise and Integral Estimates

We begin with a couple of elementary formulas.

Lemma 3.1.1. *Let $\mu \in \mathcal{M}(\mathbf{R}^N)$. Then for all $\delta > 0$*

(a)

$$\int_{|x-y|<\delta} \frac{d\mu(y)}{|x-y|^{N-\alpha}} = (N-\alpha)\int_0^\delta \frac{\mu(B(x,r))}{r^{N-\alpha}}\frac{dr}{r} + \frac{\mu(B(x,\delta))}{\delta^{N-\alpha}}, \quad 0 < \alpha < N \ ;$$

(b)

$$\int_{|x-y|<\delta} \log\frac{1}{|x-y|}\, d\mu(y) = \int_0^\delta \mu(B(x,r))\frac{dr}{r} + \mu(B(x,\delta))\log\frac{1}{\delta} \ ;$$

(c)

$$\int_{|x-y|\geq\delta} \frac{d\mu(y)}{|x-y|^{N-\alpha}} = (N-\alpha)\int_\delta^\infty \frac{\mu(B(x,r))}{r^{N-\alpha}}\frac{dr}{r} - \frac{\mu(B(x,\delta))}{\delta^{N-\alpha}}, \qquad 0<\alpha<N\ .$$

Proof. Just write $\mu(B(x,r)) = \int_{|x-y|<r} d\mu(y)$, and change the order of integration on the right hand side.

From the lemma we get the following simple but very useful pointwise estimates for Riesz potentials in terms of the maximal function Mf and fractional maximal functions $M_\alpha f$ defined in (1.1.1) and (1.1.2).

Proposition 3.1.2. *For $0<\alpha<N$, $1\leq p<\infty$, there are constants A depending on α, p, and N such that for any measurable function $f\geq 0$ and any $x\in\mathbf{R}^N$*

(a)

$$I_\alpha * f(x) \leq A\,\|f\|_p^{\alpha p/N}\, Mf(x)^{1-\alpha p/N}, \qquad 1\leq p<\frac{N}{\alpha}\ ;$$

(b)

$$I_{\alpha\theta} * f(x) \leq A\,(I_\alpha * f(x))^\theta\, Mf(x)^{1-\theta}, \qquad 0<\theta<1\ ;$$

(c)

$$I_{\alpha\theta} * f(x) \leq A\, M_\alpha f(x)^\theta\, Mf(x)^{1-\theta}, \qquad 0<\theta<1\ .$$

Proof. (a) From Lemma 3.1.1

$$\int_{|x-y|<\delta} \frac{f(y)\,dy}{|x-y|^{N-\alpha}} \leq A\,\delta^\alpha\, Mf(x)\ , \tag{3.1.1}$$

and by Hölder's inequality

$$\int_{|x-y|\geq\delta} \frac{f(y)\,dy}{|x-y|^{N-\alpha}} \leq A\,\delta^{\alpha-N/p}\,\|f\|_p\ . \tag{3.1.2}$$

The result follows by choosing

$$\delta = \delta(x) = \left(\frac{\|f\|_p}{Mf(x)}\right)^{p/N}\ .$$

(b) Taking $\alpha\theta$ instead of α and replacing (3.1.2) by

$$\int_{|x-y|\geq\delta} \frac{f(y)\,dy}{|x-y|^{N-\alpha\theta}} \leq A\,\delta^{\alpha\theta-\alpha}\, I_\alpha * f(x)\ ,$$

we obtain the result by choosing

$$\delta = \left(\frac{I_\alpha * f(x)}{Mf(x)} \right)^{1/\alpha} .$$

(c) Using Lemma 3.1.1(c) we can replace (3.1.2) by

$$\int_{|x-y|\geq\delta} \frac{f(y)\,dy}{|x-y|^{N-\alpha\theta}} \leq A\,\delta^{\alpha\theta-\alpha} M_\alpha f(x) .$$

Now choose

$$\delta = \left(\frac{M_\alpha f(x)}{Mf(x)} \right)^{1/\alpha} .$$

Remark. The inequality (b) is actually a consequence of (c). It is, in fact, easy to see that $M_\alpha f(x) \leq A\, I_\alpha * f(x)$. See Section 3.6 below.

Proposition 3.1.3. *Let* $1 \leq p < \infty$, *and* $\alpha p = N$. *There is a constant* A *depending on* α *and* p, *such that for any* $\varepsilon > 0$ *and any function* $f \in L^p_+$ *with support in the ball* $B(0, R)$ *and* $\| f \|_p = 1$,

$$\big(I_\alpha * f(x) - \varepsilon\big)_+^{p'} \leq \gamma_\alpha^{p'} \frac{\omega_{N-1}}{N} \log_+ \frac{AR^N\, Mf(x)^p}{\varepsilon^p} \qquad \textit{for } x \in B(0, R) .$$

Here γ_α *is the constant in* (1.2.6), *and* ω_{N-1} *is the area of the* $(N-1)$*-dimensional unit sphere given in* (1.2.2).

Proof. By Hölder's inequality for $x \in B(0, R)$ and $\delta < 2R$

$$\int_{|x-y|\geq\delta} \frac{f(y)\,dy}{|x-y|^{N-\alpha}} = \int_{\delta\leq|x-y|\leq 2R} \frac{f(y)\,dy}{|x-y|^{N-\alpha}}$$
$$\leq \left(\int_{\delta\leq|x-y|\leq 2R} \frac{dy}{|x-y|^N} \right)^{1/p'} = \left(\omega_{N-1} \log \frac{2R}{\delta} \right)^{1/p'} .$$

As before

$$\int_{|x-y|<\delta} \frac{f(y)\,dy}{|x-y|^{N-\alpha}} \leq A_1 \delta^\alpha Mf(x) .$$

Now choose

$$\delta^\alpha = \min\left\{ \frac{\varepsilon}{\gamma_\alpha A_1 Mf(x)}, (2R)^\alpha \right\} .$$

This gives

$$\int_{B(0,R)} \frac{f(y)\,dy}{|x-y|^{N-\alpha}} \leq \frac{\varepsilon}{\gamma_\alpha} + \left(\omega_{N-1} \log_+ \frac{2R\big(\gamma_\alpha A_1 Mf(x)\big)^{1/\alpha}}{\varepsilon^{1/\alpha}} \right)^{1/p'} .$$

Since $\alpha = N/p$ the result follows. Note that the constant γ_α appears in the proposition only because it appears in the definition (1.2.6) of I_α.

These pointwise estimates immediately give integral and weak type inequalities. The Sobolev imbedding theorem, Theorem 1.2.4, is an immediate consequence of the following theorem and its corollary.

Theorem 3.1.4. (a) *Let $f \in L^1(\mathbf{R}^N)$, and let $0 < \alpha < N$. Then*

$$\left|\{x : |I_\alpha * f(x)| > \lambda\}\right| \leq A\left(\lambda^{-1}\|f\|_1\right)^{N/(N-\alpha)} .$$

(b) *Let $f \in L^p(\mathbf{R}^N)$, $0 < \alpha < N$, $1 < p < N/\alpha$, and set $p^* = Np/(N-\alpha p)$. Then*

$$\|I_\alpha * f\|_{p^*} \leq A\,\|f\|_p \ .$$

(c) *Let $f \in L^p(\mathbf{R}^N)$, $0 < \alpha < N$, $p = N/\alpha$. Assume that* $\operatorname{supp} f \subset B(0, R)$, *and that $\|f\|_p = 1$. Then*

$$\int_{B(0,R)} \exp\left(\beta\,|I_\alpha * f|^{p'}\right) dx \leq AR^N ,$$

whenever $\beta < \beta_0 = \gamma_\alpha^{-p'} N/\omega_{N-1}$.

Here A denotes constants depending only on α, p, N, and β.

Proof. The inequalities in (a) and (b) follow immediately from Proposition 3.1.2(a), and the Hardy–Littlewood–Wiener maximal theorem, Theorem 1.1.1. The inequality in (c) follows from Proposition 3.1.3, if it is observed that for any $c < 1$ and $\varepsilon > 0$ there is a constant A such that $cx^{p'} \leq (x-\varepsilon)^{p'} + A$ for all $x \geq \varepsilon$.

Remark 1. A homogeneity argument shows that no other exponent than p^* is in general permitted in (b). See Stein [389], V.1.2.

Remark 2. By a more difficult argument one can prove that the inequality (c) is actually true for $\beta = \beta_0$. We present this proof in Section 3.2 below.

Corollary 3.1.5. *Let $f \in L^p(\mathbf{R}^N)$, $0 < \alpha \leq N$, $1 < p < N/\alpha$, and set $p^* = Np/(N-\alpha p)$. Then there is A such that for $p \leq q \leq p^*$*

$$\|G_\alpha * f\|_q \leq A\,\|f\|_p \ .$$

Proof. $G_\alpha \leq I_\alpha$ by (1.2.12), so by the theorem $\|G_\alpha * f\|_{p^*} \leq A\,\|f\|_p$. On the other hand $\|G_\alpha\|_1 = 1$ by (1.2.13), so $\|G_\alpha * f\|_p \leq \|f\|_p$ by (1.1.6). The corollary follows.

Theorem 3.1.6. *Let $0 < \alpha < N$ and let $f \in L^p(\mathbf{R}^N)$, $1 < p < N/\alpha$. Then there are constants A depending on α, p and N such that*

$$\|I_{\alpha\theta} * f\|_r \leq A\,\big\|I_\alpha * |f|\big\|_q^\theta\,\|f\|_p^{1-\theta} ,$$

and

$$\|I_{\alpha\theta} * f\|_r \leq A\,\|M_\alpha f\|_q^\theta\,\|f\|_p^{1-\theta} ,$$

for $0 < \theta < 1$, $0 < q \leq \infty$, $\frac{1}{r} = \frac{\theta}{q} + \frac{1-\theta}{p}$.

Proof. Apply the Hölder inequality and Theorem 1.1.1 to the inequalities in Proposition 3.1.2(b) and (c).

Remark 1. There is a deeper inequality than the first one, without any absolute value sign on the right hand side. See Section 3.7.1.

Remark 2. In Section 3.6 below we shall prove the deeper result that the second inequality in the theorem is true for $\theta = 1$, i.e., $\|I_\alpha * f\|_q \leq A\,\|M_\alpha f\|_q$ for $0 < q < \infty$. See Theorem 3.6.1.

The above estimates for Riesz potentials easily give corresponding estimates for derivatives of potentials, such as the following interpolation inequality.

Proposition 3.1.7. *For any multiindex ξ with $|\xi| < \alpha < N$ there is a constant A such that for any $f \in L^p(\mathbf{R}^N)$, $1 \leq p < \infty$, and almost every x*

$$\left|D^\xi(I_\alpha * f(x))\right| \leq A\, Mf(x)^{|\xi|/\alpha}\, (I_\alpha * |f|(x))^{1-|\xi|/\alpha}\ .$$

Proof. We only need to observe that almost everywhere $D^\xi(I_\alpha * f)(x) = (D^\xi I_\alpha * f)(x)$, and that there is A such that $|D^\xi I_\alpha(x)| \leq A\, I_{\alpha-|\xi|}(x)$ for all $x \neq 0$. Then choose $\theta = 1 - (|\xi|/\alpha)$ in Proposition 3.1.2(b).

We can also modify the proof to give a similar inequality for derivatives of Bessel potentials.

Proposition 3.1.8. *For any multiindex ξ with $|\xi| < \alpha < N$ there is a constant A such that for any $f \in L^p(\mathbf{R}^N)$, $1 \leq p < \infty$, and almost every x*

$$\left|D^\xi(G_\alpha * f(x))\right| \leq A\, Mf(x)^{|\xi|/\alpha}\, (G_\alpha * |f|(x))^{1-|\xi|/\alpha}\ .$$

Proof. We assume that $Mf(x) < \infty$, otherwise there is nothing to prove. We then observe that there is a constant A_1 such that $G_\alpha * |f|(x) \leq A_1 Mf(x)$. In fact, by (1.2.14), (1.2.15), and (3.1.1) there are constants A such that

$$\begin{aligned}
G_\alpha * |f|(x) &\leq A\int_{|x-y|<1} \frac{|f(y)|\,dy}{|x-y|^{N-\alpha}} + A\int_{|x-y|\geq 1} |f(y)|e^{-|x-y|/2}\,dy \\
&\leq A\, Mf(x) + A\sum_{i=1}^{\infty} e^{-i/2}\int_{i\leq|x-y|<i+1} |f(y)|\,dy \\
&\leq A\, Mf(x) + A\, Mf(x)\sum_{i=1}^{\infty}(i+1)^N e^{-i/2} \\
&= A_1\, Mf(x)\ .
\end{aligned}$$

Then, by (1.2.27), (1.2.28), (1.2.15), and (3.1.1), for any $\delta \leq 1$ we have

$$|D^{\xi}(G_\alpha * |f|)(x)| \le A\int_{|x-y|<\delta} + A\int_{\delta\le|x-y|<1} \frac{|f(y)|\,dy}{|x-y|^{N-\alpha+|\xi|}}$$

$$+ A\int_{|x-y|\ge 1} G_\alpha(x-y)|f(y)|\,dy$$

$$\le A\big(\delta^{\alpha-|\xi|}Mf(x) + \delta^{-|\xi|}G_\alpha * |f|(x) + G_\alpha * |f|(x)\big) .$$

Now choose

$$\delta^\alpha = \frac{G_\alpha * |f|(x)}{A_1\, Mf(x)} .$$

Then $\delta \le 1$, and the result follows.

3.2 A Sharp Exponential Estimate

We devote this section to the proof of a sharp inequality, which improves Theorem 3.1.4(c). The result is the following.

Theorem 3.2.1. *Let $f \in L^p(\mathbf{R}^N)$, $0 < \alpha < N$, $p = N/\alpha$. Assume that* supp $f \subset B(0, R)$, *and that $\|f\|_p = 1$. Then there is a constant $A = A(N, p)$ such that*

$$\int_{B(0,R)} \exp\left(\beta_0 |I_\alpha * f|^{p'}\right) dx \le A\,R^N$$

with $\beta_0 = \gamma_\alpha^{-p'} N/\omega_{N-1}$, where γ_α is the constant in (1.2.6).

The key to the proof is the following lemma.

Lemma 3.2.2. *Let $a(s,t)$ be a non-negative measurable function on* $[0,\infty) \times [0,\infty)$ *such that*

$$a(s,t) \le 1 \quad \text{for } 0 \le s \le t , \tag{3.2.1}$$

and

$$\sup_{t>0}\left(\int_t^\infty a(s,t)^{p'}\,ds\right)^{1/p'} = b < \infty . \tag{3.2.2}$$

Then there exists a constant $c_0 = c_0(p,b)$, such that for $\varphi(s) \ge 0$, and

$$\int_0^\infty \varphi(s)^p\,ds \le 1 ,$$

it follows that

$$\int_0^\infty e^{-F(t)}\,dt \le c_0 , \tag{3.2.3}$$

where

$$F(t) = t - \left(\int_0^\infty a(s,t)\varphi(s)\,ds\right)^{p'} .$$

Proof. We can write (3.2.3) as

$$\int_{-\infty}^{\infty} |E_\lambda| e^{-\lambda} \, d\lambda \le c_0 \ ,$$

where $E_\lambda = \{t \ge 0 : F(t) \le \lambda\}$. The result will then follow if we show that E_λ is empty for sufficiently small λ, and that there exist constants A_1 and A_2 such that $|E_\lambda| \le A_1\lambda + A_2$. So to this end we set

$$L(t) = \left(\int_t^{\infty} \varphi(s)^p \, ds\right)^{1/p} ,$$

and let $t \in E_\lambda$. Then (3.2.1) and (3.2.2) together with Hölder's inequality imply

$$t - \lambda \le \left(\int_0^t + \int_t^{\infty} a(s,t)\varphi(s) \, ds\right)^{p'} \le \left(\left(1 - L(t)^p\right)^{1/p} t^{1/p'} + bL(t)\right)^{p'} .$$

Next, we apply the binomial estimate,

$$(\alpha + \beta)^q \le \alpha^q + q2^{q-1}(\alpha^{q-1}\beta + \beta^q) \ , \tag{3.2.4}$$

which is valid for $q \ge 1$ and $\alpha, \beta \ge 0$. We then have

$$t - \lambda \le t\left(1 - L(t)^p\right)^{p'/p} + p'2^{p'-1}\left(\left(1 - L(t)^p\right)^{(p'-1)/p} t^{1/p} bL(t) + b^{p'} L(t)^{p'}\right) .$$

Now, if $p \ge 2$, then $q = p'/p \le 1$, so we can use the inequality $(1-\alpha)^q \le 1 - q\alpha$, for $0 \le \alpha \le 1$, and $0 < q \le 1$. This gives

$$t - \lambda \le t\left(1 - \tfrac{p'}{p} L(t)^p\right) + c_1 t^{1/p} L(t) + c_2$$

for some constants c_1 and c_2. Thus, if we set $\sigma = t^{1/p}L(t)$, we have

$$\tfrac{1}{p-1}\sigma^p \le \lambda + c_1\sigma + c_2 \ ,$$

or upon applying Young's inequality, $\alpha\beta \le \frac{\varepsilon^p}{p}\alpha^p + \frac{1}{p'\varepsilon^{p'}}\beta^{p'}$, to the term $c_1\sigma$,

$$tL(t)^p = \sigma^p \le c_3\lambda + c_4 \ . \tag{3.2.5}$$

Notice that (3.2.5) implies that $\lambda \ge -c_4/c_3$ if $E_\lambda \ne \emptyset$.

For $p < 2$ the argument is similar. Now use the trivial fact that $(1-\alpha)^q \le 1-\alpha$ for $0 \le \alpha \le 1$ and $q \ge 1$. Thus, (3.2.5) is valid for all $p > 1$.

To estimate $|E_\lambda|$ we now choose $t_1 < t_2$ in $E_\lambda \cap [\rho, \infty)$, and estimate $t_2 - t_1$. Again by (3.2.1), (3.2.2), and Hölder's inequality,

$$\begin{aligned} t_2 - \lambda &\le \left(\int_0^{t_1} + \int_{t_1}^{t_2} + \int_{t_2}^{\infty} a(s,t_2)\varphi(s) \, ds\right)^{p'} \\ &\le \left(t_1^{1/p'} + (t_2 - t_1)^{1/p'} L(t_1) + bL(t_2)\right)^{p'} . \end{aligned}$$

Setting $\delta = (t_2 - t_1)^{1/p'}$, we can write

$$t_1 + \delta^{p'} - \lambda \le \left(t_1^{1/p'} + (\delta + b) L(t_1) \right)^{p'} .$$

Applying (3.2.4) leads to

$$t_1 + \delta^{p'} - \lambda \le t_1 + p' 2^{p'-1} \left(t_1^{1/p} (\delta + b) L(t_1) + (\delta + b)^{p'} L(t_1)^{p'} \right) ,$$

or using (3.2.5),

$$\delta^{p'} \le \lambda + (c_5 \lambda + c_6)^{1/p} (\delta + b) + c_7 \delta^{p'} L(t_1)^{p'} + c_8 .$$

We now choose ρ sufficiently large so that

$$c_7 L(t_1)^{p'} \le \tfrac{1}{3} .$$

Notice that by (3.2.5) such a ρ can be chosen of the order $O(\lambda)$ as $\lambda \to \infty$. A further application of Young's inequality gives

$$(c_5 \lambda + c_6)^{1/p} (\delta + b) \le c_9 \lambda + \tfrac{1}{3} \delta^{p'} + c_{10} .$$

Thus

$$\tfrac{1}{3} \delta^{p'} \le c_9 \lambda + c_{10} .$$

The lemma now follows, since

$$|E_\lambda| \le \left| E_\lambda \cap [0, \rho) \right| + \left| E_\lambda \cap [\rho, \infty) \right| \le \rho + \delta^{p'} \le A_1 \lambda + A_2 .$$

Next, we define the *non-increasing rearrangement*, f^*, on $(0, \infty)$ of a function f on $\mathbf{R}^N$ by

$$f^*(t) = \inf\{ s > 0 : |E_s^f| \le t \} , \tag{3.2.6}$$

where

$$E_s^f = \{ x \in \mathbf{R}^N : |f(x)| > s \} .$$

We also define

$$f^{**}(t) = \frac{1}{t} \int_0^t f^*(s) \, ds . \tag{3.2.7}$$

Lemma 3.2.3. *If h, f, and g are measurable functions on $\mathbf{R}^N$ such that $h = f * g$, then*

$$h^{**}(t) \le t f^{**}(t) g^{**}(t) + \int_t^\infty f^*(s) g^*(s) \, ds$$

for all $t > 0$.

We omit the proof of this lemma, since it is found in W. P. Ziemer's book [438], Lemma 1.8.8, p. 30, as well as in the original paper by R. O'Neil [353].

Proof of Theorem 3.2.1. We apply Lemma 3.2.3 to the function $h = g * f$, where

$$g(x) = \beta_0^{1/p'} I_\alpha(x) = \left(\frac{N}{\omega_{N-1}}\right)^{1/p'} \frac{1}{|x|^{N-\alpha}} ,$$

and $f \geq 0$. Clearly,

$$g^*(t) = t^{-1/p'} ,$$

and

$$g^{**}(t) = p\, g^*(t) .$$

Thus we have, by the monotonicity of h^*,

$$h^*(t) \leq h^{**}(t) \leq p\, t^{-1/p'} \int_0^t f^*(s)\, ds + \int_t^{|B(0,R)|} f^*(s) s^{-1/p'}\, ds .$$

Observing that

$$\int_{B(0,R)} \exp\big(h(x)^{p'}\big)\, dx = \int_0^{|B(0,R)|} \exp\big(h^*(t)^{p'}\big)\, dt ,$$

and making the change of variables

$$t = |B(0, R)| e^{-\tau} , \tag{3.2.8}$$

we have

$$\int_{B(0,R)} \exp\big(h(x)^{p'}\big)\, dx \leq |B(0, R)| \int_0^\infty \exp\big(h^{**}(|B(0, R)| e^{-\tau})^{p'} - \tau\big)\, d\tau .$$

Now, if we set

$$\varphi(\sigma) = |B(0, R)|^{1/p} f^*(|B(0, R)| e^{-\sigma}) e^{-\sigma/p} ,$$

we find

$$h^{**}(|B(0, R)| e^{-\tau}) \leq p\, e^{\tau/p'} \int_\tau^\infty \varphi(\sigma) e^{-\sigma/p'}\, d\sigma + \int_0^\tau \varphi(\sigma)\, d\sigma$$
$$= \int_0^\infty a(\sigma, \tau) \varphi(\sigma)\, d\sigma ,$$

where

$$a(\sigma, \tau) = \begin{cases} 1, & 0 \leq \sigma \leq \tau , \\ p e^{(\tau-\sigma)/p'}, & \tau < \sigma < \infty . \end{cases}$$

The theorem now follows from Lemma 3.2.2, since

$$\int_0^\infty \varphi(\sigma)^p\, d\sigma = \int_0^{|B(0,R)|} f^*(t)^p\, dt = \int_{B(0,R)} f(x)^p\, dx = 1 .$$

3.3 Operations on Potentials

One of the most important properties of the spaces $W^{1,p}$, and one which is constantly used in applications, is the fact that they are closed under the nonlinear operations called "truncations", or more generally under contractions. These operations are even norm decreasing.

A contraction is a mapping $T : \mathbf{R} \to \mathbf{R}$ or $\mathbf{C} \to \mathbf{C}$ such that $T(0) = 0$ and $|T(x) - T(y)| \leq |x - y|$ for all x and y. It operates on functions by composition, i.e. $(T \circ u)(x) = T(u(x))$.

In the special case when T is defined by $T(x) = x$ for $|x| \leq a$, $T(x) = a$ for $x > a$ and $T(x) = -a$ for $x < -a$, for some $a > 0$, it is called a truncation.

The fundamental result is the following.

Theorem 3.3.1. *Let $O \subset \mathbf{R}^N$ be open, and let $u \in W_0^{1,p}(O)$, $1 \leq p < \infty$. Let T be a contraction. Then $T \circ u \in W_0^{1,p}(O)$ and*

$$\|T \circ u\|_{W^{1,p}} \leq \|u\|_{W^{1,p}} .$$

In fact, the chain rule applies, in the sense that

$$\nabla(T \circ u)(x) = (T' \circ u)(x)\, \nabla u(x) ,$$

where the right hand side is interpreted as 0 *wherever* $\nabla u(x) = 0$.

This theorem, although important, will be used only parenthetically in this book, so we refer to the literature for its proof. See the notes at the end of the chapter.

One reason why the theory of elliptic equations of higher order is so much less developed than the second order theory is that Theorem 3.3.1 is not true in higher order Sobolev spaces. To see that it breaks down it is enough to notice that even if $u \in C_0^\infty$, its positive part u_+ usually has derivatives with jump discontinuities. It follows that the second order (distribution) derivatives cannnot be in L^p.

The following theorem shows that it is not enough to assume that the mapping T is smooth. See also the Further Results at the end of the chapter.

Theorem 3.3.2. *Let $\alpha \geq 2$ be an integer, and let $1 \leq p < N/\alpha$ if $\alpha \geq 3$, and $1 < p < N/2$ if $\alpha = 2$. If $T \in C^\infty(\mathbf{R})$ and $T \circ u \in W^{\alpha,p}(\mathbf{R}^N)$ for all $u \in W^{\alpha,p}(\mathbf{R}^N)$, then $T(t) \equiv ct$ for some constant c.*

Proof. Let $v \in C_0^\infty(\mathbf{R}^N)$ be such that $v(x) = x_1$ for $|x| \leq 1$, and $v(x) = 0$ for $|x| \geq 2$. Choose positive numbers ε_i and C_i, $i = 1, 2, \ldots$, such that $\varepsilon_i \to 0$ and $C_i > 1$, and such that

$$\sum_{i=1}^{\infty} C_i^p \varepsilon_i^{N-\alpha p} < \infty ,$$

and

$$\sum_{i=1}^{\infty} C_i^{\alpha p-1} \varepsilon_i^{N-\alpha p} = \infty \ .$$

This is possible, since by the assumptions $\alpha p - 1 > p$.

Then choose points $\{y^i\}_1^\infty$ such that the balls $B(y^i, 2\varepsilon_i)$ are all disjoint, set $v_i(x) = C_i v(\frac{x-y^i}{\varepsilon_i})$, and define $u = \sum_1^\infty v_i$. Then

$$\sum_{|\xi|\le\alpha} \int |D^\xi u|^p \, dx \le A \sum_{i=1}^{\infty} C_i^p \varepsilon_i^{N-\alpha p} < \infty \ ,$$

and it follows that $u \in W^{\alpha,p}$.

We first prove that T must be a polynomial of degree less than α. If it is not, then there must be an interval $[a, b]$ where $T^{(\alpha)}(t) > 0$. Set

$$S_i = \{x : x \in B(y^i, \varepsilon_i), \ a < C_i(x_1 - y_1^i)/\varepsilon_i < b\} \ .$$

On S_i we have $u(x) = v_i(x) = C_i(x_1 - y_1^i)/\varepsilon_i$, and thus

$$\frac{\partial^\alpha (T \circ u)}{\partial x_1^\alpha} = (T^{(\alpha)} \circ u)(C_i/\varepsilon_i)^\alpha \ .$$

For large enough i the volume of S_i is bounded below by a constant times ε_i^N / C_i, since $\varepsilon_i / C_i \to 0$. It follows that

$$\|u\|_{W^{\alpha,p}}^p \ge \int \left| \frac{\partial^\alpha (T \circ u)}{\partial x_1^\alpha} \right|^p dx \ge A \sum_{i=1}^{\infty} C_i^{\alpha p} \varepsilon^{-\alpha p} |S_i|$$

$$\ge A \sum_{i=1}^{\infty} C_i^{\alpha p-1} \varepsilon^{N-\alpha p} = \infty \ ,$$

and thus $T \circ u \notin W^{\alpha,p}$.

Therefore T must be a polynomial. To prove that T is linear it is enough to consider functions u of the form $u(x) = |x|^{-s}\psi(x)$, where $s > 0$, $\psi(0) \ne 0$, and $\psi \in C_0^\infty$. The theorem follows easily.

In view of this negative result, the following theorem on "smooth truncation" of potentials becomes interesting. It can often serve as a substitute for Theorem 3.3.1. See also Section 3.7.

Theorem 3.3.3. *Let* $0 < \alpha < N$ *and* $1 < p < \infty$. *Let* $T \in C^k(\mathbf{R}^+)$ *for some* $k \ge \alpha$, *and suppose that* T *satisfies*

$$\sup_{t>0} |t^{i-1} T^{(i)}(t)| \le L < \infty, \qquad i = 0, 1, 2, \ldots, k \ .$$

Then $T \circ (G_\alpha * f) \in L^{\alpha,p}(\mathbf{R}^N)$ *for every* $f \in L_+^p(\mathbf{R}^N)$, *and there is a constant* A, *depending only on* α, p, *and* N *such that*

$$\|T \circ (G_\alpha * f)\|_{\alpha,p} \le AL \|G_\alpha * f\|_{\alpha,p} = AL \|f\|_p \ .$$

By far the most important case, and also the easiest one, is when α is an integer. For this reason we postpone the general case to Section 3.5, and prove only the integer case here.

Proof of Theorem 3.3.3 for α integer. The proof is a consequence of the estimates in Section 3.1.

Assume first that $f \in C_0^\infty$ and $f \geq 0$. Set $u = G_\alpha * f$, and notice that $u(x) > 0$ for all x, so that $T \circ u$ is defined. If ξ is a multiindex with $|\xi| = \alpha$, we find by the chain rule that

$$D^\xi(T \circ u) = \sum_{i=1}^{\alpha} T^{(i)} \circ u \sum c_\xi D^{\xi^1} u \cdots D^{\xi^i} u \ ,$$

where the interior sum is over all i-tuples of multiindices $\{\xi^1, \ldots, \xi^i\}$ such that $\xi^1 + \cdots + \xi^i = \xi$, and all $|\xi^j| \geq 1$. The c_ξ are coefficients, whose exact value is of no consequence to us. Thus, by assumption

$$|D^\xi(T \circ u)| \leq AL \sum_{i=1}^{\alpha} u^{1-i} \sum |D^{\xi^1} u \cdots D^{\xi^i} u| \ ,$$

For $i > 1$ we estimate these derivatives by means of Proposition 3.1.8. By the positivity of f we have

$$|D^{\xi^j} u| \leq AMf^{|\xi^j|/\alpha} u^{1-|\xi^j|/\alpha} \ .$$

Thus, since $\sum_{j=1}^{i}(1 - |\xi^j|/\alpha) = i - |\xi|/\alpha = i - 1$,

$$\sum_{i=2}^{\alpha} u^{1-i} \sum |D^{\xi^1} u \cdots D^{\xi^i} u| \leq A \sum_{i=2}^{\alpha} u^{1-i} Mf \, u^{i-1} = A \, Mf \ .$$

Taking the term with $i = 1$ into account, we obtain

$$|D^\xi(T \circ u)| \leq AL\big(|Mf| + |D^\xi u|\big) \ .$$

But we already know from Theorems 1.1.1 and 1.2.3 that $\|Mf\|_p \leq A \, \|f\|_p$, and that $\|D^\xi(G_\alpha * f)\|_p \leq A \, \|f\|_p$ for $|\xi| = \alpha$. This finishes the proof for smooth f.

Now we pass to the general case, and let f be an arbitrary function in $L_+^p(\mathbf{R}^N)$. Then there are nonnegative functions $f_i \in C_0^\infty(\mathbf{R}^N)$, $i = 1, 2, \ldots$, such that $\lim_{i\to\infty} \|f_i - f\|_p = 0$. By the first part of the proof,

$$\|T \circ (G_\alpha * f_i)\|_{\alpha,p} \leq AL\|f\|_p$$

for all sufficiently large i.

Thus, setting $T \circ (G_\alpha * f_i) = G_\alpha * g_i$, we can assume that $\{g_i\}_1^\infty$ converges weakly in L^p to an element g, with $\|g\|_p \leq AL\|f\|_p$. We have to prove that $G_\alpha * g = T \circ (G_\alpha * f)$.

The strong convergence of $\{f_i\}_1^\infty$ and the fact that $G_\alpha \in L^1$ imply, by the Minkowski inequality (1.1.6), that $\{G_\alpha * f_i\}_1^\infty$ converges strongly in L^p to $G_\alpha * f$. After extraction of a subsequence we can assume that

$$\lim_{i\to\infty} G_\alpha * f_i(x) = G_\alpha * f(x) \quad \text{a.e.}$$

But T is continuous, so it follows that

$$\lim_{i\to\infty} G_\alpha * g_i(x) = \lim_{i\to\infty} T \circ (G_\alpha * f_i)(x) = T \circ (G_\alpha * f)(x) \quad \text{a.e.}$$

On the other hand, the weak convergence of $\{g_i\}_1^\infty$ implies that the pointwise limit of $\{G_\alpha * g_i\}_1^\infty$ (which is now known to exist a.e.) is $G_\alpha * g$. In fact, setting $g_i - g = h_i$, for an arbitrary $\varepsilon > 0$

$$G_\alpha * h_i(x) = \int_{|x-y|\le\varepsilon} + \int_{|x-y|>\varepsilon} G_\alpha(x-y) h_i(y)\, dy \ .$$

By weak convergence the last term tends to zero, since G_α is in $L^{p'}$ away from the origin. Thus, by Fatou's lemma

$$\begin{aligned}\Big\| \lim_{i\to\infty} G_\alpha * h_i \Big\|_p &\le \liminf_{i\to\infty} \Big\| \textstyle\int_{|y|\le\varepsilon} G_\alpha(y) |h_i(\cdot - y)|\, dy \Big\|_p \\ &\le \sup_i \|h_i\|_p \int_{|y|\le\varepsilon} G_\alpha(y)\, dy \ ,\end{aligned}$$

which is an arbitrarily small number, and thus for a.e. x

$$G_\alpha * g(x) = \lim_{i\to\infty} G_\alpha * g_i(x) = \lim_{i\to\infty} T \circ (G_\alpha * f_i)(x) = T \circ (G_\alpha * f)(x) \ .$$

This completes the proof of the theorem.

Remark. There is a similar result for Riesz potentials of positive functions. The proof is essentially the same, only somewhat simpler, in that Proposition 3.1.7 is used instead of 3.1.8.

As a corollary we now obtain the improvement of Theorem 2.7.4 that was mentioned without proof at the end of Section 2.7.

Corollary 3.3.4. *Let $1 < p < \infty$ and $\alpha > 0$. Then there is a constant A such that for any $E \subset \mathbf{R}^N$*

$$C_{\alpha,p}(E) \le N_{\alpha,p}(E) \le A\, C_{\alpha,p}(E) \ .$$

Thus a compact set K is removable in L^p for an elliptic linear operator $\mathcal{L}$ of order α with constant coefficients if and only if $C_{\alpha,p}(K) = 0$.

Proof. It is enough to consider a compact set K. Let $\varepsilon > 0$. Then there is $f \ge 0$ such that $F = G_\alpha * f \ge 1$ on a neighborhood of K, and $\|f\|_p^p \le C_{\alpha,p}(K) + \varepsilon$ (see Proposition 2.3.5). Let $T \in C^\infty(\mathbf{R})$ be a function such that $0 \le T \le 1$, $T(t) = 0$ for $0 \le t \le \frac{1}{2}$ and $T(t) = 1$ for $t \ge 1$, and apply the theorem.

3.4 One-Sided Approximation

Consider an element $u \in W^{\alpha,p}(\mathbf{R}^N)$ for $0 < \alpha < N$ and $\alpha p \leq N$. Then u is in general unbounded, and if this is the case it is often desirable to approximate u in the $W^{\alpha,p}$-norm by bounded functions. This is always possible by means of regularization through convolution with a suitable smooth approximate identity. Sometimes, however, a more refined approximation procedure is required, where the approximating functions are pointwise bounded by the function to be approximated. The following theorem shows that this is possible.

Theorem 3.4.1. *Let $u \in W^{\alpha,p}(\mathbf{R}^N)$, where α is a positive integer, and $1 < p < \infty$. Then there exists a sequence $\{u_n\}_1^\infty$ such that:*

(a) *$u_n \in W^{\alpha,p} \cap L^\infty$, and $\operatorname{supp} u_n$ is compact;*
(b) *$|u_n(x)| \leq |u(x)|$, and $u_n(x)u(x) \geq 0$ a.e.;*
(c) *$\lim_{n\to\infty} \|u - u_n\|_{\alpha,p} = 0$.*

Proof. If $\alpha p > N$, the function u is bounded, so all that needs to be done is to multiply u by functions ζ_n, where $\zeta_n(x) = \zeta(x/n)$ and ζ is a function in C_0^∞ such that $\zeta(x) = 1$ in a neighborhood of the origin and $0 \leq \zeta \leq 1$.

Let $\alpha p \leq N$, and assume that u has compact support, if necessary by multiplying with a suitable ζ_n. We represent u as a Bessel potential, $u = G_\alpha * g$, so that $\|g\|_p \leq A\|u\|_{W^{\alpha,p}}$. Set

$$v = G_\alpha * |g| \ ,$$

and let $T \in C^\infty(\mathbf{R})$ be a function such that $0 \leq T \leq 1$, $T(t) = 1$ for $0 \leq t \leq \frac{1}{2}$ and $T(t) = 0$ for $t \geq 1$. Then set

$$u_n(x) = T\,(v(x)/n)\,u(x) \ , \qquad n = 1, 2, \ldots \ .$$

We first observe that $u_n(x) = 0$ on the set $\{x : v(x) \geq n\}$, which includes $\{x : |u(x)| \geq n\}$, so we have $|u_n(x)| < n$ a.e., and of course $u_n(x)u(x) \geq 0$. It remains to prove that $u_n \in W^{\alpha,p}$, and that u_n converges to u as n tends to ∞.

Let η be any multiindex with $0 < |\eta| = \beta \leq \alpha$. If $\eta = \eta^1 + \ldots + \eta^i$, $i > 1$, and all $|\eta^j| \geq 1$, we find by the same computation as in the proof of Theorem 3.3.3 that

$$|D^\eta T(v(x)/n)| \leq A \sum_{i=1}^{\beta} n^{-i} \sum |D^{\eta^1} v(x) \cdots D^{\eta^i} v(x)| \ .$$

By Proposition 3.1.8 we have for any multiindex η with $0 < |\eta| < \alpha$

$$|D^\eta v(x)| \leq AMg(x)^{|\eta|/\alpha} v(x)^{1-|\eta|/\alpha} \ .$$

On the open set $\{x : v(x) > n\}$ we have $D^\eta T(v(x)/n) = 0$. Thus

$$|D^\eta T(v(x)/n)| \le A \sum_{i=1}^{\beta} n^{-i} n^{i-|\eta|/\alpha} Mf(x)^{|\eta|/\alpha} \le An^{-|\eta|/\alpha} Mg(x)^{|\eta|/\alpha}$$

for $|\eta| \le \alpha - 1$, and

$$|D^\eta T(v(x)/n)| \le An^{-1}\big(Mg(x) + |D^\eta v(x)|\big)$$

for $|\eta| = \alpha$.

By Leibniz' formula we have for $|\xi| = \alpha$, if $v(x) \le n$, again using Proposition 3.1.8, that

$$\begin{aligned}
&|D^\xi u_n(x) - D^\xi u(x)| - |(1 - T\,(v(x)/n))D^\xi u(x)| \\
&\quad \le A \sum_{0<|\eta|<\alpha} |D^\eta T\,(v(x)/n)\,||D^{\xi-\eta} u(x)| + |D^\xi T\,(v(x)/n)\,||u(x)| \\
&\quad \le A \sum_{0<\beta<\alpha} n^{-\beta/\alpha} Mg(x)^{\beta/\alpha} Mg(x)^{1-\beta/\alpha} v(x)^{\beta/\alpha} \\
&\qquad + An^{-1}\big(Mg(x) + |D^\xi v(x)|\big)v(x) \\
&\quad \le A(v(x)/n)^{1/\alpha} Mg(x) + A(v(x)/n)\big(Mg(x) + |D^\xi v(x)|\big) \ .
\end{aligned}$$

If $v(x) > n$ we have $D^\xi u_n(x) = 0$. It follows that a.e.

$$\lim_{n\to\infty} |D^\xi u_n(x) - D^\xi u(x)| = 0 \ ,$$

and that

$$|D^\xi u_n(x) - D^\xi u(x)| \le |D^\xi u(x)| + A\big(Mg(x) + |D^\xi v(x)|\big) \ .$$

The functions on the right hand side all belong to L^p, so the theorem follows by dominated convergence.

Remark. In the above proof the function $T \circ (v/n)$ was chosen as a multiplier only for simplicity. The reason the proof works is that $C_{\alpha,p}(\{x : v(x) \ge n\}) \le n^{-p}\|v\|_{\alpha,p}^p = n^{-p}\|g\|_p^p$. In the proof v/n can be replaced by any $\varphi = G_\alpha * \psi$ such that $\psi \ge 0$, $\|\psi\|_p^p \le An^{-p}$, and $\varphi(x) \ge 1$ on $\{x : v(x) \ge n\}$. See Hedberg [212], Lemma 5.2.

The following consequence of Theorem 3.4.1, which is due to H. Brezis and F. E. Browder [82], was proved with applications to nonlinear partial differential equations in mind. In Chapter 9 (see Corollary 9.1.11) the result will be extended to arbitrary open sets.

Theorem 3.4.2. *Let $u \in W^{\alpha,p}(\mathbf{R}^N)$, where α is a positive integer, and $1 < p < \infty$, and let $S \in W^{-\alpha,p'}(\mathbf{R}^N) \cap L^1_{\text{loc}}(\mathbf{R}^N)$. Suppose that $S(x)u(x) \ge -|f(x)|$ a.e. for some $f \in L^1(\mathbf{R}^N)$. Then $Su \in L^1(\mathbf{R}^N)$, and*

$$\langle S, u\rangle = \int_{\mathbf{R}^N} S(x)u(x)\,dx \ ,$$

$\langle\cdot,\cdot\rangle$ denoting the duality between $W^{\alpha,p}$ and $W^{-\alpha,p'}$.

Proof. Let $\{u_n\}_1^\infty$ be the sequence defined in Theorem 3.4.1. We know that $\langle S, \varphi\rangle = \int S(x)\varphi(x)\,dx$ if φ is a test function. There are test functions that converge to u_n boundedly and in $W^{\alpha,p}$, and thus $\langle S, u_n\rangle = \int S(x)u_n(x)\,dx$.

By Theorem 3.4.1, $\lim_{n\to\infty}\langle S, u_n\rangle = \langle S, u\rangle$. On the other hand we know that $S(x)u_n(x) \ge -|f(x)|$, so by Fatou's lemma, $Su \in L^1(\mathbf{R}^N)$. But then dominated convergence gives $\lim_{n\to\infty}\int Su_n\,dx = \int Su\,dx$, since we have $|S(x)u_n(x)| \le |S(x)u(x)|$. This proves the theorem.

3.5 Operations on Potentials with Fractional Index

We now prove Theorem 3.3.3 when the index α is fractional and $p = 2$. The integral part, $[\alpha]$, of α is denoted by m. We introduce a nonlinear operator $\mathcal{D}^\alpha$, defined e.g. for $u \in \mathcal{S}$, by

$$\mathcal{D}^\alpha u(x) = \left(\int_{\mathbf{R}^N} \frac{|u(x+y) - P_x^m u(x+y)|^2}{|y|^{2\alpha}}\frac{dy}{|y|^N}\right)^{1/2},$$

where $P_x^m u$ is the corresponding Taylor polynomial of order m at x,

$$P_x^m u(x+y) = \sum_{|\beta|\le m} \frac{1}{\beta!} D^\beta u(x)\, y^\beta \ .$$

The importance of this operator is clear from the following lemma.

Lemma 3.5.1. *Let α be positive and not an integer. Then there is A such that*

$$A^{-1}\|u\|_{\alpha,2} \le \|u\|_2 + \|\mathcal{D}^\alpha u\|_2 \le A\,\|u\|_{\alpha,2} \ .$$

Proof. If $u = G_\alpha * f$, then

$$\begin{aligned}\int_{\mathbf{R}^N}(\mathcal{D}^\alpha u)^2\,dx &= \int_{\mathbf{R}^N}|y|^{-2\alpha-N}\int_{\mathbf{R}^N}|u(x+y) - P_x^m u(x+y)|^2\,dx\,dy\\ &= \frac{1}{(2\pi)^N}\int_{\mathbf{R}^N}|y|^{-2\alpha-N}\int_{\mathbf{R}^N}|\widehat{f}(\xi)|^2(1+|\xi|^2)^{-\alpha}|q(\xi\cdot y)|^2\,d\xi\,dy\\ &= \frac{1}{(2\pi)^N}\int_{\mathbf{R}^N}|\widehat{f}(\xi)|^2(1+|\xi|^2)^{-\alpha}\int_{\mathbf{R}^N}|y|^{-2\alpha-N}|q(\xi\cdot y)|^2\,dy\,d\xi \ ,\end{aligned}$$

where $q(t) = e^{it} - \sum_{j=0}^m (it)^j/j!$. Now, if we set

$$I(\xi) = \int_{\mathbf{R}^N}|y|^{-2\alpha-N}|q(\xi\cdot y)|^2\,dy \ ,$$

then $I(\xi) = |\xi|^{2\alpha}I(\omega)$, where $\omega = \xi/|\xi|$. Furthermore, $I(\omega)$ is independent of ω for $|\omega| = 1$. Hence $I(\xi) = |\xi|^{2\alpha}I(1)$, and $I(1) < \infty$, since

$$\int_{|y|\le 1}|y|^{-2\alpha-N}|q(y)|^2\,dy \le A\int_{|y|\le 1}|y|^{2(m+1-\alpha)-N}\,dy < \infty$$

for $\alpha < m + 1$, and

$$\int_{|y|\geq 1} |y|^{-2\alpha-N} |q(y)|^2 \, dy \leq A \int_{|y|\geq 1} |y|^{2(m-\alpha)-N} \, dy < \infty$$

for $\alpha > m$. Thus

$$\|\mathcal{D}^\alpha u\|_2^2 = A \int_{\mathbf{R}^N} |\widehat{f}(\xi)|^2 (1 + |\xi|^2)^{-\alpha} |\xi|^{2\alpha} \, d\xi \ ,$$

which, when added to $\|u\|_2^2$, gives a quantity comparable to $\|f\|_2^2$.

Theorem 3.3.3 for $p = 2$ now follows easily from the following pointwise estimate.

Theorem 3.5.2. *Let* $0 < \alpha < N$, *let* $T \in C^{m+1}(\mathbf{R}^+)$ *with* $m = [\alpha]$, *and suppose that*

$$\sup_{t>0} |t^{i-1} T^{(i)}(t)| \leq L < \infty \ , \quad i = 0, 1, 2, \ldots, m+1 \ .$$

Let $u = G_\alpha * f$, *where* f *is a nonnegative function in* $C_0^\infty(\mathbf{R}^N)$. *Then there is a constant* A *depending only on* α, N *and* L, *such that*

$$\mathcal{D}^\alpha (T \circ u)(x) \leq A\big(\mathcal{D}^\alpha u(x) + Mf(x)\big)$$

for all x.

It follows from this theorem, and from Lemma 3.5.1 that for smooth functions

$$\|T \circ u\|_2 + \|\mathcal{D}^\alpha (T \circ u)\|_2 \leq A\big(\|u\|_2 + \|\mathcal{D}^\alpha u\|_2 + \|f\|_2\big) \leq A \, \|u\|_{\alpha,2} \ .$$

This inequality is extended to the general case exactly as when α is an integer, and Theorem 3.3.3 follows.

It remains to prove Theorem 3.5.2. We assume, as we may, that $x = 0$, and we set

$$Q(y) = P_0^m u(y) = \sum_{|\beta|\leq m} \frac{1}{\beta!} D^\beta u(0) \, y^\beta \ ,$$

and

$$P(y) = P_0^m (T \circ u)(y) = \sum_{|\beta|\leq m} \frac{1}{\beta!} D^\beta (T \circ u)(0) \, y^\beta \ .$$

Then

$$|T \circ u - P| \leq |T \circ u - T \circ Q| + |T \circ Q - P| \leq L|u - Q| + |T \circ Q - P| \ ,$$

so that

$$\mathcal{D}^\alpha (T \circ u)(0) \leq L \, \mathcal{D}^\alpha u(0) + \left(\int_{\mathbf{R}^N} \frac{|(T \circ Q)(y) - P(y)|^2}{|y|^{2\alpha}} \frac{dy}{|y|^N} \right)^{1/2} \ .$$

By means of a series of lemmas we shall prove that the last integral is less than $A \, Mf(0)$ for a suitable A.

Lemma 3.5.3. *Let $u = G_\alpha * f$, where $f \in C_0^\infty$ and $f \geq 0$. Set*

$$R = \left(G_\alpha * f(0)/Mf(0)\right)^{1/\alpha} .$$

Then there are $\delta > 0$ and A, independent of f, such that for $|y| \leq \delta R$

$$|(T \circ Q)(y) - P(y)| \leq A\, u(0)|y|^{m+1} R^{-m-1} .$$

Proof. We first note that by the proof of Proposition 3.1.8 there is a constant A_1 such that $R \leq A_1$. It is easily seen that P is the Taylor polynomial for $T \circ Q$, and thus for $|y| \leq \delta R$

$$|(T \circ Q)(y) - P(y)| \leq A\,|y|^{m+1} \max_{|y|\leq \delta R,\, |\xi|=m+1} |D^\xi (T \circ Q)(y)| .$$

Moreover, as before

$$D^\xi (T \circ Q) = \sum_{i=1}^{m+1} (T^{(i)} \circ Q) \sum c_\xi D^{\xi^1} Q \cdots D^{\xi^i} Q ,$$

where the last sum is over i-tuples of multiindices such that $\xi^1 + \cdots + \xi^i = \xi$, and all $|\xi^j| \geq 1$.

By Proposition 3.1.8 we have

$$\begin{aligned} Q(y) &\geq u(0) - \left| \sum_{1\leq|\beta|\leq m} \frac{1}{\beta!} D^\beta u(0)\, y^\beta \right| \\ &\geq u(0) - A \sum_{1\leq|\beta|\leq m} \frac{1}{\beta!} Mf(0)^{|\beta|/\alpha} u(0)^{1-|\beta|/\alpha} |y|^{|\beta|} \\ &= u(0)\left(1 - A \sum_{1\leq|\beta|\leq m} \frac{1}{\beta!} R^{-|\beta|} |y|^{|\beta|}\right) \geq \frac{1}{2} u(0) , \end{aligned}$$

if $|y| \leq \delta R$ and δ is small enough. Then, by our assumptions on T we have

$$|T^{(i)} \circ Q(y)| \leq L|Q(y)|^{1-i} \leq L2^{1-i} u(0)^{1-i} ,$$

whenever $|y| \leq \delta R$ and $i \geq 1$. Also note that

$$|D^\xi Q(y)| \leq \sum_{|\xi|\leq|\beta|\leq m} \frac{1}{\beta!} |D^\beta u(0)|\, |y|^{|\beta|-|\xi|} \leq Au(0) R^{-|\xi|} ,$$

whenever $|y| \leq \delta R$. Hence

$$|D^\xi (T \circ Q)(y)| \leq A \sum_{i=1}^{m+1} u(0)^{1-i} u(0)^i R^{-m-1} = Au(0) R^{-m-1} ,$$

which proves the lemma.

Lemma 3.5.4. *Under the same assumptions as in the previous lemma*

$$|D^\xi(T \circ u)(0)| \le Au(0)R^{-|\xi|}$$

for $|\xi| < \alpha$.

Proof. We find again, using Proposition 3.1.8, that

$$\begin{aligned}|D^\xi(T \circ u)(0)| &\le \sum_{i=1}^{|\xi|}(T^{(i)} \circ u)(0)\sum c_\xi |D^{\xi^1}u(0)\cdots D^{\xi^i}u(0)| \\ &\le A\sum_{i=1}^{|\xi|} u(0)^{1-i}\prod_{k=1}^{i} u(0)R^{-|\xi^k|} = Au(0)R^{-|\xi|} \ .\end{aligned}$$

Lemma 3.5.5. *Under the same assumptions as in the previous lemma*

$$\left(\int_{\mathbf{R}^N}\frac{|(T \circ Q)(y) - P(y)|^2}{|y|^{2\alpha}}\frac{dy}{|y|^N}\right)^{1/2} \le A\, Mf(0) \ .$$

Proof. By Lemma 3.5.3 and the fact that $\alpha < m+1$

$$\begin{aligned}&\int_{|y|\le\delta R}\frac{|(T \circ Q)(y) - P(y)|^2}{|y|^{2\alpha}}\frac{dy}{|y|^N} \\ &\quad\le Au(0)^2R^{-2(m+1)}\int_0^{\delta R} r^{2(m+1)-2\alpha}\frac{dr}{r} \le Au(0)^2R^{-2\alpha} = A\, Mf(0)^2 \ .\end{aligned}$$

On the other hand

$$\begin{aligned}&\int_{|y|\ge\delta R}\frac{|(T \circ Q)(y) - P(y)|^2}{|y|^{2\alpha}}\frac{dy}{|y|^N} \\ &\qquad\le 2L^2\int_{|y|\ge\delta R}\frac{|Q(y)|^2}{|y|^{2\alpha}}\frac{dy}{|y|^N} + 2\int_{|y|\ge\delta R}\frac{|P(y)|^2}{|y|^{2\alpha}}\frac{dy}{|y|^N} \ .\end{aligned}$$

By Proposition 3.1.8

$$|Q(y)| \le Au(0)\sum_{|\beta|\le m} R^{-|\beta|}|y|^{|\beta|} \le Au(0)R^{-m}|y|^m$$

for $|y| \ge \delta R$, and by Lemma 3.5.4 the same inequality holds if Q is replaced by P on the left hand side. It follows that

$$\begin{aligned}\int_{|y|\ge\delta R}\frac{|(T \circ Q)(y) - P(y)|^2}{|y|^{2\alpha}}\frac{dy}{|y|^N} &\le Au(0)^2R^{-2m}\int_{\delta R}^{\infty} r^{2m-2\alpha}\frac{dr}{r} \\ &\le Au(0)^2R^{-2m}R^{2m-2\alpha} \le A\, Mf(0)^2 \ .\end{aligned}$$

This finishes the proof of the lemma and of Theorem 3.5.2, and consequently, of Theorem 3.3.3 for $p = 2$.

In order to prove Theorem 3.3.3 when the index α is fractional and $p \neq 2$ we have to replace the operator $\mathcal{D}^\alpha$ by an operator $\mathcal{S}^\alpha$, defined by

$$\mathcal{S}^\alpha u(x) = \left(\int_0^\infty \left(\frac{1}{r^{N+\alpha}} \int_{|y|\le r} |u(x+y) - P_x^m u(x+y)|\, dy\right)^2 \frac{dr}{r}\right)^{1/2},$$

where, as before, $m = [\alpha]$, the integral part of α. The reason is that Lemma 3.5.1 cannot be extended to $1 < p < \infty$, whereas for $\mathcal{S}^\alpha$ we have the following theorem of R. S. Strichartz, which gives another characterization of the space $L^{\alpha,p}$.

Theorem 3.5.6. *Let $1 < p < \infty$, and let α be positive and not an integer. Then there is A such that*

$$A^{-1}\|u\|_{\alpha,p} \le \|u\|_p + \|\mathcal{S}^\alpha u\|_p \le A\|u\|_{\alpha,p}.$$

Given this theorem, the proof of Theorem 3.3.3 in the general case is quite similar to the case $p = 2$, and will not be repeated.

Theorem 3.5.6 is a deep result, and we will postpone its proof until we have developed more powerful tools in the next chapter. See Section 4.8.

3.6 Potentials and Maximal Functions

It is an obvious fact that for any positive measure μ the corresponding Riesz potential $I_\alpha * \mu$, $0 < \alpha < N$, can be estimated below by the fractional maximal function $M_\alpha\mu$. (The fractional maximal function of a function was defined in Definition 1.1.2, and the definition can be extended to measures in the obvious way.) Indeed, for any $r > 0$

$$\int_{\mathbf{R}^N} \frac{d\mu(y)}{|x-y|^{N-\alpha}} \ge \int_{|x-y|\le r} \frac{d\mu(y)}{|x-y|^{N-\alpha}} \ge \frac{1}{r^{N-\alpha}} \int_{|x-y|\le r} d\mu(y)\ .$$

The opposite inequality is of course in general false. (Take for example $d\mu(y) = |y|^{-\alpha}\, dy$ and $x = 0$.) In view of this the following theorem of B. Muckenhoupt and R. L. Wheeden is quite surprising.

Theorem 3.6.1. *Let $1 < p < \infty$ and $0 < \alpha < N$. Then there is a constant A such that for any positive measure μ*

$$A^{-1}\|M_\alpha\mu\|_p \le \|I_\alpha * \mu\|_p \le A\,\|M_\alpha\mu\|_p\ .$$

Remark. Note that both norms are always infinite if $p \le N/(N-\alpha)$, i.e., if $\alpha p' \ge N$.

Proof. By the remark it is enough to let $\alpha p' < N$. We first suppose that μ has compact support. Then the right hand inequality is a consequence of the following so called "good λ inequality":

There exist $a > 1$ and $b > 0$ such that for any $\lambda > 0$ and any ε, $0 < \varepsilon \leq 1$,

$$\left|\{x : I_\alpha * \mu(x) > a\lambda\}\right| \leq b\varepsilon^{N/(N-\alpha)}\left|\{x : I_\alpha * \mu(x) > \lambda\}\right| + \left|\{x : M_\alpha\mu(x) > \varepsilon\lambda\}\right| . \quad (3.6.1)$$

In fact, multiplying (3.6.1) by λ^{p-1} and integrating in λ, we obtain for any positive R

$$\int_0^R \left|\{x : I_\alpha * \mu(x) > a\lambda\}\right| \lambda^{p-1}\, d\lambda \leq b\varepsilon^{N/(N-\alpha)} \int_0^R \left|\{x : I_\alpha * \mu(x) > \lambda\}\right| \lambda^{p-1}\, d\lambda + \int_0^R \left|\{x : M_\alpha\mu(x) > \varepsilon\lambda\}\right| \lambda^{p-1}\, d\lambda ,$$

or after changing variables

$$a^{-p} \int_0^{aR} \left|\{x : I_\alpha * \mu(x) > \lambda\}\right| \lambda^{p-1}\, d\lambda \leq b\varepsilon^{N/(N-\alpha)} \int_0^R \left|\{x : I_\alpha * \mu(x) > \lambda\}\right| \lambda^{p-1}\, d\lambda + \varepsilon^{-p} \int_0^{\varepsilon R} \left|\{x : M_\alpha\mu(x) > \lambda\}\right| \lambda^{p-1}\, d\lambda .$$

When μ has compact support these integrals are finite. If ε is chosen so small that $b\varepsilon^{N/(N-\alpha)} \leq \frac{1}{2}a^{-p}$, it follows that

$$a^{-p} \int_0^{aR} \left|\{x : I_\alpha * \mu(x) > \lambda\}\right| \lambda^{p-1}\, d\lambda \leq 2\varepsilon^{-p} \int_0^{\varepsilon R} \left|\{x : M_\alpha\mu(x) > \lambda\}\right| \lambda^{p-1}\, d\lambda ,$$

or after letting $R \to \infty$,

$$a^{-p} \int_{\mathbf{R}^N} (I_\alpha * \mu)^p\, dx \leq 2\varepsilon^{-p} \int_{\mathbf{R}^N} (M_\alpha\mu)^p\, dx . \quad (3.6.2)$$

If μ does not have compact support, we let μ_n be the restriction of μ to the ball $B(0, n)$ for $n = 1, 2, \ldots$. By (3.6.2), $\|I_\alpha * \mu_n\|_p \leq A\, \|M_\alpha\mu\|_p$ for all n, with an A that does not depend on n. The theorem now follows by monotone convergence.

It remains to prove the good λ inequality (3.6.1). By the lower semicontinuity of the potential, the set $\{x : I_\alpha * \mu(x) > \lambda\}$ is open. Then it has a Whitney decomposition into dyadic cubes $\{Q_i\}$ with disjoint interiors such that for every Q_i there is a point x with $\operatorname{dist}(x, Q_i) \leq 4 \operatorname{diam} Q_i$ and $I_\alpha * \mu(x) \leq \lambda$. (See Theorem 1.4.2).

Let $Q \in \{Q_i\}$, let $a > 1$, and consider the set $\{x \in Q : I_\alpha * \mu(x) > a\lambda\}$. Suppose that Q intersects the set $\{x : M_\alpha\mu(x) \le \varepsilon\lambda\}$. Let P be the ball concentric to Q with radius $6 \operatorname{diam} Q$. Denote the restriction of μ to P by μ_1, and set $\mu - \mu_1 = \mu_2$. Then by Theorem 3.1.4(a)

$$\left|\{x : I_\alpha * \mu_1(x) > a\lambda/2\}\right| \le A\left(\frac{1}{a\lambda}\int_{\mathbf{R}^N} d\mu_1\right)^{N/(N-\alpha)} .$$

Let $x_0 \in Q$ be such that $M_\alpha\mu(x_0) \le \varepsilon\lambda$, and denote by $B(x_0)$ the ball centered at x_0 with radius $8 \operatorname{diam} Q$, so that $P \subset B(x_0)$. Then

$$\int_{\mathbf{R}^N} d\mu_1 = \int_P d\mu \le \int_{B(x_0)} d\mu$$
$$\le A\, M_\alpha\mu(x_0)\,|B(x_0)|^{(N-\alpha)/N} \le A\varepsilon\lambda\,|B(x_0)|^{(N-\alpha)/N} ,$$

and

$$\left(\frac{1}{a\lambda}\int_{\mathbf{R}^N} d\mu_1\right)^{N/(N-\alpha)} \le A\left(\frac{\varepsilon}{a}\right)^{N/(N-\alpha)} |Q| .$$

It follows that there is b such that

$$\left|\{x \in Q : I_\alpha * \mu_1(x) > a\lambda/2\}\right| \le b\varepsilon^{N/(N-\alpha)}|Q| .$$

On the other hand, if x_1 is a point with $\operatorname{dist}(x_1, Q) \le 4 \operatorname{diam} Q$, then because of the choice of P, there is a constant L depending only on N such that for all $y \in P^c$ and all $x \in Q$ we have $|x_1 - y| \le L|x - y|$. Thus, if in addition $I_\alpha * \mu(x_1) \le \lambda$, then

$$I_\alpha * \mu_2(x) \le L^{N-\alpha} I_\alpha * \mu_2(x_1) \le L^{N-\alpha}\lambda .$$

Thus, if a is chosen so that $a \ge 2L^{N-\alpha}$, then $I_\alpha * \mu_2(x) \le a\lambda/2$. Hence, if $I_\alpha * \mu(x) > a\lambda$, it follows that $I_\alpha * \mu_1(x) > a\lambda/2$. In other words, either

$$Q \subset \{x : M_\alpha\mu(x) > \varepsilon\lambda\} ,$$

or

$$\{x \in Q : I_\alpha * \mu(x) > a\lambda\} \subset \{x : I_\alpha * \mu_1(x) > a\lambda/2\} .$$

In the second case it follows that

$$\left|\{x \in Q : I_\alpha * \mu(x) > a\lambda\}\right| \le b\varepsilon^{N/(N-\alpha)}|Q| .$$

Adding over all $Q \in \{Q_i\}$ we obtain the desired inequality.

We shall also need a similar inequality for Bessel potentials. For this reason we need the modified, "inhomogeneous", maximal function $M_{\alpha,\delta} f$ defined in (1.1.3).

We also define a modified Riesz kernel $I_{\alpha,\delta}$ by

$$I_{\alpha,\delta}(x) = I_\alpha(x) \quad \text{for } |x| < \delta , \qquad I_{\alpha,\delta} = 0 \quad \text{for } |x| \ge \delta . \tag{3.6.3}$$

The result is the following.

Theorem 3.6.2. *Let $0 < p < \infty$, $0 < \alpha < N$, and $\delta > 0$. Then there are positive constants A_1, A_2, and A_3 such that for any positive measure μ*

$$\|M_{\alpha,\delta}\mu\|_p \le A_1 \|I_{\alpha,\delta} * \mu\|_p \le A_2 \|G_\alpha * \mu\|_p \le A_3 \|M_{\alpha,\delta}\mu\|_p \ .$$

Proof. As in the previous theorem it is enough to consider μ with compact support, but we no longer need the restriction $p > N/(N-\alpha)$. Without loss of generality we set $\delta = 1$. We first prove that

$$\|G_\alpha * \mu\|_p \le A\,\|I_{\alpha,1} * \mu\|_p + A\,\|M_{\alpha,1}\mu\|_p \ . \tag{3.6.4}$$

By the estimates for the Bessel kernel in (1.2.14) and (1.2.15) we have

$$G_\alpha * \mu(x) \le I_{\alpha,1} * \mu(x) + A \int_{\mathbf{R}^N} e^{-|x-y|/2}\, d\mu(y) \ .$$

We denote the last integral by $I(x)$. In order to estimate it we first consider the case $p \ge 1$. We set $e^{-|x|/2} = E(x)$, so that $I(x) = E * \mu(x)$. We define $\chi_1(x) = |B(0,1)|^{-1}$ for $|x| \le 1$, and $\chi_1(x) = 0$ for $|x| > 1$. Then, clearly, there is A so that $E \le A\, E * \chi_1$, and thus

$$E * \mu \le A\, E * \chi_1 * \mu \le A\, E * M_{\alpha,1}\mu \ ,$$

since $\chi_1 * \mu \le M_{\alpha,1}\mu$. It follows by the Minkowski inequality (1.1.6) that

$$\|E * \mu\|_p \le A\,\|E * M_{\alpha,1}\mu\|_p \le A\,\|E\|_1 \|M_{\alpha,1}\mu\|_p = A\,\|M_{\alpha,1}\mu\|_p \ .$$

In the case $p < 1$ we cannot use the Minkowski inequality. We subdivide $\mathbf{R}^N$ into congruent cubes $\{Q_i\}$ of diameter 1. Then it follows from the elementary inequality $(\sum a_i)^p \le \sum a_i^p$ (see (2.6.5)) that for any x

$$I(x)^p \le \Big(\sum_i e^{-\operatorname{dist}(x,Q_i)/2}\mu(Q_i)\Big)^p \le \sum_i e^{-p \operatorname{dist}(x,Q_i)/2}\mu(Q_i)^p \ ,$$

and thus

$$\int_{\mathbf{R}^N} I(x)^p\, dx \le A \sum_i \mu(Q_i)^p \ .$$

But $Q_i \subset B(x,1)$ for any $x \in Q_i$, and thus $\mu(Q_i) \le A\, M_{\alpha,1}\mu(x)$ for all $x \in Q_i$. It follows that

$$\int_{\mathbf{R}^N} I(x)^p\, dx \le A \sum_i \int_{Q_i} M_{\alpha,1}\mu(x)^p\, dx = A \int_{\mathbf{R}^N} M_{\alpha,1}\mu(x)^p\, dx \ ,$$

which proves (3.6.4).

Thus, in order to prove the theorem it is enough to prove that

$$\|I_{\alpha,1} * \mu\|_p \le A\,\|M_{\alpha,1}\mu\|_p \ .$$

To this end we modify the proof of Theorem 3.6.1. We choose as before a Whitney cube decomposition of the set $\{x : I_{\alpha,1} * \mu(x) > \lambda\}$. However, we now modify the decomposition by further subdividing those Whitney cubes whose diameter is $> \frac{1}{8}$ into dyadic cubes with diameter $\leq \frac{1}{8}$ but $> \frac{1}{16}$. We denote the cubes in this modified decomposition by $\{Q_i\}$, and we consider a cube $Q \in \{Q_i\}$. We set $\mu = \mu_1 + \mu_2$, where μ_1 is the restriction of μ to P, the ball concentric to Q with radius $6 \operatorname{diam} Q$. With this choice of Q we have for a ball $B(x_0)$ centered at a point $x_0 \in Q$ with radius $8 \operatorname{diam} Q$ that

$$\int_{\mathbf{R}^N} d\mu_1 = \int_P d\mu \leq \int_{B(x_0)} d\mu \leq A\, M_{\alpha,1}\mu(x_0)|B(x_0)|^{(N-\alpha)/N} .$$

If $M_{\alpha,1}\mu(x_0) \leq \varepsilon\lambda$, it follows again by Theorem 3.1.4(a) that for $a > 1$ there is b such that

$$\left|\{x \in Q : I_{\alpha,1} * \mu_1(x) > a\lambda/2\}\right| \leq b\varepsilon^{N/(N-\alpha)}|Q| .$$

Next, suppose that there is a point x_1 such that $\operatorname{dist}(x_1, Q) \leq 4 \operatorname{diam} Q$ and $I_{\alpha,1} * \mu(x_1) \leq \lambda$. (This is always the case if $\operatorname{diam} Q \leq \frac{1}{16}$, by Theorem 1.4.2.) Then, if L has the same meaning as before, for $x \in Q$

$$\begin{aligned} I_{\alpha,1} * \mu_2(x) &\leq L^{N-\alpha} \int_{|x-y|<1} \frac{d\mu_2(y)}{|x_1 - y|^{N-\alpha}} \\ &\leq L^{N-\alpha} \int_{|x_1-y|<1} \frac{d\mu(y)}{|x_1 - y|^{N-\alpha}} + L^{N-\alpha} \int_{\substack{|x-y|<1 \\ |x_1-y|\geq 1}} \frac{d\mu(y)}{|x_1 - y|^{N-\alpha}} \\ &\leq L^{N-\alpha}\lambda + A\, M_{\alpha,1}\mu(x) . \end{aligned}$$

Thus, if $M_{\alpha,1}\mu(x) \leq \varepsilon\lambda$, and if a is large enough, $I_{\alpha,1} * \mu_2(x) \leq a\lambda/2$. It follows that

$$\left|\{x \in Q : I_{\alpha,1} * \mu(x) > a\lambda\}\right| \leq b\varepsilon^{N/(N-\alpha)}|Q| + \left|\{x \in Q : M_{\alpha,1}\mu(x) > \varepsilon\lambda\}\right| .$$

If there is no point x_1 such that $\operatorname{dist}(x_1, Q) \leq 4 \operatorname{diam} Q$ and $I_{\alpha,1} * \mu(x_1) \leq \lambda$, then necessarily $\operatorname{diam} Q > \frac{1}{16}$. Let $x \in Q$. Then

$$I_{\alpha,1} * \mu_2(x) = \int_{|y-x|<1} \frac{d\mu_2(y)}{|y - x|^{N-\alpha}} \leq \int_{\frac{5}{16}<|y-x|<1} \frac{d\mu(y)}{|y - x|^{N-\alpha}} ,$$

since $|y - x| \geq \frac{5}{16}$ if $y \in P^c$. Thus

$$I_{\alpha,1} * \mu_2(x) \leq (\tfrac{16}{5})^{N-\alpha} \int_{|y-x|<1} d\mu(y) \leq (\tfrac{16}{5})^{N-\alpha} M_{\alpha,1}\mu(x) .$$

If $M_{\alpha,1}\mu(x) \leq \varepsilon\lambda$, and $a \geq 2\left(\frac{16}{5}\right)^{N-\alpha}$ it follows that $I_{\alpha,1} * \mu_2(x) \leq a\lambda/2$. Thus

$$\begin{aligned} &\{x \in Q : I_{\alpha,1}\mu(x) > a\lambda\} \\ &\qquad \subset \{x \in Q : I_{\alpha,1}\mu_1(x) > a\lambda/2\} \cup \{x \in Q : M_{\alpha,1}\mu(x) > \varepsilon\lambda\} , \end{aligned}$$

and thus

$$\left|\{x \in Q : I_{\alpha,1}\mu(x) > a\lambda\}\right| \leq b\varepsilon^{N/(N-\alpha)}|Q| + \left|\{x \in Q : M_{\alpha,1}\mu(x) > \varepsilon\lambda\}\right| .$$

Now, adding over all Q_i, we obtain the inequality

$$\left|\{x : I_{\alpha,1} * \mu(x) > a\lambda\}\right| \\ \le b\varepsilon^{N/(N-\alpha)}\left|\{x : I_{\alpha,1} * \mu(x) > \lambda\}\right| + \left|\{x : M_{\alpha,1}\mu(x) > \varepsilon\lambda\}\right| ,$$

as before, and Theorem 3.6.2 follows.

We shall now look at Theorems 3.6.1 and 3.6.2 from a slightly different point of view.

As in Lemma 3.1.1 we see by a change of order of integration that

$$I_\alpha * \mu(x) = A\int_0^\infty \frac{\mu(B(x,r))}{r^{N-\alpha}}\frac{dr}{r} .$$

We want to replace this integral by a comparable sum. To this end we recall that $B_n(x)$, $n \in \mathbf{Z}$, denotes the open ball with radius 2^{-n} centered at x, and that we write B_n for $B_n(0)$. We choose a function η such that:

(a) $\operatorname{supp}\eta \subset \overline{B}_0$,
(b) η is nonnegative, bounded, and lower semicontinuous,
(c) $\eta(rx)$ is a decreasing function of $r > 0$ for any $x \in \mathbf{R}^N$.

We define η_n, $n = 0, \pm 1, \pm 2, \dots$, by setting

$$\eta_n(x) = 2^{nN}\eta(2^n x) , \tag{3.6.5}$$

so that $\operatorname{supp}\eta_n \subset B_n$ and $\int \eta_n\,dx = \int \eta\,dx$. For example we could choose η to be the characteristic function for B_0, in which case $\eta_n * \mu(x) = 2^{nN}\mu(B_n(x))$.

It is now easily seen that there is a constant A such that

$$A^{-1}\int_0^\infty \frac{\mu(B(x,r))}{r^{N-\alpha}}\frac{dr}{r} \le \sum_{n=-\infty}^{\infty} 2^{-n\alpha}\eta_n * \mu(x) \le A\int_0^\infty \frac{\mu(B(x,r))}{r^{N-\alpha}}\frac{dr}{r} .$$

In other words, there is a constant A such that

$$A^{-1}I_\alpha * \mu(x) \le \left\|\{2^{-n\alpha}\eta_n * \mu(x)\}_{-\infty}^{\infty}\right\|_{l^1} \le A\, I_\alpha * \mu(x) .$$

Similarly,

$$A^{-1}M_\alpha\mu(x) \le \left\|\{2^{-n\alpha}\eta_n * \mu(x)\}_{-\infty}^{\infty}\right\|_{l^\infty} \le A\, M_\alpha\mu(x) .$$

In the same way it follows from Lemma 3.1.1(a) that

$$A^{-1}\int_0^1 \frac{\mu(B(x,r))}{r^{N-\alpha}}\frac{dr}{r} \le \sum_{n=0}^{\infty} 2^{-n\alpha}\eta_n * \mu(x) \le A\int_0^2 \frac{\mu(B(x,r))}{r^{N-\alpha}}\frac{dr}{r} ,$$

and thus

$$A^{-1}I_{\alpha,1} * \mu(x) \le \left\|\{2^{-n\alpha}\eta_n * \mu(x)\}_0^{\infty}\right\|_{l^1} \le A\, I_{\alpha,2} * \mu(x) .$$

Also,

$$A^{-1}M_{\alpha,1}\mu(x) \le \left\|\{2^{-n\alpha}\eta_n * \mu(x)\}_0^\infty\right\|_{l^\infty} \le A\, M_{\alpha,2}\mu(x) \ .$$

For any sequence $\{a_n\}$, and $1 < q < \infty$ we have (cf. (2.6.5))

$$\|\{a_n\}\|_{l^\infty} \le \|\{a_n\}\|_{l^q} \le \|\{a_n\}\|_{l^1} \ . \tag{3.6.6}$$

This inequality is again reduced to the trivial case $\|\{a_n\}\|_{l^1} = 1$, if a_k is replaced by $a_k/\|\{a_n\}\|_{l^1}$. Thus

$$\begin{aligned} A^{-1}M_\alpha\mu(x) &\le \left\|\{2^{-n\alpha}\eta_n * \mu(x)\}_{-\infty}^\infty\right\|_{l^\infty} \le \left\|\{2^{-n\alpha}\eta_n * \mu(x)\}_{-\infty}^\infty\right\|_{l^q} \\ &\le \left\|\{2^{-n\alpha}\eta_n * \mu(x)\}_{-\infty}^\infty\right\|_{l^1} \le A\, I_\alpha * \mu(x) \ , \end{aligned}$$

and similarly in the inhomogeneous case.

The mixed $L^p(l^q)$-norm of a sequence of functions $\{f_n\}$ is defined by

$$\|\{f_n\}\|_{L^p(l^q)}^p = \|\ \|\{f_n(\cdot)\}\|_{l^q}\|_p^p = \int_{\mathbf{R}^N}\left(\sum_n |f_n(x)|^q\right)^{p/q} dx \ , \tag{3.6.7}$$

with the usual modification for $q = \infty$.

Then Theorems 3.6.1 and 3.6.2 have the following somewhat unexpected corollary.

Corollary 3.6.3. *Let* $0 < \alpha < N$, $0 < p < \infty$, $1 < q < \infty$, *and* $\delta > 0$. *Let* $\{\eta_n\}$ *be given by* (3.6.5). *There are constants A such that for all measures* $\mu \ge 0$

$$\begin{aligned} A^{-1}\|I_\alpha * \mu\|_p &\le A^{-1}\|M_\alpha\mu\|_p \le \left\|\{2^{-n\alpha}\eta_n * \mu\}_{-\infty}^\infty\right\|_{L^p(l^\infty)} \\ &\le \left\|\{2^{-n\alpha}\eta_n * \mu\}_{-\infty}^\infty\right\|_{L^p(l^q)} \le \left\|\{2^{-n\alpha}\eta_n * \mu\}_{-\infty}^\infty\right\|_{L^p(l^1)} \\ &\le A\,\|I_\alpha * \mu\|_p \ , \end{aligned}$$

and

$$\begin{aligned} A^{-1}\|G_\alpha * \mu\|_p &\le A^{-1}\|I_{\alpha,\delta} * \mu\|_p \le A^{-1}\|M_{\alpha,\delta}\mu\|_p \\ &\le \left\|\{2^{-n\alpha}\eta_n * \mu\}_0^\infty\right\|_{L^p(l^\infty)} \le \left\|\{2^{-n\alpha}\eta_n * \mu\}_0^\infty\right\|_{L^p(l^q)} \\ &\le \left\|\{2^{-n\alpha}\eta_n * \mu\}_0^\infty\right\|_{L^p(l^1)} \le A\,\|G_\alpha * \mu\|_p \ . \end{aligned}$$

We shall return to this corollary in Section 4.5 below.

3.7 Further Results

3.7.1. Let $0 < \alpha < N$ and let $f \in L^p(\mathbf{R}^N)$, $1 < p < \infty$. Then there is a constant A independent of f, such that

$$\|I_{\alpha\theta} * f\|_r \le A\,\|I_\alpha * f\|_q^\theta \|f\|_p^{1-\theta} \ ,$$

for $0 < \theta < 1$, $1 \le q \le \infty$, $\frac{1}{r} = \frac{\theta}{q} + \frac{1-\theta}{p}$. This is a stronger result than the first inequality in Theorem 3.1.6, in the sense that f appears on the right hand side instead of $|f|$. On the other hand there is no pointwise estimate corresponding to Proposition 3.1.2(b) for non-positive f. See E. Gagliardo [174], and L. Nirenberg [350] for the case with $p \geq 1$ and integral α and $\alpha\theta$, and also V. G. Maz'ya [308], Section 9.3.1. The general case is due to D. R. Adams and N. G. Meyers [23] (although stated for $f \geq 0$), and proved by complex interpolation. A similar result for periodic functions was proved earlier by I. I. Hirschman [226].

3.7.2. Let $0 < \alpha < N$ and let f be a non-negative function in $L^p(\mathbf{R}^N)$, $0 < p < \infty$. Then there is a constant A independent of f, such that

$$\|I_{\alpha\theta} * f^t\|_r \leq A \, \|I_\alpha * f\|_q^\theta \|f\|_p^{t-\theta} \ ,$$

for $0 < \theta < 1$, $0 < q \leq \infty$, $\theta < t < \theta + (1-\theta)p$, and $\frac{1}{r} = \frac{\theta}{q} + \frac{t-\theta}{p}$. See Adams and Meyers [23] for a proof by complex interpolation (for $q \geq 1$), and Hedberg [208] for a proof by the methods used in the present chapter.

3.7.3. There is another scale of spaces that is often quite useful in potential theory and partial differential equations, the Morrey–Campanato spaces $\mathcal{L}^{p,\lambda}$. We say that u belongs to $\mathcal{L}^{p,\lambda}(\mathbf{R}^N)$, $-p < \lambda \leq N$, $1 < p < \infty$, if $u \in L^p(\mathbf{R}^N)$, and

$$[u]_{p,\lambda}^p = \sup_{x,r>0} \frac{1}{|B(x,r)|^{\frac{N-\lambda}{N}}} \int_{B(x,r)} |u(y) - u_{x,r}|^p \, dy < \infty \ ,$$

where $u_{x,r} = \frac{1}{|B(x,r)|} \int_{B(x,r)} u(y) \, dy$. The norm is $\|u\|_p + [u]_{p,\lambda}$. For $0 < \lambda \leq N$, $[u]_{p,\lambda}$ is equivalent to

$$\sup_{x,r>0} \left(\frac{1}{|B(x,r)|^{\frac{N-\lambda}{N}}} \int_{B(x,r)} |u(y)|^p \, dy \right)^{1/p} ,$$

the Morrey space norm, and for $-p < \lambda < 0$ to

$$\sup_{x \neq y} \frac{|u(x) - u(y)|}{|x-y|^{-\lambda/p}} \ ,$$

the Hölder space norm. $\mathcal{L}^{p,0}(\mathbf{R}^N)$ is BMO, the space of functions of bounded mean oscillation. The definition can be extended to values of $\lambda \leq -p$ by replacing the constant $u_{x,r}$ by a polynomial. See e.g. M. Giaquinta [182], p. 65–75, and other references quoted there.

It is the imbedding properties of the Bessel potential operator $\mathcal{G}_\alpha$ that are of the most interest in potential theory. Two of these are

$$\mathcal{G}_\alpha : \mathcal{L}^{p,\lambda} \to \mathcal{L}^{p,\lambda-\alpha p} \ , \tag{3.7.1}$$

and

$$\mathcal{G}_\alpha : \mathcal{L}^{p,\lambda} \to \mathcal{L}^{\tilde{p},\lambda}, \quad \alpha p < \lambda, \quad \tilde{p} = \frac{\lambda p}{\lambda - \alpha p} \ . \tag{3.7.2}$$

See G. Stampacchia [387], and D. R. Adams [5]. The last result is an extension of the Sobolev imbedding theorem, since $\mathcal{L}^{p,N} = L^p$. Modifications in the definitions are made to accommodate Morrey–Campanato spaces over smooth domains in $\mathbf{R}^N$; see Giaquinta [182].

3.7.4. Theorem 3.3.2, which is due to B. E. J. Dahlberg [110], has lead to further research. First, the theorem leaves open the cases $W^{2,1}(\mathbf{R}^N)$, and $W^{\alpha,N/\alpha}(\mathbf{R}^N)$. But it is not difficult to prove that smooth truncation is possible in these cases. For $W^{\alpha,N/\alpha}(\mathbf{R}^N)$ this is a consequence of the Sobolev inequality. See D. R. Adams [10], and G. Bourdaud [66].

It is more interesting to consider $L^{\alpha,p}(\mathbf{R}^N)$ for $1 < p < \infty$ and $0 < \alpha < N$. Then the conclusion of Theorem 3.3.2 is true for $1 + 1/p \le \alpha < N/p$, i.e., only linear functions operate on $L^{\alpha,p}(\mathbf{R}^N)$. On the other hand, functions T on $\mathbf{R}$ such that $T(0) = 0$ and T' is of bounded variation operate on $L^{\alpha,p}(\mathbf{R}^N)$ for $\alpha < 1+1/p$, in particular truncations operate. For these results, and extensions to Besov and Lizorkin–Triebel spaces, see G. Bourdaud [66, 67, 68, 69], G. Bourdaud and Y. Meyer [71], G. Bourdaud and M. E. D. Kateb [70], and papers referred to in these articles. For example, S. Janson [232] has proved that the "Hardy–Sobolev space" $I_1(H^1)$ (which can be identified with the Lizorkin–Triebel space $F_1^{1,2}$ defined in the next chapter) is closed under truncation.

3.7.5. V. G. Maz'ya [303] (see Theorem 11) has shown that smooth truncation of *non-negative* functions in $W^{2,p}(\mathbf{R}^N)$ is always possible. In fact, let $1 < p < N/2$, let $T \in C^2(\mathbf{R})$, and suppose that T satisfies

$$\sup_{t>0} |t^{i-1}T^{(i)}(t)| \le L < \infty, \quad i = 1, 2 \ .$$

Then $T \circ u \in W^{2,p}(\mathbf{R}^N)$ for every non-negative u in $W^{2,p}(\mathbf{R}^N)$. See Maz'ya [308], Theorem 8.2.1, and D. R. Adams [6]. See also Section 7.6.3 below.

Dahlberg proved in [110] that this result cannot be extended to $W^{\alpha,p}(\mathbf{R}^N)$ for integers $\alpha > 2$. In fact, only linear functions operate on the positive cone in $W^{\alpha,p}(\mathbf{R}^N)$ for $\alpha \ge 3$, $1 < p < N/\alpha$. See also G. Bourdaud [67] for a sharpening and converse of Maz'ya's result.

3.7.6. The smooth truncation operators $u \to T \circ u$ have a natural domain of definition in $L^{\alpha,p}(\mathbf{R}^N)$. Let $\alpha > 0$ and $1 < p < \infty$. Denote by $\dot{L}^{\alpha,p}(\mathbf{R}^N)$ the space of Riesz potentials, normed by $\|u\|_{\dot{L}^{\alpha,p}} = \|f\|_p$, if $u = I_\alpha * f$. Let m be an integer such that $m \le \alpha < m+1$, and let $T \in C^m(\mathbf{R})$ satisfy T(0)=0, and $L = \max_{k \le m+1} \sup_{t\in\mathbf{R}} |T^{(k)}(t)| < \infty$. If $u \in L^{\alpha,p} \cap \dot{L}^{1,\alpha p}$, then $T \circ u \in L^{\alpha,p} \cap \dot{L}^{1,\alpha p}$, and

$$\|T \circ u\|_{\alpha,p} \le AL \left(\|u\|_{\alpha,p} + \|u\|^{\alpha}_{\dot{L}^{1,\alpha p}}\right) \ ,$$

and

$$\|T \circ u\|_{L^{1,\alpha p}} \le L \, \|u\|_{\dot{L}^{1,\alpha p}} \ .$$

Conversely, let $\alpha \ge 1$ be an integer, and suppose that u is a real-valued function on $\mathbf{R}^N$, such that $T \circ u \in L^{\alpha,p}$ for all T of the type listed. Then $u \in L^{\alpha,p} \cap \dot{L}^{1,\alpha p}$. See D. R. Adams and M. Frazier [18], and also [17].

3.8 Notes

3.1. The Sobolev inequality, Theorem 3.1.4(b) was proved in S. L. Sobolev [384]. In the one-dimensional case it had been proved earlier by G. H. Hardy and J. E. Littlewood [194]. Theorem 3.1.4(a) is due to A. Zygmund [439].

In the borderline case (c) when $p = N/\alpha$, V. I. Yudovich [436] announced that $\int_{B(0,R)} \exp(\beta |I_\alpha * f|^{p'})\, dx < \infty$ for all $f \in L^p$ and $\beta > 0$, and also for integrals over intersections of the ball with d-dimensional affine manifolds (cf. Section 7.6.4 below). N. S. Trudinger [407] (see also D. Gilbarg and N. S. Trudinger [183], Theorem 7.15, p. 162) considered the case $\alpha = 1$ and found a $\beta > 0$ such that the inequality in Theorem 3.1.4(c) is true with a constant A independent of f for $\|f\|_p \le 1$. J. A. Hempel, G. R. Morris, and N. S. Trudinger [223] proved that there is no such constant for $\beta > \beta_0$. The Trudinger inequality was extended to the case $\alpha \neq 1$ by R. S. Strichartz [393], but the present form with the correct limiting exponent is due to L. I. Hedberg [208]. The proofs given here, based on the pointwise estimates in Proposition 3.1.2, come from [208]; see also D. R. Adams [5].

See also the remarks following Sobolev's theorem, Theorem 1.2.4.

3.2. Theorem 3.2.1, which is due to D. R. Adams [14], has an interesting history. Extending the results of Trudinger et al. [407, 223] mentioned above, J. Moser [334] proved that

$$\int_{B(0,R)} \exp\left(\gamma |u|^{\frac{N}{N-1}}\right) dx \le A\, R^N$$

for all $u \in W_0^{1,N}(B(0,R))$ with $\int_{B(0,R)} |\nabla u|^N\, dx \le 1$, and $\gamma = N\omega_{N-1}^{1/(N-1)} = \gamma_N$. For $\gamma < \gamma_N$ this inequality follows from the results of [223] (see the remark on p. 372), and also from the formula (1.2.4) and Theorem 3.1.4(c) with $\alpha = 1$.

Theorem 3.2.1 implies and extends Moser's theorem, but is not implied by it, even for $\alpha = 1$. A. M. Garsia [180] gave a simpler proof of the original theorem of J. Moser [334], and the technique used to prove the crucial Lemma 3.2.2 is due to him. The change of variable (3.2.8) goes back to Moser. Lemma 3.2.3 is due to R. O'Neil [353]. In [14] it is also proved that the constant β_0 is best possible.

Long before the developments described above, A. Beurling [57] (see Théorème II, p. 30, and Lemme, p. 34) proved that if $f(z)$ is an analytic function on $|z| < 1$, satisfying $\iint_{|z|<1} |f'(z)|^2\, dx\, dy \le \pi$ and $f(0) = 0$, then its boundary values satisfy $|\{\theta : |f(e^{i\theta})| \ge \lambda\}| \le \exp(-\lambda^2 + 1)$. (It should be noted that the boundary function belongs to the space $L^{1/2,2}$ on the circle.) This inequality implies that for any $\gamma < 1$ there is A such that $\int_0^{2\pi} \exp \gamma |f(e^{i\theta})|^2\, d\theta \le A$. On the other hand there is no such A for $\gamma > 1$. S.-Y. A. Chang and D. E. Marshall [100], and D. E. Marshall [290] proved that the inequality remains true for $\gamma = 1$.

L. Carleson and S.-Y. A. Chang [93] proved the surprising result that there exists an extremal function in Moser's inequality. Sharp inequalities of the Moser type, and the associated extremal functions have important applications to geometry. See e.g. Th. P. Branson, S.-Y. A. Chang, and P. C. Yang [72]. Among other things this paper contains the following extension of Theorem 3.2.1 in the case $N = 4$: Let (M, g) be a compact Riemannian manifold without boundary, let $\Delta = -g^{-1/2}\partial_i(g^{ij}g^{1/2}\partial_j)$ be the Laplacian, and denote by V the volume measure determined by g. Let $u \in C^2(M)$, and suppose that $\int_M u\,dV = 0$, and $\int_M |\Delta u|^2\,dV \leq 1$. Then

$$\int_M \exp\bigl(32\pi^2 |u(x)|^2\bigr)\,dV \leq A\,V(M)\,,$$

and the exponent $32\pi^2$ is best possible. (This is a special case of an analogous result for general N due independently to L. Fontana [151].) See also the conference reports by S.-Y. A. Chang [99], and S.-Y. A. Chang and P. C. Yang [101]. Recently W. Beckner [49] (see also [72]) has proved a sharp inequality on the N-sphere which is somewhat related to Theorem 3.2.1.

3.3. The fundamental role played by contractions in potential theory was discovered by A. Beurling, and led to the theory of Dirichlet spaces, developed jointly with J. Deny. See A. Beurling [60] (in particular p. 238), A. Beurling and J. Deny [61, 62], and J. Deny [121]. Proofs of Theorem 3.3.1 in the case of truncations are found in e.g. J. Deny and J. L. Lions [122], Théorème 3.2, D. Gilbarg and N. S. Trudinger [183], Section 7.4, and D. Kinderlehrer and G. Stampacchia [251], Chapter II, Theorem A.1. A careful proof of the general case, due to A. Ancona, is found (along with more references) in L. Boccardo and F. Murat [64], Théorème 4.2.

Theorem 3.3.2 comes from B. E. J. Dahlberg [110]. Theorem 3.3.3 was proved in D. R. Adams [6]. See also W. Littman [274]. Corollary 3.3.4 was first proved in 1966 by V. G. Maz′ya in the case of integral α, see [301, 304], and the book [308], Section 9.3. The case of general α is due to D. R. Adams and J. C. Polking [25]. J.-P. Kahane [240] has given a proof for $p = 2$ and general $\alpha > 1$, different from the proof in Section 3.5. For $p = 2$ and $0 < \alpha \leq 1$ the result is due to A. Beurling [59] and J. Deny [119], and is a consequence of the maximum principle, or equivalently, of the fact that the space $L^{\alpha,2}$ is closed under truncation for $\alpha \leq 1$ (Theorem 3.3.1 and Lemma 3.5.1). See the notes to Section 2.7 at the end of Chapter 2 for more information on this, and for a discussion of removable singularities.

3.4. Theorem 3.4.1 was proved in L. I. Hedberg [212], Lemma 5.2. While having lunch at the Indiana University Union, Hedberg and J. R. L. Webb discovered that this lemma gave the solution to a problem on nonlinear elliptic equations that was occupying Webb. Afterwards Hedberg provided a more transparent proof of the lemma for Webb's paper [427]. Webb's problem is closely related to Theorem 3.4.2, which is due to H. Brezis and F. E. Browder [82]. They had solved the case $\alpha = 1$ in [80] and posed the problem for $\alpha = 2$ in [81]. See also F. E. Browder [84] for an application to topological degree theory. Theorem 3.4.1 has been extended to Orlicz–Sobolev spaces by A. Benkirane and J.-P. Gossez [50], and to a parabolic situation by H. Brezis and F. E. Browder [83] and R. Landes [261].

3.5. The operator $\mathcal{D}^\alpha$ (for $N = 1$) has a long history. L^p-estimates in N dimensions were first given by E. M. Stein [388]. See also E. M. Stein [389], Section V.6, pp. 161–163, where references to the earlier work are given. Theorem 3.5.6 was proved for $0 < \alpha < 1$ by R. S. Strichartz [392], who also introduced the operator $\mathcal{S}^\alpha$ in this case. The results were extended by J. C. Polking [360], and by B. E. J. Dahlberg [111], to whom Theorems 3.5.2 and 3.5.6 are due.

3.6. Theorem 3.6.1 and its proof are due to B. Muckenhoupt and R. L. Wheeden [335]. They actually proved that the same inequality is true in weighted L^p if the weight satisfies Muckenhoupt's A_∞ condition. A similar idea had been used by R. Coifman and C. Fefferman [108]. More general relations between potential and maximal operators are given by R. Kerman and E. T. Sawyer [245], and by B. Jawerth, C. Pérez, and G. Welland [234], where Theorem 3.6.2 is also found (p. 86). Our proof is different, and includes ideas by B. O. Turesson [408]. Related "two weight inequalities" have been much studied, see e.g. E. T. Sawyer [370], C. Pérez [357], I. E. Verbitsky [412], R. L. Wheeden [429], Verbitsky and Wheeden [413], and references quoted in these papers. See also the book by J. García-Cuerva and J. L. Rubio de Francia [176].

4. Besov Spaces and Lizorkin–Triebel Spaces

In this chapter we shall apply the general theory developed in Chapter 2 to the function spaces known as Besov, and Lizorkin–Triebel spaces. In Section 4.1 we define the Besov spaces $B_\alpha^{p,q}$, and present their theory in a way that suits our purposes. In Section 4.2 the Lizorkin–Triebel spaces $F_\alpha^{p,q}$ are defined, and then most of the section is devoted to a proof of the fact that this scale of spaces contains the spaces $L^{\alpha,p}$, in fact $F_\alpha^{p,2} = L^{\alpha,p}$ for $1 < p < \infty$. This result will be proved by means of a multiplier theorem of S. G. Mikhlin, whose proof is also included. In Section 4.3 we continue the presentation of these spaces in a way parallel to Section 4.1. This involves proving a rather deep theorem of J. Peetre. Now the stage is set for our application of the general nonlinear potential theory, which takes its beginning in Section 4.4. The short Section 4.5 is devoted to an important inequality of Th. H. Wolff. In Section 4.6, which is independent of most of the preceding theory in this chapter, we give a representation of the Besov, and Lizorkin–Triebel spaces by means of "smooth atoms". In Section 4.7 we apply this representation to formulate an "atomic" nonlinear potential theory, which among other things gives a new way of viewing the Wolff inequality. Finally, in Section 4.8 we use the atomic representation to give a characterization of $L^{\alpha,p}$ by means of a local approximation property. This result implies Strichartz' theorem, Theorem 3.5.6, whose proof was previously postponed.

4.1 Besov Spaces

In order to tie the potential theory of the previous chapters to the theory of Besov spaces, we shall in this section define these spaces, and prove some of their most important properties. We shall only treat those aspects of the Besov spaces that are needed for our purpose, and the reader who wants a fuller treatment of the spaces for their own sake is referred to the literature.

Let Φ belong to the Schwartz class $\mathcal{S}(\mathbf{R}^N)$, let $\widehat{\Phi}$ be its Fourier transform, and assume that

$$\operatorname{supp} \widehat{\Phi} \subset B(0,1) = B_0, \quad \text{and} \quad \widehat{\Phi}(\xi) = 1 \text{ on } B(0,\tfrac{1}{2}) = B_1 \ . \qquad (4.1.1)$$

Set

$$\Phi_n(x) = 2^{nN}\Phi(2^n x), \quad n \in \mathbf{Z} , \tag{4.1.2}$$

so that

$$\widehat{\Phi_n}(\xi) = \widehat{\Phi}(2^{-n}\xi) ,$$

and define $\{\varphi_n\}_{-\infty}^{\infty}$ by setting

$$\varphi_n(x) = \Phi_n(x) - \Phi_{n-1}(x) . \tag{4.1.3}$$

It follows that

$$\operatorname{supp}\widehat{\varphi_n} \subset B(0, 2^n) \setminus B(0, 2^{n-2}) = B_{-n}(0) \setminus B_{-n+2}(0) , \tag{4.1.4}$$

and

$$\widehat{\Phi}(\xi) + \sum_{n=1}^{\infty} \widehat{\varphi_n}(\xi) \equiv 1 . \tag{4.1.5}$$

Definition 4.1.1. Let $\alpha \in \mathbf{R}$, $0 < p \leq \infty$, and $0 < q \leq \infty$. The Besov spaces corresponding to these indices are

$$B_\alpha^{p,q} = \left\{u \in \mathcal{S}' : \|\Phi * u\|_p + \textstyle\sum_1^\infty \left(2^{n\alpha}\|\varphi_n * u\|_p\right)^q < \infty\right\}, \quad 0 < q < \infty ,$$
$$B_\alpha^{p,\infty} = \left\{u \in \mathcal{S}' : \|\Phi * u\|_p < \infty,\ \sup_{n\geq 1} 2^{n\alpha}\|\varphi_n * u\|_p < \infty\right\},$$

and their quasinorms (norms when $p, q \geq 1$) are defined by

$$\|u\|_{B_\alpha^{p,q}} = \|\Phi * u\|_p + \|\{2^{n\alpha}\|\varphi_n * u\|_p\}_1^\infty\|_{l^q}$$
$$= \|\Phi * u\|_p + \|\{2^{n\alpha}\varphi_n * u\}_1^\infty\|_{l^q(L^p)} ,$$

where the last expression also defines the mixed $l^q(L^p)$-norm.

One can also define the corresponding "homogeneous spaces". We observe that

$$\sum_{n=-\infty}^{\infty} \widehat{\varphi_n}(\xi) = 1, \quad \xi \neq 0, \quad \text{and} \quad \widehat{\varphi_n}(0) = 0, \quad n \in \mathbf{Z} .$$

Because of the last fact we have $\widehat{\varphi_n}\, D^\alpha\delta = 0$ for any derivative D^α of the Dirac measure δ at the origin, and consequently, $\varphi_n * \pi = 0$ for any polynomial π. We denote the space of all polynomials by $\mathfrak{P}$, and define the homogeneous Besov spaces as subspaces of $\mathcal{S}'/\mathfrak{P}$.

Definition 4.1.2. Let $\alpha \in \mathbf{R}$, $0 < p \leq \infty$, and $0 < q \leq \infty$. The homogeneous Besov spaces corresponding to these indices are

$$\dot{B}_\alpha^{p,q} = \left\{u \in \mathcal{S}'/\mathfrak{P} : \textstyle\sum_{-\infty}^\infty \left(2^{n\alpha}\|\varphi_n * u\|_p\right)^q < \infty\right\}, \quad 0 < q < \infty ,$$
$$\dot{B}_\alpha^{p,\infty} = \left\{u \in \mathcal{S}'/\mathfrak{P} : \sup_n 2^{n\alpha}\|\varphi_n * u\|_p < \infty\right\} ,$$

and their quasinorms are defined by

$$\|u\|_{\dot{B}_\alpha^{p,q}} = \|\{2^{n\alpha}\varphi_n * u\}_{-\infty}^\infty\|_{l^q(L^p)} .$$

In what follows we shall limit the treatment to the case $p, q \geq 1$. We also choose not to deal with the homogeneous spaces more than parenthetically. We refer to the literature, for example Chapter 3 in J. Peetre's book [356], for more information about these spaces. In the same chapter the proof of the following easy theorem is also found.

Theorem 4.1.3. *Let $\alpha \in \mathbf{R}$, $1 \leq p \leq \infty$, and $1 \leq q \leq \infty$.*

(a) *The Besov spaces $B_\alpha^{p,q}$ are Banach spaces.*
(b) *There are continuous embeddings $\mathcal{S} \subset B_\alpha^{p,q} \subset \mathcal{S}'$, and in the case $\alpha > 0$, $B_\alpha^{p,q} \subset L^p$.*
(c) *$\mathcal{S}$ is dense in $B_\alpha^{p,q}$, if $p, q < \infty$.*
(d) *The dual spaces are $(B_\alpha^{p,q})^* = B_{-\alpha}^{p',q'}$, if $p, q < \infty$, where $pp' = p + p'$, $qq' = q + q'$.*

It is a fundamental fact, due to J. Peetre, that the functions Φ and $\{\varphi_n\}_1^\infty$ in the definition can be replaced by much more general functions.

Theorem 4.1.4. *Let $\eta \in B_\alpha^{1,1}(\mathbf{R}^N)$ for some $\alpha > 0$, and set $\eta_n(x) = 2^{nN}\eta(2^n x)$ for $n = 0, 1, 2, \ldots$. Let $1 \leq p \leq \infty$, and $1 \leq q \leq \infty$. Then there is a constant A such that for any $u \in B_{-\alpha}^{p,q}$*

$$\|\{2^{-n\alpha}\eta_n * u\}_0^\infty\|_{l^q(L^p)} \leq A\,\|\eta\|_{B_\alpha^{1,1}}\|u\|_{B_{-\alpha}^{p,q}} \; .$$

Proof. We observe that for any n

$$\widehat{\Phi_n}(\xi) + \sum_{k=1}^\infty \widehat{\varphi_{k+n}}(\xi) \equiv 1 \; ,$$

so that we can write

$$\eta_n * u = \Phi_n * \eta_n * u + \sum_{k=1}^\infty \varphi_{k+n} * \eta_n * u \; .$$

Moreover

$$\operatorname{supp}\widehat{\varphi_{n-2}} \cap \operatorname{supp}\widehat{\varphi_n} = \operatorname{supp}\widehat{\varphi_n} \cap \operatorname{supp}\widehat{\varphi_{n+2}} = \emptyset \; ,$$

and thus

$$\widehat{\varphi_{n-1}}(\xi) + \widehat{\varphi_n}(\xi) + \widehat{\varphi_{n+1}}(\xi) = 1 \quad \text{for } \xi \in \operatorname{supp}\widehat{\varphi_n} \; .$$

Set $\varphi_n' = \varphi_{n-1} + \varphi_n + \varphi_{n+1}$. Then

$$\eta_n * u = \Phi_n * \eta_n * u + \sum_{k=1}^\infty \varphi_{k+n} * \eta_n * \varphi_{k+n}' * u \; ,$$

and thus

$$2^{-n\alpha}\|\eta_n * u\|_p \le 2^{-n\alpha}\|\Phi_n * \eta_n * u\|_p + 2^{-n\alpha}\sum_{k=1}^{\infty}\|\varphi_{k+n} * \eta_n\|_1\|\varphi'_{k+n} * u\|_p$$

$$= 2^{-n\alpha}\|\Phi_n * \eta_n * u\|_p + \sum_{k=1}^{\infty} 2^{k\alpha}\|\varphi_k * \eta\|_1 2^{-(n+k)\alpha}\|\varphi'_{k+n} * u\|_p$$

$$= A_n + \sum_{k=1}^{\infty} a_k b_{n+k} \ .$$

We have to estimate

$$\sum_{n=0}^{\infty} A_n^q \quad \text{and} \quad \sum_{n=0}^{\infty}\left(\sum_{k=1}^{\infty} a_k b_{n+k}\right)^q \ .$$

By Minkowski's inequality

$$\left(\sum_{n=0}^{\infty}\left(\sum_{k=1}^{\infty} a_k b_{n+k}\right)^q\right)^{1/q} \le \sum_{k=1}^{\infty} a_k \left(\sum_{n=0}^{\infty} b_{n+k}^q\right)^{1/q}$$

$$\le \sum_{k=1}^{\infty} a_k \left(\sum_{n=1}^{\infty} b_n^q\right)^{1/q} \le 3\,\|\eta\|_{B_\alpha^{1,1}}\|u\|_{B_{-\alpha}^{p,q}} \ .$$

In order to estimate $\sum_{n=0}^{\infty} A_n^q$ we observe that

$$\widehat{\Phi}(\xi) + \sum_{k=1}^{n+1}\widehat{\varphi_k}(\xi) = 1 \quad \text{on } \operatorname{supp}\widehat{\Phi_n} \ .$$

Thus

$$\Phi_n * \eta_n * u = \Phi_n * \eta_n * \left(\Phi + \sum_{k=1}^{n+1}\varphi_k\right) * u \ ,$$

and

$$\|\Phi_n * \eta_n * u\|_p \le \|\Phi_n * \eta_n\|_1\left(\|\Phi * u\|_p + \sum_{k=1}^{n+1}\|\varphi_k * u\|_p\right) = \|\Phi * \eta\|_1 \sum_{k=0}^{n+1} c_k \ .$$

We want to estimate

$$\sum_{n=0}^{\infty}\left(2^{-n\alpha}\sum_{k=0}^{n+1} c_k\right)^q \ ,$$

knowing that $\sum_{k=0}^{\infty}(2^{-k\alpha}c_k)^q < \infty$. To this end, using a well-known trick (see e.g. S. M. Nikol′skiĭ [349], 5.6(19), (20)), we choose an ε with $0 < \varepsilon < \alpha$. Hölder's inequality and summation of a geometric series give

$$\left(\sum_{k=0}^{n+1} c_k\right)^q \le \sum_{k=0}^{n+1} 2^{-k\varepsilon q} c_k^q \left(\sum_{k=0}^{n+1} 2^{k\varepsilon q'}\right)^{q/q'} \le A\, 2^{n\varepsilon q}\sum_{k=0}^{n+1} 2^{-k\varepsilon q} c_k^q \ ,$$

and it follows that

$$\begin{aligned}\sum_{n=0}^{\infty}\left(2^{-n\alpha}\sum_{k=0}^{n+1}c_k\right)^q &\le A\sum_{n=0}^{\infty}2^{n(\varepsilon-\alpha)q}\sum_{k=0}^{n+1}2^{-k\varepsilon q}c_k^q\\ &=\sum_{k=0}^{\infty}2^{-k\varepsilon q}c_k^q\sum_{n=(k-1)_+}^{\infty}2^{n(\varepsilon-\alpha)q}\\ &\le A\sum_{k=0}^{\infty}2^{-k\varepsilon q}c_k^q2^{k(\varepsilon-\alpha)q}=A\,\|\{2^{-k\alpha}c_k\}\|_{l^q}^q\ .\end{aligned}$$

This finishes the proof of the theorem.

Remark. Note that the only place the positivity of α is used is in the last estimate. In the case of homogeneous Besov spaces $\dot{B}_{-\alpha}^{p,q}$, the terms A_n are not present, and the theorem is valid for all real α. This is a result with many consequences, e.g. in approximation theory. See Peetre [356], Chapter 8, and references given there.

In order to prove a result in the converse direction we need to assume that η satisfies a "Tauberian condition" (from Wiener's Tauberian theorem; see e.g. Y. Katznelson [241], Chapter VIII.6, or K. Chandrasekharan [98], Chapter I.13), which means that $\widehat{\eta}$ satisfies a simple nonvanishing condition.

Theorem 4.1.5. *Suppose that $\eta\in L^1(\mathbf{R}^N)$ and that $|\widehat{\eta}(\xi)|\ge c>0$ on $B(0,1)$. Set $\eta_n(x)=2^{nN}\eta(2^nx)$ for $n=0,1,2,\dots$. Let $\alpha>0$, $1\le p\le\infty$, and $1\le q\le\infty$. Then there is a constant A such that for any $u\in\mathcal{S}'$*

$$\|u\|_{B_{-\alpha}^{p,q}}\le A\,\|\{2^{-n\alpha}\eta_n*u\}_0^\infty\|_{l^q(L^p)}\ .$$

Proof. By assumption $|\widehat{\eta}(\xi)|>0$ on $\operatorname{supp}\Phi$, and thus, by a theorem of Wiener (see e.g. [241], Lemma VIII.6.3, p. 228, or [98], Corollary I.11.6, p. 68), η divides Φ in the sense that there is $\Psi\in L^1$ such that

$$\widehat{\Psi}(\xi)=\frac{\widehat{\Phi}(\xi)}{\widehat{\eta}(\xi)},\qquad \xi\in\operatorname{supp}\widehat{\Phi}\ .$$

Similarly, there is ψ in L^1, and $\psi_n(x)=2^{nN}\psi(2^nx)$, $n\in\mathbf{N}$, such that

$$\widehat{\psi_n}(\xi)=\frac{\widehat{\varphi_n}(\xi)}{\widehat{\eta_n}(\xi)},\qquad \xi\in\operatorname{supp}\widehat{\varphi_n}\ ,$$

and $\|\psi_n\|_1=\|\psi\|_1<\infty$. Thus $\Phi=\eta*\Psi$ and $\|\Phi*u\|_p\le\|\Psi\|_1\|\eta*u\|_p$. In the same way $\|\varphi_n*u\|_p\le\|\psi\|_1\|\eta_n*u\|_p$. The theorem follows.

Remark. For the homogeneous spaces the above proof is valid also for $\alpha<0$. See Peetre [356], Chapter 8, for a more general result.

For positive u the condition on η in Theorem 4.1.4 can be relaxed.

Corollary 4.1.6. *Let $\eta \not\equiv 0$ be nonnegative, bounded and lower semicontinuous, with support contained in the unit ball. Set $\eta_n(x) = 2^{nN}\eta(2^n x)$ for $n = 0, 1, 2, \ldots$. Let $\alpha > 0$, $1 \le p \le \infty$, and $1 \le q \le \infty$. Then there is a constant A such that for any positive $u \in \mathcal{S}'$ (i.e. a Radon measure)*

$$A^{-1}\|u\|_{B^{p,q}_{-\alpha}} \le \|\{2^{-n\alpha}\eta_n * u\}_0^\infty\|_{l^q(L^p)} \le A\,\|u\|_{B^{p,q}_{-\alpha}}\ ,$$

and $u \in B^{p,q}_{-\alpha}$ if and only if $\|\{2^{-n\alpha}\eta_n * u\}_0^\infty\|_{l^q(L^p)} < \infty$.

Proof. The assumptions imply that $|\widehat{\eta}(\xi)| \ge c > 0$ on $B(0,1)$. Indeed,

$$|\widehat{\eta}(\xi)| = \left|\int_{|x|\le 1} \eta(x)e^{-i\langle x,\xi\rangle}\,d\xi\right|$$
$$\ge \left|\int_{|x|\le 1} \eta(x)\cos(\langle x,\xi\rangle)\,d\xi\right| \ge \cos 1 \int_{|x|\le 1} \eta(x)\,dx$$

for $|\xi| \le 1$. Thus the left hand inequality follows from Theorem 4.1.5. The right hand inequality follows from Theorem 4.1.4, if η is replaced by a smooth $\bar{\eta}$ such that $\eta \le \bar{\eta}$.

Remark 1. The assumption that $\operatorname{supp}\eta \subset B(0,1)$ has been made only for convenience. If $\eta \in L^1$ and is nonnegative, then $\widehat{\eta}(\xi) \ne 0$ on a neighborhood of 0, and the proof can easily be modified to give the same result.

Remark 2. If η is the characteristic function for $B(0,1)$ and u is a measure, we have, writing $B(x, 2^{-n}) = B_n(x)$, that $\eta_n * u(x) = 2^{nN}u(B_n(x))$. Thus, if $u \in \mathcal{M}^+(\mathbf{R}^N)$, then

$$u \in (B^{p,q}_{-\alpha})_+ \quad \text{if and only if} \quad \|\{2^{n(N-\alpha)}u(B_n(x))\}_0^\infty\|_{l^q(L^p)} < \infty\ .$$

See also Theorem 4.5.5 below.

By means of Theorems 4.1.4 and 4.1.5 we shall now show that functions in $B^{p,q}_\alpha$, $\alpha > 0$, $p > 1$, $q > 1$, admit a useful representation formula.

Theorem 4.1.7. *Let $\alpha > 0$, $1 < p \le \infty$, and $1 < q \le \infty$, and let $\{\eta_n\}_0^\infty$ satisfy the conditions in Theorems 4.1.4 and 4.1.5. Then a function (or tempered distribution) F belongs to $B^{p,q}_\alpha$ if and only if there is a function sequence $f = \{f_n\}_0^\infty \in l^q(L^p)$ such that*

$$F = \sum_{n=0}^{\infty} 2^{-n\alpha}\eta_n * f_n\ .$$

Moreover, there is a constant A, depending on η, such that

$$A^{-1}\|F\|_{B^{p,q}_\alpha} \le \inf\|f\|_{l^q(L^p)} \le A\,\|F\|_{B^{p,q}_\alpha}\ ,$$

where the infimum is taken over all such f.

Proof. First assume that $F = \sum_{n=0}^{\infty} 2^{-n\alpha}\eta_n * f_n$, where $\{f_n\}_0^\infty \in l^q(L^p)$. Let $u \in B_{-\alpha}^{p',q'}$ be arbitrary. Then by Theorem 4.1.4

$$
\begin{aligned}
|\langle u, F\rangle| &\le \sum |\langle u, 2^{-n\alpha}\eta_n * f_n\rangle| = \sum |\langle 2^{-n\alpha}\eta_n * u, f_n\rangle| \\
&\le \sum \|2^{-n\alpha}\eta_n * u\|_{p'}\|f_n\|_p \le \|\{2^{-n\alpha}\eta_n * u\}\|_{l^{q'}(L^{p'})}\|\{f_n\}\|_{l^q(L^p)} \\
&\le A\,\|\eta\|_{B_\alpha^{1,1}}\|u\|_{B_{-\alpha}^{p',q'}}\|\{f_n\}\|_{l^q(L^p)} \ .
\end{aligned}
$$

It follows that $F \in (B_{-\alpha}^{p',q'})^* = B_\alpha^{p,q}$.

To prove the converse we let $F \in B_\alpha^{p,q}$. We have

$$
F = \Phi * F + \sum_{n=1}^{\infty} \varphi_n * F \ .
$$

By Wiener's theorem there are functions Ψ and ψ_n, $n = 1, 2, \dots$, in L^1, such that $\widehat{\eta}\,\widehat{\Psi} = 1$ on $\operatorname{supp}\widehat{\Phi}$, $\widehat{\eta_n}\,\widehat{\psi_n} = 1$ on $\operatorname{supp}\widehat{\varphi_n}$, and $\|\psi_n\|_1 = \|\psi_1\|_1 \le A$. (See e.g. Y. Katznelson [241], or K. Chandrasekharan [98], loc. cit.)

Define $\{f_n\}_0^\infty$ by

$$
\widehat{f_0} = \widehat{\Psi}\widehat{\Phi}\widehat{F}, \quad \text{and} \quad \widehat{f_n} = 2^{n\alpha}\widehat{\psi_n}\widehat{\varphi_n}\widehat{F}, \quad n = 1, 2, \dots \ .
$$

Then

$$
\widehat{F} = \widehat{\Phi}\widehat{F} + \sum_{n=1}^{\infty}\widehat{\varphi_n}\widehat{F} = \widehat{\eta}\widehat{\Psi}\widehat{\Phi}\widehat{F} + \sum_{n=1}^{\infty}\widehat{\eta_n}\widehat{\psi_n}\widehat{\varphi_n}\widehat{F} = \sum_{n=0}^{\infty} 2^{-n\alpha}\widehat{\eta_n}\widehat{f_n} \ ,
$$

and thus

$$
F = \sum_{n=0}^{\infty} 2^{-n\alpha}\eta_n * f_n \ .
$$

Moreover

$$
\|f_0\|_p \le \|\Psi\|_1\|\Phi * F\|_p \ ,
$$

and

$$
\|f_n\|_p \le 2^{n\alpha}\|\psi_n\|_1\|\varphi_n * F\|_p \le A\,2^{n\alpha}\|\varphi_n * F\|_p \ ,
$$

which proves that $f = \{f_n\}_0^\infty \in l^q(L^p)$, and

$$
\|f\|_{l^q(L^p)} \le A\,\|F\|_{B_\alpha^{p,q}} \ .
$$

4.2 Lizorkin–Triebel Spaces

If one makes a permutation in Definition 4.1.1, one obtains a different scale of spaces, the Lizorkin–Triebel spaces.

Definition 4.2.1. Let $\alpha \in \mathbf{R}$, $0 < p < \infty$, and $0 < q \leq \infty$. The Lizorkin-Triebel spaces corresponding to these indices are

$$F_\alpha^{p,q} = \left\{u \in \mathcal{S}' : \|\Phi * u\|_p + \left\|\left(\sum_1^\infty |2^{n\alpha}\varphi_n * u|^q\right)^{1/q}\right\|_p < \infty\right\}, \quad 0 < q < \infty ,$$

$$F_\alpha^{p,\infty} = \left\{u \in \mathcal{S}' : \|\Phi * u\|_p < \infty, \ \left\|\sup_{n\geq 1} 2^{n\alpha}|\varphi_n * u|\right\|_p < \infty\right\} ,$$

and their quasinorms (norms for $p, q \geq 1$) are defined by

$$\begin{aligned}\|u\|_{F_\alpha^{p,q}} &= \|\Phi * u\|_p + \left\|\|\{2^{n\alpha}\varphi_n * u\}_1^\infty\|_{l^q}\right\|_p \\ &= \|\Phi * u\|_p + \left\|\{2^{n\alpha}\varphi_n * u\}_1^\infty\right\|_{L^p(l^q)} .\end{aligned}$$

As before one can also define the corresponding "homogeneous spaces", $\dot{F}_\alpha^{p,q}$, by extending the sum to $-\infty$.

The main importance of these spaces is that they include many classical function spaces as special cases. For example, $F_0^{p,2}$ coincides with L^p for $1 < p < \infty$, with the Hardy space H^p for $0 < p \leq 1$, and the definition can be extended to the case $p = \infty$ so that $F_0^{\infty,2}$ can be identified with the space of functions of bounded mean oscillation, BMO. (See the notes at the end of the chapter.) An obvious fact is that $F_\alpha^{p,p} = B_\alpha^{p,p}$ for $p < \infty$. For us their importance lies in the following theorem, which will give us a useful new representation of Bessel potentials.

Theorem 4.2.2. *The Lizorkin–Triebel space $F_\alpha^{p,2}$ coincides with $L^{\alpha,p}$ for $\alpha \in \mathbf{R}$ and $1 < p < \infty$, and there are constants A_1 and A_2 such that*

$$A_1\|u\|_{\alpha,p} \leq \|u\|_{F_\alpha^{p,2}} \leq A_2\|u\|_{\alpha,p} .$$

This highly non-trivial result is a theorem of so called Littlewood–Paley type. (See e.g. Chapter IV in Stein [389].) We shall give a self-contained proof, except for a reference to the standard estimate of singular integral operators given as Theorem 1.1.5.

The proof uses the so called Rademacher functions $\{r_n(t)\}_0^\infty$, $0 < t < 1$. These functions are defined by

$$\begin{aligned} r_0(t) &= 1 && \text{for } 0 < t \leq \tfrac{1}{2} , \\ r_0(t) &= -1 && \text{for } \tfrac{1}{2} < t \leq 1 , \\ r_0(t+1) &= r_0(t) && \text{for } t \in \mathbf{R} , \\ r_n(t) &= r_0(2^n t) && \text{for } n \in \mathbf{N} . \end{aligned}$$

Clearly the Rademacher functions form an orthonormal system in $L^2(0,1)$, and in particular

$$\int_0^1 \left|\sum_{n=0}^m a_n r_n(t)\right|^2 dt = \sum_{n=0}^m |a_n|^2 \tag{4.2.1}$$

for all constants a_n. They satisfy the following important inequality of A. Ya. Khinchin.

Lemma 4.2.3. *Let $0 < p < \infty$, and let $R_m(t) = \sum_{n=0}^m a_n r_n(t)$. Then there are constants A_1 and A_2, only depending on p, such that*

$$A_1 \| R_m \|_{L^p(0,1)} \leq \| R_m \|_{L^2(0,1)} \leq A_2 \| R_m \|_{L^p(0,1)} \ .$$

Proof. We shall only prove the special case

$$\left(\sum_{n=0}^m a_n^2\right)^{1/2} = \| R_m \|_{L^2(0,1)} \leq \sqrt{3}\, \| R_m \|_{L^1(0,1)} , \quad a_n \in \mathbf{R} , \tag{4.2.2}$$

which is all that is needed for the proof of Theorem 4.2.2. (See also the notes at the end of the chapter.)

We first prove

$$\int_0^1 R_m(t)^4\, dt \leq 3\left(\sum_{n=0}^m a_n^2\right)^2 . \tag{4.2.3}$$

This inequality is easily proved by means of induction. For $m = 0$ the inequality is obvious. We prove that (4.2.3) is true for R_{m+1}, given that it is true for R_m. We have

$$\begin{aligned}\int_0^1 R_{m+1}^4\, dt &= \int_0^1 R_m^4\, dt + 4a_{m+1} \int_0^1 r_{m+1} R_m^3\, dt + 6a_{m+1}^2 \int_0^1 r_{m+1}^2 R_m^2\, dt \\ &\quad + 4a_{m+1}^3 \int_0^1 r_{m+1}^3 R_m\, dt + a_{m+1}^4 \int_0^1 r_{m+1}^4\, dt \ ,\end{aligned}$$

which by means of the induction hypothesis (4.2.3) and the orthogonality relations gives

$$\int_0^1 R_{m+1}^4\, dt \leq 3\left(\sum_{n=0}^m a_n^2\right)^2 + 6a_{m+1}^2 \sum_{n=0}^m a_n^2 + a_{m+1}^4 \leq 3\left(\sum_{n=0}^{m+1} a_n^2\right)^2 ,$$

as required.

Hölder's inequality, and (4.2.3) now give

$$\begin{aligned}\sum_{n=0}^m a_n^2 &= \int_0^1 R_m^2\, dt = \int_0^1 R_m^{4/3} R_m^{2/3}\, dt \leq \left(\int_0^1 R_m^4\, dt\right)^{1/3} \left(\int_0^1 |R_m|\, dt\right)^{2/3} \\ &\leq 3^{1/3} \left(\sum_{n=0}^m a_n^2\right)^{2/3} \left(\int_0^1 |R_m|\, dt\right)^{2/3} ,\end{aligned}$$

and (4.2.2) follows.

The second main ingredient in the proof of Theorem 4.2.2 is the following well-known multiplier theorem of S. G. Mikhlin and L. Hörmander.

Theorem 4.2.4. *Let $m \in L^\infty(\mathbf{R}^N)$, and suppose that m satisfies*

$$\frac{1}{R^N}\int_{\frac{1}{2}R\le|\xi|\le 2R} |R^{|\sigma|}D^\sigma m(\xi)|^2\,d\xi \le B^2\ , \quad 0<R<\infty\ , \tag{4.2.4}$$

for all multiindices σ with $|\sigma|\le\kappa$, where κ is the least integer such that $\kappa>\frac{1}{2}N$. Then, for every p with $1<p<\infty$ there is a constant A, depending only on p and N, such that for all $f\in L^p(\mathbf{R}^N)$

$$\|\mathcal{F}^{-1}(m\mathcal{F}(f))\|_p \le A\,B\,\|f\|_p\ . \tag{4.2.5}$$

Proof. We write $m=\sum_{-\infty}^{\infty} m_n$, where $m_n = m\widehat{\varphi_n}$, and $\widehat{\varphi_n}$ is given by (4.1.3). Let k_n be the inverse Fourier transform of m_n, and set $M_n=\sum_{-n}^{n} m_j$, and $K_n=\sum_{-n}^{n} k_j$. We shall show that K_n satisfies the "Hörmander condition" for the validity of the Calderón–Zygmund estimate (see Theorem 1.1.5),

$$\int_{|x|\ge 2t} |K_n(x-y)-K_n(x)|\,dx \le A\,B\ , \quad |y|\le t,\quad n\in\mathbf{N}\ , \tag{4.2.6}$$

for a constant A independent of n.

By assumption (see (4.1.3)) $\operatorname{supp} m_j \subset B(0,2^j)\setminus B(0,2^{j-2})$. The Leibniz formula and (4.2.4) give

$$2^{-jN}\int_{\mathbf{R}^N}\sum_{|\sigma|\le\kappa}\left|2^{j|\sigma|}D^\sigma m_j\right|^2 d\xi \le A\,B^2\ ,$$

whence by Plancherel's formula

$$\int_{\mathbf{R}^N}(1+2^{2j}|x|^2)^\kappa |k_j|^2\,dx \le A\,B^2 2^{jN}\ .$$

By the Cauchy–Schwarz inequality

$$\int_{\mathbf{R}^N}|k_j|\,dx \le A\,B\left(2^{jN}\int_{\mathbf{R}^N}\left(1+2^{2j}|x|^2\right)^{-\kappa}dx\right)^{1/2} \le A\,B\ , \tag{4.2.7}$$

since the last integral is finite for $2\kappa>N$. This implies that $\|m_j\|_\infty\le A\,B$, and thus $\|M_n\|_\infty\le A\,B$, since at most two $m_j(\xi)$ are different from 0 at any point ξ. It follows that the operation $f\mapsto\mathcal{F}^{-1}(M_n\mathcal{F}(f))$ is bounded in L^2, independently of n.

In order to estimate $\int_{|x|\ge 2t}|k_j(x+y)-k_j(x)|\,dx$ for $|y|\le t$ we first note that the Cauchy–Schwarz inequality also gives

$$\int_{|x|\ge t}|k_j|\,dx \le A\,B\left(2^{jN}\int_{|x|\ge t}\left(2^{2j}|x|^2\right)^{-\kappa}dx\right)^{1/2} \le A\,B\,(2^j t)^{(N-2\kappa)/2}\ ,$$

and thus for $|y|\le t$

$$\int_{|x|\ge 2t}|k_j(x+y)-k_j(x)|\,dx \le A\,B\,(2^j t)^{(N-2\kappa)/2}\ .$$

This is a good estimate for $2^j t \geq 1$, since $N < 2\kappa$, but for $2^j t < 1$ we need a better inequality. We can obtain such an estimate by observing that $\operatorname{supp} \widehat{k_j} \subset B(0, 2^j)$, and thus $\widehat{k_j} = \widehat{\Phi_{j+1} k_j}$ with Φ_{j+1} given by (4.1.2). Writing

$$k_j(x+y) - k_j(x) = \int_{\mathbf{R}^N} \big(\Phi_{j+1}(x+y-z) - \Phi_{j+1}(x-z)\big)\, k_j(z)\, dz \ ,$$

and using (4.2.7) and the fact that $\Phi \in \mathcal{S}$, we easily find

$$\int_{\mathbf{R}^N} |k_j(x+y) - k_j(x)|\, dx \leq \int_{\mathbf{R}^N} |\Phi(x + 2^{j+1} y) - \Phi(x)|\, dx \int_{\mathbf{R}^N} |k_j(z)|\, dz \leq A\, B\, 2^{j+1} |y|$$

for $|y| \leq 2^{-j}$. Thus, for $|y| \leq t$,

$$\int_{|x| \geq 2t} |K_n(x+y) - K_n(x)|\, dx \leq A\, B\, t \sum_{2^j \leq 1/t} 2^j + A\, B\, t^{(N-2\kappa)/2} \sum_{2^j > 1/t} 2^{j(N-2\kappa)/2} \ ,$$

which clearly implies (4.2.6).

It follows from Theorem 1.1.5 that the mapping

$$f \mapsto \mathcal{F}^{-1}(M_n \mathcal{F}(f)) = K_n * f$$

is bounded in each L^p, $1 < p < \infty$, independently of n. Now, $\|M_n\|_\infty$ is bounded, and $M_n(\xi) \to m(\xi)$ for $\xi \neq 0$, so $M_n \to m$ in $\mathcal{S}'$, as $n \to \infty$. The Fourier transformation is an isomorphism of $\mathcal{S}'$, so K_n converges in $\mathcal{S}'$ to a distribution k with $\widehat{k} = m$. The norm of $f \mapsto k * f$ as a mapping on L^p is

$$\begin{aligned} \sup_{\|f\|_p \leq 1} \|k * f\|_p &= \sup\Big\{ \big|\textstyle\int_{\mathbf{R}^N} (k * f) g\, dx\big| : f, g \in \mathcal{S},\ \|f\|_p \leq 1,\ \|g\|_{p'} \leq 1 \Big\} \\ &= \sup\Big\{ |\langle k, f * g\rangle| : f, g \in \mathcal{S},\ \|f\|_p \leq 1,\ \|g\|_{p'} \leq 1 \Big\} \ , \end{aligned}$$

and since $\langle k, f * g\rangle = \lim_{N\to\infty} \langle K_N, f * g\rangle$ for $f, g \in \mathcal{S}$, the theorem follows.

For our application of the multiplier theorem, we need the following lemma.

Lemma 4.2.5. *Let $\alpha \in \mathbf{R}$, let Φ and $\{\varphi_n\}_1^\infty$ be functions satisfying* (4.1.1)–(4.1.3), *and let $\{r_n\}_0^\infty$ be the Rademacher functions. Then* (4.2.4) *is satisfied by*

$$m(\xi) = \frac{1}{\widehat{\Phi}^2(\xi) + \sum_{n=1}^\infty \widehat{\varphi_n}^2(\xi)} \ ,$$

and for all $t \in [0, 1]$ by

$$m(\xi) = \frac{r_0(t)\widehat{\Phi}(\xi) + \sum_{n=1}^\infty 2^{n\alpha} r_n(t) \widehat{\varphi_n}(\xi)}{(1 + |\xi|^2)^{\alpha/2}} \ ,$$

Here the constant B appearing in (4.2.4) *depends only on α and does not depend on t.*

It is enough to prove that m satisfies

$$|D^\sigma m(\xi)| \le B\,|\xi|^{-|\sigma|} \quad , \quad \xi \in \mathbf{R}^N \setminus \{0\} \ ,$$

for $|\sigma| \le \kappa$, $\kappa > \frac{1}{2}N$. We omit this elementary verification.

Proof of Theorem 4.2.2. We shall prove the inequalities

$$A_1 \|u\|_{\alpha,p} \le \left\| \left(|\Phi * u|^2 + \textstyle\sum_1^\infty 2^{2n\alpha} |\varphi_n * u|^2 \right)^{1/2} \right\|_p \le A_2 \|u\|_{\alpha,p} \, . \tag{4.2.8}$$

It is enough to prove these inequalities for $u \in L^p$ with $\operatorname{supp} \widehat{u}$ compact, since such functions are dense both in $L^{\alpha,p}$ and in $F_\alpha^{p,2}$. We observe that under this assumption the infinite sum in (4.2.8) is actually finite.

We first prove the right hand estimate. Let $u = G_\alpha * f$, i.e.,

$$\widehat{f}(\xi) = (1 + |\xi|^2)^{\alpha/2} \widehat{u}(\xi) \ .$$

Then by Lemma 4.2.5, for each $t \in (0,1)$

$$\left\| r_0(t)\Phi * u + \textstyle\sum_{n=1}^\infty r_n(t) 2^{n\alpha} \varphi_n * u \right\|_p \le A \, \|f\|_p = A\, \|u\|_{\alpha,p} \ ,$$

and thus

$$\int_0^1 \left\| r_0(t)\Phi * u + \textstyle\sum_{n=1}^\infty r_n(t) 2^{n\alpha} \varphi_n * u \right\|_p dt \le A\, \|u\|_{\alpha,p} \ .$$

Applying Khinchin's inequality (4.2.2), and Minkowski's inequality (1.1.5), this immediately gives

$$\begin{aligned}
&\left\| \left(|\Phi * u|^2 + \textstyle\sum_1^\infty 2^{n\alpha} |\varphi_n * u|^2 \right)^{1/2} \right\|_p \\
&\qquad \le \sqrt{3} \left\| \int_0^1 \left| r_0(t)\Phi * u + \textstyle\sum_{n=1}^\infty r_n(t) 2^{n\alpha} \varphi_n * u \right| dt \right\|_p \\
&\qquad \le \sqrt{3} \int_0^1 \left\| r_0(t)\Phi * u + \textstyle\sum_{n=1}^\infty r_n(t) 2^{n\alpha} \varphi_n * u \right\|_p dt \\
&\qquad \le A\, \|u\|_{\alpha,p} \ ,
\end{aligned}$$

as claimed.

The converse inequality follows easily by duality. In fact, let

$$h = \mathcal{F}^{-1}\left((\widehat{\Phi}^2 + \textstyle\sum_{n=1}^\infty \widehat{\varphi}_n{}^2) \widehat{f} \right) \ . \tag{4.2.9}$$

Then Lemma 4.2.5 gives

$$\|f\|_p \le A\, \|h\|_p \ . \tag{4.2.10}$$

Let $g \in L^{p'}(\mathbf{R}^N)$ be a function such that $\|g\|_{p'} = 1$, $\operatorname{supp} \widehat{g}$ is compact, and

$$\int_{\mathbf{R}^N} g(x) h(x)\, dx \ge \tfrac{1}{2} \|h\|_p \ . \tag{4.2.11}$$

If a function v is defined by $\widehat{v}(\xi) = (1+|\xi|^2)^{\alpha/2}\widehat{g}(\xi)$, i.e., $g = G_\alpha * v$, and $u = G_\alpha * f$ as above, so that $\widehat{u}\,\widehat{v} = \widehat{f}\,\widehat{g}$, it follows from (4.2.9)–(4.2.11) that

$$\begin{aligned}\|u\|_{\alpha,p} = \|f\|_p &\le A\int_{\mathbf{R}^N} hg\,dx = A\int_{\mathbf{R}^N} \widehat{h}\,\widehat{g}\,d\xi \\ &= A\int_{\mathbf{R}^N} \widehat{f}\,\widehat{g}\left(\widehat{\Phi}^2 + \textstyle\sum_{n=1}^{\infty}\widehat{\varphi}_n{}^2\right)d\xi \\ &= A\int_{\mathbf{R}^N}\left((\widehat{u}\,\widehat{\Phi})(\widehat{v}\,\widehat{\Phi}) + \textstyle\sum_{n=1}^{\infty}(2^{n\alpha}\widehat{u}\,\widehat{\varphi}_n)(2^{-n\alpha}\widehat{v}\,\widehat{\varphi}_n)\right)d\xi\ ,\end{aligned}$$

which by Plancherel's formula and the Cauchy and Hölder inequalities gives

$$\begin{aligned}\|u\|_{\alpha,p} \le A\left\|\left(|\Phi * u|^2 + \textstyle\sum_1^\infty 2^{2n\alpha}|\varphi_n * u|^2\right)^{1/2}\right\|_p \\ \times\left\|\left(|\Phi * v|^2 + \textstyle\sum_1^\infty 2^{-2n\alpha}|\varphi_n * v|^2\right)^{1/2}\right\|_{p'}\ .\end{aligned}$$

But by the part of (4.2.8) already proved

$$\left\|\left(|\Phi * v|^2 + \textstyle\sum_1^\infty 2^{-2n\alpha}|\varphi_n * v|^2\right)^{1/2}\right\|_{p'} \le A\,\|v\|_{-\alpha,p'} = A\,\|g\|_{p'} \le A\ .$$

This proves the theorem.

4.3 Lizorkin–Triebel Spaces, Continued

The main goal of this section is to prove a representation theorem analogous to Theorem 4.1.7. This turns out to be more difficult than in the case of Besov spaces.

For the purpose of stating and proving the next two theorems it is practical to define a kind of weighted Besov spaces. However, our use of these spaces will be transient.

Definition 4.3.1. Let $\alpha \in \mathbf{R}$, $1 \le p \le \infty$, and $1 \le q \le \infty$. Let $\lambda \ge 0$. The space $B^{p,q}_{\alpha;\lambda}(\mathbf{R}^N)$ consists of the distributions $u \in \mathcal{S}'$ such that the norm

$$\|u\|_{B^{p,q}_{\alpha;\lambda}} = \left\|(1+|x|)^\lambda \Phi * u\right\|_p + \left\|\{2^{n\alpha}(1+2^n|x|)^\lambda \varphi_n * u\}_1^\infty\right\|_{l^q(L^p)}\ .$$

is finite.

Theorem 4.1.4 now has the following counterpart, also due to J. Peetre.

Theorem 4.3.2. *Let $\eta \in B^{1,1}_{\alpha;N}(\mathbf{R}^N)$ for some $\alpha > 0$, and denote $\eta_n(x) = 2^{nN}\eta(2^n x)$ for $n = 0, 1, 2, \ldots$. Let $1 < p < \infty$, and $1 < q \le \infty$. Then there is a constant A such that for any $u \in F^{p,q}_{-\alpha}$*

$$\left\|\{2^{-n\alpha}\eta_n * u\}_0^\infty\right\|_{L^p(l^q)} \le A\,\|\eta\|_{B^{1,1}_{\alpha;N}}\|u\|_{F^{p,q}_{-\alpha}}\ .$$

The proof of Theorem 4.3.2 depends on the following Theorem 4.3.4, which in itself is of great interest. In order to state the theorem we need to define a maximal function introduced by J. Peetre.

Definition 4.3.3. For any continuous function u on $\mathbf{R}^N$ the maximal function u^{**} is defined by

$$u^{**}(x) = \sup_{y \in \mathbf{R}^N} \frac{|u(x-y)|}{(1+|y|)^N} ,$$

and for the sequence $\{\eta_n\}_0^\infty$ a sequence $\{\eta_n^{**}u\}_0^\infty$ is defined by

$$\eta_n^{**}u(x) = \sup_{y \in \mathbf{R}^N} \frac{|\eta_n * u(x-y)|}{(1+2^n|y|)^N} .$$

Theorem 4.3.4. *Let* $\eta \in B^{1,1}_{\alpha+N;N}(\mathbf{R}^N)$ *for some* $\alpha > 0$*, and define* $\{\eta_n\}_0^\infty$ *as before. Let* $1 < p < \infty$*, and* $1 < q \le \infty$*. Then there is a constant A such that for any* $u \in F^{p,q}_{-\alpha}$

$$\left\| \{2^{-n\alpha}\eta_n^{**}u\}_0^\infty \right\|_{L^p(l^q)} \le A \, \|\eta\|_{B^{1,1}_{\alpha+N;N}} \|u\|_{F^{p,q}_{-\alpha}} . \tag{4.3.1}$$

The proof of Theorem 4.3.4 depends on the maximal theorem of C. Fefferman and E. M. Stein, Theorem 1.1.2, which extends the Hardy–Littlewood–Wiener maximal theorem (Theorem 1.1.1) to l^q-valued functions.

We also need the following elementary lemma.

Lemma 4.3.5. *For any* $u \in C^1(\mathbf{R}^N)$ *and* $0 < \delta \le 1$

$$u^{**}(x) \le \delta^{-N} Mu(x) + 2^N \delta (\nabla u)^{**}(x) \quad \textit{for all } x .$$

Proof. For any $x, z \in \mathbf{R}^N$ we have by the mean value theorem

$$\begin{aligned} |u(x-z)| &\le \min_{|x-z-y|\le\delta} |u(y)| + \delta \max_{|x-z-y|\le\delta} |\nabla u(y)| \\ &\le |B_\delta|^{-1} \int_{|x-z-y|\le\delta} |u(y)|\,dy + \delta \max_{|x-z-y|\le\delta} |\nabla u(y)| . \end{aligned}$$

By the definitions of Mu and $(\nabla u)^{**}$

$$|u(x-z)| \le \delta^{-N}(\delta+|z|)^N Mu(x) + \delta(\nabla u)^{**}(x)(1+\delta+|z|)^N ,$$

which gives the result.

Proof of Theorem 4.3.4. As in the proof of Theorem 4.1.4 we can write

$$\eta_n * u = \Phi_n * \eta_n * u + \sum_{k=1}^{\infty} \varphi'_{k+n} * \eta_n * \varphi_{k+n} * u = A_n + \sum_{k=1}^{\infty} B_{nk} .$$

Using the inequalities $1+a+b \le (1+a)(1+b)$ and $1+ab \le (1+a)(1+b)$, valid for $a, b \ge 0$, we find

$$
\begin{aligned}
|B_{nk}(x-z)| &\le \int_{\mathbf{R}^N} |\varphi'_{k+n} * \eta_n(y)|\,|\varphi_{k+n} * u(x-z-y)|\,dy \\
&\le \varphi^{**}_{k+n}u(x) \int_{\mathbf{R}^N} |\varphi'_{k+n} * \eta_n(y)|(1+2^{k+n}|y+z|)^N\,dy \\
&\le \varphi^{**}_{k+n}u(x)(1+2^{k+n}|z|)^N \int_{\mathbf{R}^N} |\varphi'_k * \eta(2^n y)|2^{nN}(1+2^{k+n}|y|)^N\,dy \\
&\le \varphi^{**}_{k+n}u(x)(1+2^n|z|)^N(1+2^k)^N \int_{\mathbf{R}^N} |\varphi'_k * \eta(y)|(1+2^k|y|)^N\,dy \ .
\end{aligned}
\tag{4.3.2}
$$

Setting

$$
(1+2^k)^{\alpha+N} \int_{\mathbf{R}^N} |\varphi'_k * \eta(y)|(1+2^k|y|)^N\,dy = t_k \ ,
$$

it follows that

$$
\widetilde{B}_{nk}(x) = \sup_{z\in\mathbf{R}^N} \frac{2^{-n\alpha}|B_{nk}(x-z)|}{(1+2^n|z|)^N} \le 2^{-(n+k)\alpha}\varphi^{**}_{k+n}u(x)t_k \ .
$$

It is easily seen that

$$
\sum_{k=1}^{\infty} t_k \le A\,\|\eta\|_{B^{1,1}_{\alpha+N;N}} < \infty \ ,
$$

which by assumption is finite, and thus, by Minkowski's inequality,

$$
\begin{aligned}
\left(\sum_{n=0}^{\infty}\left(\sum_{k=1}^{\infty}\widetilde{B}_{nk}(x)\right)^q\right)^{1/q} &\le A\sum_{k=1}^{\infty} t_k \left(\sum_{n=k}^{\infty}(2^{-n\alpha}\varphi^{**}_n u(x))^q\right)^{1/q} \\
&\le A\,\|\eta\|_{B^{1,1}_{\alpha+N;N}}\left\|\{2^{-n\alpha}\varphi^{**}_n u(x)\}_0^\infty\right\|_{l^q} \ .
\end{aligned}
$$

The previous estimate is valid for all $\alpha \in \mathbf{R}$, but in order to estimate the term A_n we will need the assumption $\alpha > 0$. As in the proof of Theorem 4.1.4 we can write

$$
\Phi_n * \eta_n * u = \Phi_n * \eta_n * \left(\Phi + \sum_{k=1}^{n+1}\varphi_k\right) * u \ .
$$

Using the inequality

$$
|\Phi * u(x-z-y)| \le (\Phi * u)^{**}(x)(1+|z|)^N(1+|y|)^N
$$

and the analogous inequalities for $|\varphi_k * u(x-z-y)|$ we find

$$
\begin{aligned}
|A_n(x-z)| \le & \int_{\mathbf{R}^N} |\Phi_n * \eta_n(y)|\,|\Phi * u(x-z-y)|\,dy \\
& + \sum_{k=1}^{n+1}\int_{\mathbf{R}^N} |\Phi_n * \eta_n(y)|\,|\varphi_k * u(x-z-y)|\,dy \\
\le & \ (\Phi * u)^{**}(x)(1+|z|)^N \int_{\mathbf{R}^N} |\Phi_n * \eta_n(y)|(1+|y|)^N\,dy \\
& + \sum_{k=1}^{n+1}\varphi^{**}_k u(x)(1+2^k|z|)^N \int_{\mathbf{R}^N} |\Phi_n * \eta_n(y)|(1+2^k|y|)^N\,dy \ .
\end{aligned}
$$

For $0 \le k \le n+1$ we have

$$\int_{\mathbf{R}^N} |\Phi_n * \eta_n(y)|(1+2^k|y|)^N\,dy = \int_{\mathbf{R}^N} |\Phi * \eta(2^n y)|2^{nN}(1+2^k|y|)^N\,dy$$

$$\le \int_{\mathbf{R}^N} |\Phi * \eta(y)|(1+2|y|)^N\,dy = C \le \|\eta\|_{B^{1,1}_{\alpha+N;N}}\,,$$

which gives

$$\frac{2^{-n\alpha}|A_n(x-z)|}{(1+2^n|z|)^N}$$
$$\le \frac{C2^{-n\alpha}}{(1+2^n|z|)^N}\left((\Phi * u)^{**}(x)(1+|z|)^N + \sum_{k=1}^{n+1} \varphi_k^{**}u(x)(1+2^k|z|)^N\right)$$
$$\le AC2^{-n\alpha}\left((\Phi * u)^{**}(x) + \sum_{k=1}^{n+1} \varphi_k^{**}u(x)\right) = AC2^{-n\alpha}\sum_{k=0}^{n+1} c_k\,. \tag{4.3.3}$$

It follows as in the proof of Theorem 4.1.4 that

$$\sum_{n=0}^{\infty}\left(\sup_{z\in\mathbf{R}^N} \frac{2^{-n\alpha}|A_n(x-z)|}{(1+2^n|z|)^N}\right)^q \le AC^q \sum_{n=0}^{\infty}(2^{-k\alpha}c_k)^q\,.$$

Putting the estimates of B_{nk} and A_n together we obtain

$$\|\{2^{-n\alpha}\eta_n^{**}u(x)\}_0^\infty\|_{l^q}$$
$$\le A\,\|\eta\|_{B^{1,1}_{\alpha+N;N}}\left((\Phi * u)^{**}(x) + \|\{2^{-n\alpha}\varphi_n^{**}u(x)\}_0^\infty\|_{l^q}\right)\,. \tag{4.3.4}$$

Thus we have reduced the problem to proving

$$\int_{\mathbf{R}^N} \left((\Phi * u)^{**} + \|\{2^{-n\alpha}\varphi_n^{**}u\}_0^\infty\|_{l^q}\right)^p dx$$
$$\le A\int_{\mathbf{R}^N} \left((\Phi * u) + \|\{2^{-n\alpha}\varphi_n * u\}_0^\infty\|_{l^q}\right)^p dx\,. \tag{4.3.5}$$

We first note that (4.3.4) applies to the derivatives of φ, and thus there is a constant A such that

$$\|\{2^{-n\alpha}(\nabla\varphi)_n^{**}u(x)\}_0^\infty\|_{l^q} \le A\big((\Phi * u)^{**}(x) + \|\{2^{-n\alpha}\varphi_n^{**}u(x)\}_0^\infty\|_{l^q}\big)\,.$$

The above estimates for the special case $n = 0$ applied to $\nabla\Phi$ also give

$$(\nabla\Phi * u)^{**}(x) \le A\big((\Phi * u)^{**}(x) + \|\{2^{-n\alpha}\varphi_n^{**}u(x)\}_0^\infty\|_{l^q}\big)\,.$$

All the previous estimates are still true if the exponent N in the definition of u^{**} is replaced by an arbitrary $\lambda > 0$. But now we apply Lemma 4.3.5, where it is important that the exponent is N. We obtain

$$\varphi_n^{**}u \le \delta^{-N}M(\varphi_n * u) + 2^N\delta(\nabla\varphi)_n^{**}u, \quad \delta \le 1\,,$$

and

$$(\Phi * u)^{**} \le \delta^{-N} M(\Phi * u) + 2^N \delta (\nabla \Phi * u)^{**}, \quad \delta \le 1 ,$$

Combining these inequalities we have

$$\begin{aligned}(\Phi * u)^{**} &+ \|\{2^{-n\alpha}\varphi_n^{**}u\}_0^\infty\|_{l^q} \\ &\le A\Big(\delta^{-N} M(\Phi * u) + \delta^{-N}\|\{2^{-n\alpha}M(\varphi_n * u)\}_0^\infty\|_{l^q} \\ &\quad + \delta(\nabla\Phi * u)^{**} + \delta\|\{2^{-n\alpha}(\nabla\varphi)_n^{**}u\}_0^\infty\|_{l^q}\Big) \\ &\le A\Big(\delta^{-N} M(\Phi * u) + \delta^{-N}\|\{2^{-n\alpha}M(\varphi_n * u)\}_0^\infty\|_{l^q} \\ &\quad + \delta(\Phi * u)^{**} + \delta\|\{2^{-n\alpha}\varphi_n^{**}u\}_0^\infty\|_{l^q}\Big) .\end{aligned}$$

By choosing δ small enough we obtain

$$(\Phi * u)^{**} + \|\{2^{-n\alpha}\varphi_n^{**}u\}_0^\infty\|_{l^q} \le A\big(M(\Phi * u) + \|\{2^{-n\alpha}M(\varphi_n * u)\}_0^\infty\|_{l^q}\big) ,$$

at least if we know that the left side is finite, which is certainly true if $u \in \mathcal{S}$. We now appeal to the Fefferman–Stein theorem (Theorem 1.1.2) to obtain

$$\begin{aligned}\|(\Phi * u)^{**}\|_p &+ \|\{2^{-n\alpha}\varphi_n^{**}u\}_0^\infty\|_{L^p(l^q)}^p \\ &\le A\big(\|\Phi * u\|_p + \|\{2^{-n\alpha}M(\varphi_n * u)\}_0^\infty\|_{L^p(l^q)}\big) \le A\|u\|_{F_{-\alpha}^{pq}} .\end{aligned}$$

The theorem follows from the fact that $\mathcal{S}$ is dense in $F_{-\alpha}^{p,q}$.

We can now at last prove Theorem 4.3.2.

Proof of Theorem 4.3.2. Applying the estimates (4.3.2) and (4.3.3) in the proof of Theorem 4.3.4 for $z = 0$, and setting

$$(1 + 2^k)^\alpha \int_{\mathbf{R}^N} |\varphi_k' * \eta(y)|(1 + 2^k|y|)^N \, dy = t_k ,$$

we find

$$2^{-n\alpha}|B_{nk}(x)| \le 2^{-(n+k)\alpha}(\varphi_{k+n}^{**}u)(x)t_k ,$$

and

$$2^{-n\alpha}|A_n(x)| \le AC2^{-n\alpha}\left((\Phi * u)^{**}(x) + \sum_{k=1}^{n+1}\varphi_k^{**}u(x)\right) .$$

It follows as in (4.3.4) that

$$\begin{aligned}\|\{2^{-n\alpha}&\eta_n * u(x)\}_0^\infty\|_{l^q} \\ &\le A\,\|\eta\|_{B_{\alpha;N}^{1,1}}\big((\Phi * u)^{**}(x) + \|\{2^{-n\alpha}\varphi_n^{**}u(x)\}_0^\infty\|_{l^q}\big) .\end{aligned}$$

In proving Theorem 4.3.4 we proved that the L^p-norm of the right-hand side is majorized by $\|u\|_{F_{-\alpha}^{p,q}}$, and this finishes the proof of Theorem 4.3.2.

Conversely we have the following analogue of Theorem 4.1.5.

Theorem 4.3.6. *Suppose that $\eta \in L^1(\mathbf{R}^N)$, that $|\eta(x)| \leq A\,(1+|x|)^{-N-1}$, and that $|\widehat{\eta}(\xi)| \geq c > 0$ on $B(0,1)$. Set $\eta_n(x) = 2^{nN}\eta(2^n x)$ for $n = 0, 1, 2, \ldots$. Let $\alpha > 0$, $1 < p < \infty$, and $1 < q \leq \infty$. Then there is a constant A such that for any $u \in \mathcal{S}'$*

$$\|u\|_{F^{p,q}_{-\alpha}} \leq A\,\|\{2^{-\alpha n}\eta_n * u\}_0^\infty\|_{L^p(l^q)} .$$

We need an elementary lemma.

Lemma 4.3.7. *Let $\eta \in L^1(\mathbf{R}^N)$, and suppose that $|\eta(x)| \leq (1+|x|)^{-N-1}$ for almost all x. Then for any integrable u and all x*

$$|\eta * u(x)| \leq A\,Mu(x) .$$

Proof. We assume without loss of generality that u and η are nonnegative, and let $x = 0$. Then

$$\begin{aligned}\eta * u(0) &\leq \int_{|y|\leq 1} u\,dy + \sum_{n=1}^{\infty} 2^{-n(N+1)} \int_{2^n \leq |y| \leq 2^{n+1}} u\,dy \\ &\leq A\,Mu(0) \sum_{n=0}^{\infty} 2^{-n} = A\,Mu(0) .\end{aligned}$$

Proof of Theorem 4.3.6. As in the proof of Theorem 4.1.5 there are Ψ and ψ_n, $\psi_n(x) = 2^{nN}\psi(2^n x)$, such that $\Phi = \eta * \Psi$ and $\varphi_n = \eta_n * \psi_n$. The assumption on η implies that we can choose Ψ and ψ so that $(1+|x|)^{N+1}\Psi(x)$ and $(1+|x|)^{N+1}\psi(x)$ are bounded. This implies that

$$|\varphi_n * u(x)| = |\psi_n * \eta_n * u(x)| \leq A\,M(\eta_n * u)(x) ,$$

and similarly for $\Phi * u$. By the Fefferman–Stein theorem (Theorem 1.1.2)

$$\|\Phi * u\|_p + \|\{2^{-n\alpha}\varphi_n * u\}_1^\infty\|_{L^p(l^q)} \leq A\,\|\{2^{-n\alpha}\eta_n * u\}_0^\infty\|_{L^p(l^q)} ,$$

which proves the theorem.

Corollary 4.3.8. *Let $\eta \not\equiv 0$ be nonnegative, bounded and lower semicontinuous, with support contained in the unit ball. Set $\eta_n(x) = 2^{nN}\eta(2^n x)$ for $n = 0, 1, 2, \ldots$. Let $\alpha > 0$, $1 < p < \infty$, and $1 < q \leq \infty$. Then there is a constant A such that for any positive $u \in \mathcal{S}'$ (i.e. a Radon measure)*

$$A^{-1}\|u\|_{F^{p,q}_{-\alpha}} \leq \|\{2^{-n\alpha}\eta_n * u\}_0^\infty\|_{L^p(l^q)} \leq A\,\|u\|_{F^{p,q}_{-\alpha}} ,$$

and $u \in F^{p,q}_{-\alpha}$ if and only if $\|\{2^{-n\alpha}\eta_n * u\}_0^\infty\|_{L^p(l^q)} < \infty$.

Proof. As the proof of Corollary 4.1.6.

As in the second remark following Corollary 4.1.6 it follows that if $u \in \mathcal{M}^+(\mathbf{R}^N)$, then

$$u \in (F^{p,q}_{-\alpha})_+ \quad \text{if and only if} \quad \left\| \{2^{n(N-\alpha)} u(B_n(x))\}_0^\infty \right\|_{L^p(l^q)} < \infty \ .$$

In particular, by Theorem 4.2.2,

$$u \in (L^{-\alpha,p})_+ \quad \text{if and only if} \quad \left\| \{2^{n(N-\alpha)} u(B_n(x))\}_0^\infty \right\|_{L^p(l^2)} < \infty \ .$$

If these observations are combined with the Muckenhoupt–Wheeden inequality in the form of Corollary 3.6.3, we obtain the following remarkable result. See also Theorem 4.5.5 below.

Corollary 4.3.9. *Let $\alpha > 0$ and $1 < p < \infty$. Then for all q, $1 < q \leq \infty$, the positive cones $(F^{p,q}_{-\alpha})_+$ and $(L^{-\alpha,p})_+ = (F^{p,2}_{-\alpha})_+$ coincide, and*

$$\begin{aligned} A^{-1} \|u\|_{F^{p,q}_{-\alpha}} &\leq \left\| \{2^{n(N-\alpha)} u(B_n(x))\}_0^\infty \right\|_{L^p(l^\infty)} \\ &\leq \left\| \{2^{n(N-\alpha)} u(B_n(x))\}_0^\infty \right\|_{L^p(l^q)} \\ &\leq \left\| \{2^{n(N-\alpha)} u(B_n(x))\}_0^\infty \right\|_{L^p(l^1)} \leq A \, \|u\|_{F^{p,q}_{-\alpha}} \end{aligned}$$

for all $u \in \mathcal{M}^+(\mathbf{R}^N)$.

Finally, we have a representation theorem corresponding to Theorem 4.1.7.

Theorem 4.3.10. *Let $\alpha > 0$, $1 < p < \infty$, and $1 < q < \infty$, and let $\{\eta_n\}_0^\infty$ satisfy the conditions in Theorems 4.3.2 and 4.3.6. Then a function (or tempered distribution) F belongs to $F^{p,q}_\alpha$ if and only if there is a function sequence $f = \{f_n\}_0^\infty \in L^p(l^q)$ such that*

$$F = \sum_{n=0}^{\infty} 2^{-n\alpha} \eta_n * f_n \ .$$

Moreover, there is a constant A, depending on η, such that

$$A^{-1} \|F\|_{F^{p,q}_\alpha} \leq \inf \|f\|_{L^p(l^q)} \leq A \, \|F\|_{F^{p,q}_\alpha} \ ,$$

where the infimum is taken over all such f.

Proof. Suppose that F has a representation as in the theorem. Let $u \in F^{p',q'}_{-\alpha}$ be arbitrary. Then

$$\langle u, F \rangle = \sum_{n=0}^{\infty} \langle u, 2^{-n\alpha} \eta_n * f_n \rangle = \sum_{n=0}^{\infty} \langle 2^{-n\alpha} \eta_n * u, f_n \rangle \ ,$$

and by Theorem 4.3.2

$$\begin{aligned} |\langle u, F \rangle| &\leq \int_{\mathbf{R}^N} \sum_{n=0}^{\infty} 2^{-n\alpha} |\eta_n * u| \, |f_n| \, dx \\ &\leq \int_{\mathbf{R}^N} \left\| \{2^{-n\alpha} \eta_n * u(x)\}_0^\infty \right\|_{l^{q'}} \|f(x)\|_{l^q} \, dx \\ &\leq \left\| \{2^{-n\alpha} \eta_n * u\}_0^\infty \right\|_{L^{p'}(l^{q'})} \|f\|_{L^p(l^q)} \\ &\leq A \, \|\eta\|_{B^{1,1}_{\alpha;N}} \|u\|_{F^{p',q'}_{-\alpha}} \|f\|_{L^p(l^q)} \ . \end{aligned}$$

It follows that $F \in (F^{p',q'}_{-\alpha})^* = F^{p,q}_\alpha$.

Conversely, if $F \in F_\alpha^{p,q}$, we can as in the proof of Theorem 4.1.7 write

$$F = \sum_0^\infty 2^{-n\alpha} \eta_n * f_n \ ,$$

where $f_0 = \Psi * \Phi * F$, and $f_n = 2^{n\alpha} \psi_n * \varphi_n * F$. We can assume that $(1+|x|)^{N+1}\Psi(x)$ and $(1+|x|)^{N+1}\psi(x)$ are bounded, and it follows as before from the Fefferman–Stein theorem (Theorem 1.1.2) that $f = \{f_n\}_0^\infty \in L^p(l^q)$ and

$$\|f\|_{L^p(l^q)} \le A \, \|F\|_{F_\alpha^{p,q}} \ .$$

4.4 More Nonlinear Potentials

In this section we shall see that thanks to the representation theorems 4.1.7 and 4.3.10 we can obtain a theory of capacities associated to the Besov and Lizorkin–Triebel spaces as a special case of the general theory of Sections 2.3 and 2.5. In the next section we then apply the estimates of Section 3.6 to obtain an inequality of Th. H. Wolff.

We consider the inhomogeneous case, the homogeneous case being similar and slightly simpler, and define a measure ν on $\mathbf{M} = \mathbf{R}^N \times \mathbf{N}$ by

$$\nu = m \otimes \sum_0^\infty \delta_n \ , \tag{4.4.1}$$

where m is Lebesgue measure, and δ_n denotes unit mass at n. Thus $f \in L^p(\nu)$ means $f = \{f_n\}_0^\infty$, with

$$\|f\|_{L^p(\nu)}^p = \sum_0^\infty \|f_n\|_p^p < \infty \ .$$

In other words, $f = \{f_n\}_0^\infty \in L^p(l^p) = l^p(L^p)$, and $\|f\|_{L^p(\nu)} = \|f\|_{l^p(L^p)}$.

We shall also need the mixed norms

$$\|f\|_{l^q(L^p)(\nu)} = \|f\|_{l^q(L^p)} \text{ and } \|f\|_{L^p(l^q)(\nu)} = \|f\|_{L^p(l^q)} \ .$$

We define a kernel h_α on $\mathbf{R}^N \times \mathbf{M} = \mathbf{R}^N \times \mathbf{R}^N \times \mathbf{N}$ by

$$h_\alpha(x, y, n) = 2^{-n\alpha} \eta_n(x - y) \ , \tag{4.4.2}$$

where $\eta_0 = \eta \not\equiv 0$ is a nonnegative function in $C_0^\infty(B_0)$, and $\eta_n(x) = 2^{nN}\eta(2^n x)$. If f is ν-measurable and nonnegative, and $\mu \in \mathcal{M}^+(\mathbf{R}^N)$, the corresponding potentials $\mathcal{H}_\alpha f$ and $\check{\mathcal{H}}_\alpha \mu$ are everywhere well defined and given by

$$\mathcal{H}_\alpha f = \sum_0^\infty 2^{-n\alpha} \eta_n * f_n \ ,$$

and

$$\check{\mathcal{H}}_\alpha \mu = \{2^{-n\alpha} \eta_n * \mu\}_0^\infty \ .$$

The following proposition is now an immediate consequence of Theorems 4.1.4, 4.1.5, and 4.1.7 in the Besov case, and of Theorems 4.3.2, 4.3.6, and 4.3.10 in the Lizorkin–Triebel case.

Proposition 4.4.1. *Let $\alpha > 0$, $1 < p < \infty$, and $1 < q < \infty$. Let $\{\eta_n\}_0^\infty$ be as above. Then the Besov space $B_\alpha^{p,q}$ consists of the functions F that have a representation $F = \mathcal{H}_\alpha f$, where $f \in l^q(L^p)$, the dual space $B_{-\alpha}^{p',q'}$ consists of the distributions $u \in \mathcal{S}'$ such that $\check{\mathcal{H}}_\alpha u \in l^{q'}(L^{p'})$, and there is equivalence in norms.*

In the same way the Lizorkin–Triebel space $F_\alpha^{p,q}$ consists of the functions $F = \mathcal{H}_\alpha f$, where $f \in L^p(l^q)$, and the dual space $F_{-\alpha}^{p',q'}$ consists of the distributions $u \in \mathcal{S}'$ such that $\check{\mathcal{H}}_\alpha u \in L^{p'}(l^{q'})$, with equivalence in norms.

We now use these potential representations to define capacities, denoted $C(\,\cdot\,; B_\alpha^{p,q})$ and $C(\,\cdot\,; F_\alpha^{p,q})$, as in Definition 2.3.3.

Definition 4.4.2. Let $1 < p < \infty$, $1 < q < \infty$, and $\alpha > 0$. Let $E \subset \mathbf{R}^N$ be arbitrary. Then

$$C(E; B_\alpha^{p,q}) = \inf\{\, \|f\|_{l^q(L^p)}^q : f \geq 0,\ \mathcal{H}_\alpha f(x) \geq 1 \text{ on } E \,\} \ ,$$

and

$$C(E; F_\alpha^{p,q}) = \inf\{\, \|f\|_{L^p(l^q)}^p : f \geq 0,\ \mathcal{H}_\alpha f(x) \geq 1 \text{ on } E \,\} \ .$$

Remark. The capacity associated to $B_\alpha^{p,q}$ is often defined as $C(E; B_\alpha^{p,q})^{p/q}$. See e.g. D. R. Adams [15]. Our choice has the advantage that (4.4.11) below is valid, but there are also good reasons for the other choice; mainly that both $C(\,\cdot\,; B_\alpha^{p,q})^{p/q}$ and $C(\,\cdot\,; F_\alpha^{p,q})$ are $(N - \alpha p)$-dimensional set functions. See also Section 5.6.7 below.

The following proposition is a counterpart to Proposition 2.3.13.

Proposition 4.4.3. *There is a constant A such that for any compact set K*

$$A^{-1}C(K; B_\alpha^{p,q}) \leq \inf\{\, \|F\|_{B_\alpha^{p,q}}^q : F \in \mathcal{S},\ F(x) \geq 1 \text{ on } K \,\} \leq A\, C(K; B_\alpha^{p,q}) \ ,$$

and similarly for $C(K; F_\alpha^{p,q})$.

The proof is similar to the proof of Proposition 2.3.13, and is left to the reader.

The mutual energy of a ν-measurable f and a $\mu \in \mathcal{M}(\mathbf{R}^N)$ is

$$\mathcal{E}_{h_\alpha}(\mu, f) = \sum_{n=0}^{\infty} 2^{-n\alpha} \int_{\mathbf{R}^N} (\eta_n * f_n)\, d\mu = \sum_{n=0}^{\infty} 2^{-n\alpha} \int_{\mathbf{R}^N} (\eta_n * \mu) f_n \, dx \ ,$$

and clearly

$$|\mathcal{E}_{h_\alpha}(\mu, f)| \leq \|\check{\mathcal{H}}_\alpha \mu\|_{l^{q'}(L^{p'})} \|f\|_{l^q(L^p)} \ ,$$

and

$$|\mathcal{E}_{h_\alpha}(\mu, f)| \le \|\check{\mathcal{H}}_\alpha \mu\|_{L^{p'}(l^{q'})} \|f\|_{L^p(l^q)} \ .$$

If $\mathcal{H}_\alpha f \ge 1$ on a compact set K and if $\mu \in \mathcal{M}^+(K)$, then

$$\mu(K) \le \int \mathcal{H}_\alpha f \, d\mu = \mathcal{E}_{h_\alpha}(\mu, f) \le \|\check{\mathcal{H}}_\alpha \mu\|_{l^{q'}(L^{p'})} \|f\|_{l^q(L^p)} \ .$$

and thus

$$\sup\left\{ \frac{\mu(K)}{\|\check{\mathcal{H}}_\alpha \mu\|_{l^{q'}(L^{p'})}} : \mu \in \mathcal{M}^+(K) \right\} \le C(K; B_\alpha^{p,q})^{1/q} \ ,$$

and

$$\sup\left\{ \frac{\mu(K)}{\|\check{\mathcal{H}}_\alpha \mu\|_{L^{p'}(l^{q'})}} : \mu \in \mathcal{M}^+(K) \right\} \le C(K; F_\alpha^{p,q})^{1/p} \ .$$

Now assume that $p = q$. Then by the dual definition of capacity, Theorem 2.5.1,

$$\begin{aligned} C(K; B_\alpha^{p,p})^{1/p} &= C(K; F_\alpha^{p,p})^{1/p} \\ &= \sup\left\{ \frac{\mu(K)}{\|\check{\mathcal{H}}_\alpha \mu\|_{L^{p'}(l^{p'})}} : \mu \in \mathcal{M}^+(K) \right\} \ . \end{aligned} \tag{4.4.3}$$

By Theorem 2.5.3, there is an extremal function $f = f^K$, which is of the form

$$f^K = (\check{\mathcal{H}}_\alpha \mu^K)^{p'-1} = \left\{ \left(2^{-\alpha n} \eta_n * \mu^K \right)^{p'-1} \right\}_{n=0}^{\infty} \ ,$$

where $\mu^K \in \mathcal{M}^+(K)$, and

$$\mu^K(K) = C(K; B_\alpha^{p,p}) \ .$$

For any $\mu \in \mathcal{M}^+(\mathbf{R}^N)$ the corresponding nonlinear potential is

$$\begin{aligned} \mathcal{V}_{\alpha,p}^\mu = \mathcal{H}_\alpha(\check{\mathcal{H}}_\alpha \mu)^{p'-1} &= \sum_0^\infty 2^{-n\alpha} \eta_n * \left(2^{-n\alpha} \eta_n * \mu \right)^{p'-1} \\ &= \sum_0^\infty 2^{-n\alpha p'} \eta_n * (\eta_n * \mu)^{p'-1} \ . \end{aligned} \tag{4.4.4}$$

For the extremal μ^K the generalized energy is

$$\begin{aligned} \int_K \mathcal{V}_{\alpha,p}^{\mu^K} d\mu^K &= \sum_0^\infty 2^{-n\alpha p'} \int_K \eta_n * (\eta_n * \mu^K)^{p'-1} \, d\mu^K \\ &= \sum_0^\infty \|2^{-n\alpha} \eta_n * \mu^K\|_{p'}^{p'} = \|\check{\mathcal{H}}_\alpha \mu^K\|_{l^{p'}(L^{p'})}^{p'} \\ &= \|f^K\|_{l^p(L^p)}^p = C(K; B_\alpha^{p,p}) \ . \end{aligned} \tag{4.4.5}$$

We shall study the potentials $\mathcal{V}_{\alpha,p}^\mu$ more closely in Section 6.3.

We now return to the case of general $p, q > 1$. It is easy to see that Theorem 2.5.1 extends to spaces with mixed norm, so we have

$$C(K; B_\alpha^{p,q})^{1/q} = \sup\left\{\frac{\mu(K)}{\|\check{\mathcal{H}}_\alpha\mu\|_{l^{q'}(L^{p'})}} : \mu \in \mathcal{M}^+(K)\right\} , \tag{4.4.6}$$

and similarly

$$C(K; F_\alpha^{p,q})^{1/p} = \sup\left\{\frac{\mu(K)}{\|\check{\mathcal{H}}_\alpha\mu\|_{L^{p'}(l^{q'})}} : \mu \in \mathcal{M}^+(K)\right\} . \tag{4.4.7}$$

Clearly $C(\cdot\,; F_\alpha^{p,2})$ is equivalent to the Bessel capacity $C_{\alpha,p}$ (by Theorem 4.3.2). It is quite remarkable that $C_{\alpha,p}$ is also equivalent to the capacities $C(\cdot\,; F_\alpha^{p,q})$ for $1 < q < \infty$, and thus, in particular, to $C(\cdot\,; B_\alpha^{p,p})$.

In fact, the following proposition is an immediate consequence of the dual characterization (4.4.7) of capacity, and of Corollary 4.3.9, i.e., of the Muckenhoupt–Wheeden inequality (Theorem 3.6.2 and Corollary 3.6.3).

Proposition 4.4.4. *Let $1 < p < \infty$ and $0 < p \leq N/\alpha$. Then there is a constant A such that for all $E \subset \mathbf{R}^N$ and $1 < q < \infty$*

$$A^{-1}C_{\alpha,p}(E) \leq C(E; F_\alpha^{p,q}) \leq A\, C_{\alpha,p}(E) .$$

We shall now briefly see how the above formulas for nonlinear potentials and energies should be modified in the case of $p \neq q$. Using the conditions for equality in Hölder's and Minkowski's inequalities, it is easily seen that, in the case of the Besov space $B_\alpha^{p,q}$, the extremal $f = f^K$ now has the form

$$\begin{aligned} f^K &= \left\{(\check{\mathcal{H}}_\alpha\mu^K)_n^{p'-1}\|(\check{\mathcal{H}}_\alpha\mu^K)_n\|_{p'}^{q'-p'}\right\}_{n=0}^\infty \\ &= \left\{\left(2^{-n\alpha}\eta_n * \mu^K\right)^{p'-1}\|2^{-n\alpha}\eta_n * \mu^K\|_{p'}^{q'-p'}\right\}_{n=0}^\infty , \end{aligned} \tag{4.4.8}$$

Thus, for any positive measure μ it is natural to define the nonlinear potential

$$\begin{aligned} \mathcal{V}_{\alpha,p,q}^\mu &= \sum_0^\infty 2^{-n\alpha}\eta_n * \left(2^{-n\alpha}\eta_n * \mu\right)^{p'-1}\|2^{-n\alpha}\eta_n * \mu\|_{p'}^{q'-p'} \\ &= \sum_0^\infty 2^{-n\alpha q'}\eta_n * (\eta_n * \mu)^{p'-1}\|\eta_n * \mu\|_{p'}^{q'-p'} , \end{aligned} \tag{4.4.9}$$

and then the energy is

$$\begin{aligned} \int_{\mathbf{R}^N}\mathcal{V}_{\alpha,p,q}^\mu\,d\mu &= \sum_0^\infty 2^{-n\alpha q'}\|\eta_n * \mu\|_{p'}^{q'-p'}\int_{\mathbf{R}^N}\eta_n * (\eta_n * \mu)^{p'-1}\,d\mu \\ &= \sum_0^\infty 2^{-n\alpha q'}\|\eta_n * \mu\|_{p'}^{q'-p'}\int_{\mathbf{R}^N}(\eta_n * \mu)^{p'}\,dx \\ &= \left\|\{2^{-n\alpha}\eta_n * \mu\}_{n=0}^\infty\right\|_{l^{q'}(L^{p'})}^{q'} = \left\|\check{\mathcal{H}}_\alpha\mu\right\|_{l^{q'}(L^{p'})}^{q'} . \end{aligned} \tag{4.4.10}$$

Thus, for the extremal μ^K,

$$\int_K \mathcal{V}^{\mu^K}_{\alpha,p,q}\,d\mu^K = \|\check{\mathcal{H}}_\alpha\mu^K\|^{q'}_{l^{q'}(L^{p'})} = \|f^K\|^q_{l^q(L^p)} = C(K;B^{p,q}_\alpha)\ . \quad (4.4.11)$$

Now consider the Lizorkin–Triebel space $F^{p,q}_\alpha$. Then the extremal $f^K = \{f^K_n\}_0^\infty$ is given by

$$\begin{aligned} f^K_n(x) &= (\check{\mathcal{H}}_\alpha\mu^K)_n(x)^{q'-1}\|\check{\mathcal{H}}_\alpha\mu^K(x)\|^{p'-q'}_{l^{q'}} \\ &= \left(2^{-n\alpha}\eta_n * \mu^K(x)\right)^{q'-1}\left\|\{2^{-n\alpha}\eta_n * \mu^K(x)\}^\infty_{n=0}\right\|^{p'-q'}_{l^{q'}}\ , \end{aligned} \quad (4.4.12)$$

and nonlinear potentials are defined by

$$\begin{aligned} \mathcal{V}^\mu_{\alpha,p,q} &= \sum_0^\infty 2^{-n\alpha}\eta_n * \left(\left(2^{-n\alpha}\eta_n * \mu\right)^{q'-1}\left\|\{2^{-n\alpha}\eta_n * \mu\}^\infty_{n=0}\right\|^{p'-q'}_{l^{q'}}\right) \\ &= \sum_0^\infty 2^{-n\alpha p'}\eta_n * \left((\eta_n * \mu)^{q'-1}\left\|\{\eta_n * \mu\}^\infty_{n=0}\right\|^{p'-q'}_{l^{q'}}\right)\ . \end{aligned} \quad (4.4.13)$$

Then the energy is

$$\begin{aligned} \int_{\mathbf{R}^N}\mathcal{V}^\mu_{\alpha,p,q}\,d\mu &= \sum_0^\infty 2^{-n\alpha p'}\int_{\mathbf{R}^N}\eta_n * \left((\eta_n * \mu)^{q'-1}\left\|\{\eta_n * \mu\}^\infty_{n=0}\right\|^{p'-q'}_{l^{q'}}\right)d\mu \\ &= \sum_0^\infty 2^{-n\alpha p'}\int_{\mathbf{R}^N}(\eta_n * \mu)^{q'}\left\|\{\eta_n * \mu\}^\infty_{n=0}\right\|^{p'-q'}_{l^{q'}}\,dx \\ &= \left\|\{2^{-n\alpha}\eta_n * \mu\}^\infty_{n=0}\right\|^{p'}_{L^{p'}(l^{q'})} = \left\|\check{\mathcal{H}}_\alpha\mu\right\|^{p'}_{L^{p'}(l^{q'})}\ . \end{aligned} \quad (4.4.14)$$

Thus, for the extremal μ^K,

$$\int_K \mathcal{V}^{\mu^K}_{\alpha,p,q}\,d\mu^K = \left\|\check{\mathcal{H}}_\alpha\mu^K\right\|^{p'}_{L^{p'}(l^{q'})} = \|f^K\|^p_{L^p(l^q)} = C(K;F^{p,q}_\alpha)\,. \quad (4.4.15)$$

4.5 An Inequality of Wolff

We return to the case $p = q$, and we make the following further assumption on η (see (4.4.2)):

$$\eta(x) \le 1 \text{ for } |x| \le 1, \text{ and } \eta(x) \ge \tfrac{1}{2} \text{ for } |x| \le \tfrac{1}{2}\ . \quad (4.5.1)$$

Then

$$\eta_n * (\eta_n * \mu)^{p'-1}(x) \le 2^{nN(p'-1)}\mu(B_{n-1}(x))^{p'-1}\int_{\mathbf{R}^N}\eta_n\,dy\ ,$$

and

$$\eta_n * (\eta_n * \mu)^{p'-1}(x) \geq 2^{1-p'} 2^{nN(p'-1)} \mu(B_{n+2}(x))^{p'-1} \int_{B_{n+2}(x)} \eta_n \, dy \ ,$$

since $B_{n+1}(y) \supset B_{n+2}(x)$ for $|x-y| < 2^{-n-2}$. It follows from (4.4.4) that

$$A^{-1} \sum_{n=2}^{\infty} \left(2^{n(N-\alpha p)} \mu(B_n(x))\right)^{p'-1} \leq \mathcal{V}^{\mu}_{\alpha,p}(x) = \sum_{n=0}^{\infty} 2^{-n\alpha p'} \eta_n * (\eta_n * \mu)^{p'-1}$$

$$\leq A \sum_{n=-1}^{\infty} \left(2^{n(N-\alpha p)} \mu(B_n(x))\right)^{p'-1} \ . \quad (4.5.2)$$

We make the following definition.

Definition 4.5.1. Let $\mu \in \mathcal{M}^+(\mathbf{R}^N)$, $1 < p < \infty$, and $0 < \alpha p \leq N$. Then

$$W^{\mu}_{\alpha,p}(x) = \sum_{n=0}^{\infty} \left(2^{n(N-\alpha p)} \mu(B_n(x))\right)^{p'-1}.$$

The last inequality then gives

$$A^{-1} W^{\mu}_{\alpha,p}(x) - A^{-1} \mu(B_0(x))^{p'-1} \leq \mathcal{V}^{\mu}_{\alpha,p}(x)$$

$$\leq A\, W^{\mu}_{\alpha,p}(x) + A\, \mu(B_{-1}(x))^{p'-1} \ . \quad (4.5.3)$$

The following important theorem is now a consequence of the above and Corollary 3.6.3.

Theorem 4.5.2 (Wolff's Inequality). *Let* $1 < p < \infty$ *and* $0 < \alpha p \leq N$. *Then there is a constant* A *such that for any* $\mu \in \mathcal{M}^+(\mathbf{R}^N)$

$$A^{-1} \int_{\mathbf{R}^N} (G_\alpha * \mu)^{p'} \, dx = A^{-1} \int_{\mathbf{R}^N} V^{\mu}_{\alpha,p} \, d\mu$$

$$\leq \int_{\mathbf{R}^N} W^{\mu}_{\alpha,p} \, d\mu \leq A \int_{\mathbf{R}^N} (G_\alpha * \mu)^{p'} \, dx \ .$$

Proof. By Corollary 3.6.3

$$\int_{\mathbf{R}^N} (G_\alpha * \mu)^{p'} \, dx \approx \left\| \{2^{-n\alpha} \eta_n * \mu\}_{n=0}^{\infty} \right\|_{L^{p'}(l^{p'})} = \int_{\mathbf{R}^N} \mathcal{V}^{\mu}_{\alpha,p} d\mu \ .$$

But $\mathcal{V}^{\mu}_{\alpha,p}(x)$ is comparable to $W^{\mu}_{\alpha,p}(x)$ by (4.5.3).

There is a homogeneous variant of Theorem 4.5.2, which we record for completeness.

Definition 4.5.3. Let $\mu \in \mathcal{M}^+(\mathbf{R}^N)$, $1 < p < \infty$, and $0 < \alpha p < N$. Then

$$\dot{W}^{\mu}_{\alpha,p}(x) = \sum_{n=-\infty}^{\infty} \left(2^{n(N-\alpha p)} \mu(B_n(x))\right)^{p'-1} \ .$$

Theorem 4.5.4. *Let* $1 < p < \infty$ *and* $0 < \alpha p < N$. *Then there is a constant* A *such that for any* $\mu \in \mathcal{M}^+(\mathbf{R}^N)$

$$A^{-1}\int_{\mathbf{R}^N}(I_\alpha * \mu)^{p'}\,dx = A^{-1}\int_{\mathbf{R}^N}\dot{V}^{\mu}_{\alpha,p}\,d\mu \le \int_{\mathbf{R}^N}\dot{W}^{\mu}_{\alpha,p}\,d\mu \le A\int_{\mathbf{R}^N}(I_\alpha * \mu)^{p'}\,dx\ .$$

The proof is completely analogous to the proof of Theorem 4.5.2.

The following characterization of the positive cones in the dual Besov and Bessel potential spaces is an easy consequence of Corollary 4.3.9 and of Corollary 4.1.6 (see Remark 2).

Theorem 4.5.5. *Let* $\alpha > 0$ *and* $1 < p < \infty$. *Let* $\mu \in \mathcal{M}^+(\mathbf{R}^N)$. *Then* $\mu \in (L^{-\alpha,p'})_+ = (F^{p',q'}_{-\alpha})_+$, $1 < q' \le \infty$, *if and only if*

$$\int_{\mathbf{R}^N} W^{\mu}_{\alpha,p}\,d\mu = \int_{\mathbf{R}^N}\sum_{n=0}^{\infty}\left(2^{n(N-\alpha p)}\mu(B_n(x))\right)^{p'-1}d\mu(x) < \infty\ ,$$

$\mu \in (B^{p',q'}_{-\alpha})_+$, $1 \le q' < \infty$, *if and only if*

$$\sum_{n=0}^{\infty}\int_{\mathbf{R}^N}\left(\left(2^{n(N-\alpha p)}\mu(B_n(x))\right)^{p'-1}d\mu(x)\right)^{p'/q'} < \infty\ ,$$

and $\mu \in (B^{p',\infty}_{-\alpha})_+$ *if and only if*

$$\sup_{n\ge 0}\int_{\mathbf{R}^N}\left(2^{n(N-\alpha p)}\mu(B_n(x))\right)^{p'-1}d\mu(x) < \infty\ .$$

Remark. It is easily seen that instead of the sums in Definitions 4.5.1 and 4.5.3 we could equally well have used integrals in the following way:

$$W^{\mu}_{\alpha,p}(x) = \int_0^1\left(\frac{\mu(B(x,t))}{t^{N-\alpha p}}\right)^{p'-1}\frac{dt}{t},\quad \alpha p \le N\ , \tag{4.5.4}$$

$$\dot{W}^{\mu}_{\alpha,p}(x) = \int_0^\infty\left(\frac{\mu(B(x,t))}{t^{N-\alpha p}}\right)^{p'-1}\frac{dt}{t},\quad \alpha p < N\ . \tag{4.5.5}$$

The same is of course true for Theorem 4.5.5.

4.6 An Atomic Decomposition

As we have seen in Definitions 4.1.1 and 4.2.1, the elements in the Besov space $B_\alpha^{p,q}$, or in the corresponding Lizorkin–Triebel space $F_\alpha^{p,q}$ can be represented as

$$f = \Phi * f + \sum_1^\infty \varphi_n * f \ , \tag{4.6.1}$$

where the functions Φ, Φ_n, and φ_n satisfy the conditions (4.1.1)–(4.1.5). We shall now decompose this representation further in the case when $\alpha > 0$, and $p, q > 1$.

We set

$$x_{nk} = 2^{-n}k \quad \text{for } n \in \mathbf{N} \text{ and } k \in \mathbf{Z}^N \ , \tag{4.6.2}$$

and we define dyadic cubes Q_{nk} by saying that

$$x \in Q_{nk} \ , \ \text{if } x = x_{nk} + 2^{-n}y, \text{ where } \ 0 \le y_i < 1 \text{ for } 1 \le i \le N \ . \tag{4.6.3}$$

The characteristic function for Q_{nk} is denoted χ_{nk}.

For any cube Q we denote its side by $l(Q)$, and for $\lambda > 0$ we denote by λQ the cube concentric to Q with side $\lambda l(Q)$. We also write $\widetilde{Q}_{nk}$ for $3Q_{nk}$, and we denote the characteristic function of $\widetilde{Q}_{nk}$ by $\widetilde{\chi}_{nk}$.

Definition 4.6.1. Let S be a nonnegative integer. A function a_{nk} in $C^S(\mathbf{R}^N)$ is called a C^S *atom* for a cube Q_{nk} if it satisfies the following conditions:

$$\operatorname{supp} a_{nk} \subset \widetilde{Q}_{nk} \ ; \tag{4.6.4}$$

$$\sup_x 2^{-n|\gamma|} |D^\gamma a_{nk}(x)| \le 1 \qquad \text{for all } |\gamma| \le S \ . \tag{4.6.5}$$

Theorem 4.6.2. *Let f belong to $B_\alpha^{p,q}$ or to $F_\alpha^{p,q}$ with $\alpha > 0$, $1 < p \le \infty$ ($1 < p < \infty$ in the case of $F_\alpha^{p,q}$), and $1 < q \le \infty$, and let S be a nonnegative integer. Then there exist C^S atoms $\{a_{nk}\}$ for the dyadic cubes $\{Q_{nk}\}$, and constants $\{s_{nk}\}$ such that*

$$f(x) = \sum_{n=0}^\infty 2^{-n\alpha} u_n(x) \quad \textit{with} \quad u_n(x) = \sum_{k\in\mathbf{Z}^N} s_{nk} a_{nk}(x) \ , \tag{4.6.6}$$

and, denoting $s_n(x) = \sum_{k\in\mathbf{Z}^N} s_{nk}\chi_{nk}(x)$, there are constants A independent of f such that, if $f \in B_\alpha^{p,q}$,

$$\left\| \{s_n\}_0^\infty \right\|_{l^q(L^p)} = \left(\sum_{n=0}^\infty \left(\sum_{k\in\mathbf{Z}^N} 2^{-nN} |s_{nk}|^p \right)^{q/p} \right)^{1/q} \le A \, \|f\|_{B_\alpha^{p,q}} \ , \tag{4.6.7}$$

or, if $f \in F_\alpha^{p,q}$,

$$\left\| \{s_n\}_0^\infty \right\|_{L^p(l^q)} = \left\| \left(\sum_{Q_{nk}\ni x} |s_{nk}|^q \right)^{1/q} \right\|_p \le A \, \|f\|_{F_\alpha^{p,q}} \ . \tag{4.6.8}$$

Proof. We write the representation (4.6.1) as

$$f = \sum_{n=0}^{\infty} 2^{-n\alpha} u_n \ .$$

We let $\eta \in C_0^\infty(Q_{00})$, $\int \eta\, dx = 1$, $\eta_n(x) = 2^{nN}\eta(2^n x)$, and we define $\eta_{nk} = \eta_n * \chi_{nk}$, so that $\operatorname{supp} \eta_{nk} \subset \widetilde{Q}_{nk}$, and $\sum_{k\in \mathbf{Z}^N} \eta_{nk} \equiv 1$. We set $b_{nk} = \eta_{nk} u_n$,

$$s_{nk} = \max_{x\in \widetilde{Q}_{nk},\ |\gamma|\le S} 2^{-n|\gamma|} \left| D^\gamma b_{nk}(x) \right| \ ,$$

and

$$a_{nk}(x) = \frac{b_{nk}(x)}{s_{nk}} \ .$$

Then clearly the $\{a_{nk}\}$ are C^S atoms, and

$$f = \sum_{n=0}^{\infty} 2^{-n\alpha} \sum_{k\in \mathbf{Z}^N} s_{nk} a_{nk} \ .$$

We observe that $\widehat{\Phi_{n+1}}(\xi) = 1$ on $\operatorname{supp} \widehat{u_n}$, so that

$$u_n = \Phi_{n+1} * u_n \ .$$

Thus, for $x \in \widetilde{Q}_{nk}$ and $|\gamma| \le S$

$$\begin{aligned} |D^\gamma u_n(x)| &\le |D^\gamma \Phi_{n+1}| * |u_n|(x) \\ &\le A\, 2^{n(N+|\gamma|)} \int_{\mathbf{R}^N} \frac{|u_n(y)|}{(1+2^n|y-x|)^{N+1}}\, dy \\ &\le A\, 2^{n|\gamma|} \inf_{y\in Q_{nk}} M u_n(y) \ , \end{aligned}$$

where the last inequality is proved in the same way as Lemma 4.3.7. By the Leibniz formula

$$s_{nk} \le A \max_{x\in \widetilde{Q}_{nk},\ |\gamma|\le S} 2^{-n|\gamma|} |D^\gamma u_n(x)| \ .$$

Hence $s_{nk} \le A\, M u_n(x)$ for all $x \in Q_{nk}$, and thus

$$s_n(x) = \sum_{k\in \mathbf{Z}^N} s_{nk} \chi_{nk}(x) \le A\, M u_n(x) \ .$$

The inequalities (4.6.7), and (4.6.8) now follow from the Hardy–Littlewood and the Fefferman–Stein maximal theorems, respectively.

Remark. An alternative approach would be to obtain the atomic representations from the representations in Theorems 4.1.7 and 4.3.10 by setting $b_{nk} = \eta_n * (f_n \chi_{nk})$.

In the converse direction we have the following theorem.

Theorem 4.6.3. *Let α, p, q be as in Theorem 4.6.2, and let $S > \alpha$ be an integer. Let $\{a_{nk}\}$ be a sequence of C^S atoms for the cubes $\{Q_{nk}\}$ and let the coefficients $\{s_{nk}\}$ be such that the left side in (4.6.7) is finite. Then the function f defined by*

$$f(x) = \sum_{n=0}^{\infty} 2^{-n\alpha} \sum_{k \in \mathbf{Z}^N} s_{nk} a_{nk}(x)$$

belongs to $B_\alpha^{p,q}$, and with $s_n(x) = \sum_{k \in \mathbf{Z}^N} s_{nk} \chi_{nk}(x)$,

$$\|f\|_{B_\alpha^{p,q}} \le A \left\| \{s_n\}_0^\infty \right\|_{l^q(L^p)} = A \left(\sum_{n=0}^{\infty} \left(\sum_{k \in \mathbf{Z}^N} 2^{-nN} |s_{nk}|^p \right)^{q/p} \right)^{1/q} . \tag{4.6.9}$$

Similarly, if (4.6.8) is finite, then f belongs to $F_\alpha^{p,q}$, and

$$\|f\|_{F_\alpha^{p,q}} \le A \left\| \{s_n\}_0^\infty \right\|_{L^p(l^q)} = A \left\| \left(\sum_{Q_{nk} \ni x} |s_{nk}|^q \right)^{1/q} \right\|_p . \tag{4.6.10}$$

The proof requires two technical lemmas.

Lemma 4.6.4. *Let Φ and $\{\varphi_n\}_0^\infty$ satisfy (4.1.1)–(4.1.3), and let a_{mk} be a C^S atom for Q_{mk}. Then*

$$|\Phi * a_{mk}(x)| \le \frac{A\, 2^{-mN}}{\left(1 + |x - x_{mk}|\right)^{N+1}} \quad \text{for } m \ge 0 \ , \tag{4.6.11}$$

$$|\varphi_n * a_{mk}(x)| \le A \frac{A\, 2^{-(m-n)N}}{\left(1 + 2^n |x - x_{mk}|\right)^{N+1}} \quad \text{for } 0 \le n \le m \ , \tag{4.6.12}$$

and

$$|\varphi_n * a_{mk}(x)| \le \frac{A\, 2^{-(n-m)S}}{\left(1 + 2^m |x - x_{mk}|\right)^{N+1}} \quad \text{for } 0 \le m \le n \ . \tag{4.6.13}$$

Proof. We first prove (4.6.11). We can assume that $k = 0$. Then

$$\begin{aligned} |\Phi * a_{m0}(x)| &\le \int_{|y| \le |x|/2} + \int_{|y| > |x|/2} |\Phi(x - y) a_{m0}(y)|\, dy \\ &\le \max_{|y| \ge |x|/2} |\Phi(y)| \int_{\mathbf{R}^N} |a_{m0}(y)|\, dy + \max_y |\Phi(y)| \int_{|y| > |x|/2} |a_{m0}(y)|\, dy \\ &\le \frac{A\, 2^{-mN}}{(1 + |x|)^{N+1}} \ , \end{aligned}$$

since $\Phi \in \mathcal{S}$, and $a_{m0}(y) = 0$ off $\widetilde{Q}_{m0}$.

To prove (4.6.12) we observe that

$$|\varphi_n * a_{mk}(x)| = 2^{nN} \int_{\mathbf{R}^N} \varphi_0\big(2^n(x-y)\big) a_{mk}(y)\,dy$$
$$= \int_{\mathbf{R}^N} \varphi_0(2^n x - y) a_{mk}(2^{-n}y)\,dy \ .$$

Thus it is enough to estimate $|\varphi_0 * a_{m-n,0}(2^n x)|$. But by the first part of the proof,

$$|\varphi_0 * a_{m-n,0}(2^n x)| \le \frac{A\,2^{-(m-n)N}}{\big(1+2^n|x|\big)^{N+1}} \ ,$$

which gives the desired result.

In proving (4.6.13) we observe in the same way that

$$|\varphi_n * a_{mk}(x)| = \int_{\mathbf{R}^N} \varphi_{n-m}(2^m x - y) a_{mk}(2^{-m}y)\,dy \ ,$$

so it is enough to estimate $|\varphi_{n-m} * a_{00}(2^m x)|$. Using the fact that the moments of φ_n of all orders are 0, we find

$$\varphi_n * a_{00}(x) = \int_{\mathbf{R}^N} \varphi_n(y) a_{00}(x-y)\,dy = \int_{\mathbf{R}^N} \varphi_n(y) R(x,y)\,dy \ ,$$

where

$$R(x,y) = a_{00}(x-y) - \sum_{|\beta|\le S-1} \frac{(-y)^\beta}{\beta!} D^\beta a_{00}(x) \ .$$

Here $R(x,y)=0$ if neither x nor $x-y$ belong to $\widetilde{Q}_{00}$, and for all x and y

$$|R(x,y)| \le A\,|y|^S \max_z \sum_{|\beta|=S} |D^\beta a_{00}(z)| \le A\,|y|^S \ .$$

It follows that

$$\int_{|y|\le|x|/2} |\varphi_n(y) R(x,y)|\,dy = 0 \quad \text{for } |x| \ge 4\sqrt{N} \ ,$$

and in general

$$\int_{|y|\le|x|/2} |\varphi_n(y)R(x,y)|\,dy \le A \int_{\mathbf{R}^N} |\varphi_n(y)||y|^S\,dy$$
$$= A\,2^{-nS} \int_{\mathbf{R}^N} |\varphi_0(y)||y|^S\,dy \ ,$$

which is finite by the assumption that $\varphi_0 \in \mathcal{S}$. Moreover,

$$\int_{|y|\geq|x|/2} |\varphi_n(y)R(x,y)|\,dy \leq A\,2^{-nS}\int_{|y|\geq 2^{n-1}|x|} |\varphi_0(y)||y|^S\,dy$$
$$\leq \frac{A\,2^{-nS}}{\left(1+2^n|x|\right)^{N+1}} ,$$

where the last inequality again comes from the fact that $\varphi_0 \in \mathcal{S}$. Collecting the estimates, we have

$$|\varphi_n * a_{00}(x)| \leq \frac{A\,2^{-nS}}{(1+|x|)^{N+1}} ,$$

which proves (4.6.13).

Lemma 4.6.5. *Let* $s_n(x) = \sum_{k\in\mathbf{Z}^N} s_{nk}\chi_{nk}(x)$, *where* $n \geq 0$ *and* $s_{nk} \geq 0$. *Let* l *be a nonnegative integer. Then*

$$\sum_{k\in\mathbf{Z}^N} \frac{s_{nk}}{\left(1+2^{n-l}|x-x_{nk}|\right)^{N+1}} \leq A\,2^{lN} Ms_n(x) .$$

Proof. We can assume that $x \in Q_{n0}$. Set $B_0 = \{k \in \mathbf{Z}^N : |k| \leq 2^l\}$, and $B_m = \{k \in \mathbf{Z}^N : 2^{l+m-1} < |k| \leq 2^{l+m}\}$ for $m = 1, 2, \ldots$. Then

$$\sum_{k\in B_m} \frac{s_{nk}}{\left(1+2^{n-l}|x-x_{nk}|\right)^{N+1}} \leq A\,2^{-m(N+1)} \sum_{k\in B_m} s_{nk}$$
$$= A\,2^{-m(N+1)}2^{nN}\int_{\mathbf{R}^N}\sum_{k\in B_m} s_{nk}\chi_{nk}(y)\,dy$$
$$\leq A\,2^{-m}2^{lN} Ms_n(x) .$$

The lemma follows by summing over m.

Proof of Theorem 4.6.3. Let $f = \sum_{m=0}^{\infty} 2^{-m\alpha}\sum_{k\in\mathbf{Z}^N} s_{mk}a_{mk}$. By (4.6.11) and Lemma 4.6.5

$$|\Phi * f(x)| \leq A\sum_{m=0}^{\infty} 2^{-m\alpha}\sum_{k\in\mathbf{Z}^N}\frac{2^{-mN}s_{mk}}{\left(1+|x-x_{mk}|\right)^{N+1}} \leq A\sum_{m=0}^{\infty}2^{-m\alpha}Ms_m(x) .$$

Similarly, for $n \geq 1$, by (4.6.13) and Lemma 4.6.5 with $l = 0$ for $m \leq n$, and by (4.6.12) and Lemma 4.6.5 with $l = m - n$ for $m > n$,

$$2^{n\alpha}|\varphi_n * f(x)| \leq A\sum_{m=0}^{n} 2^{(n-m)\alpha}\sum_{k\in\mathbf{Z}^N}\frac{2^{-(n-m)S}s_{mk}}{(1+2^m|x-x_{mk}|)^{N+1}}$$
$$+ A\sum_{m=n+1}^{\infty} 2^{(n-m)\alpha}\sum_{k\in\mathbf{Z}^N}\frac{2^{-(m-n)N}s_{mk}}{\left(1+2^n|x-x_{mk}|\right)^{N+1}}$$
$$\leq A\sum_{m=0}^{n}2^{-(n-m)(S-\alpha)}Ms_m(x) + A\sum_{m=n+1}^{\infty}2^{(n-m)\alpha}Ms_m(x) .$$

By the Hardy–Littlewood maximal theorem

$$2^{n\alpha}\|\varphi_n * f\|_p \leq A\left(\sum_{m=0}^{n} 2^{-(n-m)(S-\alpha)}\|s_m\|_p + \sum_{m=n+1}^{\infty} 2^{(n-m)\alpha}\|s_m\|_p\right)$$
$$= A\sum_{m=0}^{\infty} c_{n-m}\|s_m\|_p \ ,$$

and similarly

$$\|\Phi * f\|_p \leq A\sum_{m=0}^{\infty} 2^{-m\alpha}\|s_m\|_p = A\sum_{m=0}^{\infty} c_{-m}\|s_m\|_p \ .$$

Here the assumption $S > \alpha > 0$ implies that $\sum_{n=-\infty}^{\infty}|c_n| < \infty$. It follows from the inequality $\|c * d\|_{l^q} \leq \|c\|_{l^1}\|d\|_{l^q}$ (Minkowski's inequality) that

$$\left(\|\Phi * f\|_p^q + \sum_{n=1}^{\infty}(2^{n\alpha}\|\varphi_n * f\|_p)^q\right)^{1/q} \leq A\left(\sum_{m=0}^{\infty}\|s_m\|_p^q\right)^{1/q} ,$$

which proves (4.6.9).

Turning to (4.6.10), we have, again using Minkowski's inequality,

$$\left(|\Phi * f(x)|^q + \sum_{n=1}^{\infty}|2^{n\alpha}\varphi_n * f(x)|^q\right)^{1/q} \leq A\left(\sum_{m=0}^{\infty} Ms_m(x)^q\right)^{1/q} .$$

The desired inequality now follows from the Fefferman–Stein theorem.

4.7 Atomic Nonlinear Potentials

We shall now give an "atomic modification" of the theory in Section 4.4, and show that all information concerning capacities of sets is contained in the spaces of dyadic sequences $\{s_{nk}\}$. Again, we consider only the inhomogeneous case.

We set $\mathbf{M} = \mathbf{N} \times \mathbf{Z}^N$, and we define a measure ν on $\mathbf{M}$ by

$$\nu = \sum_{(n,k)\in\mathbf{M}} 2^{-nN}\delta_{nk} \ , \tag{4.7.1}$$

where δ_{nk} is unit mass at (n, k). Compared to (4.4.1) this means that for each $n \in \mathbf{N}$ we replace Lebesgue measure in each dyadic cube Q_{nk} (see (4.6.3)) by a point measure of the same mass.

The functions on $\mathbf{M}$ are sequences $s = \{s_{nk}\}_{(n,k)\in\mathbf{M}}$, and defining

$$s_n(x) = \sum_{k\in\mathbf{Z}^N} s_{nk}\chi_{nk}(x) \ ,$$

we have

$$\|s\|_{L^p(\nu)} = \sum_{(n,k)\in\mathbf{M}} 2^{-nN}|s_{nk}|^p = \left\|\{s_n\}_0^\infty\right\|_{l^p(L^p)} .$$

The mixed norms are defined by

$$\|s\|_{l^q(L^p)(\nu)} = \left\|\{s_n\}_0^\infty\right\|_{l^q(L^p)} \quad \text{and} \quad \|s\|_{L^p(l^q)(\nu)} = \left\|\{s_n\}_0^\infty\right\|_{L^p(l^q)} .$$

The kernel $\mathfrak{h}_\alpha$ is now defined on $\mathbf{R}^N \times \mathbf{M}$ by

$$\mathfrak{h}_\alpha(x, n, k) = 2^{n(N-\alpha)}\chi_{nk}(x) , \tag{4.7.2}$$

where, as before, χ_{nk} is the characteristic function for Q_{nk}. Thus, compared to (4.4.2), for each $n \in \mathbf{N}$ the weight function $\eta_n(\cdot - y)$ is replaced by the normalized characteristic function $2^{nN}\chi_{nk}$ if $y \in Q_{nk}$.

If $s = \{s_{nk}\}_{(n,k)\in\mathbf{M}}$, and $\mu \in \mathcal{M}^+(\mathbf{R}^N)$, the corresponding potentials are

$$\mathfrak{H}_\alpha s(x) = \sum_{(n,k)\in\mathbf{M}} \mathfrak{h}_\alpha(x, n, k)s_{nk}2^{-nN}$$

$$= \sum_{(n,k)\in\mathbf{M}} 2^{-n\alpha}s_{nk}\chi_{nk}(x) = \sum_{n=0}^{\infty} 2^{-n\alpha}s_n(x) , \tag{4.7.3}$$

and, denoting $\mu(Q_{nk}) = \mu_{nk}$,

$$\check{\mathfrak{H}}_\alpha\mu = \left\{\left(\check{\mathfrak{H}}_\alpha\mu\right)_{nk}\right\}_{(n,k)\in\mathbf{M}} , \tag{4.7.4}$$

where

$$\left(\check{\mathfrak{H}}_\alpha\mu\right)_{nk} = \int_{\mathbf{R}^N} \mathfrak{h}_\alpha(x, n, k)\, d\mu(x) = 2^{n(N-\alpha)}\mu_{nk} . \tag{4.7.5}$$

Denoting

$$\mu_n(x) = \sum_{k\in\mathbf{Z}^N} \mu_{nk}\chi_{nk}(x) ,$$

and

$$(\check{\mathfrak{H}}_\alpha\mu)_n(x) = \sum_{k\in\mathbf{Z}^N} (\check{\mathfrak{H}}_\alpha\mu)_{nk}\chi_{nk}(x) = 2^{n(N-\alpha)}\mu_n(x) ,$$

we find for the mutual energy

$$\mathcal{E}_{\mathfrak{h}_\alpha}(\mu, s) = \int_{\mathbf{R}^N} \mathfrak{H}_\alpha s\, d\mu = \sum_{(n,k)\in\mathbf{M}} 2^{-n\alpha}s_{nk}\mu_{nk}$$

$$= \sum_{(n,k)\in\mathbf{M}} (\check{\mathfrak{H}}_\alpha\mu)_{nk}s_{nk}2^{-nN} = \int_{\mathbf{R}^N} \sum_{n=0}^{\infty}(\check{\mathfrak{H}}_\alpha\mu)_n\, s_n\, dx . \tag{4.7.6}$$

For $\alpha > 0$ we denote by $b_\alpha^{p,q}$ the space of functions that have a potential representation

$$\mathfrak{H}_\alpha s(x) = \sum_{(n,k)\in\mathbf{M}} 2^{-n\alpha} s_{nk}\chi_{nk}(x) = \sum_{n=0}^{\infty} 2^{-n\alpha} s_n(x) ,$$

such that

$$\left\| \{s_n\}_0^\infty \right\|_{l^q(L^p)} < \infty ,$$

and similarly, by $f_\alpha^{p,q}$ the space of potentials $\mathfrak{H}_\alpha s$ such that

$$\left\| \{s_n\}_0^\infty \right\|_{L^p(l^q)} < \infty .$$

As in Definition 4.4.2 we can define capacities $C(\,\cdot\,; b_\alpha^{p,q})$ and $C(\,\cdot\,; f_\alpha^{p,q})$.

Definition 4.7.1. Let $1 < p < \infty$, $1 < q < \infty$, and $\alpha > 0$. Let $E \subset \mathbf{R}^N$ be arbitrary. Then

$$C(E; b_\alpha^{p,q}) = \inf\left\{ \left\| \{s_n\}_0^\infty \right\|_{l^q(L^p)}^q : s \geq 0,\ \mathfrak{H}_\alpha s(x) \geq 1 \text{ on } E \right\} ,$$

and

$$C(E; f_\alpha^{p,q}) = \inf\left\{ \left\| \{s_n\}_0^\infty \right\|_{L^p(l^q)}^p : s \geq 0,\ \mathfrak{H}_\alpha s(x) \geq 1 \text{ on } E \right\} .$$

By the dual definition, Theorem 2.5.1, as in (4.4.6) and (4.4.7)

$$C(K; b_\alpha^{p,q})^{1/q} = \sup\left\{ \frac{\mu(K)}{\|\check{\mathfrak{H}}_\alpha \mu\|_{l^{q'}(L^{p'})}} : \mu \in \mathcal{M}^+(K) \right\} , \tag{4.7.7}$$

and similarly

$$C(K; f_\alpha^{p,q})^{1/p} = \sup\left\{ \frac{\mu(K)}{\|\check{\mathfrak{H}}_\alpha \mu\|_{L^{p'}(l^{q'})}} : \mu \in \mathcal{M}^+(K) \right\} \tag{4.7.8}$$

for compact K and $1 < p, q < \infty$.

The following proposition is an easy consequence of Theorems 4.6.2 and 4.6.3.

Proposition 4.7.2. *There is a constant A such that for any set E*

$$A^{-1} C(E; B_\alpha^{p,q}) \leq C(E; b_\alpha^{p,q}) \leq A\, C(E; B_\alpha^{p,q}) ,$$

and

$$A^{-1} C(E; F_\alpha^{p,q}) \leq C(E; f_\alpha^{p,q}) \leq A\, C(E; F_\alpha^{p,q}) .$$

Combining this with Proposition 4.4.4 we immediately obtain the following important result.

Proposition 4.7.3. *Let $0 < \alpha < N$ and $1 < p < \infty$. Then there is a constant A such that for all $E \subset \mathbf{R}^N$ and $1 < q < \infty$*

$$A^{-1} C_{\alpha,p}(E) \leq C(E; f_\alpha^{p,q}) \leq A\, C_{\alpha,p}(E) .$$

We now assume that $p = q$. Then, by Theorem 2.5.3, for a compact K, the extremal sequence $s = s^K$ is of the form

$$s^K = (\check{\mathfrak{H}}_\alpha \mu^K)^{p'-1} = \left\{\left(2^{n(N-\alpha)}\mu^K_{nk}\right)^{p'-1}\right\}_{(n,k)\in\mathbf{M}} ,$$

where $\mu^K \in \mathcal{M}^+(K)$ and

$$\mu^K(K) = C(K; b^{p,p}_\alpha) .$$

For any $\mu \in \mathcal{M}^+(\mathbf{R}^N)$ the corresponding nonlinear potential is

$$\begin{aligned}\mathfrak{V}^\mu_{\alpha,p}(x) &= \mathfrak{H}_\alpha(\check{\mathfrak{H}}_\alpha\mu)^{p'-1}(x) = \sum_{(n,k)\in\mathbf{M}} 2^{-n\alpha}\left(2^{n(N-\alpha)}\mu_{nk}\right)^{p'-1}\chi_{nk}(x)\\ &= \sum_{(n,k)\in\mathbf{M}} 2^{-n\alpha p'}\left(2^{nN}\mu_{nk}\right)^{p'-1}\chi_{nk}(x) . \qquad (4.7.9)\end{aligned}$$

For the extremal μ^K

$$\begin{aligned}\int_K \mathfrak{V}^{\mu^K}_{\alpha,p}\,d\mu^K &= \sum_{(n,k)\in\mathbf{M}} \left(2^{n(N-\alpha)}\mu^K_{nk}\right)^{p'} 2^{-nN}\\ &= \|\check{\mathcal{H}}_\alpha\mu^K\|^{p'}_{L^{p'}(\nu)} = \|s^K\|^p_{L^p(\nu)} = C(K; B^{p,p}_\alpha) .\end{aligned}$$

If we rewrite $\mathfrak{V}^\mu_{\alpha,p}$ as

$$\mathfrak{V}^\mu_{\alpha,p}(x) = \sum_{(n,k)\in\mathbf{M}} \left(2^{n(N-\alpha p)}\mu_{nk}\right)^{p'-1}\chi_{nk}(x) ,$$

we see that we have arrived at an atomic version of the potential $W^\mu_{\alpha,p}$ of Definition 4.5.1. Thus, *by using atoms, we have gained that the analogue of* $W^\mu_{\alpha,p}$ *appears naturally as the nonlinear potential* $\mathfrak{V}^\mu_{\alpha,p}$.

Now consider general $p, q > 1$. For the mutual energy we find by (4.7.6) in the Besov case

$$\begin{aligned}\mathcal{E}_{\mathfrak{h}_\alpha}(\mu, s) &= \sum_{n=0}^\infty \int_{\mathbf{R}^N} (\check{\mathfrak{H}}_\alpha\mu)_n(x)\, s_n(x)\,dx \le \sum_{n=0}^\infty \|(\check{\mathfrak{H}}_\alpha\mu)_n\|_{p'}\|s_n\|_p\\ &\le \left\|\{(\check{\mathfrak{H}}_\alpha\mu)_n\}_0^\infty\right\|_{l^{q'}(L^{p'})}\left\|\{s_n\}_0^\infty\right\|_{l^q(L^p)} ,\end{aligned}$$

and in the Lizorkin–Triebel case similarly

$$\begin{aligned}\mathcal{E}_{\mathfrak{h}_\alpha}(\mu, s) &= \int_{\mathbf{R}^N} \sum_{n=0}^\infty (\check{\mathfrak{H}}_\alpha\mu)_n(x)\, s_n(x)\,dx\\ &\le \int_{\mathbf{R}^N} \left\|\{(\check{\mathfrak{H}}_\alpha\mu)_n(x)\}_0^\infty\right\|_{l^{q'}}\left\|\{s_n(x)\}_0^\infty\right\|_{l^q}dx\\ &\le \left\|\{(\check{\mathfrak{H}}_\alpha\mu)_n\}_0^\infty\right\|_{L^{p'}(l^{q'})}\left\|\{s_n\}_0^\infty\right\|_{L^p(l^q)} .\end{aligned}$$

In the first case we have equality in both applications of Hölder's inequality if μ is positive, and

$$s_n(x) = (\check{\mathfrak{H}}_\alpha\mu)_n(x)^{p'-1}\|(\check{\mathfrak{H}}_\alpha\mu)_n\|_{p'}^{q'-p'}$$
$$= 2^{n(N-\alpha)(q'-1)}\mu_n(x)^{p'-1}\|\mu_n\|_{p'}^{q'-p'} ,$$

and thus

$$s_{nk} = 2^{n(N-\alpha)(q'-1)}\mu_{nk}^{p'-1}\left\|\mu_n\right\|_{p'}^{q'-p'} .$$

Thus, the natural definition of a nonlinear potential is

$$\mathfrak{V}^\mu_{\alpha,p,q}(x) = \sum_{n=0}^{\infty} 2^{-n\alpha}(\check{\mathfrak{H}}_\alpha\mu)_n(x)^{p'-1}\|(\check{\mathfrak{H}}_\alpha\mu)_n\|_{p'}^{q'-p'}$$
$$= \sum_{(n,k)\in\mathbf{M}} 2^{-n\alpha}\left\|2^{n(N-\alpha)}\mu_n\right\|_{p'}^{q'-p'}\left(2^{n(N-\alpha)}\mu_{nk}\right)^{p'-1}\chi_{nk}(x)$$
$$= \sum_{n=0}^{\infty} 2^{-n\alpha}\left\|2^{n(N-\alpha)}\mu_n\right\|_{p'}^{q'-p'}\left(2^{n(N-\alpha)}\mu_n(x)\right)^{p'-1} .$$

Then

$$\int_{\mathbf{R}^N}\mathfrak{V}^\mu_{\alpha,p,q}\,d\mu = \sum_{(n,k)\in\mathbf{M}}\left\|2^{n(N-\alpha)}\mu_n\right\|_{p'}^{q'-p'}\left(2^{n(N-\alpha)}\mu_{nk}\right)^{p'}2^{-nN}$$
$$= \sum_{n=0}^{\infty}\left\|2^{n(N-\alpha)}\mu_n\right\|_{p'}^{q'} = \left\|\check{\mathfrak{H}}_\alpha\mu\right\|_{l^{q'}(L^{p'})}^{q'} .$$

In the Lizorkin–Triebel case we have equality if $\mu \geq 0$, and

$$s_{nk} = (\check{\mathfrak{H}}_\alpha\mu)_{nk}^{q'-1}2^{nN}\int_{Q_{nk}}\left\|\check{\mathfrak{H}}_\alpha\mu(y)\right\|_{l^{q'}}^{p'-q'}dy .$$

Notice that, compared to (4.4.12), the function $\left\|\check{\mathfrak{H}}_\alpha\mu(x)\right\|_{l^{q'}}^{p'-q'}$ has been replaced by its average over Q_{nk} in order to make s_{nk} a constant. Nonlinear potentials are defined by

$$\mathfrak{V}^\mu_{\alpha,p,q}(x) = \sum_{(n,k)\in\mathbf{M}} 2^{n(N-\alpha)}(\check{\mathfrak{H}}_\alpha\mu)_{nk}^{q'-1}\chi_{nk}(x)\int_{\mathbf{R}^N}\left\|\check{\mathfrak{H}}_\alpha\mu(y)\right\|_{l^{q'}}^{p'-q'}\chi_{nk}(y)\,dy .$$

In fact, this gives

$$\int_{\mathbf{R}^N}\mathfrak{V}^\mu_{\alpha,p,q}\,d\mu = \sum_{(n,k)\in\mathbf{M}}(\check{\mathfrak{H}}_\alpha\mu)_{nk}^{q'}\int_{\mathbf{R}^N}\left\|\check{\mathfrak{H}}_\alpha\mu(y)\right\|_{l^{q'}}^{p'-q'}\chi_{nk}(y)\,dy$$
$$= \int_{\mathbf{R}^N}\left\|\check{\mathfrak{H}}_\alpha\mu(y)\right\|_{l^{q'}}^{p'-q'}\sum_{(n,k)\in\mathbf{M}}(\check{\mathfrak{H}}_\alpha\mu)_{nk}^{q'}\chi_{nk}(y)\,dy$$
$$= \int_{\mathbf{R}^N}\left\|\check{\mathfrak{H}}_\alpha\mu(y)\right\|_{l^{q'}}^{p'}dy = \left\|\check{\mathfrak{H}}_\alpha\right\|_{L^{p'}(l^{q'})}^{p'} .$$

Finally, we remark that it has certain advantages to replace the cubes Q_{nk} in the above by expanded cubes λQ_{nk}, $\lambda > 1$, for example by the cubes $3Q_{nk} = \widetilde{Q}_{nk}$. This means that we change (4.7.2) to

$$\mathfrak{h}_\alpha(x, n, k) = 2^{n(N-\alpha)}\widetilde{\chi}_{nk}(x) \ . \tag{4.7.10}$$

Then (4.7.3) becomes

$$\mathfrak{H}_\alpha s(x) = \sum_{(n,k)\in\mathbf{M}} 2^{-n\alpha} s_{nk}\widetilde{\chi}_{nk}(x) \ , \tag{4.7.11}$$

and (4.7.4) is replaced by

$$\check{\mathfrak{H}}_\alpha \mu = \left\{(\check{\mathfrak{H}}_\alpha \mu)_{nk}\right\}_{(n,k)\in\mathbf{M}} = \left\{2^{n(N-\alpha)}\widetilde{\mu}_{nk}\right\}_{(n,k)\in\mathbf{M}} \ , \tag{4.7.12}$$

with

$$\widetilde{\mu}_{nk} = \mu(\widetilde{Q}_{nk}) \ . \tag{4.7.13}$$

Thus, for $p = q$,

$$\mathfrak{V}^\mu_{\alpha,p}(x) = \sum_{(n,k)\in\mathbf{M}} 2^{-n\alpha p'}(2^{-nN}\widetilde{\mu}_{nk})^{p'-1}\widetilde{\chi}_{nk}(x) \ . \tag{4.7.14}$$

The advantages are two:

(a) If the cubes $\widetilde{Q}_{nk}$ are chosen to be open, the potentials $\mathfrak{H}_\alpha s$ are lower semicontinuous on $\mathbf{R}^N$ for nonnegative sequences s, being limits of increasing sequences of lower semicontinuous functions.

(b) The functions $W^\mu_{\alpha,p}$ and $\mathfrak{V}^\mu_{\alpha,p}$ are directly comparable. We have, in fact, the following inequality:

$$W^\mu_{\alpha,p}(x) \le \mathfrak{V}^\mu_{\alpha,p}(x) \le A\,\mu(B_{-m}(x))^{p'-1} + A\,W^\mu_{\alpha,p}(x) \ , \tag{4.7.15}$$

if the integer m is chosen so that $2^{-m}3\sqrt{N} \le 1$. This is easily seen, because if $x \in Q_{nk}$, then $B_n(x) \subset \widetilde{Q}_{nk}$, and thus

$$\begin{aligned} W^\mu_{\alpha,p}(x) &= \sum_{n=0}^{\infty} 2^{-n\alpha p'}\left(2^{-nN}\mu(B_n(x))\right)^{p'-1} \\ &\le \sum_{(n,k)\in\mathbf{M}} 2^{-n\alpha p'}(2^{-nN}\widetilde{\mu}_{nk})^{p'-1}\chi_{nk}(x) \le \mathfrak{V}^\mu_{\alpha,p}(x) \ . \end{aligned}$$

On the other hand, because of the way m was chosen, $\widetilde{Q}_{nk} \subset B_{n-m}(x)$ if $x \in \widetilde{Q}_{nk}$, and thus

$$\begin{aligned} \mathfrak{V}^\mu_{\alpha,p}(x) &\le A\sum_{n=0}^{\infty} 2^{-n\alpha p'}\left(2^{-nN}\mu(B_{n-m}(x))\right)^{p'-1} \\ &\le A\,\mu(B_{-m}(x))^{p'-1} + A\,W^\mu_{\alpha,p}(x) \ . \end{aligned}$$

At the same time the capacities and energies remain equivalent. In fact,

$$\int_{\mathbf{R}^N} \mathfrak{V}^{\mu}_{\alpha,p}\, d\mu = \sum_{n=0}^{\infty} 2^{-n\alpha p'} 2^{-nN(p'-1)} \sum_{k\in\mathbf{Z}^N} (\widetilde{\mu}_{nk})^{p'} ,$$

and, denoting the l^∞ norm on $\mathbf{Z}^N$ by $\|\cdot\|_\infty$,

$$\sum_{k\in\mathbf{Z}^N} (\widetilde{\mu}_{nk})^{p'} \le \sum_{k\in\mathbf{Z}^N} 3^{N(p'-1)} \sum_{\|l-k\|_\infty\le 1} (\mu_{nl})^{p'} \le 3^{Np'} \sum_{l\in\mathbf{Z}^N} (\mu_{nk})^{p'} .$$

These observations are easily extended to general $p, q > 1$, using the Fefferman–Stein theorem in the Lizorkin–Triebel case.

4.8 A Characterization of $L^{\alpha,p}$

With Theorem 4.2.2 and the results of Section 4.6 at our disposition, we can now prove Theorem 3.5.6, which characterizes the spaces $L^{\alpha,p}$ in terms of an approximation property.

For a given $\alpha > 0$ we first define the *local degree of approximation* of a function u on a measurable set F by

$$\mathcal{E}(u, F) = \min \int_F |u - \pi|\, dx ,$$

where the minimum is taken over all polynomials π of degree $\le m = [\alpha]$, the integer part of α, and we define a new operator $\mathcal{E}^\alpha$ by

$$\mathcal{E}^\alpha u(x) = \left(\int_0^1 \left(\frac{\mathcal{E}(u, B(x,r))}{r^{N+\alpha}} \right)^2 \frac{dr}{r} \right)^{1/2} .$$

We shall prove the following theorem, which contains Theorem 3.5.6.

Theorem 4.8.1. *Let $1 < p < \infty$, and let α be positive and not an integer. Then there are A_1, and A_2 such that*

$$\|u\|_{\alpha,p} \le A_1\big(\|u\|_p + \|\mathcal{E}^\alpha u\|_p\big) \le A_1\big(\|u\|_p + \|\mathcal{S}^\alpha u\|_p\big) \le A_2\|u\|_{\alpha,p} .$$

Proof. We use the notation from Section 4.6. We first observe that if Q is a cube in $\mathbf{R}^N$ with sidelength 1, then for all polynomials π of degree $\le [\alpha] = m$ and all multiindices β

$$\max_{y\in Q} |D^\beta \pi(y)| \le A \int_Q |\pi(x)|\, dx , \tag{4.8.1}$$

where A is a constant depending only on N and m. This well-known inequality is an immediate consequence of the fact that on a finite dimensional space all norms are equivalent, which follows from the compactness of the unit ball in a finite dimensional normed space.

Let $(n,k) \in \mathbf{Z} \times \mathbf{Z}^N$, and let a_{nk} be an atom in the sense of Definition 4.6.1. It follows from (4.8.1) that there is another constant A, depending only on N and m, such that if π is a polynomial of degree $\leq m$ satisfying

$$A \int_{2Q_{nk}} |\pi(x)|\, dx \leq 2^{-nN} \;, \tag{4.8.2}$$

then πa_{nk} is also an atom.

We now prove the first inequality in the theorem. It is easily seen that for a suitable n_0 depending on N

$$A\, \mathcal{E}^\alpha u(x)^2 \geq \sum_{n=n_0}^{\infty} \sum_{k \in \mathbf{Z}^N} \left(2^{n(\alpha+N)} \mathcal{E}(u, 9Q_{nk})\right)^2 \chi_{nk}(x) \;. \tag{4.8.3}$$

In order to apply Theorems 4.2.2 and 4.6.3 we shall find a representation of u as $u = \sum_{n=0}^{\infty} \sum_{k \in \mathbf{Z}^N} u_{nk}$, where $\operatorname{supp} u_{nk} \subset 2Q_{nk}$, and

$$\|D^\beta u_{nk}\|_\infty \leq A\, 2^{n(|\beta|+N)} \mathcal{E}(u, 9Q_{nk}) \quad \text{for } 0 \leq |\beta| \leq m+1 \;. \tag{4.8.4}$$

Let $\varphi \in C_0^\infty(2Q_{00})$ be a function such that $\sum_{k \in \mathbf{Z}^N} \varphi(x-k) \equiv 1$. Denote by π_{nk} a polynomial of degree $\leq m$ such that

$$\|u - \pi_{nk}\|_{L^1(3Q_{nk})} \leq 2\mathcal{E}(u, 3Q_{nk}) \;. \tag{4.8.5}$$

Set $U_n = \sum_{k \in \mathbf{Z}^N} \pi_{nk} \varphi_{nk}$, where $\varphi_{nk}(x) = \varphi(2^n x - k)$.

We first observe that $\lim_{n\to\infty} U_n(x) = u(x)$ for a.e. x. In fact, we can assume that $\mathcal{E}^\alpha u(x) < \infty$, and that x is a Lebesgue point for u. We can also assume, without loss of generality, that $u(x) = 0$, and thus

$$\lim_{n\to\infty,\, Q_{nk} \ni x} 2^{nN} \int_{Q_{nk}} |u(y)|\, dy = 0 \;. \tag{4.8.6}$$

We claim that under these assumptions $\lim_{n\to\infty} U_n(x) = 0$. Suppose that $x \in Q_{nk}$ for some k. Then $\varphi_{nl}(x) \neq 0$ only if $Q_{nk} \subset 3Q_{nl}$. But then

$$\begin{aligned}\int_{Q_{nk}} |\pi_{nl}|\, dy &\leq \int_{Q_{nk}} |u - \pi_{nl}|\, dy + \int_{Q_{nk}} |u|\, dy \\ &\leq 2\mathcal{E}(u, 3Q_{nl}) + \int_{Q_{nk}} |u|\, dy \leq 2\mathcal{E}(u, 9Q_{nk}) + \int_{Q_{nk}} |u|\, dy \;.\end{aligned}$$

Here the right hand side is $o(2^{-nN})$, as $n \to \infty$, by (4.8.3) and (4.8.6), and thus $|\pi_{nl}(x)| = o(1)$ by (4.8.1), which implies the claim.

We write $U_n - U_{n-1} = u_n$ for $n \geq 1$, so that $u = U_0 + \sum_{n=1}^{\infty} u_n$. Then

$$\begin{aligned} u_n &= \sum_{k \in \mathbf{Z}^N} \pi_{nk}\varphi_{nk} - \sum_{l \in \mathbf{Z}^N} \pi_{n-1,l}\varphi_{n-1,l} \\ &= \sum_{l \in \mathbf{Z}^N} \varphi_{n-1,l} \sum_{k \in \mathbf{Z}^N} \pi_{nk}\varphi_{nk} - \sum_{k \in \mathbf{Z}^N} \varphi_{nk} \sum_{l \in \mathbf{Z}^N} \pi_{n-1,l}\varphi_{n-1,l} \\ &= \sum_{l \in \mathbf{Z}^N} \sum_{k \in \mathbf{Z}^N} \varphi_{nk}\varphi_{n-1,l}(\pi_{nk} - \pi_{n-1,l}) \;. \end{aligned}$$

But $\operatorname{supp}\varphi_{nk} \subset 2Q_{nk}$, so in the last sum, for each k only the 3^N terms with $2Q_{nk} \cap 2Q_{n-1,l} \neq \emptyset$ give any contribution. We set

$$\sum_{l \in \mathbf{Z}^N} \varphi_{nk}\varphi_{n-1,l}(\pi_{nk} - \pi_{n-1,l}) = u_{nk} \ ,$$

so that $u_n = \sum_{k \in \mathbf{Z}^N} u_{nk}$ with $\operatorname{supp} u_{nk} \subset 2Q_{nk}$. The inequality (4.8.4) for $n \geq 1$ is an easy consequence of (4.8.1), and of the estimate

$$\|\pi_{nk} - \pi_{n-1,l}\|_{L^1(2Q_{nk})} \\ \leq \|u - \pi_{nk}\|_{L^1(3Q_{nk})} + \|u - \pi_{n-1,l}\|_{L^1(3Q_{n-1,l})} \leq 4\mathcal{E}(u, 9Q_{nk}) \ ,$$

which follows from (4.8.5) if $2Q_{nk} \cap 2Q_{n-1,l} \neq \emptyset$.

Writing

$$U_0 = \sum_{k \in \mathbf{Z}^N} \pi_{0k}\varphi_{0k} = \sum_{k \in \mathbf{Z}^N} u_{0k} \ ,$$

and observing that (4.8.5) gives

$$\|\pi_{0k}\|_{L^1(3Q_{0k})} \leq \|u\|_{L^1(3Q_{0k})} + 2\mathcal{E}(u, 3Q_{0k}) \leq 3\|u\|_{L^1(3Q_{0k})} \ ,$$

we find

$$\|D^\beta u_{0k}\|_\infty \leq A\, \|u\|_{L^1(3Q_{0k})} \quad \text{for } 0 \leq |\beta| \leq m+1 \ .$$

The first inequality in the theorem follows easily by means of Theorems 4.2.2 and 4.6.3 from this estimate and (4.8.4).

The second inequality is obvious from the definition of local degree of approximation.

In order to prove the last inequality we let

$$u = \sum_{n=0}^{\infty} u_n = \Phi_0 * u + \sum_{n=1}^{\infty} \varphi_n * u$$

be a representation of u that satisfies the conditions of Definition 4.2.1. For any u_n such that $\operatorname{supp}\widehat{u_n} \subset B(0, 2^n)$ we have the estimate

$$|D^\beta u_n(x)| \leq A\, 2^{n|\beta|} \min_{y \in B(x, 2^{-n})} Mu_n(y) \ . \tag{4.8.7}$$

See the proof of Theorem 4.6.2. We clearly have

$$S^\alpha u(x)^2 \leq A \sum_{j=-\infty}^{\infty} \left(2^{j(\alpha+N)} \int_{|y| \leq 2^{-j}} |u(x+y) - P_x^m u(x+y)|\, dy \right)^2 .$$

Easy estimates using Taylor's formula and (4.8.7) give

$$2^{jN} \int_{|y| \leq 2^{-j}} |u(x+y) - P_x^m u(x+y)|\, dy \\ \leq A \sum_{0 \leq n < j} 2^{(n-j)(m+1)} Mu_n(x) + A \sum_{\max\{j,0\}}^{\infty} 2^{(n-j)m} Mu_n(x) \ .$$

Writing $Mu_n(x) = c_n$, and choosing $\varepsilon > 0$ so that $m + \varepsilon < \alpha$ (cf. the last part of the proof of Theorem 4.1.3), we find by Cauchy's inequality

$$\sum_{j=-\infty}^{\infty} 2^{2j\alpha}\left(\sum_{\max\{j,0\}}^{\infty} 2^{(n-j)m} c_n\right)^2$$
$$\leq A \sum_{j=-\infty}^{\infty} 2^{2j\alpha - 2\varepsilon \max\{j,0\}} \sum_{\max\{j,0\}}^{\infty} 2^{2(n-j)m + 2n\varepsilon} c_n^2$$
$$= A \sum_{n=0}^{\infty} 2^{2n(m+\varepsilon)} c_n^2 \sum_{j=-\infty}^{n} 2^{2j(\alpha-m) - 2\varepsilon \max\{j,0\}} \leq A \sum_{n=0}^{\infty} 2^{2n\alpha} c_n^2 \; .$$

Similarly, choosing $\varepsilon > 0$ so that $\alpha + \varepsilon < m + 1$,

$$\sum_{j=0}^{\infty} 2^{2j\alpha}\left(\sum_{n=0}^{j-1} 2^{(n-j)(m+1)} c_n\right)^2$$
$$\leq A \sum_{j=0}^{\infty} 2^{2j\alpha} 2^{-2j(\alpha+\varepsilon)} \sum_{n=0}^{j-1} 2^{2n(\alpha+\varepsilon)} c_n^2$$
$$= A \sum_{n=0}^{\infty} 2^{2n(\alpha+\varepsilon)} c_n^2 \sum_{j=n+1}^{\infty} 2^{-2j\varepsilon} \leq A \sum_{n=0}^{\infty} 2^{2n\alpha} c_n^2 \; .$$

It follows that

$$S^\alpha u(x)^2 \leq A \sum_{j=0}^{\infty} 2^{2j\alpha} Mu_j(x)^2 \; .$$

The Fefferman–Stein theorem (Theorem 1.1.2) and Theorem 4.2.2 give the desired result.

4.9 Notes

4.1. Besov spaces appear naturally as spaces of traces of functions in Sobolev spaces. They were introduced by O. V. Besov [55, 56]. References to earlier work are found in e.g. the book of S. M. Nikol′skiĭ [349]. See also E. M. Stein [389], Section V.5. The treatment of Besov spaces given here is inspired by J. Peetre's book [356]. The important Theorems 4.1.4 and 4.1.5 are special cases of results found in [356], Chapter 8, Theorems 1 and 2, and Remark 1. Peetre attributes the use of Wiener's theorem in this context to H. S. Shapiro [376]. The case $p = q = 2$ is found already in L. Hörmander's book [228], Corollary 2.4.1.

4.2–4.3. The Lizorkin–Triebel spaces for $1 < p, q < \infty$ were introduced simultaneously by P. I. Lizorkin [277] (see also P. I. Lizorkin [276]), and H. Triebel [402].

The proof of Theorem 4.2.2 given here, depending on a simple case of Khinchin's inequality and Mikhlin's multiplier theorem, is a modification of arguments found in E. M. Stein [389], Section IV.5, and was shown to the authors by

Yu. V. Netrusov. See also H. Triebel [405], 2.5.6, p. 88, where the proof is based on a matrix multiplier theorem. A short proof of the Khinchin inequality, Lemma 4.2.3, is found in Stein [389], Appendix D, p. 276. See also A. Zygmund [440], Chapter V, Theorem 8.4, p. 213.

The multiplier theorem, Theorem 4.2.4, and its proof are taken from L. Hörmander [227], (Theorem 2.5, p. 120). The result goes back to J. Marcinkiewicz [288], who proved a corresponding multiplier theorem for Fourier series, and S. G. Mikhlin [325, 326, 327]. Mikhlin's conditions involve maximum norms of derivatives up to the order N, and are actually sufficient for our application. A different proof of Hörmander's result is given by Stein [389], Theorem IV:3, p. 96.

The theory was extended to $p > 0$ and $q > 0$ by J. Peetre [355], who also observed the equivalence of $F_0^{p,2}$ with the Hardy space H^p for $0 < p \leq 1$. The definition of $F_0^{\infty,2}$ and its identification with BMO, i.e the space of functions of bounded mean oscillation, are found in M. Frazier and B. Jawerth [155]. Our exposition draws heavily on Peetre's paper. In particular, Definition 4.3.3, and Theorems 4.3.2, and 4.3.4 are adapted from [355], where these results are proved for all $p > 0$ and $q > 0$, but in the homogeneous case. Peetre also shows that the Mikhlin–Hörmander multiplier theorem (Theorem 4.2.4) is an easy consequence of Theorem 4.3.4. Much more information on the Lizorkin–Triebel spaces is found in the books [404, 405, 406] by H. Triebel.

Corollary 4.3.9 was first observed by Per Nilsson in 1983 in conversations with the authors. See L. I. Hedberg and Th. H. Wolff [219], p. 175, and D. R. Adams [15]. The result has been extended to $p > 0$ and $q > 0$ by Yu. V. Netrusov [341], and B. Jawerth, C. Pérez, and G. Welland [234]. For a different approach, see W. S. Cohn and I. E. Verbitsky [109], Theorem 2.

4.4. The connection between Besov spaces and nonlinear potentials was apparently first noticed in D. R. Adams [15]. Nonlinear "Besov potentials" were also introduced there. Some of the material in this section appeared in L. I. Hedberg [217], and some appears here for the first time. See also T. Sjödin [381].

4.5. Wolff's inequality, Theorem 4.5.2, was first proved in L. I. Hedberg and Th. H. Wolff [219]. (Because of the fact that the theorem was published in a joint paper, it has sometimes erroneously been attributed to the authors jointly. See [219], p. 166.) See also the notes to Chapter 9. Wolff's proof was direct, and quite complicated. Simpler proofs were given by Per Nilsson (see [219], p. 174) and J. L. Lewis (unpublished), before it was observed in D. R. Adams [12] that the inequality is a consequence of the Muckenhoupt–Wheeden inequality, Theorem 3.6.1. The characterization of the positive cone in the dual Besov space (Theorem 4.5.5) is due to D. R. Adams [15].

4.6. The representation of elements in Besov and Lizorkin–Triebel spaces by means of "smooth atoms" is a part of the much more comprehensive theory of M. Frazier and B. Jawerth [153, 154, 155]. See also Yu. V. Netrusov [337, 338], where this representation was discovered independently, and used in proving embedding theorems. Our proofs are adapted mainly from [153] and [154], and our exposition has benefited greatly from remarks by Netrusov. A nice survey of many related results is given by M. Frazier, B. Jawerth, and G. Weiss [156].

A significant omission in our presentation, but one which gives some simplification of the proofs, is that we treat only the case $p, q > 1$. In order to extend the theory to general positive p and q one has to modify the definition of atoms so that they have a certain number of moments equal to zero.

The work of Frazier and Jawerth predated the theory of wavelets, which we have also chosen to omit here, as well as the related theory of multiresolution analyses. We refer to the excellent expositions of Y. Meyer [317], I. Daubechies [115], and others.

4.7. An "atomic" nonlinear potential theory appeared in L. I. Hedberg and Th. H. Wolff [219], in fact the basic inequality ([219], p. 170) proved by Wolff is atomic. The connection with Besov and Lizorkin–Triebel spaces was, however, not observed (and is published here for the first time), nor was it noticed that the theory could be subsumed under the general theory of B. Fuglede and N. G. Meyers. The simple, but sometimes very useful Proposition 4.7.2 has been systematically used by Yu. V. Netrusov in [341]–[345].

4.8. Theorem 4.8.1, and its interesting proof were communicated to the authors by Yu. V. Netrusov in November 1993. The original proof of Theorem 3.5.6 of R. S. Strichartz [392] depends on a Banach space valued version of Theorem 1.1.5. Using atoms, Strichartz [394] has proved an extension of Theorem 3.5.6 to the spaces $F_\alpha^{p,2}$ (denoted $I_\alpha(H^p)$) for $0 < p \leq 1$, and $\alpha > N(1/p - 1)$. See also the notes to Chapter 3.

5. Metric Properties of Capacities

Many problems have definitive solutions in terms of capacities, but the latter have the drawback that their geometrical meaning is not transparent. For this reason we devote most of this chapter to comparing the (α, p)-capacities $C_{\alpha,p}$ for $1 < p < \infty$ and $0 < \alpha p \le N$ to the more geometric quantities known as Hausdorff measures. As we now know, $C_{\alpha,p}$ is associated not only to the Sobolev spaces and Bessel potential spaces $L^{\alpha,p}$, but also to the Besov spaces $B_\alpha^{p,p}$ and the Lizorkin–Triebel spaces $F_\alpha^{p,q}$, $1 < q < \infty$.

In Section 5.1, after having estimated the capacities for balls, we define Hausdorff measures, and prove some of their most important properties, including a classic theorem of O. Frostman. We give upper and lower estimates for capacities in terms of Hausdorff measures. These are proved to be sharp in Section 5.4 by means of estimates for Cantor sets proved in Section 5.3. In Section 5.2 we give an estimate for capacities under Lipschitz mappings, which as an easy consequence has a boundary value result in halfspaces. In Section 5.5, finally, we investigate the relations between (α, p)-capacities for different α and p.

5.1 Comparison Theorems

In order to give a more concrete idea of the properties of (α, p)-capacities we prove some comparison theorems. We recall from Section 2.6 that a function g on $\mathbf{R}^N$ is called a radially decreasing convolution kernel, if $g(x) = g_0(|x|)$, where g_0 is a non-negative, lower semi-continuous, non-increasing function on $\mathbf{R}^+$ for which $\int_0^1 g_0(t)t^{N-1}\,dt < \infty$.

Proposition 5.1.1. *Let $p > 1$ and let g be a radially decreasing convolution kernel on $\mathbf{R}^N$ such that $g \notin L^{p'}(\mathbf{R}^N)$, and $\int_{|x|\ge 1} g^{p'}\,dx < \infty$. Let B_r be a ball in $\mathbf{R}^N$ with radius $r > 0$. Then there is a constant $A > 0$ such that for all $r > 0$*

$$A^{-1}C_{g,p}(B_r) \le \min\left\{ r^N\left(\int_{|x|\le 2r} g\,dx\right)^{-p}, \left(\int_{|x|\ge 2r} g^{p'}\,dx\right)^{1-p}\right\}$$
$$\le A\,C_{g,p}(B_r)\ .$$

Proof. We assume that the balls B_r are centered at the origin. In order to prove the first inequality we let μ be an arbitrary measure in $\mathcal{M}^+(B_r)$ that satisfies $\mu(B_r) = 1$. If $y \in B_r$ and $x \notin B_r$, then $|x - y| \le 2|x|$, and thus

$$\int_{\mathbf{R}^N}\left(\int_{B_r} g(x-y)\,d\mu(y)\right)^{p'} dx \geq \int_{|x|\geq r} g(2x)^{p'}\,dx = 2^{-N}\int_{|x|\geq 2r} g(x)^{p'}\,dx.$$

On the other hand, by Hölder's inequality,

$$\int_{\mathbf{R}^N}\left(\int_{B_r} g(x-y)\,d\mu(y)\right)^{p'} dx \geq \int_{B_{2r}}\left(\int_{B_r} g(x-y)\,d\mu(y)\right)^{p'} dx$$
$$\geq |B_{2r}|^{1-p'}\left(\int_{B_{2r}}\int_{B_r} g(x-y)\,d\mu(y)\,dx\right)^{p'} = D\ .$$

Hence, setting $x-y=z$, and taking into account that $\operatorname{supp}\mu\subset B_r$ and $B_r+B_r\subset B_{2r}$,

$$D \geq |B_{2r}|^{1-p'}\left(\int_{B_r}\int_{B_r} g(z)\,d\mu(y)\,dz\right)^{p'} = |B_{2r}|^{1-p'}\left(\int_{B_r} g(z)\,dz\right)^{p'},$$

which by the monotonicity of g gives

$$D \geq |B_{2r}|^{1-p'}\left(\int_{B_r} g(2z)\,dz\right)^{p'} = |B_{2r}|^{1-p'}\left(2^{-N}\int_{B_{2r}} g(z)\,dz\right)^{p'}.$$

By Theorem 2.5.1 these inequalities imply

$$C_{g,p}(B_r)^{1/p} \leq A\,\min\left\{ r^{N/p}\left(\int_{B_{2r}} g(x)\,dx\right)^{-1}, \left(\int_{|x|\geq 2r} g(x)^{p'}\,dx\right)^{-1/p'}\right\},$$

which gives the required upper estimate.

In order to prove the lower estimate of $C_{g,p}$ we specialize μ to be Lebesgue measure restricted to B_r, normalized so that $\mu(B_r)=1$. Then for $|x|\leq 3r$, again by monotonicity,

$$\int_{B_r} g(x-y)\,d\mu(y) \leq \frac{1}{|B_r|}\int_{|y-x|\leq 4r} g(x-y)\,dy$$
$$= \frac{1}{|B_r|}\int_{|y|\leq 4r} g(y)\,dy \leq \frac{2^N}{|B_r|}\int_{|y|\leq 2r} g(y)\,dy\ .$$

For $|x|\geq 3r$ we have $|x-y|\geq |x|-|y|\geq 2|x|/3$, so that

$$\int_{|x|\geq 3r}\left(\int_{B_r} g(x-y)\,d\mu(y)\right)^{p'} dx$$
$$\leq \int_{|x|\geq 3r} g(2x/3)^{p'}\,dx = (\tfrac{3}{2})^N\int_{|x|\geq 2r} g(x)^{p'}\,dx\ .$$

Thus, by the inequality $a+b\leq 2\max\{a,b\}$,

$$\int_{\mathbf{R}^N}\left(\int_{B_r} g(x-y)\,d\mu(y)\right)^{p'} dx$$
$$\leq 2\max\left\{3^N|B_r|\left(\frac{2^N}{|B_r|}\int_{|x|\leq 2r} g(x)\,dx\right)^{p'},\ (\tfrac{3}{2})^N\int_{|x|\geq 2r} g(x)^{p'}\,dx\right\}.$$

Theorem 2.5.1 now gives the desired estimate.

For Riesz and Bessel capacities we can say more.

Proposition 5.1.2. *Let B_r be a ball in $\mathbf{R}^N$ with radius $r > 0$. Then $\dot{C}_{\alpha,p}(B_r) > 0$ for $\alpha p < N$, and*

$$\dot{C}_{\alpha,p}(B_r) = r^{N-\alpha p}\dot{C}_{\alpha,p}(B_1).$$

Remark. We know already from Proposition 2.6.1 that $\dot{C}_{\alpha,p}(B_r) = 0$ for all $E \subset \mathbf{R}^N$ if $\alpha p \geq N$.

Proof. Lebesgue measure m restricted to B_r gives a potential $I_\alpha * \mu$ that belongs to $L^{p'}(\mathbf{R}^N)$ if $\alpha p < N$, and thus $\dot{C}_{\alpha,p}(B_r) > 0$.

Now assume that the balls B_r are centered at the origin, and let $h \geq 0$ be a function in $L^p(\mathbf{R}^N)$ such that $f(x) = I_\alpha * h(x) \geq 1$ on B_1. Then, if f_r is defined by $f_r(x) = f(x/r)$, we have $f_r(x) \geq 1$ on B_r , and it is easily seen that $f_r = I_\alpha * (r^{-\alpha}h_r)$. But $\int (r^{-\alpha}h_r)^p\,dx = r^{N-\alpha p}\int h^p\,dx$, which gives the proposition.

Proposition 5.1.3. *Let $p > 1$ and $\alpha p = N$. Then for any $c > 1$ there is $A > 0$ (depending on c) so that for $0 < r \leq 1$*

$$A^{-1}\left(\log\frac{c}{r}\right)^{1-p} \leq C_{\alpha,p}(B_r) \leq A\left(\log\frac{c}{r}\right)^{1-p}.$$

Remark. The proof will show that the right inequality is valid for all $r < c$.

Proof. It follows from Proposition 5.1.1 and (1.2.14) that

$$C_{\alpha,p}(B_r) \leq A\left(\int_{2r\leq|x|\leq 2c}|x|^{-N}\,dx\right)^{1-p} = A\left(\log\frac{c}{r}\right)^{1-p}.$$

On the other hand, by Proposition 5.1.1 and (1.2.15)

$$C_{\alpha,p}(B_r) \geq A^{-1}\left(\int_{2r\leq|x|\leq 2c}|x|^{-N}\,dx + \int_{|x|\geq 2c} e^{-p'|x|/2}\,dx\right)^{1-p}$$
$$\geq A^{-1}\left(\log\frac{c}{r}\right)^{1-p}.$$

Proposition 5.1.4. *Let $\alpha p < N$.*
(a) *For all $E \subset \mathbf{R}^N$*

$$\dot{C}_{\alpha,p}(E) \leq C_{\alpha,p}(E).$$

(b) *For each $R > 0$ there is A such that*

$$C_{\alpha,p}(E) \leq A\,\dot{C}_{\alpha,p}(E)$$

for all $E \subset \mathbf{R}^N$ with diameter at most R.

Remark. An inequality that is independent of the diameter of E is given in Section 5.6.1.

Proof. The first part of the proposition is an immediate consequence of the fact that $G_\alpha \le I_\alpha$.

For the proof of the second part we recall that in Section 3.6 we defined a modified kernel $I_{\alpha,\delta}$ for $\delta > 0$ by

$$I_{\alpha,\delta}(x) = I_\alpha(x) \quad \text{for } |x| < \delta, \qquad I_{\alpha,\delta} = 0 \quad \text{for } |x| \ge \delta .$$

Let $\delta > 0$ and let $h \ge 0$ be a function such that $I_\alpha * h(x) \ge 1$ on a set E with diameter at most R. Assume that $\|h\|_p^p \le 2\dot{C}_{\alpha,p}(E) \le A R^{N-\alpha p}$. Then for any x

$$|I_\alpha * h(x) - I_{\alpha,\delta} * h(x)| \le \|h\|_p \, \|I_\alpha - I_{\alpha,\delta}\|_{p'} \le A \, R^{(N-\alpha p)/p} \, \delta^{-(N-\alpha p)/p} .$$

Thus, if δ is chosen large enough, we have $I_{\alpha,\delta} * h(x) \ge \frac{1}{2}$ on E. But for any $\delta > 0$ there is A such that $I_{\alpha,\delta}(x) \le A G_\alpha(x)$ for all x; see (1.2.14) and (1.2.24). Then $G_\alpha * (Ah)(x) \ge \frac{1}{2}$ on E, and the proposition follows.

We next recall the definition of *Hausdorff measure*. Let $h(r)$ be an increasing function, defined ($\le +\infty$) for $r \ge 0$, and satisfying $h(0) = 0$. Let $E \subset \mathbf{R}^N$, and consider coverings of E by countable unions of (open or closed) balls $\{B(x_i, r_i)\}_{i=1}^\infty$ with radii $\{r_i\}_1^\infty$. Then for any ρ, $0 < \rho \le \infty$, a set function $\Lambda_h^{(\rho)}$ is defined by

$$\Lambda_h^{(\rho)}(E) = \inf \sum_1^\infty h(r_i) ,$$

where the infimum is taken over all such coverings with $\sup_i r_i \le \rho$. Clearly $\Lambda_h^{(\rho)}(E)$ is a decreasing function of ρ, so $\lim_{\rho \to 0} \Lambda_h^{(\rho)}(E)$ exists ($\le +\infty$), and we can define

$$\Lambda_h(E) = \lim_{\rho \to 0} \Lambda_h^{(\rho)}(E) .$$

This is the Hausdorff measure of E with respect to the function h. If $h(r) = r^\alpha$, we write Λ_α for Λ_{r^α}.

The set function $\Lambda_h^{(\infty)}$ will often be more useful for us than the Hausdorff measure itself. It is sometimes called the *Hausdorff content* or the *Hausdorff capacity*.

Proposition 5.1.5. $\Lambda_h^{(\infty)}(E) = 0$ *if and only if* $\Lambda_h(E) = 0$.

Proof. We always have $\Lambda_h^{(\infty)}(E) \le \Lambda_h(E)$, so it is enough to prove the "only if" part. Assume that E is such that $\Lambda_h(E) > 0$. Choose a number c so that $0 < c < \Lambda_h(E)$, and $\rho > 0$ so small that $\Lambda_h^{(\rho)}(E) > c$. Then $\sum h(r_i) > c$ for all coverings of E by balls $\{B(x_i, r_i)\}$ such that $\sup_i r_i \le \rho$. For all other coverings $\sum h(r_i) \ge h(\rho) > 0$, because $r_i \ge \rho$ for some i. Thus $\Lambda_h^{(\infty)}(E) \ge \min\{c, h(\rho)\} > 0$.

Proposition 5.1.6. *Let Λ_{h_1} and Λ_{h_2} be Hausdorff measures defined by functions h_1 and h_2 such that $h_1(t) = o(h_2(t))$ as $t \to 0$. Then*

$$\Lambda_{h_1}(E) > 0 \quad \Longrightarrow \quad \Lambda_{h_2}(E) = \infty \ .$$

Proof. Let $\varepsilon > 0$ and choose ρ so small that $h_1(t) \le \varepsilon h_2(t)$ for $t \le \rho$. Let $\{B(x_i, r_i)\}$ be a covering of E with all $r_i \le \rho$. Then $\sum h_1(r_i) \le \varepsilon \sum h_2(r_i)$. Thus also $\Lambda_{h_1}^{(\rho)}(E) \le \varepsilon \Lambda_{h_2}^{(\rho)}(E)$, and consequently, taking the limit as $\rho \to 0$, $\Lambda_{h_1}(E) \le \varepsilon \Lambda_{h_2}(E)$.

Corollary 5.1.7. *For all $E \subset \mathbf{R}^N$*

$$\sup\{\, \alpha \in \mathbf{R}^+ : \Lambda_\alpha(E) = \infty \,\} = \inf\{\, \beta \in \mathbf{R}^+ : \Lambda_\beta(E) = 0 \,\}.$$

The number defined in the corollary is called the *Hausdorff dimension* of E.

We have so far assumed only that the function h is increasing and satisfies $h(0) = 0$. The following proposition shows that in all interesting cases we can also assume that the function $h(r)r^{-N}$ is decreasing.

Proposition 5.1.8. *Let $h(r)$ be an increasing function for $r \ge 0$, such that $h(0) = 0$.*
(a) *If* $\liminf_{r\to 0} h(r)r^{-N} = 0$, *then* $\Lambda_h^{(\rho)}(E) = 0$ *for all* $E \subset \mathbf{R}^N$.
(b) *If* $\liminf_{r\to 0} h(r)r^{-N} > 0$, *then there is an increasing function $h^*(r)$ for $r \ge 0$, such that $h^*(0) = 0$, h^* is continuous, $h^*(r)r^{-N}$ is decreasing, and there is $A > 0$ such that for all $E \subset \mathbf{R}^N$ and all $\rho > 0$*

$$A^{-1} \Lambda_h^{(\rho)}(E) \le \Lambda_{h^*}^{(\rho)}(E) \le A\, \Lambda_h^{(\rho)}(E) \ . \tag{5.1.1}$$

Proof. (a) Assume that $\liminf_{r\to 0} h(r)r^{-N} = 0$. It enough to prove that $\Lambda_h^{(\rho)}(Q) = 0$ for the unit cube Q. For a constant A_N depending only on the dimension N, and any $r > 0$, Q can be covered by $A_N r^{-N}$ balls of radius r. It follows that $\Lambda_h^{(\rho)}(Q) \le A_N \inf_{0<r\le\rho} r^{-N} h(r)$, which proves the claim.

(b) We define h^* by

$$\frac{h^*(r)}{r^N} = \inf_{0<t\le r} \frac{h(t)}{t^N} \ .$$

The condition $\liminf_{r\to 0} h(r)r^{-N} > 0$ guarantees that $h^*(r) > 0$ for $r > 0$. Clearly $h^*(r)r^{-N}$ is increasing, and $h^*(r) \le h(r)$. It is also easy to see that h^* is increasing. In fact, let $0 < r < R$, and let $\varepsilon > 0$ be arbitrary. Then there is t, $r \le t \le R$, such that $h(t)t^{-N} < (1+\varepsilon)h^*(R)R^{-N}$, and consequently $h^*(r) \le h(r) \le h(t) < (1+\varepsilon)h^*(R)(t/R)^N \le (1+\varepsilon)h^*(R)$.

The continuity of h^* now follows from the fact that both $h^*(r)$ and $r^N/h^*(r)$ are increasing.

The right hand inequality in (5.1.1) (with $A = 1$) follows immediately from the inequality $h^* \le h$.

To prove the left hand inequality, finally, we let $\{B(x_i, r_i)\}_1^\infty$ be a covering of the set E such that all $r_i \le \rho$, and $\sum_i h^*(r_i) < \Lambda_{h^*}^{(\rho)}(E) + \varepsilon$, where

$\varepsilon > 0$ is arbitrary. Then, as in the proof of (a), each of the balls $B(x_i, r_i)$ can be covered by $A_N(r_i/r)^N$ balls of radius r for any $r \le r_i$. Thus, $\Lambda_h^{(\rho)}(E) \le A_N \sum_i \inf_{0<r\le r_i}(r_i/r)^N h(r) = A_N \sum_i h^*(r_i)$, and the inequality follows.

We are now ready for our first comparison theorem.

Theorem 5.1.9. *Let $p > 1$ and $0 < \alpha p \le N$, and let $E \subset \mathbf{R}^N$. Set $h(r) = r^{N-\alpha p}$, if $\alpha p < N$, and $h(r) = \left(\log_+ \frac{2}{r}\right)^{1-p}$, if $\alpha p = N$. Then there is A independent of E such that*

$$C_{\alpha,p}(E) \le A\,\Lambda_h^{(1)}(E) \ ,$$

and moreover

$$\Lambda_h(E) < \infty \quad \Longrightarrow \quad C_{\alpha,p}(E) = 0 \ .$$

Proof. We first prove the first statement. Let $E \subset \bigcup_1^\infty B(x_i, r_i)$, where the covering is chosen so that $r_i \le 1$ for all i, and $\sum h(r_i) \le 2\Lambda_h^{(1)}(E)$. Then, by Propositions 5.1.2, 5.1.3, and 5.1.4(b)

$$C_{\alpha,p}(E) \le \sum_1^\infty C_{\alpha,p}(B(x_i, r_i)) \le A \sum_1^\infty h(r_i) \le A\,\Lambda_h^{(1)}(E) \ .$$

For the second statement we need two lemmas.

Lemma 5.1.10. *Let $g(x) = g_0(|x|)$ be a radially decreasing convolution kernel. Let $\mu \in \mathcal{M}^+$ and suppose that $g * \mu \in L^p$, where $p \ge 1$. Then there is a kernel $\overline{g}$ with the same properties, such that*

(a) $\overline{g} * \mu \in L^p$;
(b) $\overline{g}(x) \ge g(x)$;
(c) $g(x) = o(\overline{g}(x))$, *as* $|x| \to 0$.

Proof. Set

$$v_i(x) = \int_{2^{-i} \le |x-y| < 2^{-i+1}} g(x-y)\,d\mu(y), \quad n = 1, 2, \dots \ .$$

Note that $\int_{y=x} g(x-y)\,d\mu(y) = 0$ for almost all x, since $\mu(\{x\}) > 0$ for at most a countable number of points x. Thus $g * \mu(x) = \sum_{-\infty}^\infty v_i(x)$ in the L^p-sense.

We claim that there are $\{a_i\}_1^\infty$ such that $a_i \ge 1$, $a_i \nearrow \infty$ as $i \to \infty$, and $\sum_1^\infty a_i v_i \in L^p$.

We observe that $\lim_{k\to\infty} \left\|\sum_k^\infty v_i\right\|_p = 0$ by dominated convergence, since $\sum_k^\infty v_i \le g * \mu \in L^p$, and

$$\sum_k^\infty v_i(x) = \int_{0<|x-y|<2^{-k+1}} g(x-y)\,d\mu(y) \ ,$$

which tends to 0, as k tends to infinity.

It follows that there is a subsequence $\{k_j\}_{j=1}^{\infty}$ such that

$$\sum_{j=1}^{\infty}\left\|\sum_{i=k_j}^{\infty} v_i\right\|_p < \infty .$$

But then there are $\{b_j\}_{j=1}^{\infty}$ such that $b_j \geq 1$, $b_j \nearrow \infty$, and

$$\sum_{j=1}^{\infty} b_j\left\|\sum_{i=k_j}^{\infty} v_i\right\|_p < \infty .$$

Set $a_i = b_j$ for $k_j \leq i < k_{j+1}$ and $a_i = 1$ for $i < k_1$. Then we have

$$\left\|\sum_{i=k_1}^{\infty} a_i v_i\right\|_p \leq \sum_{j=1}^{\infty}\left\|\sum_{i=k_j}^{k_{j+1}-1} a_i v_i\right\|_p \leq \sum_{j=1}^{\infty} b_j\left\|\sum_{i=k_j}^{\infty} v_i\right\|_p < \infty .$$

The lemma follows if we set $\overline{g}(x) = a_i g(x)$ for $2^{-i} \leq |x| < 2^{-i+1}$, $i = 1, 2, \ldots,$ and $\overline{g}(x) = g(x)$ for $|x| \geq 1$.

Lemma 5.1.11. *Let $p > 1$ and $\alpha p \leq N$. Let $\overline{g}$ be a radially decreasing convolution kernel such that $\overline{g}(x) \geq G_\alpha(x)$ and $G_\alpha(x) = o(\overline{g}(x))$, as $|x| \to 0$. Then*

$$\lim_{r\to 0}\frac{C_{\overline{g},p}(B_r)}{C_{\alpha,p}(B_r)} = 0 .$$

Proof. Set $\sup_{|x|\leq\delta} G_\alpha(x)/\overline{g}(x) = m(\delta)$, so that $\lim_{\delta\to 0} m(\delta) = 0$. First let $\alpha p < N$. By Proposition 5.1.1, the property (1.2.14) of G_α, and Propositions 5.1.2 and 5.1.4

$$\begin{aligned} C_{\overline{g},p}(B_r) &\leq A\, r^N\left(\int_{|x|\leq 2r}\overline{g}\,dx\right)^{-p} \leq A\, r^N m(2r)^p\left(\int_{|x|\leq 2r} G_\alpha\,dx\right)^{-p} \\ &\leq A\, m(2r)^p\, r^{N-\alpha p} \leq A\, m(2r)^p\, C_{\alpha,p}(B_r) , \end{aligned}$$

which proves the lemma in this case.

For $\alpha p = N$ we choose $\varepsilon > 0$ and δ, $0 < \delta \leq 1$, so that $m(2\delta) < \varepsilon$. Proposition 5.1.1 and (1.2.14) give for $r < \delta$

$$\begin{aligned} C_{\overline{g},p}(B_r) &\leq A\left(\int_{2r\leq|x|\leq 2\delta}\overline{g}^{p'}\,dx\right)^{1-p} \\ &\leq A\, m(2\delta)^p\left(\int_{2r\leq|x|\leq 2\delta}|x|^{-N}\,dx\right)^{1-p} < A\,\varepsilon^p\left(\log\frac{\delta}{r}\right)^{1-p} . \end{aligned}$$

Together with Proposition 5.1.3 this implies

$$\limsup_{r\to 0}\frac{C_{\overline{g},p}(B_r)}{C_{\alpha,p}(B_r)} \leq A\,\varepsilon^p ,$$

and the lemma follows.

End of proof of Theorem 5.1.9. Assume that $C_{\alpha,p}(E) > 0$. First assume that E is compact. Then there is a nonzero $\mu \in \mathcal{M}^+(E)$ such that $G_\alpha * \mu \in L^{p'}$. By Lemma 5.1.10 there is a kernel $\overline{g}$ such that $G_\alpha(x) = o(\overline{g}(x))$ as $|x| \to 0$, and $\overline{g} * \mu \in L^{p'}$, which implies that $C_{\overline{g},p}(E) > 0$. But for every covering of E

$$C_{\overline{g},p}(E) \le \sum_i C_{\overline{g},p}(B(x_i,r_i)) \le \sup_i \frac{C_{\overline{g},p}(B(x_i,r_i))}{C_{\alpha,p}(B(x_i,r_i))} \sum_i C_{\alpha,p}(B(x_i,r_i))$$
$$\le \sup_i \frac{C_{\overline{g},p}(B(x_i,r_i))}{C_{\alpha,p}(B(x_i,r_i))} \sum_i h(r_i) \ .$$

If we consider coverings with all $r_i \le \rho$, it follows that

$$C_{\overline{g},p}(E) \le \sup_{r\le\rho} \frac{C_{\overline{g},p}(B_r)}{C_{\alpha,p}(B_r)} \Lambda_h^{(\rho)}(E) \ .$$

Lemma 5.1.11 now gives a contradiction.

We then observe that in order to prove the theorem for general E, it is enough to prove it when E is a set of type G_δ. In fact, it is easy to see that for any E there is a G_δ-set E' such that $E \subset E'$ and $\Lambda_h(E) = \Lambda_h(E')$. Then, if E is G_δ and $C_{\alpha,p}(E) > 0$, there is a compact $K \subset E$ such that $C_{\alpha,p}(K) > 0$. This is a special case of Choquet's theorem, Theorem 2.3.11. The theorem follows.

The converse estimate is a deeper result, and depends on the following theorem of O. Frostman [158].

Theorem 5.1.12. *Let h be an increasing function on* $[0,\infty)$ *such that* $h(0) = 0$, *and let $E \subset \mathbf{R}^N$ be a compact set. Then*

$$\mu(E) \le \Lambda_h^{(\infty)}(E)$$

for all $\mu \in \mathcal{M}^+(E)$ such that $\mu(B(x,r)) \le h(r)$ for all balls $B(x,r)$. Furthermore, there is a constant $A > 0$, depending only on N, and a $\mu \in \mathcal{M}^+(E)$, satisfying $\mu(B(x,r)) \le h(r)$ for all $B(x,r)$, such that

$$\Lambda_h^{(\infty)}(E) \le A\,\mu(E) \ .$$

Proof. The first inequality is obvious. In fact, if $E \subset \bigcup B(x_i,r_i)$, then $\mu(E) \le \sum \mu(B(x_i,r_i)) \le \sum h(r_i)$.

In order to construct a measure satisfying the opposite inequality we subdivide $\mathbf{R}^N$ for each integer n into a mesh $\mathcal{Q}_n$ of cubes with side 2^{-n}. Here the cubes in each $\mathcal{Q}_n$ are half-open and disjoint, and $\mathcal{Q}_{n+1}$ is obtained from $\mathcal{Q}_n$ by subdividing each cube into 2^N cubes with half the side. Let $\mathcal{Q} = \{\mathcal{Q}_n\}_{n=-\infty}^{\infty}$.

Suppose that $E \subset Q_0 \in \mathcal{Q}_0$. Fix a positive integer n, and let μ_n be a measure such that μ_n has constant density and has mass equal to $h(2^{-n})$ on each $Q_n \in \mathcal{Q}_n$ that intersects E.

We now modify μ_n in the following way. Consider the cubes $Q_{n-1} \in \mathcal{Q}_{n-1}$. If $\mu_n(Q_{n-1}) > h(2^{-n+1})$ for some Q_{n-1} we reduce its mass uniformly on Q_{n-1}

until it equals $h(2^{-n+1})$. If on the other hand $\mu_n(Q_{n-1}) \le h(2^{-n+1})$, we leave μ_n unchanged on Q_{n-1}. This way we obtain a new measure μ_{n1}

We repeat this procedure with μ_{n1}, obtaining μ_{n2}, and after n such steps we have obtained μ_{nn}. Then $\mu_{nn}(Q_\nu) \le h(2^{-\nu})$ for each $Q_\nu \in \mathcal{Q}_\nu$, $\nu = 0, 1, 2, \ldots, n$.

Now let $n \to \infty$. Then $\{\mu_{nn}\}_1^\infty$ has a subsequence that converges weakly to a measure μ, and clearly $\operatorname{supp}\mu \subset E$. Moreover, $\mu(Q_\nu) \le 3^N h(2^{-\nu})$ if $Q_\nu \in \mathcal{Q}_\nu$, $\nu = 0, 1, 2, \ldots$. Indeed, if χ_ν is a continuous function with support in $3Q_\nu$ (the cube concentric to Q_ν with three times the side) such that $0 \le \chi_\nu(x) \le 1$ on $3Q_\nu$ and $\chi_\nu(x) = 1$ on Q_ν, then

$$\mu(Q_\nu) \le \int \chi_\nu \, d\mu = \lim_{i\to\infty} \int \chi_\nu \, d\mu_{n_i n_i} \le 3^N h(2^{-\nu}) \ .$$

On the other hand, for any n each $x \in E$ belongs to some (or several) $Q^{(j)} \in \mathcal{Q}_{n_j}$, $0 \le n_j \le n$, such that $\mu_{nn}(Q^{(j)}) = h(2^{-n_j})$. These cubes are either disjoint or contained in one another. We obtain a disjoint covering, $E \subset \bigcup Q^{(j)}$, such that

$$\mu_{nn}(Q_0) = \sum_j \mu_{nn}(Q^{(j)}) = \sum_j h(2^{-n_j}) \ge \inf \sum_i h(2^{-n_i}) \ ,$$

where the infimum is taken over all finite or denumerable coverings of E with $Q^{(i)} \in \mathcal{Q}$. The right hand side is independent of n, and letting $n \to \infty$ it follows that also

$$\mu(Q_0) = \mu(E) \ge \inf \sum_i h(2^{-n_i}) \ .$$

To finish the proof it remains only to replace the cubes by balls. Suppose that $E \subset \bigcup_{j=1}^\infty Q^{(j)}$, where $Q^{(j)} \in \mathcal{Q}_{n_j}$. Then there is a constant A_N such that each $Q^{(j)}$ is contained in the union of A_N balls with radius 2^{-n_j}. Thus $\Lambda_h^{(\infty)}(E) \le A_N \inf \sum h(2^{-n_j})$, where the infimum is taken as above, and thus

$$\Lambda_h^{(\infty)}(E) \le A_N \mu(E) \ .$$

On the other hand $\mu(Q_N) \le 3^N h(2^{-N})$ for all $Q_n \in \mathcal{Q}_n$. It follows that if $2^{-n} \le r < 2^{-n+1}$, then any ball $B(x, r)$ is contained in the union of 5^N cubes Q_n, and thus

$$\mu(B(x, r)) \le 5^N 3^N h(2^{-n}) \le 5^N 3^N h(r) \ .$$

Thus $\nu = 15^{-N}\mu$ satisfies $\nu(B(x, r)) \le h(r)$ for all balls $B(x, r)$, and $\Lambda_h^{(\infty)}(E) \le A\,\nu(E)$. The theorem is proved.

We can now give a lower estimate for $C_{\alpha,p}$ in terms of Hausdorff content.

Theorem 5.1.13. *Let $p > 1$ and $0 < \alpha p \le N$, and let h be an increasing function on $[0, \infty)$ such that $h(0) = 0$, and*

$$\int_0^1 \left(\frac{h(r)}{r^{N-\alpha p}}\right)^{1/(p-1)} \frac{dr}{r} < \infty \ .$$

Let $E \subset \mathbf{R}^N$ be compact and satisfy $\Lambda_h^{(\infty)}(E) > 0$, choose δ, $0 \le \delta \le 1$, so that

$$h(\delta) \le \Lambda_h^{(\infty)}(E) \ ,$$

and set

$$H = \int_0^\delta \left(\frac{h(r)}{r^{N-\alpha p}} \right)^{p'-1} \frac{dr}{r} + \Lambda_h^{(\infty)}(E)^{p'-1} \int_\delta^1 \left(\frac{1}{r^{N-\alpha p}} \right)^{p'-1} \frac{dr}{r} \ .$$

Then there is a constant $A > 0$, independent of h and E, such that

$$\Lambda_h^{(\infty)}(E) \le A\, H^{p-1} C_{\alpha,p}(E) \ . \tag{5.1.2}$$

In particular

$$C_{\alpha,p}(E) = 0 \implies \Lambda_h(E) = 0.$$

Remark. The theorem can be extended to general sets. In fact, there is a G_δ-set E' such that $E \subset E'$, $C_{\alpha,p}(E) = C_{\alpha,p}(E')$, and $\Lambda_h^{(\infty)}(E) = \Lambda_h^{(\infty)}(E')$. One can prove that the set function $\Lambda_h^{(\infty)}$ satisfies the assumptions in Choquet's theorem, Theorem 2.3.11, and thus $\Lambda_h^{(\infty)}(E') = \sup_{K \subset E'} \Lambda_h^{(\infty)}(K)$, the supremum being taken over compact sets. See e.g. L. Carleson [92], Theorem II.2 or C. A. Rogers [365], Chapter 2:7.

Proof. We prove the theorem by combining Theorem 5.1.12 with Wolff's inequality, Theorem 4.5.2. Let E be compact with $\Lambda_h^{(\infty)}(E) > 0$, and let $\mu \in \mathcal{M}^+(E)$ be a measure as in Theorem 5.1.12 such that $\mu(B(x,r)) \le h(r)$ for all balls, and $A^{-1}\Lambda_h^{(\infty)}(E) \le \mu(E) \le \Lambda_h^{(\infty)}(E)$.

By Theorem 4.5.2

$$\int_{\mathbf{R}^N} (G_\alpha * \mu)^{p'} \, dx \le A \int_{\mathbf{R}^N} W_{\alpha,p}^\mu \, d\mu \ .$$

For all x, defining $W_{\alpha,p}^\mu$ by (4.5.4),

$$W_{\alpha,p}^\mu(x) = \int_0^1 \left(\frac{\mu(B(x,r))}{r^{N-\alpha p}} \right)^{p'-1} \frac{dr}{r} \le H \ .$$

Thus, $\|G_\alpha * \mu\|_{p'} \le A\, H^{1/p'} \mu(E)^{1/p'}$, and Theorem 2.5.1 gives

$$C_{\alpha,p}(E)^{1/p} \ge \frac{\mu(E)}{\|G_\alpha * \mu\|_{p'}} \ge \frac{A\,\mu(E)^{1-1/p'}}{H^{1/p'}} \ge \frac{A\,\Lambda_h^{(\infty)}(E)^{1/p}}{H^{1/p'}} \ ,$$

which finishes the proof.

Corollary 5.1.14. *Let $E \subset \mathbf{R}^N$ be compact. Let $p, q > 1$ and $0 < \beta q \le \alpha p \le N$. Set $h(r) = r^{N-\beta q}$ if $\beta q < N$, and $h(r) = (\log_+ \frac{2}{r})^{1-q}$ if $\beta q = N$. Then there are constants A independent of E such that*

$$\Lambda_h^{(\infty)}(E)^{N-\alpha p} \le A\, C_{\alpha,p}(E)^{N-\beta q}, \quad \textit{if } \beta q < \alpha p < N \ , \tag{5.1.3}$$

$$\left(1 + \log_+ \frac{1}{\Lambda_h^{(\infty)}(E)}\right)^{1-p} \le A\, C_{\alpha,p}(E), \quad \textit{if } \beta q < \alpha p = N \ , \tag{5.1.4}$$

$$\Lambda_h^{(\infty)}(E)^{p-1} \le A\, C_{\alpha,p}(E)^{q-1}, \quad \textit{if } \beta q = \alpha p = N,\ p < q \ . \tag{5.1.5}$$

Proof. The corollary follows from Theorem 5.1.13 by a simple computation. We write $\Lambda_h^{(\infty)}(E) = \Lambda$.

In the first case, with $\delta^{N-\beta q} = \min\{\Lambda, 1\}$, we find

$$H \le A\, \delta^{(\alpha p - \beta q)(p'-1)} + A\, \Lambda^{p'-1} \delta^{-(N-\alpha p)(p'-1)} = A\, \Lambda^{\frac{(\alpha p - \beta q)(p'-1)}{N-\beta q}} \ ,$$

if $\Lambda \le 1$. For $\Lambda \ge 1$, H is bounded independently of Λ. Thus, (5.1.2) gives

$$\Lambda \le A\, \Lambda^{\frac{\alpha p - \beta q}{N - \beta q}} C_{\alpha,p}(E) \ ,$$

and (5.1.3) follows.

In the second case, with δ as above,

$$H \le A\, \delta^{(N-\beta q)(p'-1)} + \Lambda^{p'-1} \log \tfrac{1}{\delta} \le A\, \Lambda^{p'-1}\big(1 + \log \tfrac{1}{\Lambda}\big) \ ,$$

if $\Lambda \le 1$, and thus for all $\Lambda > 0$ by (5.1.2)

$$\Lambda \le A\, \Lambda \big(1 + \log^+ \tfrac{1}{\Lambda}\big)^{p-1} C_{\alpha,p}(E) \ ,$$

which gives (5.1.4).

In the third case, if $h(\delta) = \big(\log \frac{2}{\delta}\big)^{1-q} = \Lambda$, and $\delta \le 1$, we have

$$\begin{aligned} H &= \int_0^\delta \big(\log \tfrac{2}{r}\big)^{(1-q)(p'-1)} \frac{dr}{r} + \Lambda^{p'-1} \int_\delta^1 \frac{dr}{r} \\ &\le A\, \big(\log \tfrac{2}{\delta}\big)^{(p-q)(p'-1)} + \Lambda^{p'-1} \log \tfrac{2}{\delta} = A\, \Lambda^{p'-q'} \ . \end{aligned}$$

Thus, for all $\Lambda > 0$, by (5.1.2)

$$\Lambda \le A\, \Lambda^{(p'-q')(p-1)} C_{\alpha,p}(E) \ ,$$

and this proves (5.1.5).

Corollary 5.1.15. *Let $M \subset \mathbf{R}^N$ be a manifold of dimension $N - d$, where $0 < d < N$. Then $C_{\alpha,p}(M) = 0$ if and only if $\alpha p \le d$.*

The proof is immediate.

5.2 Lipschitz Mappings and Capacities

Here we investigate the behavior of (α, p)-capacity with respect to a Lipschitz map. Our main result is the following theorem.

Theorem 5.2.1. *Let $\alpha > 0$, and $1 < p \leq N/\alpha$. Assume that $E \subset \mathbf{R}^N$, and that $\Phi : E \to \mathbf{R}^N$ is a Lipschitz mapping, i.e. there is a constant L such that Φ satisfies*

$$|\Phi(x) - \Phi(y)| \leq L\,|x - y|$$

for all $x, y \in E$. Then there is a constant A, which depends only on α, p, N, and L, such that

$$C_{\alpha,p}(\Phi(E)) \leq A\, C_{\alpha,p}(E)\ .$$

Remark. It is worth noting that by a theorem of M. D. Kirszbraun [252], any Lipschitz mapping from a subset of $\mathbf{R}^N$ to $\mathbf{R}^N$ can be extended to all of $\mathbf{R}^N$ without increasing the Lipschitz constant. (See also G. J. Minty [329] for a simple proof and references, and H. Federer [141], Theorem 2.10.43, p. 201.) Thus, it would not be any restriction in the above theorem to assume that Φ is defined on $\mathbf{R}^N$.

The proof of Theorem 5.2.1 is based on estimating the "energy" $\|G_\alpha * \mu\|_{p'}^{p'}$ by means of the Wolff inequality, Theorem 4.5.2. We begin by establishing a lemma.

Lemma 5.2.2. *Let $K \subset \mathbf{R}^N$ be compact, and let $\Phi : K \to \mathbf{R}^N$ be continuous. Then, for any measure $\mu \in \mathcal{M}^+(\Phi(K))$ there is a measure $\mu^* \in \mathcal{M}^+(K)$, such that*

$$\int_K f \circ \Phi\, d\mu^* = \int_{\Phi(K)} f\, d\mu \tag{5.2.1}$$

for all $f \in C(\Phi(K))$, i.e., $\mu^*\big(\Phi^{-1}(E)\big) = \mu(E)$ for all Borel $E \subset \Phi(K)$.

Proof. Given a measure $\mu \in \mathcal{M}^+(\Phi(K))$, set

$$p(f) = \|\mu\|_1 \sup_{x \in K} f(x) \quad \text{for } f \in C(K)\ .$$

Then on the subspace $\{\, f_* \circ \Phi : f_* \in C(\Phi(K))\,\}$ of $C(K)$ the linear functional λ, defined by $\lambda(f_* \circ \Phi) = \int_{\Phi(K)} f_*\, d\mu$, satisfies

$$\lambda(f_* \circ \Phi) \leq \|\mu\|_1 \sup_{y \in \Phi(K)} f_*(y) = \|\mu\|_1 \sup_{x \in K} f_* \circ \Phi(x) = p(f_* \circ \Phi)\ .$$

Hence, by the Hahn–Banach theorem (see e.g. Rudin [368], Theorem 3.2), there exists an extension μ^* of λ to the whole of $C(K)$, such that $\mu^*(f) \leq p(f)$ for all f in $C(K)$. But clearly $\mu^* \in \mathcal{M}^+(K)$, since $p(f) \leq 0$ when $f \leq 0$, so $\mu^*(f) = -\mu^*(-f) \geq 0$ for $f \geq 0$. The lemma follows.

Proof of Theorem 5.2.1. We estimate $\int_{\mathbf{R}^N}(G_\alpha * \mu^*)^{p'}\,dx$ using Theorem 4.5.2 and (4.5.4). Writing $y = \Phi(x)$, we have for $\mu \in \mathcal{M}^+(\Phi(K))$,

$$\begin{aligned}\int_{\Phi(K)} W^\mu_{\alpha,p}(y)\,d\mu(y) &= \int_K W^\mu_{\alpha,p}(\Phi(x))\,d\mu^*(x)\\ &= \int_K\int_0^1 \left(\frac{\mu^*\big(\Phi^{-1}(B(\Phi(x),t))\big)}{t^{N-\alpha p}}\right)^{p'-1}\frac{dt}{t}\,d\mu^*(x)\\ &\geq \int_K\int_0^1 \left(\frac{\mu^*(B(x,t/L))}{t^{N-\alpha p}}\right)^{p'-1}\frac{dt}{t}\,d\mu^*(x)\\ &= L^{(\alpha p-N)(p'-1)}\int_K\int_0^{1/L} \left(\frac{\mu^*(B(x,t))}{t^{N-\alpha p}}\right)^{p'-1}\frac{dt}{t}\,d\mu^*(x)\ ,\end{aligned}$$

since $\Phi^{-1}(B(\Phi(x),t)) \supset B(x,t/L)$ for all x and all $t > 0$. If $L \leq 1$ the inner integral is greater than $W^{\mu^*}_{\alpha,p}(x)$, and thus

$$\int_{\Phi(K)} W^\mu_{\alpha,p}(y)\,d\mu(y) \geq L^{(\alpha p-N)(p'-1)}\int_K W^{\mu^*}_{\alpha,p}(x)\,d\mu^*(x)\ , \tag{5.2.2}$$

and by Theorem 4.5.2

$$\|G_\alpha * \mu^*\|_{p'} \leq A\,L^{(N-\alpha p)/p}\,\|G_\alpha * \mu\|_{p'}\ , \tag{5.2.3}$$

where μ and μ^* are related as in Lemma 5.2.2, and A is independent of L. If $L \geq 1$ we have to use Theorem 3.6.2 with $\delta = 1/L$ in combination with Theorem 4.5.2 to conclude that there is a constant $A > 0$ (depending on L) such that

$$\int_K\int_0^{1/L}\left(\frac{\mu^*(B(x,t))}{t^{N-\alpha p}}\right)^{p'-1}\frac{dt}{t}\,d\mu^*(x) \geq A^{-1}\int_K W^{\mu^*}_{\alpha,p}(x)\,d\mu^*(x)\ , \tag{5.2.4}$$

which again implies (5.2.3), but with A depending on L.

Choosing a test measure μ for $C_{\alpha,p}(\Phi(K))$ as in Theorem 2.5.1, we get a measure μ^* with the same total mass testing $C_{\alpha,p}(K)$ with (5.2.3) holding. Hence the result follows for compact sets K. For E the countable union of compact sets, one has a sequence of compact sets $K_j \nearrow E$, and then $\Phi(K_j) \nearrow \Phi(E)$, hence by Proposition 2.3.12, $C_{\alpha,p}(\Phi(E)) \leq A\,C_{\alpha,p}(E)$. For the general case, let G be an open set containing E; then $C_{\alpha,p}(\Phi(E)) \leq C_{\alpha,p}(\Phi(G)) \leq A\,C_{\alpha,p}(G)$. The result is now a consequence of the fact that $C_{\alpha,p}$ is an outer capacity, i.e., Proposition 2.3.5 holds.

Replacing $W^\mu_{\alpha,p}$ in the above proof by $\dot{W}^\mu_{\alpha,p}$, we immediately obtain the following corollary.

Corollary 5.2.3. *Under the assumptions of Theorem 5.2.1, with $p < \alpha/N$, there is a constant A, independent of L, such that*

$$\dot{C}_{\alpha,p}(\Phi(E)) \leq A\,L^{N-\alpha p}\,\dot{C}_{\alpha,p}(E)\ .$$

Theorem 5.2.1 allows us to draw conclusions about the continuity properties of functions in $L^{\alpha,p}$. We will partly anticipate some results proved in Section 6.1.

Let M be an affine subspace contained in $\mathbf{R^N}$, i.e. a translate of a (linear) subspace of $\mathbf{R}^N$. Also set $M^\perp$ to be the largest subspace of $\mathbf{R}^N$ orthogonal to M. For $a \in M$, we write $a + M^\perp$ for the corresponding affine subspace of $\mathbf{R}^N$ orthogonal to M. Since orthogonal projections of $\mathbf{R}^N$ onto such M, denoted P_M, are contractions of $\mathbf{R}^N$ into itself, we can apply Theorem 5.2.1. The result is the following.

Theorem 5.2.4. *Let M be an affine subspace of $\mathbf{R}^N$. For any $f \in L^{\alpha,p}(\mathbf{R}^N)$ there is a set $F \subset M$ such that $C_{\alpha,p}(F) = 0$, and the restriction of $G_\alpha * f$ to $a + M^\perp$ is continuous for all $a \in M \setminus F$.*

Proof. Choose a sequence $\{f_n\}_1^\infty$ in $C_0^\infty(\mathbf{R}^N)$ such that f_n converges to f in $L^p(\mathbf{R}^N)$. Clearly $G_\alpha * f_n \in C(\mathbf{R}^N)$. By Proposition 2.3.8 there is a subsequence $\{f_{n_i}\}_{i=1}^\infty$ such that $G_\alpha * f_{n_i}(x)$ converges to $G_\alpha * f(x)$ (α, p)-q.e. on $\mathbf{R}^N$, uniformly outside an open set G of arbitrarily small (α, p)-capacity. Thus, the convergence is uniform on $a + M^\perp$ for all $a \in M \setminus P_M G$, and by Theorem 5.2.1, $C_{\alpha,p}(P_M G)$ can be made as small as we please. The result follows.

Remark. In order for this theorem to have any content, it is necessary that $C_{\alpha,p}(M) > 0$. By Corollary 5.1.15 this means that we must have $\alpha p > d$, where $N - d$ is the dimension of M.

Corollary 5.2.5. *If $u \in W^{m,p}(\mathbf{R}_+^N)$, $m \in \mathbf{Z}_+$, $mp > 1$, where $\mathbf{R}_+^N$ is the upper half space $\{x : x_N > 0\}$, then the boundary values $\lim_{x_N \to 0} u(x)$ exist (m, p)-q.e. (in the sense of $\mathbf{R}^N$) on $\mathbf{R}^{N-1} = \{x : x_N = 0\}$.*

Proof. This follows immediately through extending u to the lower half space by Theorem 1.2.2.

Remark. More general versions of the last corollary can easily be proved by replacing Theorem 1.2.2 by the more general extension theorem of A. P. Calderón mentioned in Section 1.2. See also Corollary 6.2.3 below.

5.3 The Capacity of Cantor Sets

For Cantor sets it is possible to give a necessary and sufficient condition for their (α, p)-capacity to vanish.

Let $\mathcal{L} = \{l_k\}_{k=0}^\infty$ be a decreasing sequence such that $0 < 2l_{k+1} < l_k$ for $k = 0, 1, \ldots$. Without loss of generality we can assume that $l_0 = 1$. Let E_0 be a closed interval of length l_0, and let E_1 be the set obtained by removing an open interval of length $l_0 - 2l_1$ in the middle, so that E_1 consists of two closed intervals of length l_1. Then remove an interval of length $l_1 - 2l_2$ in the middle of each of these intervals, to obtain E_2 consisting of 2^2 intervals of length l_2. Continuing like this we obtain after k steps a set E_k consisting of 2^k intervals of length l_k. Denote the Cartesian product of N copies of E_k by $E_k^{(N)}$, and set

$$E_{\mathcal{L}} = \bigcap_{k=0}^{\infty} E_k^{(N)} \ . \tag{5.3.1}$$

Then $E_{\mathcal{L}}$ is called the *Cantor set* corresponding to $\mathcal{L}$.

For the sake of comparison we begin by estimating the Hausdorff measure of $E_{\mathcal{L}}$.

Theorem 5.3.1. *Let $\mathcal{L}$ and $E_{\mathcal{L}}$ be as above. Let $h(r)$ be an increasing function, defined for $r \geq 0$, satisfying $h(0) = 0$. Then there is $A > 0$ such that*

$$A^{-1} \liminf_{k\to\infty} 2^{kN} h(l_k) \leq \Lambda_h(E_{\mathcal{L}}) \leq A \liminf_{k\to\infty} 2^{kN} h(l_k) \ .$$

Proof. In one direction the theorem is obvious. In fact, by its construction, $E_{\mathcal{L}}$ can be covered by 2^{kN} cubes of side l_k, and thus by $A\,2^{kN}$ balls of radius l_k. It follows that $\Lambda_h^{(\rho)}(E_{\mathcal{L}}) \leq A\,2^{kN} h(l_k)$, if $\rho \geq l_k$, and thus $\Lambda_h(E_{\mathcal{L}}) \leq A \liminf_{k\to\infty} 2^{kN} h(l_k)$.

In order to prove the converse inequality we assume that

$$\liminf_{k\to\infty} 2^{kN} h(l_k) = c > 0 \ ,$$

and we let k_0 be a number such that $2^{kN} h(l_k) \geq c/2$ for all $k \geq k_0$. As in the easy part of Theorem 5.1.12 it suffices in order to get a lower estimate for $\Lambda_h(E_{\mathcal{L}})$ to find a positive measure μ supported by $E_{\mathcal{L}}$ such that $\mu(B(x,r)) \leq h(r)$, $r \leq \rho$, for some $\rho > 0$. In fact, it then follows that $\mu(E_{\mathcal{L}}) \leq \Lambda_h^{(\rho)}(E_{\mathcal{L}}) \leq \Lambda_h(E_{\mathcal{L}})$.

We let $\mu_{\mathcal{L}}$ be the unit measure on $E_{\mathcal{L}}$ obtained by the following well-known construction (which generalizes the construction of the singular Lebesgue function in one dimension). Define for each k a measure $\mu_k \in \mathcal{M}^+(E_k^{(N)})$, so that $\mu_k(E_k^{(N)}) = 1$, and μ_k distributes the mass 2^{-kN} uniformly on each of the 2^{kN} cubes with side l_k that constitute $E_k^{(N)}$. A subsequence, still denoted $\{\mu_k\}_1^\infty$, converges weakly to a measure $\mu_{\mathcal{L}}$. (It is in fact easy to see that the entire sequence $\{\mu_k\}_1^\infty$ converges weakly to $\mu_{\mathcal{L}}$.)

Then clearly $\mu_{\mathcal{L}} \in \mathcal{M}^+(E_{\mathcal{L}})$, $\mu_{\mathcal{L}}(E_{\mathcal{L}}) = 1$, and for any of the cubes Q_k of $E_k^{(N)}$ we have $\mu_{\mathcal{L}}(Q_k) = 2^{-kN}$. Thus, by the assumption, $\mu_{\mathcal{L}}(Q_k) = 2^N 2^{-(k+1)N} \leq 2^{N+1} c^{-1} h(l_{k+1})$ for $k \geq k_0$.

Now consider any ball $B(x,r)$ with radius r, $r \leq l_{k_0}$. Fix an r and determine k so that $l_{k+1} \leq r < l_k$. Then $B(x,r)$ can intersect at most 3^N of the cubes Q_k from $E_k^{(N)}$, so $\mu_{\mathcal{L}}(B(x,r)) \leq 3^N 2^{N+1} c^{-1} h(l_{k+1}) \leq 3^N 2^{N+1} c^{-1} h(r)$. Thus the measure $\mu = c\,3^{-N} 2^{-N-1} \mu_{\mathcal{L}}$ satisfies $\mu(B(x,r)) \leq h(r)$ for $r \leq l_{k_0}$, and the theorem follows.

The main result of this section is the following theorem.

Theorem 5.3.2. *Let $\mathcal{L}$ and $E_{\mathcal{L}}$ be as above. Then $C_{\alpha,p}(E_{\mathcal{L}}) > 0$ if and only if*

$$\sum_{k=0}^{\infty} \left(2^{-kN} l_k^{\alpha p - N}\right)^{p'-1} < \infty, \quad \text{if } \alpha p < N \ ,$$

and if and only if

$$\sum_{k=0}^{\infty} 2^{-kN(p'-1)} \log \frac{1}{l_k} < \infty, \quad \textit{if } \alpha p = N \ .$$

Remark. If h is defined as in Theorem 5.1.9, i.e., $h(r) = r^{N-\alpha p}$, if $\alpha p < N$, and $h(r) = \left(\log_+ \frac{2}{r}\right)^{1-p}$, if $\alpha p = N$, then Theorem 5.3.1 says that $\Lambda_h(E_{\mathcal{L}}) > 0$ if and only if

$$\limsup_{k\to\infty} 2^{-kN} l_k^{\alpha p-N} < \infty, \quad \text{if } \alpha p < N \ ,$$

and if and only if

$$\limsup_{k\to\infty} 2^{-kN(p'-1)} \log \frac{1}{l_k} < \infty, \quad \text{if } \alpha p = N \ .$$

Before proving the theorem we give it an alternative formulation. Let $h_{\mathcal{L}}$ be an increasing function on $(0, \infty)$ such that

$$h_{\mathcal{L}}(l_k) = 2^{-kN}, \quad k = 0, 1, 2, \dots \ .$$

Theorem 5.3.3. *Let* $0 < \alpha p \le N$. *Then* $C_{\alpha,p}(E_{\mathcal{L}}) > 0$ *if and only if*

$$\int_0^1 \left(\frac{h_{\mathcal{L}}(r)}{r^{N-\alpha p}}\right)^{p'-1} \frac{dr}{r} < \infty \ .$$

Proof of the equivalence of Theorems 5.3.2 and 5.3.3. If $\alpha p < N$ we obtain on the one hand

$$\begin{aligned}\int_0^1 \left(\frac{h_{\mathcal{L}}(r)}{r^{N-\alpha p}}\right)^{p'-1} \frac{dr}{r} &\le \sum_{k=1}^{\infty} 2^{-(k-1)N(p'-1)} \int_{l_k}^{l_{k-1}} r^{(\alpha p-N)(p'-1)} \frac{dr}{r} \\ &\le A \sum_{k=1}^{\infty} \left(2^{-kN} l_k^{\alpha p-N}\right)^{p'-1} \ ,\end{aligned}$$

and on the other

$$\begin{aligned}\int_0^1 \left(\frac{h_{\mathcal{L}}(r)}{r^{N-\alpha p}}\right)^{p'-1} \frac{dr}{r} &\ge \sum_{k=1}^{\infty} 2^{-kN(p'-1)} \int_{l_k}^{l_{k-1}} r^{(\alpha p-N)(p'-1)} \frac{dr}{r} \\ &= A \sum_{k=1}^{\infty} 2^{-kN(p'-1)} \left(l_k^{(\alpha p-N)(p'-1)} - l_{k-1}^{(\alpha p-N)(p'-1)}\right) \\ &\ge A \sum_{k=1}^{\infty} \left(2^{-kN} l_k^{\alpha p-N}\right)^{p'-1} \ .\end{aligned}$$

Here the last inequality follows from a summation by parts. The proof for $\alpha p = N$ is similar.

One of the implications in Theorem 5.3.3 depends on the following result, which is interesting in its own right.

Theorem 5.3.4. *Let $E \subset \mathbf{R}^N$ be compact (or Suslin). Let $\mathcal{A}(r)$ be the minimal number of balls of radius r required to cover E. Then*

$$C_{\alpha,p}(E) \leq A\left(\int_0^1 \left(\mathcal{A}(r)r^{N-\alpha p}\right)^{1-p'} \frac{dr}{r}\right)^{1-p} ,$$

where A is a constant independent of E. In particular, $C_{\alpha,p}(E) = 0$, if

$$\int_0^1 \left(\mathcal{A}(r)r^{N-\alpha p}\right)^{1-p'} \frac{dr}{r} = \infty .$$

Proof. Let $\mu \in \mathcal{M}^+(E)$ be such that $\|G_\alpha * \mu\|_{p'} < \infty$. As in the proof of Corollary 3.6.3 we have

$$(G_\alpha * \mu(x))^{p'} \geq A\,\mathcal{I}_{\alpha,1}\mu(x)^{p'} \geq A\sum_{k=1}^{\infty}\left(2^{k(N-\alpha)}\mu(B(x,2^{-k}))\right)^{p'} .$$

Thus

$$\|G_\alpha * \mu\|_{p'}^{p'} \geq A\sum_{k=1}^{\infty} 2^{k(N-\alpha)p'} \int_{\mathbf{R}^N} \left(\mu(B(x,2^{-k}))\right)^{p'} dx .$$

Suppose that $E \subset \bigcup_{j=1}^{\mathcal{A}_k} B_k^{(j)}$, where $\mathcal{A}_k = \mathcal{A}(2^{-k})$, and $B_k^{(j)}$ are balls of radius 2^{-k}. Then there is a number A_N such that no point belongs to more than A_N of these balls. A_N can be chosen as the minimal number of balls of radius 1 that are needed to cover a ball of radius 2. In fact, all balls of radius 2^{-k} that contain a point x are contained in the ball $B(x, 2^{-k+1})$, which can be covered by A_N balls of radius 2^{-k}. Thus, if x were contained in $A_N + 1$ of the balls $B_k^{(j)}$, this would contradict the minimality of $\mathcal{A}_k$. It follows that

$$\begin{aligned}\int_{\mathbf{R}^N} \left(\mu(B(x,2^{-k})\right)^{p'} dx &\geq A_N^{-1}\sum_{j=1}^{\mathcal{A}_{k+1}} \int_{B_{k+1}^{(j)}} \left(\mu(B(x,2^{-k})\right)^{p'} dx \\ &\geq A\,A_N^{-1}2^{-kN}\sum_{j=1}^{\mathcal{A}_{k+1}}\left(\mu(B_{k+1}^{(j)})\right)^{p'} .\end{aligned}$$

But by Hölder's inequality

$$\mu(E) \leq \sum_{j=1}^{\mathcal{A}_{k+1}} \mu(B_{k+1}^{(j)}) \leq \left(\sum_{j=1}^{\mathcal{A}_{k+1}}\left(\mu(B_{k+1}^{(j)})\right)^{p'}\right)^{1/p'} \mathcal{A}_{k+1}^{1/p} .$$

Thus

$$\sum_{j=1}^{\mathcal{A}_{k+1}}\left(\mu(B_{k+1}^{(j)})\right)^{p'} \geq \mu(E)^{p'}\mathcal{A}_{k+1}^{1-p'} ,$$

and

$$\|G_\alpha * \mu\|_{p'}^{p'} \geq A\,\mu(E)^{p'} \sum_{k=1}^{\infty} 2^{k(N-\alpha)p'} 2^{-kN} \mathcal{A}_{k+1}^{1-p'}$$

$$= A\,\mu(E)^{p'} \sum_{k=1}^{\infty} \left(2^{k(N-\alpha p)} \mathcal{A}_{k+1}\right)^{1-p'} .$$

For $2^{-k-1} \leq r \leq 2^{-k}$ we clearly have $\mathcal{A}(r) \geq \mathcal{A}_k \geq A_N^{-1} \mathcal{A}_{k+1}$. It follows that

$$\frac{\|G_\alpha * \mu\|_{p'}^{p'}}{\mu(E)^{p'}} \geq A \int_0^1 \left(\mathcal{A}(r) r^{N-\alpha p}\right)^{1-p'} \frac{dr}{r} ,$$

and the result follows from Theorem 2.5.1.

Proof of Theorem 5.3.3. It is possible to cover $E_\mathcal{L}$ by 2^{kN} balls of radius $\frac{1}{2} l_k \sqrt{N}$, since $E_k^{(N)}$ consists of 2^{kN} cubes of side l_k. Thus $\mathcal{A}(r) \leq 2^{kN} = h_\mathcal{L}(l_k)^{-1}$ if $\frac{1}{2} l_k \sqrt{N} \leq r \leq \frac{1}{2} l_{k-1} \sqrt{N}$. It follows as in Theorem 5.3.3 that

$$\int_0^{\frac{1}{2}\sqrt{N}} \left(\mathcal{A}(r) r^{N-\alpha p}\right)^{1-p'} \frac{dr}{r} \geq A \int_0^1 \left(\frac{h_\mathcal{L}(r)}{r^{N-\alpha p}}\right)^{p'-1} \frac{dr}{r} .$$

Together with Theorem 5.3.4 this shows that $C_{\alpha,p}(E_\mathcal{L}) = 0$ if the integral in the theorem diverges.

In the converse direction the theorem follows easily from Wolff's inequality (Theorem 4.5.2). Let $\mu_\mathcal{L}$ be the measure constructed in the proof of Theorem 5.3.1, so that $\mu_\mathcal{L} \in \mathcal{M}^+(E_\mathcal{L})$, $\mu_\mathcal{L}(E_\mathcal{L}) = 1$, and for any of the cubes Q of $E_k^{(N)}$ we have $\mu_\mathcal{L}(Q) = 2^{-kN}$. It follows that for any ball $B(x, r)$ with radius $l_{k+1} \leq r \leq l_k$ we have $\mu_\mathcal{L}(B(x, r)) \leq 2^N 2^{-kN} = 2^{2N} h_\mathcal{L}(l_{k+1}) \leq 2^{2N} h_\mathcal{L}(r)$. The result now follows from Theorem 4.5.2 as in the proof of Theorem 5.1.13.

5.4 Sharpness of Comparison Theorems

In this section we show that Theorems 5.1.9 and 5.1.13 are sharp.

Theorem 5.4.1. *Let* $0 < \alpha p \leq N$, $c(r) = r^{N-\alpha p}$ *if* $\alpha p < N$, *and* $c(r) = \left(\log_+ \frac{2}{r}\right)^{1-p}$ *if* $\alpha p = N$. *Let* h *be an increasing positive function on* $(0, \infty)$ *such that*

$$\liminf_{r \to 0} \frac{h(r)}{c(r)} = 0 .$$

Then there is a compact set $E \subset \mathbf{R}^N$ *such that* $\Lambda_h(E) = 0$ *and* $C_{\alpha,p}(E) > 0$.

Proof. We only consider the case $\alpha p < N$, since the case $\alpha p = N$ is analogous. It follows from Theorems 5.3.1 and 5.3.2 that it suffices to construct a sequence of positive numbers $\{l_k\}_0^\infty$ such that

(a) $0 < 2l_{k+1} < l_k$;
(b) $\liminf_{k\to\infty} 2^{kN} h(l_k) = 0$;
(c) $\sum_{k=0}^{\infty} \left(2^{-kN} l_k^{\alpha p - N}\right)^{p'-1} < \infty$.

Choose a sequence $\{a_i\}_0^\infty$ such that $a_0 = 1$, $0 < 2a_{i+1} < a_i$, and

$$\frac{h(a_i)}{a_i^{N-\alpha p}} < 2^{-2i} \ . \tag{5.4.1}$$

Then define integers $m(i)$ so that

$$2^{-(m(i)+1)N} < 2^{-i} a_i^{N-\alpha p} \le 2^{-m(i)N} \ . \tag{5.4.2}$$

Let β be such that $1 < \beta < N/(N - \alpha p)$, and set $N - (N - \alpha p)\beta = \gamma$. Define the sequence $\{l_k\}_0^\infty$ by

$$l_k = a_i 2^{(m(i)-k)\beta} \quad \text{for } m(i) \le k < m(i+1), \quad i = 0, 1, \dots \ .$$

By the choice of β we can clearly also assume that $a_i 2^{(m(i)-m(i+1))\beta} > a_{i+1}$.

Then (a) is satisfied, since $\beta > 1$, and (b) is satisfied, since

$$h\big(l_{m(i)}\big)2^{m(i)N} = h(a_i)2^{m(i)N} \le h(a_i)2^i a_i^{\alpha p-N} \le 2^{-i} \ .$$

Here we have used the second inequality in (5.4.2), and (5.4.1).

Finally, (c) is implied by the estimate

$$\begin{aligned}\sum_{k=m(i)}^{m(i+1)-1} \left(2^{-kN} l_k^{\alpha p-N}\right)^{p'-1} &\le \left(a_i 2^{m(i)\beta}\right)^{(\alpha p-N)(p'-1)} \sum_{k=m(i)}^{\infty} 2^{-k\gamma(p'-1)} \\ &\le A\left(a_i^{\alpha p-N} 2^{-m(i)N}\right)^{p'-1} \le A\, 2^{(N-i)(p'-1)} \ ,\end{aligned}$$

where the final inequality is the first inequality in (5.4.2).

Theorem 5.4.2. *Let h be an increasing nonnegative function on $[0, \infty)$, and let $0 < \alpha p \le N$. If*

$$\int_0^1 \left(\frac{h(r)}{r^{N-\alpha p}}\right)^{p'-1} \frac{dr}{r} = \infty \ ,$$

then there is a compact set $E \subset \mathbf{R}^N$ such that $\Lambda_h(E) > 0$ and $C_{\alpha,p}(E) = 0$.

Proof. This is a corollary of Theorem 5.3.3. By Proposition 5.1.8 we can assume that $h(2r) \le 2^N h(r)$ for all r. It is no loss of generality to assume that the inequality is strict. We define a sequence $\mathcal{L} = \{l_k\}_0^\infty$ such that $h(l_k) = 2^{-kN}$. It follows that $2l_{k+1} < l_k$, so there is a Cantor set $E_\mathcal{L}$ as in (5.3.1). Then $C_{\alpha,p}(E_\mathcal{L}) = 0$ by Theorem 5.3.3, but $\Lambda_h^{(\infty)}(E_\mathcal{L}) \ge 2^{-2N}\mu_\mathcal{L}(E_\mathcal{L}) > 0$, if $\mu_\mathcal{L}$ is the measure constructed in the proof of Theorem 5.3.1.

5.5 Relations Between Different Capacities

We shall show that there is a natural ordering in the two-parametric family of (α, p)-capacities, $0 < \alpha p \le N$, and that no two capacities in the family are equivalent. More precisely we have the following theorem.

Theorem 5.5.1. *Let $E \subset \mathbf{R}^N$ be an arbitrary set with $\operatorname{diam} E \le 1$. There are constants A such that*

(a)

$$C_{\beta,q}(E)^{1/(N-\beta q)} \le A\, C_{\alpha,p}(E)^{1/(N-\alpha p)} \quad \textit{for } 0 < \beta q < \alpha p < N\,;$$

(b)

$$C_{\beta,q}(E) \le A\, C_{\alpha,p}(E) \quad \textit{for } \beta q = \alpha p < N, \quad p < q\,;$$

(c)

$$\left(\log \frac{A}{C_{\beta,q}(E)}\right)^{-1} \le A\, C_{\alpha,p}(E)^{p'-1} \quad \textit{for } 0 < \beta q < \alpha p = N\,;$$

(d)

$$C_{\beta,q}(E)^{q'-1} \le A\, C_{\alpha,p}(E)^{p'-1} \quad \textit{for } \beta q = \alpha p = N, \quad p < q\,.$$

If the Bessel capacity is replaced by Riesz capacity in (a) *and* (b), *the restriction on the diameter of E can be removed.*

Moreover, in all cases (a), (b), (c), *and* (d) *there exist sets E such that $C_{\beta,q}(E) = 0$ but $C_{\alpha,p}(E) > 0$.*

Remark. Taking E to be a ball with radius r one sees that the exponents in the theorem are the right ones. See Propositions 5.1.2, 5.1.3, and 5.1.4.

Proof. It is no restriction to prove the theorem for compact sets. This is because of the capacitability of open sets and the fact that our capacities are outer.

In all cases except (b) the inequalities follow directly from Theorem 5.1.9 and Corollary 5.1.14.

To prove the theorem in the case (b) we let $\mu \in \mathcal{M}^+(E)$ be the (β, q)-capacitary measure for E. Then $\mu(E) = C_{\beta,q}(E)$ and $V^{\mu}_{\beta,q}(x) \le 1$ on $\operatorname{supp}\mu$. By the remark following Proposition 2.6.9, and Definition 4.5.1, we have

$$W^{\mu}_{\beta,q}(x) = \sum_{n=0}^{\infty} \left(2^{n(N-\beta q)}\mu(B_n(x))\right)^{q'-1} \le A\, V^{\mu}_{\beta,q}(x) \le A\,.$$

The fact that $(\sum a_n^s)^{1/s}$ is a decreasing function of s for $s > 0$ (cf. (2.6.5)) gives

$$\left(\sum_{n=0}^{\infty} \left(2^{n(N-\beta q)}\mu(B_n(x))\right)^{q'-1}\right)^{q-1} \ge \left(\sum_{n=0}^{\infty} \left(2^{n(N-\alpha p)}\mu(B_n(x))\right)^{p'-1}\right)^{p-1}$$

and thus $W^{\mu}_{\alpha,p}(x) \le A$ on supp μ. It follows from Wolff's inequality (Theorem 4.5.2) that

$$\frac{\mu(E)}{\|G_\alpha * \mu\|_{p'}} \ge A\mu(E)^{1/p} ,$$

whence by Theorem 2.5.1

$$A\, C_{\alpha,p}(E) \ge \mu(E) = C_{\beta,q}(E) .$$

This proves (b).

To prove the final statement of the theorem it is enough to apply Theorem 5.3.2. In cases (a) and (c) we choose l_k so that $l_k^{N-\beta q} = 2^{-kN}$, and in case (d) we choose l_k so that $\log(1/l_k) = 2^{kN(q'-1)}$. Then $C_{\beta,q}(E_{\mathcal{L}}) = 0$ but $C_{\alpha,p}(E_{\mathcal{L}}) > 0$.

The case (b) is again somewhat more delicate. But if we construct $\mathcal{L}$ so that

$$\sum_{k=0}^{\infty} \left(2^{-kN} l_k^{\beta q-N}\right)^{q'-1} = \infty ,$$

and

$$\sum_{k=0}^{\infty} \left(2^{-kN} l_k^{\alpha p-N}\right)^{p'-1} < \infty ,$$

then $C_{\beta,q}(E_{\mathcal{L}}) = 0$, and $C_{\alpha,p}(E_{\mathcal{L}}) > 0$, and in order to achieve this all we have to do is to choose l_k so that

$$\left(2^{-kN} l_k^{\beta q-N}\right)^{q'-1} = \frac{1}{k} .$$

Then

$$\left(2^{-kN} l_k^{\alpha p-N}\right)^{p'-1} = \left(2^{-kN} l_k^{\beta q-N}\right)^{p'-1} = k^{-\frac{p'-1}{q'-1}} ,$$

and $p' - 1 > q' - 1$. This completes the proof of Theorem 5.5.1.

Remark. An alternative proof of (a), (c), (d), that does not depend on Theorems 5.1.12 and 5.1.13, follows simply by noting that if the potential $V^{\mu}_{\beta,q}(x)$ is bounded, then the potential $W^{\mu}_{\beta,q}(x)$ is bounded. Indeed, if μ is the (β, q)-capacitary measure for E, and $\beta q < N$, we have by Proposition 2.6.9

$$\int_0^2 \left(\frac{\mu(B(x,r))}{r^{N-\beta q}}\right)^{q'-1} \frac{dr}{r} \le A\, V^{\mu}_{\beta,q}(x) \le A ,$$

for $x \in E$. This implies that

$$\left(\frac{\mu(B(x,r))}{(2r)^{N-\beta q}}\right)^{q'-1} \int_r^{2r} \frac{dt}{t} \le A \qquad \text{for all } r \le 1 ,$$

and thus $\mu(B(x,r)) \le A\, r^{N-\beta q}$. For $\beta q = N$ we have similarly

$$\int_0^2 \mu(B(x,r))^{q'-1} \frac{dr}{r} \le A\, V^{\mu}_{\beta,q}(x) \le A$$

for $x \in E$, which gives

$$\mu(B(x,r))^{q'-1}\int_r^2 \frac{dt}{t} \le A \quad \text{for } r \le 1 \ ,$$

and thus $\mu(B(x,r)) \le A\big(\log \frac{2}{r}\big)^{1-q}$. In both cases the result now follows if $W^{\mu}_{\alpha,p}(x)$ is estimated by the same computation as in the proof of Corollary 5.1.14.

5.6 Further Results

5.6.1. Let $\alpha > 0$, $1 < p < N/\alpha$. Then there is a constant A, such that for all $E \subset \mathbf{R}^N$

$$\dot{C}_{\alpha,p}(E) \le C_{\alpha,p}(E) \le A\left(\dot{C}_{\alpha,p}(E) + \dot{C}_{\alpha,p}(E)^{N/(N-\alpha p)}\right) \ .$$

See D. R. Adams [8]. His proof of the right inequality depends on the "smooth truncation" results in Section 3.3. The following simpler proof was shown to the authors by Yu. V. Netrusov.

Let $\dot{C}_{\alpha,p}(E) = \eta > 0$. Then there is $f \in L^p(\mathbf{R}^N)$ such that $\|f\|_p^p \le 2\eta$, and $I_\alpha * f \ge \chi_E$. Denote by I'_α the function defined by $I'_\alpha(x) = 0$ for $|x| < 1$, and $I'_\alpha(x) = I_\alpha(x)$ for $|x| \ge 1$. Then, by (1.2.10) and (1.2.11), there is $c > 0$ such that $I_\alpha \le cG_\alpha + I'_\alpha$. If $E_1 = \{x : cG_\alpha * f \ge \frac{1}{2}\}$, and $E_2 = \{x : I'_\alpha * f \ge \frac{1}{2}\}$, this implies that $E \subset E_1 \cup E_2$. It is easy to show that there is $c_1 > 0$ such that $G_\alpha * I'_\alpha \ge c_1 I'_\alpha$. Thus $E_2 \subset E_3 = \{x : G_\alpha * (I'_\alpha * f) \ge \frac{1}{2}c_1\}$. From the facts that $I'_\alpha * f$ is continuous, and $\|G_\alpha\|_1 < \infty$, it follows that there are positive constants c_2 and c_3, such that $E_3 \subset E_4 = \{x : G_\alpha * ((I'_\alpha * f)\chi_{E_5}) \ge c_2\}$, where $E_5 = \{x : I'_\alpha * f \ge c_3\}$. To finish the proof it is enough to estimate $\|(I'_\alpha * f)\chi_{E_5}\|_p$. Applying the Sobolev inequality with $1/p^* = 1/p - \alpha/N$, we find $\|(I'_\alpha * f)\chi_{E_5}\|_p^p \le c_3^{p-p^*}\|I'_\alpha * f\|_{p^*}^{p^*} \le A\|f\|_p^{p^*} \le A\eta^{p^*/p} = A\eta^{N/(N-\alpha p)}$, and the desired inequality follows.

5.6.2. The capacities $\dot{C}_{\alpha,p}$ fail to be additive on disjoint sets. Indeed, the open and closed balls as well as their boundary, the unit sphere, have the same $\dot{C}_{1,2}$ capacity. It is thus of some interest to know what kind of decompositions of E into countable disjoint unions E_k are possible so that

$$\dot{C}_{\alpha,p}(E) \ge A\sum_k \dot{C}_{\alpha,p}(E_k)$$

for some constant A independent of the decomposition. This property is known as quasiadditivity. A related problem was studied by M. Brelot [76]. In N. S. Landkof [266] (Lemma 5.5), and Adams [8], it is shown that E_k can be taken to be the part of E contained in spherical shells centered at a fixed point. An extension of these results appears in H. Aikawa [31], where the fixed point is replaced by a fixed set of dimension less than $n - \alpha p$ for $\alpha p < n$. See also H. Aikawa [32], and H. Aikawa and A. A. Borichev [33].

5.6.3. By Frostman's theorem, Theorem 5.1.12, we have for compact $K \subset \mathbf{R}^N$ and $0 < \alpha < N$

$$\Lambda_{N-\alpha}^{(\infty)}(K) > 0 \qquad \Longleftrightarrow \qquad \exists \mu \in \mathcal{M}^+(K),\ 0 < \|M_\alpha \mu\|_\infty < \infty\ .$$

On the other hand, by Corollary 3.6.3 we have for $p > 1$, $0 < \alpha p \leq N$

$$C_{\alpha,p}(K) > 0 \qquad \Longleftrightarrow \qquad \exists \mu \in \mathcal{M}^+(K),\ 0 < \|M_\alpha \mu\|_{p'} < \infty\ .$$

Thus, in some sense the counterpart of $C_{\alpha,p}$ for $p = 1$ is $\Lambda_{N-\alpha}^{(\infty)}$. This point of view is pursued in D. R. Adams [13]. See also Sections 6.5.1, and 7.6.9–7.6.11 below.

5.6.4. Theorem 5.4.1 can be modified to show that for $0 < \alpha p \leq N$ there exists a compact $E \subset \mathbf{R}^N$ such that $C_{\alpha,p}(E) > 0$ but $\Lambda_h(E) = 0$ for every increasing nonnegative h such that

$$\int_0^1 \left(\frac{h(r)}{r^{N-\alpha p}} \right)^{p'-1} \frac{dr}{r} < \infty\ .$$

Thus no complete description of (α, p)-capacity in terms of Hausdorff measure is possible. See L. Carleson [92], Theorem IV:5, for the classical case.

5.6.5. There is a more general result than Theorem 5.4.2, that goes back to S. J. Taylor [395]. A set E is said to have positive *lower spherical h-density* at a point x if

$$\liminf_{r \to \infty} \frac{\Lambda_h(B(x,r) \cap E)}{h(2r)} > 0\ .$$

Let h be continuous, increasing, and satisfy

$$\int_0^1 \left(\frac{h(r)}{r^{N-\alpha p}} \right)^{p'-1} \frac{dr}{r} = \infty\ .$$

Let E be a Borel set such that E has positive lower spherical h-density at every $x \in E$, and such that $0 < \Lambda_h(E) < \infty$. Then, if μ is defined by $\mu(A) = \Lambda_h(A \cap E)$, it follows that $W_{\alpha,p}^{\mu}(x) = \infty$ on E. We shall prove later (Theorem 6.3.12) that this implies $C_{\alpha,p}(E) = 0$. The set $E_{\mathcal{L}}$ constructed in the proof of Theorem 5.4.2 is an example of a set satisfying the conditions. See H. Wallin [424] for the case $\alpha = 1$, $p = N$.

5.6.6. An alternative approach to Theorem 5.5.1(b) is via the Morrey–Campanato spaces (see Section 3.7.3). Here the key is to note that if μ^K is the (α, p)-capacitary measure of Theorem 2.2.7 for a compact set $K \subset \mathbf{R}^N$, then $G_\alpha * \mu^K \in \mathcal{L}^{p',\alpha p}$ for $\alpha p \leq N$; see D. R. Adams [10]. Hence, Theorem 3.1.6 implies that $G_\beta * \mu^K \in L^{q'}$, $\alpha p = \beta q < N$, $0 < \alpha < \beta$. Actually, one can even get $G_\beta * \mu^K \in \mathcal{L}^{q',\beta q}$; see D. R. Adams [5].

5.6.7. We have seen in Proposition 4.4.4 that the "Lizorkin–Triebel capacities" $C(\cdot\,; F_\alpha^{p,q})$ are equivalent to (α, p)-capacity for all $q > 1$. For the "Besov capacities" $C(E; B_\alpha^{p,q})$ the situation is different, and relations between different Besov capacities have been investigated by D. R. Adams [11, 15], and Yu. V. Netrusov [342], [344], and [345] (Proposition 1.1).

The result is the following: If $C(E; B_\alpha^{p,q})$, $1 < p, q < \infty$, is the "Besov capacity", defined in Definition 4.4.2, then $C(E; B_\alpha^{p,q}) = 0$ implies $C(E; B_\beta^{r,s}) = 0$ if and only if

(a) $\beta r < \alpha p$;
(b) $\beta r = \alpha p = N$, with $\frac{p}{q'} < \frac{r}{s'}$, or $\frac{p}{q'} = \frac{r}{s'}$, when $r \le p$;
(c) $\beta r = \alpha p < N$, with $\frac{p}{q'} \le \frac{r}{s'}$, when $r \le p$, and $\frac{q}{p} \le \frac{s}{r}$, when $p \le r$.

Netrusov [342, 344], has also studied the Besov capacities $C(\cdot\,, B_\alpha^{p,q})$ for $0 < p < \infty$, $0 < q \le \infty$. He has obtained an equivalence of $C(\cdot\,, B_\alpha^{p,q})$ with an interesting variant of Hausdorff content (see Section 5.1) when $0 < p < \infty$ and $0 < q < 1$. In particular, he shows that $C(\cdot\,, B_{d/p}^{p,\theta p})$ and $C(\cdot\,, B_{d/r}^{r,\theta r})$ are equivalent when $0 < p < r < \infty$, $0 < \theta < \infty$, $0 < d < N$, and $\theta r < 1$.

5.7 Notes

5.1. Most of the results in Sections 5.1, 5.3, and 5.4, are extensions to the nonlinear case of results exposed in the case $p = 2$ in L. Carleson [92] and in S. J. Taylor [395]. We refer to these sources for references to the older literature. More information on Hausdorff measures is found in e.g. C. A. Rogers [365] and K. J. Falconer [139], and in the more recent texts by W. P. Ziemer [438] and by L. C. Evans and R. F. Gariepy [137]. Proposition 5.1.8 was shown to us by Yu. V. Netrusov.

Propositions 5.1.3, and 5.1.4, and Theorem 5.1.9 are due to N. G. Meyers [318]. Theorem 5.1.12 was proved by O. Frostman in his thesis [158], and used by him to prove Theorem 5.1.13 for $p = 2$. In our situation, Theorem 5.1.13 is due to V. P. Havin and V. G. Maz′ya [203]. A weaker lower estimate was proved by Yu. G. Reshetnyak [362]; see also the book by V. M. Gol′dshteĭn and Yu. G. Reshetnyak [184].

5.2. In spite of its simplicity, Theorem 5.2.1 (which seems to be published here for the first time) has a long history. It is well known that classical capacities decrease under contractions, i.e. Lipschitz mappings with Lipschitz constant ≤ 1. See G. C. Evans [135], p. 232, M. Brelot [73], and N. Aronszajn and K. T. Smith [40], p. 440. An elegant proof in a general situation was given by B. Fuglede [171]. Our proof follows [171], the main difference being that instead of estimating the energy $\|G_\alpha * \mu\|_{p'}^{p'} = \int V_{\alpha,p}^\mu \, d\mu$, which is difficult, we estimate the equivalent quantity $\int W_{\alpha,p}^\mu \, d\mu$.

Whether the capacities $C_{\alpha,p}$ or $\dot{C}_{\alpha,p}$ actually decrease under contractions for $p \ne 2$ seems to be an open question. In the case when the mapping is an orthogonal projection this was established at about the same time by N. G. Meyers [319] and Yu. G. Reshetnyak [363]. As a consequence both these authors obtained the limit theorem, Theorem 5.2.4. Their proofs, as well as those of Evans and Aronszajn–Smith, were also based on Lemma 5.2.2, which Aronszajn–Smith and Meyers proved by means of the Hahn–Banach theorem as above, whereas Evans and Reshetnyak gave constructive proofs. Other examples of contractions that decrease (α, p)-capacity have been given by H. Aikawa [29].

Note that our proof shows that the equivalent (α, p)-capacities defined by means of an energy $\int W^{\mu}_{\alpha,p}\, d\mu$ decrease under all contractions. The value of the constant in Theorem 5.2.1 is of course without importance if one is only interested in the consequence that null sets are carried into null sets.

Corollary 5.2.5 extends a result of A. Beurling [58], and J. Deny [119], p. 175. It is interesting that this problem led Beurling to introduce the notion of outer capacity (also introduced at about the same time by others); cf. [119], p. 176. See also L. Carleson [92], Theorem V.3, p. 55, and H. Wallin [423].

5.3. Theorems 5.3.2, 5.3.3, and 5.3.4 are due to Havin and Maz′ya [203]. Theorem 5.3.2 extends a theorem of M. Ohtsuka [352]. Theorem 5.3.1 was proved (differently, and under a restriction on the growth of h; cf. Proposition 5.1.8) by A. Besicovitch and S. J. Taylor [54] for $N = 1$, and their proof was extended to arbitrary N by D. R. Adams and N. G. Meyers [23].

5.4. Theorem 5.4.1 in the generality given here is due to K. Hatano [198], who obtained it as a consequence of an extension of Theorem 5.3.2 to more general sets of Cantor type. It was pointed out to the authors by Yu. V. Netrusov that the result follows already from Theorems 5.3.1 and 5.3.2. The case $p = 2$ is in Carleson [92], Theorem IV:4, and in S. J. Taylor [395], and for $p > 2$ the result is found in H. Wallin [424], Theorem 4.4 and Remark 4.2. H. Aikawa [30] has proved similar results for more general kernels by replacing Wolff's inequality by an inequality of R. Kerman and E. T. Sawyer [245]. Theorem 5.4.2 is due to Havin and Maz′ya [203].

5.5. In the case when $p = 2$ or $q = 2$, i.e., when one of the capacities is classical, most of the results of Theorem 5.5.1 are due to B. Fuglede [160], Theorem A, p. 198, with proofs and references to older work given in Fuglede [162]. Here sets of (α, p)-capacity zero were seen as sets of infinities of Riesz potentials $I_\alpha * f$ for $f \in L^p_+$. Inspired by J. Serrin [374] (see the notes to Chapter 2), H. Wallin [422] compared the capacity $C'_{1,p}$ to classical capacities, and found similar results in this special case. The relations between general (α, p)-capacities were clarified by Adams and Meyers [23]. Theorem 5.5.1 includes improvements from Adams and Hedberg [20], and Adams [10].

6. Continuity Properties

A statement such as "f belongs to an L^p space" can be understood in different ways. The strict interpretation is that f is an equivalence class of functions, the equivalence relation being equality almost everywhere. But one can also think of some representative of this equivalence class, perhaps defined at all points outside a set of measure zero. In particular, if a continuous function is identified with an element in L^p, the identification means that one representative of the corresponding equivalence class is singled out, and this distinguished element is usually thought of as belonging to L^p.

The same is true if f is an element of one of the function spaces studied in this book, $f \in L^{\alpha,p}(\mathbf{R}^N)$, say. However, if $\alpha p > N$, the S. L. Sobolev imbedding theorem (Theorem 1.2.4) tells us that every equivalence class contains a continuous function. Even when $\alpha p \leq N$, there are trace theorems of Sobolev and others that give the existence of distinguished elements in the equivalence classes, so that restrictions to some sets of zero Lebesgue measure, such as submanifolds of $\mathbf{R}^N$, can be defined. See e.g. Sobolev [385], and R. A. Adams [26].

In Section 6.1 we show that in the case $\alpha p \leq N$ the equivalence classes have representatives with a property that is known as *quasicontinuity*, and we show that the notion of trace is meaningful on arbitrary sets of positive (α, p)-capacity. In Section 6.2 we investigate the Lebesgue points of functions in $L^{\alpha,p}$, and show that for the quasicontinuous representatives quasi all points are Lebesgue points in a strong sense.

In addition to the quasicontinuity studied in Section 6.1, functions in $L^{\alpha,p}$ enjoy another, subtler continuity property, called *fine continuity*. This is a concept with deep roots in classical potential theory, and it is central in axiomatic potential theory.

Fine continuity is closely related to the concept of a *thin set*, which is the subject of Section 6.3. We review some of the basic properties of thin sets in classical potential theory, and we then generalize the theory to the L^p situation. The main result of the section is Theorem 6.3.11, which gives necessary and sufficient conditions for sets to be (α, p)-thin, analogous to Brelot's characterization in the classical case (Theorem 6.3.2). Extensions of the so called Kellogg and Choquet properties follow easily. It is interesting that in order to obtain a satisfactory theory for (α, p)-thin sets, one has to go beyond the spaces $L^{\alpha,p}$, and apply the potential theory for the Besov spaces $B_\alpha^{p,p}$ developed in Section 4.4.

The results of Section 6.3 are then applied in Section 6.4 to extend a number of classical results about fine continuity and fine topology.

6.1 Quasicontinuity

We first make a general definition.

Definition 6.1.1. Let C be a capacity on $\mathbf{R}^N$, and let the function f be defined C-quasieverywhere on $\mathbf{R}^N$ or on some open subset. Then f is said to be C-quasicontinuous if for every $\varepsilon > 0$ there is an open set G such that $C(G) < \varepsilon$ and $f|_{G^c} \in C(G^c)$. In other words, the restriction of f to the complement of G is continuous in the induced topology.

We already know that if $f \in L^{\alpha,p}(\mathbf{R}^N)$, then f can be represented as $f = G_\alpha * g$, $g \in L^p$. The following result was essentially proved in Chapter 2.

Proposition 6.1.2. *If $g \in L^p(\mathbf{R}^N)$, $1 < p < \infty$, then the potential $G_\alpha * g$, $\alpha > 0$, is (α, p)-quasicontinuous. Thus every element in $L^{\alpha,p}(\mathbf{R}^N)$ has an (α, p)-quasicontinuous representative.*

Proof. By Proposition 2.3.7 we know that $G_\alpha * g(x)$ is well defined and finite (α, p)-q.e. Let $\{g_i\}_1^\infty$ be a sequence of functions in C_0^∞ that converges to g in L^p. Then $G_\alpha * g_i \in \mathcal{S}$, and by Proposition 2.3.8 there is a subsequence $\{i_n\}_1^\infty$ such that $G_\alpha * g_{i_n}(x)$ converges to $G_\alpha * g(x)$ (α, p)-quasieverywhere, and uniformly outside an open set of arbitrarily small capacity. The proposition follows.

Denote the normalized characteristic function for the unit ball by χ, i.e., $\chi(x) = \chi(x, B(0, 1))/|B(0, 1)|$, and define χ_r for $r > 0$ by $\chi_r(x) = r^{-N}\chi(x/r)$. Then

$$\frac{1}{|B(x,r)|}\int_{B(x,r)} G_\alpha * g\,dy = \chi_r * G_\alpha * g(x) = G_\alpha * \chi_r * g(x)\ .$$

Moreover $\chi_r * g$ is continuous and converges to g in L^p. It follows as in the above proof that for a suitable subsequence $\{r_i\}_1^\infty$

$$\lim_{r_i\to 0}\frac{1}{|B(x,r_i)|}\int_{B(x,r_i)} G_\alpha * g\,dy = G_\alpha * g(x)$$

(α, p)-quasieverywhere, and uniformly outside an open set of arbitrarily small capacity.

However, we can also easily prove the following stronger result.

Proposition 6.1.3. *Let $f = G_\alpha * g \in L^{\alpha,p}(\mathbf{R}^N)$, $1 < p < \infty$, $\alpha > 0$. Then*

$$\lim_{r\to 0}\frac{1}{|B(x,r)|}\int_{B(x,r)} f(y)\,dy = G_\alpha * g(x)\ ,$$

whenever $G_\alpha * |g|(x) < \infty$, i.e., (α, p)-q.e.

Proof. We assume that $\alpha p \le N$, since otherwise there is nothing to prove. It is easily seen, using the estimates for the Bessel kernel in Chapter 1, that there is a constant A such that $\chi_r * G_\alpha \le A\,G_\alpha$ for all $r \le 1$. In fact, if $|x| \le 2$ and $r \le \frac{1}{2}|x|$, then by (1.2.14)

$$\frac{1}{|B(x,r)|}\int_{B(x,r)} G_\alpha(y)\,dy \le G_\alpha(\tfrac{1}{2}x) \le A\,G_\alpha(x)\ .$$

Similarly, if $|x| \ge 2$ and $r \le 1$, then by (1.2.16)

$$\frac{1}{|B(x,r)|}\int_{B(x,r)} G_\alpha(y)\,dy \le \max_{|x-y|\le 1} G_\alpha(y) \le A\,G_\alpha(x)\ .$$

And if $|x| \le 2$ and $r \ge \frac{1}{2}|x|$, then by (1.2.14)

$$\begin{aligned}\frac{1}{|B(x,r)|}\int_{B(x,r)} G_\alpha(y)\,dy &\le \frac{1}{|B(x,r)|}\int_{B(0,3r)} G_\alpha(y)\,dy\\ &\le A\,r^{\alpha-N} \le A\,(\tfrac{1}{2}|x|)^{\alpha-N} \le A\,G_\alpha(x)\ .\end{aligned}$$

It follows by Lebesgue's theorem that

$$\lim_{r\to 0}\frac{1}{|B(x,r)|}\int_{B(x,r)} f(y)\,dy = \lim_{r\to 0}\int_{\mathbf{R}^N} (\chi_r * G_\alpha)(y)\,g(x-y)\,dy = G_\alpha * g(x)\ ,$$

wherever $G_\alpha * |g|(x) < \infty$.

The idea of a quasicontinuous representative would not be very interesting if an element in a function space could have several different such representatives. For this reason the following is an important result.

Theorem 6.1.4. *Let f_1 and f_2 be (α, p)-quasicontinuous functions, $\alpha > 0$, $1 < p < \infty$, and suppose that $f_1(x) = f_2(x)$ almost everywhere. Then $f_1(x) = f_2(x)$ (α, p)-quasieverywhere.*

Remark 1. Notice that we are not assuming that the functions are in $L^{\alpha,p}$. The theorem is valid for (α, p)-quasicontinuous functions in general.

Remark 2. A different proof of the theorem will be given in Section 6.4. See the remark following Corollary 6.4.7.

Proof. Suppose that $f = f_1 - f_2$ is (α, p)-quasicontinuous and that $f(x) = 0$ almost everywhere. We shall prove that $f(x) = 0$ (α, p)-quasieverywhere.

By the definition of quasicontinuity there are open sets $\{O_n\}_1^\infty$ such that $\lim_{n\to\infty} C_{\alpha,p}(O_n) = 0$ and $f|_{O_n^c} \in C(O_n^c)$. Then there are also $\psi_n \in L_+^p$ such that $\lim_{n\to\infty}\|\psi_n\|_p = 0$ and $\varphi_n(x) = G_\alpha * \psi_n(x) \ge 1$ everywhere on O_n. According to Proposition 2.3.8 we can assume that $\lim_{n\to\infty}\varphi_n(x) = 0$ (α, p)-q.e.

Let x be such a point. By Theorem 6.1.3 we can also assume that

$$\lim_{r\to 0}\frac{1}{|B(x,r)|}\int_{B(x,r)} \varphi_n\,dy = \varphi_n(x)$$

for all n. Then, if n is large enough, we have

$$\limsup_{r\to 0}\frac{|O_n\cap B(x,r)|}{|B(x,r)|}\le \limsup_{r\to 0}\frac{1}{|B(x,r)|}\int_{O_n\cap B(x,r)}\varphi_n\,dy$$

$$\le \lim_{r\to 0}\frac{1}{|B(x,r)|}\int_{B(x,r)}\varphi_n\,dy=\varphi_n(x)<1\ .$$

Thus, we see that $F_r = B(x,r)\setminus O_n$ has positive Lebesgue measure for all $r > 0$. By the assumption this implies that there are $y_r \in F_r$ for all $r > 0$ such that $f(y_r) = 0$. But $x \in O_n^c$ and $f \in C(O_n^c)$, so $f(x) = 0$. The theorem follows.

Now let $f \in L^{\alpha,p}$ for some $\alpha > 1$, and assume that f is (α, p)-quasicontinuous. Let Df denote any of the first order partial derivatives of f, taken in the sense of distributions. Then $Df \in L^{\alpha-1,p}$, so we can assume that Df is $(\alpha-1, p)$-quasicontinuous. The following corollary will be used in Chapter 9.

Corollary 6.1.5. *Let f, g, and fg belong to $L^{\alpha,p}$, $\alpha > 1$. Then*

$$D(fg)(x) = f(x)Dg(x) + Df(x)g(x) \quad (\alpha-1, p)\text{-}q.e.$$

Proof. The distribution derivatives of f, g, and fg are almost everywhere equal to pointwise derivatives. (See L. Schwartz [373], Ch. II, Théorème V.) Thus the two sides are equal almost everywhere, by the ordinary product rule. But both sides are $(\alpha-1, p)$-quasicontinuous, since the sums and products of quasicontinuous functions are clearly quasicontinuous.

Theorem 6.1.4 makes it possible to extend the notion of trace of a function in $L^{\alpha,p}$ to arbitrary sets. We make a formal definition.

Definition 6.1.6. If $f \in L^{\alpha,p}(\mathbf{R}^N)$, $1 < p \le N/\alpha$, and $E \subset \mathbf{R}^N$ is an arbitrary set, then the trace of f on E, denoted $f|_E$, is the restriction to E of any (α, p)-quasicontinuous representative $\tilde{f}$ of f.

Similarly, if $f \in L^{\alpha,p}(\mathbf{R}^N)$, and β is a multiindex with $|\beta| < \alpha$, then $D^\beta f|_E$ is the restriction to E of any $(\alpha-|\beta|, p)$-quasicontinuous representative $\widetilde{D^\beta f}$ of $D^\beta f$, where $D^\beta f$ is defined as the distribution derivative of f and considered as an element in $L^{\alpha-|\beta|,p}(\mathbf{R}^N)$.

6.2 Lebesgue Points

If $f \in L^1_{\text{loc}}(\mathbf{R}^N)$, a point $x \in \mathbf{R}^N$ is called a *Lebesgue point* for f if

$$\lim_{r\to 0}\frac{1}{|B(x,r)|}\int_{B(x,r)}|f(y)-f(x)|\,dy=0\ ,$$

and by a theorem of Lebesgue almost every point is a Lebesgue point. Also, if $f \in L^p_{\text{loc}}(\mathbf{R}^N)$ for some p, $1 \le p < \infty$, then almost every x is a Lebesgue point in the strong sense that

$$\lim_{r\to 0} \frac{1}{|B(x,r)|} \int_{B(x,r)} |f(y) - f(x)|^p \, dy = 0 \ .$$

See e.g. Stein [389], Section I.5.7.

We recall that if $f \in L^{\alpha,p}(\mathbf{R}^N)$, $1 < p < \infty$, $0 < \alpha p \le N$, then by Sobolev's theorem (Theorem 3.1.4) $f \in L^q_{\text{loc}}$ for all q, $1 \le q \le p^*$, where $p^* = Np/(N - \alpha p)$, if $\alpha p < N$, and for all q, $1 \le q < \infty$, if $\alpha p = N$. We shall now prove for such f that (α, p)-quasi all points are Lebesgue points in the sense of L^q.

Theorem 6.2.1. *Let $f = G_\alpha * g \in L^{\alpha,p}(\mathbf{R}^N)$, $1 < p < \infty$, $0 < \alpha p \le N$. Let q be as above. Then (α, p)-quasievery x is a Lebesgue point for f in the L^q-sense, i.e.*

$$\lim_{r\to 0} \frac{1}{|B(x,r)|} \int_{B(x,r)} f(y) \, dy = \widetilde{f}(x) \quad \textit{exists},$$

and

$$\lim_{r\to 0} \frac{1}{|B(x,r)|} \int_{B(x,r)} |f(y) - \widetilde{f}(x)|^q \, dy = 0 \ .$$

Moreover, the convergence is uniform outside an open set of arbitrarily small (α, p)-capacity, $\widetilde{f}$ is an (α, p)-quasicontinuous representative for f, and

$$\widetilde{f}(x) = G_\alpha * g(x) \quad (\alpha, p)\text{-q.e.}$$

The proof follows the standard pattern, i.e., it depends on a density argument and on a weak type estimate. The latter, which is of independent interest, is the content of the following lemma. The case $q = p$ will play an important role in Chapter 9. As usual, M denotes the Hardy–Littlewood maximal operator.

Lemma 6.2.2. *Suppose that $1 < p < \infty$, $0 < \alpha p \le N$, and let q be as above. Let $f = G_\alpha * g$, where $g \in L^p_+$. Set*

$$E_\lambda = \{ x : M(f^q)(x) > \lambda^q \} \ .$$

Then there is a constant A, independent of f, such that

$$C_{\alpha,p}(E_\lambda) \le \frac{A}{\lambda^p} \|g\|_p^p$$

for all $\lambda \ge \|g\|_p$ (for all $\lambda > 0$ if $q = 1$).

We first prove the theorem, assuming the lemma. We shall see that the lemma has a simple proof if $q = 1$, and that this is enough for the proof of the theorem in that case.

Proof of Theorem 6.2.1. Let $f = G_\alpha * g$, $g \in L^p$. We already know from Proposition 6.1.3 that $\widetilde{f}(x)$ exists (α, p)-q.e. and equals $G_\alpha * g(x)$, but we prefer to give an independent proof, which at the same time gives uniformity of convergence.

We define χ_r for $r > 0$ as a normalized characteristic function as in Section 6.1. Let $\varepsilon > 0$ and choose $g_0 \in \mathcal{S}$ so that $\|g - g_0\|_p < \varepsilon$. Then $f_0 = G_\alpha * g_0 \in \mathcal{S}$, and thus $\lim_{r\to 0} \chi_r * f_0(x) = f_0(x)$ for all x.

For $\delta > 0$ we define

$$\Omega_\delta f(x) = \sup_{0<r<\delta} (\chi_r * f)(x) - \inf_{0<r<\delta} (\chi_r * f)(x) \ .$$

It follows that

$$\Omega_\delta f(x) \le \Omega_\delta (f - f_0)(x) + \Omega_\delta f_0(x) \ .$$

By uniform continuity, we can choose δ so small that $\Omega_\delta f_0(x) < \varepsilon$ for all x. Moreover,

$$|\chi_r * (f - f_0)(x)| \le M(f - f_0)(x) \ ,$$

so

$$\Omega_\delta f(x) \le M(f - f_0)(x) + \varepsilon \ .$$

If $\varepsilon < \frac{1}{2}\lambda$ this implies that

$$\{x : \Omega_\delta f(x) > \lambda\} \subset \{x : M(f - f_0)(x) > \tfrac{1}{2}\lambda\} \ ,$$

and thus, by the lemma applied to $G_\alpha * |g - g_0|$ for $q = 1$, there is A so that

$$C_{\alpha,p}\big(\{x : \Omega_\delta f(x) > \lambda\}\big) \le A(1/\lambda)^p \|g - g_0\|_p^p \le A(\varepsilon/\lambda)^p \ .$$

Now choose $\lambda = 2^{-n}$, and $\varepsilon = 4^{-n}$ for $n = 1, 2, \ldots$, and denote the corresponding δ by δ_n. Set

$$E_n = \{x : \Omega_{\delta_n} f(x) > 2^{-n}\} \ ,$$

so that

$$C_{\alpha,p}(E_n) \le A\, 2^{-np} \ .$$

If $F_m = \bigcup_{n=m}^{\infty} E_n$, it follows that

$$C_{\alpha,p}(F_m) \le A \sum_{n=m}^{\infty} 2^{-np} \ ,$$

which tends to 0 as m tends to ∞, whence

$$C_{\alpha,p}\left(\bigcap\nolimits_{m=1}^{\infty} F_m\right) = 0 \ .$$

If $x \notin F_m$, we see that $\Omega_\delta f(x) \le 2^{-n}$ for $\delta \le \delta_n$ for all $n \ge m$. It follows that $\lim_{r\to 0} \chi_r * f(x) = \widetilde{f}(x)$ exists if $x \notin \bigcap_{m=1}^{\infty} F_m$, and uniformly outside F_m for any m, which proves the first part of the theorem.

For the second part we modify the proof slightly and define

$$\Omega_{q,\delta}(f - \widetilde{f}(x))(x) = \sup_{0<r\le\delta} \big((\chi_r * |f - \widetilde{f}(x)|^q)(x)\big)^{1/q} \ .$$

We choose $\varepsilon > 0$, g_0, and $f_0 = G_\alpha * g_0$ as before. Then $\widetilde{f_0} = f_0$, and as before we can choose δ so small that $\Omega_{q,\delta}(f_0 - f_0(x))(x) < \varepsilon$ for all x. We find

$$\begin{aligned}\Omega_{q,\delta}(f-\widetilde{f}(x))(x) &\le \Omega_{q,\delta}\big(f-f_0-(\widetilde{f}(x)-f_0(x))\big)(x)+\Omega_{q,\delta}(f_0-f_0(x))(x)\\ &\le \sup_{0<r\le\delta}\big((\chi_r*|f-f_0|^q)(x)\big)^{1/q}+|\widetilde{f}(x)-f_0(x)|+\varepsilon\\ &\le \big(M|f-f_0|^q(x)\big)^{1/q}+|\widetilde{f}(x)-f_0(x)|+\varepsilon\ .\end{aligned}$$

Thus, if $\varepsilon < \frac{1}{3}\lambda$,

$$\begin{aligned}&\{x:\Omega_{q,\delta}(f-\widetilde{f}(x))(x)>\lambda\}\\ &\qquad\subset\{x:M|f-f_0|^q(x)>(\tfrac{1}{3}\lambda)^q\}\cup\{x:|\widetilde{f}(x)-f_0(x)|>\tfrac{1}{3}\lambda\}\ .\end{aligned}$$

But $|\widetilde{f}(x)-f_0(x)|\le G_\alpha*|g-g_0|(x)$ (α,p)-q.e. Hence, by the lemma applied to $G_\alpha*|g-g_0|$, and by the definition of capacity, there is A such that

$$C_{\alpha,p}\big(\{x:\Omega_{q,\delta}(f-\widetilde{f}(x))(x)>\lambda\}\big)\le A(1/\lambda)^p\|g-g_0\|_p^p\le A(\varepsilon/\lambda)^p\ .$$

The proof is now concluded as before.

Corollary 6.2.3. *Let $f\in L^{\alpha,p}(\mathbf{R}^N)$ and denote by $u(x,t)$ the Poisson integral of f, $u(x,t)=P_t*f(x)$, $t>0$, $x\in\mathbf{R}^N$. Then the nontangential limit, $\lim_{(x,t)\to(x_0,0)}u(x,t)=\widetilde{f}(x_0)$, exists for (α,p)-q.e. x_0.*

Proof. We only have to observe that there is a constant A such that if (x,t) belongs to a nontangential cone with vertex at x_0, then $|P_t*f(x)|\le A\,G_\alpha*Mf(x_0)$, and then apply the same proof ($q=1$). See Stein [389], Theorem VII.1, for details.

Proof of Lemma 6.2.2. We prove the lemma in the following steps: $q=1$, $q=p$, $q=p^*<\infty$, and finally $q<\infty$, if $\alpha p=N$.

Step 1. $q=1$. Clearly $\chi_r*f(x)=\chi_r*G_\alpha*g(x)=G_\alpha*\chi_r*g(x)\le G_\alpha*Mg(x)$, and thus $Mf(x)=\sup_{r>0}\chi_r*f(x)\le G_\alpha*Mg(x)$.

It follows that $\{x:Mf(x)>\lambda\}\subset\{x:G_\alpha*Mg(x)>\lambda\}$, and by the definition of capacity $C_{\alpha,p}(E_\lambda)\le\lambda^{-p}\|Mg\|_p^p$ for all $\lambda>0$. The conclusion now follows from the Hardy–Littlewood–Wiener maximal theorem, Theorem 1.1.1.

Step 2. $q=p$. Suppose that $x_0\in E_\lambda$, and let $r>0$ be such that

$$\frac{1}{|B(x_0,r)|}\int_{B(x_0,r)}(G_\alpha*g)^p\,dx>\lambda^p\ . \tag{6.2.1}$$

We observe that by Minkowski's inequality (1.1.6)

$$|B(x_0,r)|<\frac{1}{\lambda^p}\int_{\mathbf{R}^N}(G_\alpha*g)^p\,dx\le\frac{1}{\lambda^p}\|G_\alpha\|_1^p\|g\|_p^p=\frac{1}{\lambda^p}\|g\|_p^p\ ,$$

so $|B(x_0,r)|<1$ if $\lambda\ge\|g\|_p$.

Set $g=g'+g''$, where $g'(x)=0$ for $|x-x_0|>2r$, and $g'(x)=g(x)$ for $|x-x_0|\le 2r$. Then

$$\lambda|B(x_0,r)|^{1/p} < \left(\int_{B(x_0,r)} (G_\alpha * g)^p\, dx\right)^{1/p}$$
$$\le \left(\int_{B(x_0,r)} (G_\alpha * g')^p\, dx\right)^{1/p} + \left(\int_{B(x_0,r)} (G_\alpha * g'')^p\, dx\right)^{1/p} ,$$

so that either

$$\frac{1}{|B(x_0,r)|}\int_{B(x_0,r)} (G_\alpha * g')^p\, dx > \left(\frac{\lambda}{2}\right)^p , \tag{6.2.2}$$

or

$$\frac{1}{|B(x_0,r)|}\int_{B(x_0,r)} (G_\alpha * g'')^p\, dx > \left(\frac{\lambda}{2}\right)^p . \tag{6.2.3}$$

But by Lemma 3.1.1, for any $x \in B(x_0,r)$

$$G_\alpha * g'(x) \le \int_{B(x,3r)} G_\alpha(x-y)\, g'(y)\, dy \le A\, Mg'(x)\, r^\alpha . \tag{6.2.4}$$

By the Hardy–Littlewood–Wiener theorem (Theorem 1.1.1)

$$\int_{B(x_0,r)} (Mg')^p\, dx \le A\, \|g'\|_p^p = A\int_{B(x_0,2r)} g^p\, dx ,$$

so (6.2.2) implies

$$A\int_{B(x_0,2r)} g^p\, dx > r^{N-\alpha p}\lambda^p . \tag{6.2.5}$$

We note that if $N = \alpha p$, then this cannot occur if $\lambda > A\|g\|_p$, since always

$$\int_{B(x_0,r)} g^p\, dx \le \int_{\mathbf{R}^N} g^p\, dx .$$

If on the other hand (6.2.3) holds, then we claim that

$$A\, G_\alpha * g(x_0) > \lambda . \tag{6.2.6}$$

In fact, for arbitrary x_1 and x_2 in $B(x_0,r)$ and y outside of $B(x_0,2r)$ we have

$$\tfrac{1}{3}|x_2-y| \le |x_1-y| \le 3|x_2-y| ,$$

and

$$|x_2-y| - 2r \le |x_1-y| \le |x_2-y| + 2r .$$

Thus, $G_\alpha(x_1-y) \le A\, G_\alpha(x_2-y)$ by the estimates (1.2.14) and (1.2.16) for Bessel kernels, so that for any $x_1 \in B(x_0,r)$

$$G_\alpha * g''(x_1) \le A \inf_{x\in B(x_0,r)} G_\alpha * g''(x) \le A \inf_{x\in B(x_0,r)} G_\alpha * g(x) .$$

Hence, (6.2.3) implies that

$$\lambda \le A \inf_{x \in B(x_0,r)} G_\alpha * g(x) \ ,$$

which implies (6.2.6).

Let U be the set of all $x \in E_\lambda$ such that (6.2.5) holds, i. e.

$$A \int_{B(x,2r)} g^p \, dx > r^{N-\alpha p} \lambda^p \ .$$

Then, by (6.2.6), $A\, G_\alpha * g(x) > \lambda$ everywhere on $E_\lambda \setminus U$, so

$$C_{\alpha,p}(E_\lambda \setminus U) \le \frac{A}{\lambda^p} \|g\|_p^p \ .$$

By Theorem 1.4.1 there are disjoint balls $\{B(x_i, 2r_i)\}_1^\infty$, satisfying

$$A \int_{B(x_i,2r_i)} g^p \, dx > r_i^{N-\alpha p} \lambda^p \ ,$$

such that

$$U \subset \bigcup_1^\infty B(x_i, 10r_i) \ .$$

We can assume that $N > \alpha p$, so that

$$C_{\alpha,p}(U) \le \sum_1^\infty C_{\alpha,p}(B(x_i, 10r_i)) \le A \sum_1^\infty r_i^{N-\alpha p}$$

$$\le \frac{A}{\lambda^p} \sum_1^\infty \int_{B(x_i,2r_i)} g^p \, dx \le \frac{A}{\lambda^p} \int_{\mathbf{R}^N} g^p \, dx \ .$$

The lemma follows, since $C_{\alpha,p}(E_\lambda) \le C_{\alpha,p}(E_\lambda \setminus U) + C_{\alpha,p}(U)$.

Step 3. $q = p^*$, $\alpha p < N$. As in the previous proof we have (6.2.2) and (6.2.3) with p replaced by p^*. In (6.2.4) we now apply Proposition 3.1.2 to obtain

$$G_\alpha * g'(x) \le I_\alpha * g'(x) \le A \|g'\|_p^{\alpha p/N} Mg'(x)^{1-\alpha p/N} \ ,$$

that is,

$$(G_\alpha * g'(x))^{p^*} \le A \|g'\|_p^{p^*-p} Mg'(x)^p \ .$$

Thus, (6.2.2) now implies, by the Hardy–Littlewood–Wiener theorem,

$$\lambda^{p^*} < \frac{A}{r^N} \|g'\|_p^{p^*-p} \int_{B(x_0,r)} (Mg')^p \, dx \le \frac{A}{r^N} \|g'\|_p^{p^*} \ ,$$

which is (6.2.5), i.e.,

$$A \int_{B(x_0,2r)} g^p \, dx > r^{N-\alpha p} \lambda^p \ .$$

From this point on the proof is unchanged.

Step 4. $q < \infty$, $\alpha p = N$. We let $q < \infty$ and determine $\alpha' < \alpha$ by requiring $q = Np/(N - \alpha' p)$. Let $x_0 \in E_\lambda$ and $r \leq 1$. Then by Proposition 3.1.2(a)

$$G_\alpha * g'(x) \leq I_\alpha * g'(x) \leq A\, r^{\alpha-\alpha'} I_{\alpha'} * g'(x) \\ \leq A\, r^{\alpha-\alpha'} \|g'\|_p^{\alpha' p/N} Mg'(x)^{1-\alpha' p/N}$$

for all $x \in B(x_0, r)$, and thus

$$(G_\alpha * g'(x))^q \leq A r^{(\alpha-\alpha')q} \|g'\|_p^{q-p} Mg'(x)^p \ .$$

Applying (6.2.2) as before we find

$$\lambda^q < \frac{A r^{(\alpha-\alpha')q}}{r^N} \|g'\|_p^{q-p} \int_{B(x_0,r)} (Mg')^p\, dx \leq \frac{A r^{(\alpha-\alpha')q}}{r^N} \|g'\|_p^q = A\, \|g'\|_p^q \ ,$$

since, by assumption, $(\alpha - \alpha')q = \alpha p = N$. Again, this is impossible if $\lambda > A\, \|g'\|_p$, so (6.2.6) holds everywhere on E_λ, and the lemma follows.

Remark. Using Lemma 6.2.2 it is also easy to prove differentiability in an L^q sense for functions in $L^{\alpha,p}$. See W. P. Ziemer [438], Chapter 3. We shall return to this subject in Chapter 10, and by a different approach, due to Yu. V. Netrusov, we shall prove a differentiability theorem (Theorem 10.1.4), which gives information that is not available by the methods of the present chapter.

6.3 Thin Sets

One of the important ideas in classical potential theory is the concept of a *thin set*. This is the generalization to arbitrary sets of the notion of an *irregular set* for the Dirichlet problem. If $\Omega \subset \mathbf{R}^N$ is a region, then a boundary point a is said to be irregular if there are continuous boundary data such that the classical Dirichlet problem is not solvable in a neighborhood of a, and then Ω^c is said to be thin at a. Otherwise, a is a *regular point* for Ω

Closely related is the *fine topology*, which is the coarsest topology in $\mathbf{R}^N$ such that all subharmonic functions are continuous. A *fine neighborhood* of a can, in fact, be defined as an ordinary neighborhood minus a set that is thin at a.

We do not assume that the reader is familiar with this classical theory. However, by way of motivation, we start by recalling a few of its fundamental facts. They will be subsumed under the more general theory that follows.

There are many equivalent ways of defining a thin set, but the following one suits our purposes.

Definition 6.3.1. Let $E \subset \mathbf{R}^N$ be an arbitrary set. Then E is *thin* at a point a if there exists a positive measure μ such that

$$G_2 * \mu(a) < \liminf_{x \to a,\, x \in E \setminus \{a\}} G_2 * \mu(x) \ .$$

If E is not thin at a it is said to be *thick* there.

If $a \notin \overline{E}$, the definition is interpreted as meaning that E is thin at a. Note that $G_2 * \mu$ is a lower semicontinuous function, so that

$$G_2 * \mu(a) \le \liminf_{x \to a,\ x \in E \setminus \{a\}} G_2 * \mu(x)$$

always holds.

The set where E is thin is denoted $e(E)$, and its complement is denoted $b(E)$. Clearly

$$E^0 \subset b(E) \subset \overline{E} \quad \text{and} \quad (\overline{E})^c \subset e(E) \subset (E^0)^c \ .$$

The main result characterizing thin sets is the following.

Theorem 6.3.2. *Let $E \subset \mathbf{R}^N$, $N \ge 2$, and let $a \in \mathbf{R}^N$. The following statements are equivalent:*

(a) *E is thin at a.*

(b) *Let G be a neighborhood of a and let μ be the capacitary measure for $E \cap G$. If G is sufficiently small, then*

$$V_{1,2}^{\mu}(a) = G_2 * \mu(a) < 1 \ .$$

(c)
$$\int_0^1 \frac{C_{1,2}(E \cap B(a,r))}{r^{N-2}} \frac{dr}{r} < \infty \ .$$

The last condition can be written in the equivalent form,

(c′)
$$\sum_{n=0}^{\infty} 2^{n(N-2)} C_{1,2}(E \cap B_n(a)) < \infty \ .$$

In the special case when $E = \Omega^c$ and Ω is a domain in $\mathbf{R}^N$, $N \ge 2$, we have Wiener's theorem [433].

Theorem 6.3.3 (Wiener's Criterion). *A point $a \in \partial\Omega$ is regular for the Dirichlet problem for the Laplace equation in Ω if and only if*

$$\int_0^1 \frac{C_{1,2}(\Omega^c \cap B(a,r))}{r^{N-2}} \frac{dr}{r} = \infty \ .$$

An important result is the following so called *Kellogg property*. As we shall see later, it is an easy consequence of Theorem 6.3.2.

Theorem 6.3.4. $C_{1,2}(e(E) \cap E) = 0$ *for all $E \subset \mathbf{R}^N$.*

There is also the more general *Choquet property*.

Theorem 6.3.5. *For any $E \subset \mathbf{R}^N$ and any $\varepsilon > 0$ there is an open G such that $e(E) \subset G$, and $C_{1,2}(E \cap G) < \varepsilon$.*

There are rather straightforward generalizations of these results (with the exception of Theorem 6.3.3) to the case of $(\alpha, 2)$-capacities. Our purpose is to produce an analogue of Theorem 6.3.2 (see Theorem 6.3.11 below) for the (α, p)-capacities for $1 < p \le N/\alpha$. Generalizations of Theorems 6.3.4 and 6.3.5 will then be easy consequences of this result.

The earliest nonlinear Wiener condition for boundary regularity is due to V. G. Maz'ya [302]. He considered a class of quasilinear equations of second order, including the so called p-Laplace equation,

$$\Delta_p u = \operatorname{div}\left(\nabla u |\nabla u|^{p-2}\right) = 0 \ ,$$

and defined regular boundary points for solutions in $W^{1,p}$. His main result contains the sufficiency part of the following theorem.

Theorem 6.3.6. *A point $a \in \partial\Omega$ is regular for the Dirichlet problem for the p-Laplace equation $\Delta_p u = 0$, $1 < p \leq N$, in a region $\Omega \subset \mathbf{R}^N$ if and only if*

$$\int_0^1 \left(\frac{C_{1,p}\left(\Omega^c \cap B(a,r)\right)}{r^{N-p}}\right)^{p'-1} \frac{dr}{r} = \infty \ .$$

The necessity part was proved almost a quarter of a century later by T. Kilpeläinen and J. Malý [250]. To prove these results would fall outside the scope of this book, but we make some further remarks in Section 6.5.5 at the end of the chapter.

A natural definition of (α, p)-thinness would seem to be the following statement about a set $E \subset \mathbf{R}^N$ and a point $a \in \mathbf{R}^N$.

(A) There exists a positive measure μ such that

$$V^{\mu}_{\alpha,p}(a) < \liminf_{x \to a,\ x \in E\setminus\{a\}} V^{\mu}_{\alpha,p}(x) \ .$$

Here $V^{\mu}_{\alpha,p}$ is the nonlinear potential $V^{\mu}_{\alpha,p} = G_\alpha * (G_\alpha * \mu)^{p'-1}$.

The natural generalization of the integral condition for thinness is the following.

(B) $$\int_0^1 \left(\frac{C_{\alpha,p}(E \cap B(a,t))}{r^{N-\alpha p}}\right)^{p'-1} \frac{dr}{r} < \infty \ .$$

Unfortunately (A) and (B) are not in general equivalent. It turns out that the good choice is given by (B), and we make the following definition.

Definition 6.3.7. Let $E \subset \mathbf{R}^N$ and let $1 < p \leq N/\alpha$. Then E is (α, p)-*thin* at a point $a \in \mathbf{R}^N$ if

$$\int_0^1 \left(\frac{C_{\alpha,p}(E \cap B(a,r))}{r^{N-\alpha p}}\right)^{p'-1} \frac{dr}{r} < \infty \ , \tag{6.3.1}$$

or equivalently

$$\sum_{n=0}^{\infty} \left(2^{n(N-\alpha p)} C_{\alpha,p}(E \cap B_n(a))\right)^{p'-1} < \infty \ . \tag{6.3.2}$$

If E is not (α, p)-thin at a it is said to be (α, p)-*thick* there. The set of points where E is (α, p)-thin is denoted by $e_{\alpha,p}(E)$, and its complement by $b_{\alpha,p}(E)$.

Remark. The above definition has the small defect that it is not invariant under change of scale. A modified definition, in terms of a relative capacity, which has this invariance is given in Chapter 8. See Definition 8.2.3, and Corollary 8.2.6.

An explanation of the situation is that the family of potentials $V^{\mu}_{\alpha,p}$ is not sufficiently rich to characterize the (α, p)-thin sets. Moreover, if we take the point of view that the basic objects of study are the function spaces $L^{\alpha,p}$ (or $W^{\alpha,p}$), then the potentials $V^{\mu}_{\alpha,p}$ are not really intrinsic to the situation. Indeed, the spaces can be normed in many different ways, and each different norm gives a different, but equivalent definition of (α, p)-capacity. But the extremal functions in the definition of capacity have a form that depends strongly on the norm that is actually being used, and this makes them less suitable for defining thinness.

However, we shall see that with the definition of an equivalent (α, p)-capacity, $C(E; B^{p,p}_{\alpha}) = C(E; F^{p,p}_{\alpha})$, that was given in Definition 4.4.2 (see Proposition 4.4.4), the extremals have the desired properties, and we shall use this result in extending Theorems 6.3.2, 6.3.4, and 6.3.5. Indeed, the "right" nonlinear potentials in this context are the "Besov potentials" $\mathcal{V}^{\mu}_{\alpha,p}$ defined in (4.4.4), or the "atomic" $\mathfrak{V}^{\mu}_{\alpha,p}$ defined in (4.7.14), and the closely related $W^{\mu}_{\alpha,p}$ defined in Definition 4.5.1. We shall therefore now take a fresh look at the potentials $\mathcal{V}^{\mu}_{\alpha,p}$.

We recall that by (4.4.4)

$$\mathcal{V}^{\mu}_{\alpha,p} = \mathcal{H}_{\alpha}(\check{\mathcal{H}}_{\alpha}\mu)^{p'-1} = \sum_{n=0}^{\infty} 2^{-n\alpha p'} \eta_n * (\eta_n * \mu)^{p'-1} \ ,$$

where $\eta_0 = \eta \not\equiv 0$ is a nonnegative smooth function with support in the unit ball, and $\eta_n(x) = 2^{nN}\eta(2^n x)$. Furthermore, if η in addition satisfies $\max_x \eta(x) \le 1$, and $\eta(x) \ge \frac{1}{2}$ for $|x| \le \frac{1}{2}$, then by (4.5.3)

$$\begin{aligned} A^{-1} W^{\mu}_{\alpha,p}(x) - A^{-1}\mu(B_0(x))^{p'-1} &\le \mathcal{V}^{\mu}_{\alpha,p}(x) \\ &\le A\, W^{\mu}_{\alpha,p}(x) + A\, \mu(B_{-1}(x))^{p'-1} \ , \end{aligned} \tag{6.3.3}$$

where

$$W^{\mu}_{\alpha,p}(x) = \sum_{n=0}^{\infty} \left(2^{n(N-\alpha p)}\mu(B_n(x))\right)^{p'-1} .$$

We denote the capacity $C(\,\cdot\,; F^{p,p}_{\alpha}) = C(\,\cdot\,; B^{p,p}_{\alpha})$ defined in Definition 4.4.2 by $\mathcal{C}_{\alpha,p}$, and then Proposition 4.4.4 gives

$$A^{-1} C_{\alpha,p}(E) \le \mathcal{C}_{\alpha,p}(E) \le A\, C_{\alpha,p}(E) \tag{6.3.4}$$

for all $E \subset \mathbf{R}^N$.

Our next observation is that the assumptions in Theorem 2.5.6 are satisfied in this situation. We formulate this as a lemma.

Lemma 6.3.8. *Let* $\varphi = \{\varphi_n\}_0^{\infty} \in C_0^{\infty}(\mathbf{R}^N \times \mathbf{N})$. *Then* $\mathcal{H}_{\alpha}\varphi$ *is continuous on* $\mathbf{R}^N$ *and has compact support.*

Proof. We have

$$\mathcal{H}_\alpha \varphi = \sum_0^\infty 2^{-n\alpha} \eta_n * \varphi_n \ .$$

The assumption means that there is R such that $\varphi_n(y) = 0$ for $n > R$, or $|y| > R$. But then the lemma is obvious, since η_n has compact support.

It follows that the following consequence of Theorem 2.5.6 holds.

Theorem 6.3.9. *Let $E \subset \mathbf{R}^N$, let $1 < p < \infty$, $0 < \alpha \leq N/p$, and suppose that $C_{\alpha,p}(E) < \infty$. Then there is a $\mu^E \in \mathcal{M}^+(\overline{E})$, the capacitary measure for E, such that*

$$\mathcal{V}_{\alpha,p}^{\mu^E}(x) \geq 1 \quad (\alpha, p)\text{-q.e. on } E \ ,$$
$$\mathcal{V}_{\alpha,p}^{\mu^E}(x) \leq 1 \quad \text{on } \operatorname{supp} \mu^E \ ,$$

and

$$C_{\alpha,p}(E) = \mu^E(\overline{E}) = \int \mathcal{V}_{\alpha,p}^{\mu^E} d\mu^E \ .$$

Another consequence of Lemma 6.3.8 is that potentials are quasicontinuous. Let ν be the measure on $\mathbf{R}^N \times \mathbf{N}$ defined by (4.4.1).

Proposition 6.3.10. *Let $f = \{f_n\}_0^\infty \in L^p(\nu)$, $1 < p < \infty$. Then $\mathcal{H}_\alpha f$ is (α, p)-quasicontinuous. In particular, $\mathcal{V}_{\alpha,p}^\mu$ is (α, p)-quasicontinuous for any $\mu \in \mathcal{M}^+(\mathbf{R}^N)$ such that $\check{\mathcal{H}}_\alpha \mu \in L^{p'}(\nu)$.*

Proof. When we have Lemma 6.3.8 the proof is the same as for Proposition 6.1.2, again using Proposition 2.3.8.

We shall now justify our claim that Definition 6.3.7 is a good definition of thinness by proving that the nonlinear potentials $\mathcal{V}_{\alpha,p}^\mu$ have the property (A) above. This is the content of the following theorem, which generalizes part of Theorem 6.3.2. In the course of the proof we also generalize the remaining part of Theorem 6.3.2; see Proposition 6.3.14.

Theorem 6.3.11. *Let $E \subset \mathbf{R}^N$ and let $1 < p \leq N/\alpha$. Then E is (α, p)-thin at a point $a \in \overline{E}$ if and only if there is a $\mu \in \mathcal{M}^+(\mathbf{R}^N)$ such that*

$$\mathcal{V}_{\alpha,p}^\mu(a) < \liminf_{x \to a,\ x \in E \setminus \{a\}} \mathcal{V}_{\alpha,p}^\mu(x) \ .$$

Remark 1. By Proposition 2.3.2 we know that $\mathcal{V}_{\alpha,p}^\mu$ is lower semicontinuous. Thus

$$\mathcal{V}_{\alpha,p}^\mu(a) \leq \liminf_{x \to a,\ x \in E \setminus \{a\}} \mathcal{V}_{\alpha,p}^\mu(x) \ .$$

for all E.

Remark 2. In much the same way one can prove that $W_{\alpha,p}^\mu$ has the property of Theorem 6.3.11, i.e., E is (α, p)-thin at a point $a \in \overline{E}$ if and only if there is a $\mu \in \mathcal{M}^+(\mathbf{R}^N)$ such that

$$W^{\mu}_{\alpha,p}(a) < \liminf_{x \to a,\ x \in E \setminus \{a\}} W^{\mu}_{\alpha,p}(x) \ ,$$

and similarly for $\dot{W}^{\mu}_{\alpha,p}$ if $\alpha p < N$. We omit the proofs of these results, but we note that they easily imply the classical Theorem 6.3.2 and its extension to the linear potentials $V^{\mu}_{\alpha,2} = G_{2\alpha} * \mu$ for $0 < 2\alpha \leq N$, and $\dot{V}^{\mu}_{\alpha,2} = I_{2\alpha} * \mu$ for $0 < 2\alpha < N$. In order to see this it is enough to observe that by Lemma 3.1.1 e.g. $I_{2\alpha} * \mu = A\, \dot{W}^{\mu}_{\alpha,2}$.

In order to prove the theorem we need the following three propositions. The first one is needed for the proof of sufficiency and the remaining two for the necessity part.

Proposition 6.3.12. *Let $\mu \in \mathcal{M}^+(\mathbf{R}^N)$ with $\mu(\mathbf{R}^N) < \infty$. Then there are constants A_1 and A_2 such that*

$$C_{\alpha,p}\left(\{ x : \mathcal{V}^{\mu}_{\alpha,p}(x) > \lambda \}\right) \leq \frac{A_1 \mu(\mathbf{R}^N)}{\lambda^{p-1}} \ , \tag{6.3.5}$$

and

$$C_{\alpha,p}\left(\{ x : W^{\mu}_{\alpha,p}(x) > \lambda \}\right) \leq \frac{A_2 \mu(\mathbf{R}^N)}{\lambda^{p-1}} \tag{6.3.6}$$

for all $\lambda > 0$. In particular $\mathcal{V}^{\mu}_{\alpha,p}(x) < \infty$, and $W^{\mu}_{\alpha,p}(x) < \infty$ (α, p)-q.e.

Remark. This proposition is false if $\mathcal{V}^{\mu}_{\alpha,p}$ is replaced by $V^{\mu}_{\alpha,p}$, for $1 < p \leq 2 - \alpha/N$. In fact, in this case $V^{\mu}_{\alpha,p} \equiv \infty$, if μ is a point mass. On the other hand it is easy to verify that if $\mu = \delta$, the Dirac measure at the origin, then $W^{\delta}_{\alpha,p}(x) \leq A\,|x|^{-(N-\alpha p)(p'-1)}$ (if $N > \alpha p$), and by Propositions 5.1.2–5.1.4 this gives $C_{\alpha,p}(\{ x : W^{\delta}_{\alpha,p}(x) > \lambda \}) \leq A\,\lambda^{1-p}$, and similarly for $\mathcal{V}^{\delta}_{\alpha,p}$.

Proof. Let K be a compact subset of the open set $\{ x : \mathcal{V}^{\mu}_{\alpha,p}(x) > \lambda \}$. By Theorem 6.3.9 there is a positive measure γ on K such that $\gamma(K) = C_{\alpha,p}(K)$, and $\mathcal{V}^{\gamma}_{\alpha,p}(x) \leq 1$, when x belongs to $\operatorname{supp}\gamma$. It is clear that we obtain an equivalent capacity by using the nonlinear potential

$$\sum_{n=m}^{\infty} 2^{-n\alpha p'} \eta_n * (\eta_n * \mu)^{p'-1}$$

for any integer m. (See (4.4.5), Corollary 3.6.3, and Theorem 3.6.2.) Thus, there is a constant A_1, and a measure $\gamma \in \mathcal{M}^+(K)$ such that

$$\sum_{n=-6}^{\infty} 2^{-n\alpha p'} \eta_n * (\eta_n * \gamma)^{p'-1} \leq 1 \quad \text{for } x \in \operatorname{supp}\gamma \ ,$$

and $C_{\alpha,p}(K) \leq A_1 \gamma(K)$. Now let $x \in \operatorname{supp}\gamma$ and define

$$M_\gamma \mu(x) = \sup_{n \in \mathbf{Z}} \frac{\mu(B_{n+3}(x))}{\gamma(B_n(x))} \ .$$

Then, for $x \in \operatorname{supp}\gamma$, by two applications of (4.5.2), and by the assumption on γ,

$$\begin{aligned}
\lambda < \mathcal{V}^{\mu}_{\alpha,p}(x) &\le A_2 \sum_{n=-1}^{\infty} \left(2^{n(N-\alpha p)}\mu(B_n(x))\right)^{p'-1} \\
&\le A_2 M_\gamma \mu(x)^{p'-1} \sum_{n=-1}^{\infty} \left(2^{n(N-\alpha p)}\gamma(B_{n-3}(x))\right)^{p'-1} \\
&= A_3 M_\gamma \mu(x)^{p'-1} \sum_{n=-4}^{\infty} \left(2^{n(N-\alpha p)}\gamma(B_n(x))\right)^{p'-1} \\
&\le A_4 M_\gamma \mu(x)^{p'-1} \sum_{n=-6}^{\infty} 2^{-n\alpha p'} \eta_n * (\eta_n * \gamma)^{p'-1}(x) \\
&\le A_4 M_\gamma \mu(x)^{p'-1} .
\end{aligned}$$

Thus

$$\operatorname{supp}\gamma \subset \left\{x : A_4 M_\gamma \mu(x)^{p'-1} > \lambda\right\} = \left\{x : M_\gamma \mu(x) > (\lambda/A_4)^{p-1}\right\} .$$

By Theorem 1.4.1 we can cover $\operatorname{supp}\gamma$ by a union of balls $\{B_{n_i}(x_i)\}$ such that the balls $\{B_{n_i+3}(x_i)\}$ are disjoint, and

$$\frac{\mu(B_{n_i+3}(x_i))}{\gamma(B_{n_i}(x_i))} > (\lambda/A_4)^{p-1}.$$

It follows that

$$\begin{aligned}
C_{\alpha,p}(K) &\le A_1\gamma(K) \le A_1 \sum_i \gamma(B_{n_i}(x_i)) \\
&< A_1(A_4/\lambda)^{p-1} \sum_i \mu(B_{n_i+3}(x_i)) \le A_1(A_4/\lambda)^{p-1}\mu(\mathbf{R}^N) ,
\end{aligned}$$

which proves the first part of the proposition. The proof for $W^{\mu}_{\alpha,p}$ is almost the same, with the difference that one application of (4.5.2) is enough.

Proposition 6.3.13. *Let E be a Borel (or Suslin) set, suppose that E satisfies* $0 < C_{\alpha,p}(E) < \infty$, *and let* μ^E *be its capacitary measure. Then*

$$\mu^E(S) \le C_{\alpha,p}(S \cap E)$$

for any open set S.

Proof. By Proposition 2.3.12 Suslin sets are capacitable. Thus, there is a sequence $\{F_n\}_1^\infty$ of compact subsets of E such that $\lim_{n\to\infty} C_{\alpha,p}(F_n) = C_{\alpha,p}(E)$. By the proof of Theorem 2.5.6 we can assume that the corresponding capacitary measures μ_n converge weak* as measures to μ^E. Let K be a compact subset of S and denote the restriction $\mu_n|_K$ by σ_n. Then σ_n satisfies $\mathcal{V}^{\sigma_n}_{\alpha,p}(x) \le 1$ on $\operatorname{supp}\sigma_n \subset K \cap F_n$, so by Theorem 2.5.5

$$\sigma_n(K) = \sigma_n(K \cap F_n) \leq C_{\alpha,p}(K \cap F_n) \leq C_{\alpha,p}(S \cap E) \ ,$$

and thus

$$\mu_n(S) = \sup_{K \subset S} \sigma_n(K) \leq C_{\alpha,p}(S \cap E) \ .$$

The weak* convergence of $\{\mu_n\}$ to μ^E gives

$$\mu^E(S) \leq C_{\alpha,p}(S \cap E) \ .$$

In fact, by the theory of measures, $\mu^E(S) = \sup_\varphi \int \varphi \, d\mu^E$, the supremum being taken over all $\varphi \in C_0^\infty(S)$ such that $\varphi \leq 1$ on S. But for every such φ

$$\int \varphi \, d\mu^E = \lim_{n\to\infty} \int \varphi \, d\mu_n \leq \lim_{n\to\infty} \mu_n(S) \leq C_{\alpha,p}(S \cap E) \ ,$$

which proves the proposition.

Remark. The proposition is not true for closed sets S. For example, if E is open and $S = \partial E$, then $S \cap E = \emptyset$. But in the classical case, i.e., when $\alpha = 1$, $p = 2$, and in many other cases, the measure μ^E is carried by ∂E, and consequently $\mu^E(S) = \mu^E(\overline{E}) > 0$.

Proposition 6.3.14. *Let $E \subset \mathbf{R}^N$, $a \in \overline{E} \cap e_{\alpha,p}(E)$, and $\varepsilon > 0$. Then, for every set V contained in a sufficiently small ball $B_m(a)$, the capacitary measure $\mu^{E \cap V} = \mu$ satisfies*

$$\mathcal{V}^\mu_{\alpha,p}(a) < \varepsilon \ .$$

Proof. First suppose that E and V are Suslin. (All we need for proving Theorem 6.3.11 is the case when E and V are open.) By (6.3.3) we have

$$\mathcal{V}^\mu_{\alpha,p}(a) \leq A \sum_{n=-1}^{\infty} \left(2^{n(N-\alpha p)} \mu(B_n(a))\right)^{p'-1} \ .$$

Let $\varepsilon' > 0$, and choose m so large that

$$\sum_{n=m+1}^{\infty} \left(2^{n(N-\alpha p)} C_{\alpha,p}(E \cap B_n(a))\right)^{p'-1} < \varepsilon' \ . \qquad (6.3.7)$$

Suppose that $V \subset B_m(a)$. For all n we then have

$$\mu(B_n(a)) \leq C_{\alpha,p}(E \cap B_n(a))$$

by Proposition 6.3.13, and

$$\mu(B_n(a)) \leq C_{\alpha,p}(E \cap V) \leq C_{\alpha,p}(E \cap B_m(a)) \ ,$$

by Theorem 6.3.9. It follows that

$$\mathcal{V}^\mu_{\alpha,p}(a) \leq A\varepsilon' + AC_{\alpha,p}(E \cap B_m(a))^{p'-1} \sum_{n=0}^{m} 2^{n(N-\alpha p)(p'-1)} < \varepsilon \ ,$$

if we first choose $\varepsilon' < \varepsilon/(2A)$, and then make m large enough. Here we use the observation that for $\alpha p < N$ the convergence of the sum in (6.3.7) implies that

$$\lim_{m\to\infty}\left(2^{m(N-\alpha p)}\mathcal{C}_{\alpha,p}(E\cap B_m(a))\right)^{p'-1} = 0 \ ,$$

and that for $\alpha p = N$ the convergence of (6.3.7) together with the monotonicity of $\mathcal{C}_{\alpha,p}(E\cap B_m(a))$ implies that

$$\lim_{m\to\infty} m\big(\mathcal{C}_{\alpha,p}(E\cap B_m(a))\big)^{p'-1} = 0 \ .$$

In the case of general sets E and V there are G_δ sets $\{F_n\}_0^\infty$, such that $E\cap B_n(a)\subset F_n$, and $\mathcal{C}_{\alpha,p}(E\cap B_n(a)) = \mathcal{C}_{\alpha,p}(F_n)$. Set

$$F = \bigcap_0^\infty (F_n\cup B_n(a)^c) \ .$$

Then F is a G_δ set, $F\supset E$, and

$$\mathcal{C}_{\alpha,p}(E\cap B_n(a)) = \mathcal{C}_{\alpha,p}(F\cap B_n(a)) \quad \text{for all } n = 0,\ 1,\ \ldots \ ,$$

and consequently F is (α, p)-thin at a. We now choose a G_δ set H so that $E\cap V\subset H$, and $\mathcal{C}_{\alpha,p}(E\cap V) = \mathcal{C}_{\alpha,p}(H)$. We can assume that $H\subset F$. As in the proof of Theorem 2.5.6 the capacitary potentials corresponding to the sets $E\cap V$ and H are equal. The result now follows from the case of G_δ sets, applied to F and H.

Proof of Theorem 6.3.11. Let $a\in \overline{E}$ and suppose that there is a $\mu\in\mathcal{M}^+(\mathbf{R}^N)$ such that $\mathcal{V}^\mu_{\alpha,p}(a) < \infty$, and

$$\liminf_{x\to a,\ x\in E\setminus\{a\}} \mathcal{V}^\mu_{\alpha,p}(x) - \mathcal{V}^\mu_{\alpha,p}(a) = \eta > 0 \ .$$

We claim that this implies that E is (α, p)-thin at a, i.e. that

$$\sum_{n=0}^{\infty}\left(2^{n(N-\alpha p)}\mathcal{C}_{\alpha,p}(E\cap B_n(a))\right)^{p'-1} < \infty \ .$$

We first observe that the assumptions imply that for any $\varepsilon > 0$, $M > 0$, and integer $m > 0$ there is a $\gamma\in\mathcal{M}^+(B_m(a))$ such that

$$\mathcal{V}^\gamma_{\alpha,p}(a) < \varepsilon \ ,$$

and

$$\liminf_{x\to a,\ x\in E\setminus\{a\}} \mathcal{V}^\gamma_{\alpha,p}(x) \ge M \ .$$

To see this let $\varepsilon > 0$, $M > 0$, and $m > 0$ be given, and denote the restriction of μ to $B_m(a)$ by μ_m. By assumption (see (4.4.2)), $\operatorname{supp}\eta_n\subset B_n$. It follows that $\eta_n * \mu_m(x) = \eta_n * \mu(x)$ for $x\in B_{m+1}(a)$ if $n\ge m+1$, and consequently, that

$\eta_n * (\eta_n * \mu_m)^{p'-1}(x) = \eta_n * (\eta_n * \mu)^{p'-1}(x)$ for $x \in B_{m+2}(a)$ if $n \geq m+2$. Thus, for $x \in B_{m+2}(a)$ we have

$$\mathcal{V}^{\mu}_{\alpha,p}(x) - \mathcal{V}^{\mu_m}_{\alpha,p}(x) = \sum_{n=0}^{m+1} 2^{-n\alpha p'} \eta_n * (\eta_n * \mu)^{p'-1}(x) \ ,$$

which is clearly a continuous function of x. This shows that

$$\liminf_{x \to a,\ x \in E \setminus \{a\}} \mathcal{V}^{\mu_m}_{\alpha,p}(x) - \mathcal{V}^{\mu_m}_{\alpha,p}(a) = \liminf_{x \to a,\ x \in E \setminus \{a\}} \mathcal{V}^{\mu}_{\alpha,p}(x) - \mathcal{V}^{\mu}_{\alpha,p}(a) = \eta \ .$$

On the other hand, (6.3.3) and the fact that $\mathcal{V}^{\mu}_{\alpha,p}(a) < \infty$ imply that

$$W^{\mu}_{\alpha,p}(a) = \sum_{n=0}^{\infty} \Big(2^{n(N-\alpha p)} \mu(B_n(a))\Big)^{p'-1} < \infty \ . \tag{6.3.8}$$

Another application of (6.3.3) shows that $\mathcal{V}^{\mu_m}_{\alpha,p}(a) \to 0$ as $m \to \infty$. Now choose m so large that $\mathcal{V}^{\mu_m}_{\alpha,p}(a) < \eta\varepsilon/M$ and set $\gamma = (M/\eta)^{p-1}\mu_m$. Clearly, γ has the desired properties.

We also observe that by choosing m large, we can make the total mass of γ as small as we please. Applying (6.3.3) again, we then see that if M is suitably chosen, γ can be chosen so that

$$W^{\gamma}_{\alpha,p}(a) < \varepsilon \ ,$$

and

$$\liminf_{x \to a,\ x \in E \setminus \{a\}} W^{\gamma}_{\alpha,p}(x) \geq 1 \ .$$

The idea of the proof is now to use Proposition 6.3.12 to obtain an estimate $\mathcal{C}_{\alpha,p}(E \cap B_{n+1}(a)) \leq A\,\gamma(B_n(a))$ for sufficiently large n, and then apply (6.3.8) to show thinness.

Let γ be the measure constructed above, let $l > m$, and denote the restriction of γ to $B_l(a)$ by γ_l. We set $\gamma = \gamma_l + \gamma_l'$. If $x \in B_{l+1}(a)$ we have $\gamma_l'(B_k(x)) = 0$ for $k \geq l+1$, and $\gamma(B_k(x)) = \gamma(B_k(a)) = \gamma(B_m(a))$ for $k \leq m-1$. Thus, for $x \in B_{l+1}(a)$,

$$\begin{aligned}
W^{\gamma_l'}_{\alpha,p}(x) &= \sum_{k=0}^{l} \Big(2^{k(N-\alpha p)} \gamma_l'(B_k(x))\Big)^{p'-1} \\
&\leq \sum_{k=m}^{l} \Big(2^{k(N-\alpha p)} \gamma_l'(B_k(x))\Big)^{p'-1} + W^{\gamma_l'}_{\alpha,p}(a) \\
&\leq \sum_{k=m}^{l} \Big(2^{k(N-\alpha p)} \gamma_l'(B_{k-1}(a))\Big)^{p'-1} + W^{\gamma_l'}_{\alpha,p}(a) \\
&= 2^{(N-\alpha p)(p'-1)} \sum_{k=m-1}^{l-1} \Big(2^{k(N-\alpha p)} \gamma_l'(B_k(a))\Big)^{p'-1} + W^{\gamma_l'}_{\alpha,p}(a) \\
&\leq A_1 W^{\gamma_l'}_{\alpha,p}(a) < A_1 \varepsilon \ .
\end{aligned}$$

But $W^{\gamma}_{\alpha,p}(x) \le A_2\left(W^{\gamma_l}_{\alpha,p}(x) + W^{\gamma_l'}_{\alpha,p}(x)\right)$, so on $E \cap B_{l+1}(a) \setminus \{a\}$ we have

$$W^{\gamma_l}_{\alpha,p}(x) \ge A_2^{-1} W^{\gamma}_{\alpha,p}(x) - W^{\gamma_l'}_{\alpha,p}(x) > \frac{1}{2A_2} - A_1\varepsilon > \frac{1}{4A_2} \ ,$$

if l is so large that $W^{\gamma}_{\alpha,p}(x) \ge \frac{1}{2}$, and if $A_1\varepsilon < 1/(4A_2)$. Thus

$$E \cap B_{l+1}(a) \subset \left\{ x : W^{\gamma_l}_{\alpha,p}(x) > 1/(4A_2) \right\} \cup \{a\} \ .$$

By Proposition 6.3.12 applied to γ_l there is a constant A such that

$$\mathcal{C}_{\alpha,p}(E \cap B_{l+1}(a)) \le A\gamma_l(\mathbf{R}^N) = A\gamma(B_l(a)) \ .$$

That E is thin at a now follows from (6.3.8).

For the proof of the converse we let $a \in \overline{E}$, and we suppose that E is (α, p)-thin at a. We can assume that E is open and that $a \notin E$. In fact, it is easy to see that there is an open set containing $E \setminus \{a\}$ which is also thin at a. We can choose this set as the union of suitable open neighborhoods of the sets $\{x : x \in E, 2^{-n-1} \le |x - a| < 2^{-n}\}$, $n \in \mathbf{Z}$.

By Proposition 6.3.14 there is, for any $\varepsilon > 0$, an open neighborhood G of a and a measure μ such that $\mathcal{V}^{\mu}_{\alpha,p}(a) < \varepsilon$ and $\mathcal{V}^{\mu}_{\alpha,p}(x) \ge 1$ (α, p)-q. e. on $E \cap G$. But the complement of a set of capacity zero is thick at all points, and consequently the part of the theorem that has already been proved can be applied to all points in the open set $E \cap G$. (Cf. Proposition 2.6.7.) It follows that $\mathcal{V}^{\mu}_{\alpha,p}(x) \ge 1$ everywhere on $E \cap G$, and this finishes the proof of the theorem.

Corollary 6.3.15. *Let $1 < p \le N/\alpha$, let $E \subset \mathbf{R}^N$ and let $a \in \overline{E}$.*

If $a \in \overline{E} \cap e_{\alpha,p}(E)$ and if $\varepsilon > 0$, then there is $r > 0$ such that for every neighborhood G of a with $G \subset B(a, r)$ the capacitary potential $\mathcal{V}^{\mu}_{\alpha,p}$ for $E \cap G$ satisfies $\mathcal{V}^{\mu}_{\alpha,p}(a) < \varepsilon$.

If $a \in b_{\alpha,p}(E)$, then for any neighborhood G of a the capacitary potential for $E \cap G$ satisfies $\mathcal{V}^{\mu}_{\alpha,p}(a) \ge 1$.

Proof. The first part is contained in Proposition 6.3.14. If $a \in b_{\alpha,p}(E)$, the capacitary potential for $E \cap G$ satisfies $\mathcal{V}^{\mu}_{\alpha,p}(x) \ge 1$ (α, p)-q.e. on $E \cap G$. If the exceptional set of capacity zero where $\mathcal{V}^{\mu}_{\alpha,p}(x) < 1$ is removed, the remaining subset of $E \cap G$ is still (α, p)-thick at a, and it follows from the theorem that $\mathcal{V}^{\mu}_{\alpha,p}(a) \ge 1$.

Corollary 6.3.15 has the following immediate consequence.

Corollary 6.3.16. *Let $1 < p \le N/\alpha$ and let $E \subset \mathbf{R}^N$. Let $\{B_n\}_1^\infty$ be an enumeration of the open balls in $\mathbf{R}^N$ that have rational centers and radii and intersect E. Let $\mathcal{V}_n$ be the $\mathcal{C}_{\alpha,p}$-capacitary potential for $E \cap B_n$, and set $A_n = \{x \in \overline{E} \cap B_n : \mathcal{V}_n(x) < 1\}$. (In particular $A_n = \overline{E} \cap B_n$ if $\mathcal{C}_{\alpha,p}(E \cap B_n) = 0$.) Then*

$$e_{\alpha,p}(E) \cap \overline{E} = \bigcup_1^\infty A_n \ ,$$

and thus $e_{\alpha,p}(E)$ is a Borel set.

The extension of the Kellogg property, Theorem 6.3.4, is now an easy consequence of Corollary 6.3.16

Corollary 6.3.17 (Kellogg Property). *Let* $1 < p \leq N/\alpha$ *and let* $E \subset \mathbf{R}^N$. *Then*

$$C_{\alpha,p}\left(e_{\alpha,p}(E) \cap E\right) = 0 \ .$$

Proof. Corollary 6.3.16 gives the inclusion $e_{\alpha,p}(E) \cap E \subset \cup_1^\infty (A_n \cap E)$, and $C_{\alpha,p}(A_n \cap E) = \mathcal{C}_{\alpha,p}(A_n \cap E) = 0$ for all n by Theorem 6.3.9 and Proposition 4.4.4.

We finally extend the Choquet property, Theorem 6.3.5.

Theorem 6.3.18 (Choquet Property). *Let* $1 < p \leq N/\alpha$ *and let* $E \subset \mathbf{R}^N$. *For every* $\varepsilon > 0$ *there is an open set* G *such that*

$$e_{\alpha,p}(E) \subset G \quad \textit{and} \quad C_{\alpha,p}(E \cap G) < \varepsilon \ .$$

Proof. Again using Corollary 6.3.16 we have

$$e_{\alpha,p}(E) = (\overline{E})^c \bigcup \left(\bigcup_1^\infty A_n\right) \ .$$

Let $\varepsilon > 0$. Since $\mathcal{V}_n(x) \geq 1$ (α, p)-q.e. on $E \cap B_n$, and since $\mathcal{V}_n$ is quasicontinuous according to Proposition 6.3.10, there is an open set U_n with $\mathcal{C}_{\alpha,p}(U_n) < \varepsilon 2^{-n}$ such that $\mathcal{V}_n|_{U_n^c} \in C(U_n^c)$ and $\mathcal{V}_n(x) \geq 1$ on $E \cap B_n \cap U_n^c$. Set $F = E \setminus (\cup_1^\infty U_n)$. Then $\overline{F} \cap (\cup_1^\infty U_n) = \emptyset$.

We shall see that $G = (\overline{F})^c$ has the required properties. In fact, $\mathcal{V}_n$ is continuous on $\overline{F}$ and $\mathcal{V}_n(x) \geq 1$ on $F \cap B_n$. Thus $\mathcal{V}_n(x) \geq 1$ on $\overline{F} \cap B_n$, and thus $\overline{F} \cap A_n = \emptyset$. It follows that $e_{\alpha,p}(E) \subset G$. Moreover,

$$G \cap E \subset F^c \cap E \subset \bigcup_1^\infty U_n \ ,$$

so $\mathcal{C}_{\alpha,p}(G \cap E) < \varepsilon$. The theorem follows from Proposition 4.4.4.

Remark 1. The equality in Corollary 6.3.16 is not really needed in the proofs of the Kellogg and Choquet properties. It is enough to have inclusion of $e_{\alpha,p}(E)$, and this inclusion is a consequence of Proposition 6.3.14.

Remark 2. Note that although we have formulated the Kellogg and Choquet properties in terms of the capacity $C_{\alpha,p}$, Definition 6.3.7 shows that these properties are true for any (α, p)-capacity equivalent to $C_{\alpha,p}$. The proof given here, on the other hand, uses very strongly properties of the special nonlinear potentials $\mathcal{V}^{\mu}_{\alpha,p}$, in particular the fact that for any E the set $e_{\alpha,p}(E)$ can be described by means of a denumerable family of such potentials (Corollary 6.3.16). There is a more direct proof of the Kellogg property due to Th. H. Wolff (see [219], Lemma 2), but for the Choquet property no essentially different proof seems to be known.

6.4 Fine Continuity

As in the classical theory it is useful to define an (α, p)-fine topology associated to the concept of (α, p)-thinness.

Definition 6.4.1. Let $x \in \mathbf{R}^N$. Then a set $U \subset \mathbf{R}^N$ is called an *(α, p)-fine neighborhood* of x if $x \in U$ and U^c is (α, p)-thin at x.

A set $G \subset \mathbf{R}^N$ is *(α, p)-finely open* if it is an (α, p)-fine neighborhood of each of its points.

A set $F \subset \mathbf{R}^N$ is *(α, p)-finely closed* if F^c is (α, p)-finely open.

This definition defines a topology. This follows from the obvious facts that the union of a finite number of (α, p)-thin sets, and any intersection of (α, p)-thin sets, are also (α, p)-thin.

In most of the remainder of this section we drop the (α, p), so we write *fine* instead of *(α, p)-fine*, and $e(E)$ and $b(E)$ for $e_{\alpha,p}(E)$ and $b_{\alpha,p}(E)$ (see Definition 6.3.7).

We formulate a few consequences of the definition:

The *fine closure* of a set E, denoted $\widetilde{E}$, is $E \cup b(E)$. A point x is a *finely isolated point* of E if $x \in E \cap e(E)$. The *fine interior* of E is $E \cap e(E^c)$.

A (real valued) function f that is defined on a set F is *finely continuous* at a point $x \in F$ if $\{ y : y \in F,\ a < f(y) < b \} \cup F^c$ is a fine neighborhood of x for all (a, b) such that $f(x) \in (a, b)$.

More explicitly, we can formulate the following definition.

Definition 6.4.2. A (real valued) function f that is defined on a set F is finely continuous at a point $x \in F$ if $\{ y : y \in F,\ |f(y) - f(x)| \geq \varepsilon \}$ is thin at x for all $\varepsilon > 0$.

We say that two sets are *equivalent* if their symmetric difference has capacity zero. It is clear from Definition 6.3.7 that equivalent sets are thin at the same points. It follows that if f is defined quasieverywhere on an open set G, i.e., if f is defined on $F \subset G$, and $G \setminus F$ has capacity zero, then f is finely continuous at $x \in F$ if and only if $\{ y : y \in F,\ |f(y) - f(x)| \geq \varepsilon \} \cup F^c$ is thin at x for all $\varepsilon > 0$.

The meaning of fine continuity is further clarified by the following proposition.

Proposition 6.4.3. *Let a function f be defined on a set F and finely continuous at a point $x \in F \cap b(F)$. Then there is a set $E \subset F$ such that E is thin at x and*

$$\lim_{y \to x,\ y \in F \setminus E} f(y) = f(x)\ .$$

The proof depends on a lemma.

Lemma 6.4.4. *Let $\{E_n\}_1^\infty$ be a sequence of sets, each of which is (α, p)-thin at a point x. Then there are positive numbers $\{r_n\}_1^\infty$ such that*

$$\bigcup_1^\infty (E_n \cap B(x, r_n))$$

is (α, p)-thin at x.

Proof. Choose r_n so small that

$$\int_0^1 \left(\frac{C_{\alpha,p}(E_n \cap B(x, r_n) \cap B(x, r))}{r^{N-\alpha p}} \right)^{p'-1} \frac{dr}{r} < 2^{-n} .$$

The lemma follows from the subadditivity of capacity and the inequalities

$$\left(\sum_1^\infty a_n\right)^{p'-1} \le \sum_1^\infty a_n^{p'-1} \quad \text{for } 2 \le p < \infty ,$$

$$\left(\sum_1^\infty a_n\right)^{p'-1} \le \sum_1^\infty (na_n)^{p'-1} \left(\sum_1^\infty n^{-1/(2-p)}\right)^{(p'-2)} \quad \text{for } 1 < p < 2 ,$$

valid for all positive a_n.

Proof of Proposition 6.4.3. Suppose that f is defined on F and finely continuous at $x \in F$. Then, for each n, $n = 1, 2, \ldots$, the set

$$E_n = \{ y : y \in F, \ |f(y) - f(x)| \ge n^{-1} \}$$

is thin at x. Choose r_n as in the lemma, and set

$$E = \bigcup_1^\infty (E_n \cap B(x, r_n)) .$$

Then E is thin at x, and for $y \in (F \setminus E) \cap B(x, r_n)$ we have $|f(y) - f(x)| < n^{-1}$.

The main results of this section are the following two theorems.

Theorem 6.4.5. *Every (α, p)-quasicontinuous function is (α, p)-finely continuous (α, p)-q.e.*

Theorem 6.4.6. *Every function that is defined and (α, p)-finely continuous (α, p)-q.e. is (α, p)-quasicontinuous.*

We give two proofs of the first theorem. The second one will be given after the proof of Proposition 6.4.13 below.

First proof of Theorem 6.4.5. Let f be the given function. That f is quasicontinuous implies that there is a decreasing sequence of open sets, G_n, $n = 1, 2, 3, \ldots$, such that $C_{\alpha,p}(G_n) \to 0$ and the restriction of f to G_n^c is continuous. Denote by $\mathcal{V}_n$ the capacitary potential for G_n. By Proposition 2.3.8 we can assume that $\lim_{n\to\infty} \mathcal{V}_n(x) = 0$ (α, p)-q.e. Let a be such a point. On the other hand $\mathcal{V}_n(x) \ge 1$ everywhere on G_n by Corollary 6.3.15. It follows, again from Corollary

6.3.15, that G_n is (α, p)-thin at a for all sufficiently large n. This clearly implies that f is (α, p)-finely continuous at a.

We postpone the second proof of Theorem 6.4.5 and the proof of Theorem 6.4.6 until after the proof of Proposition 6.4.13 below. Theorem 6.4.5 has the following corollary.

Corollary 6.4.7. *Let f be a function which is (α, p)-finely continuous (α, p)-q.e. Suppose that h is an (α, p)-quasicontinuous function such that $f(x) = h(x)$ almost everywhere. Then $f(x) = h(x)$ (α, p)-q.e., and thus f is (α, p)-quasicontinuous.*

Proof. By the theorem h is (α, p)-finely continuous (α, p)-q.e. Let a be a point where both f and h are (α, p)-finely continuous. Suppose that $|f(a) - h(a)| = c > 0$. Then $|f(x) - h(x)| > c/2$ outside a set that is thin at a, which contradicts the assumption that $f(x) - h(x) = 0$ a.e.

Remark. If both f and h are assumed to be quasicontinuous, the same argument gives another proof for Theorem 6.1.4.

It is useful to introduce the so called *quasitopology* associated to our capacities.

Definition 6.4.8. A set $E \subset \mathbf{R}^N$ is (α, p)-*quasiopen* if for any $\varepsilon > 0$ there is an open set G with $C_{\alpha,p}(G) < \varepsilon$ such that $E \setminus G$ is open in G^c in the relative topology.

A set is (α, p)-*quasiclosed* if its complement is (α, p)-quasiopen.

The definition does not define a topology, because arbitrary unions of quasiopen sets do not have to be quasiopen. On the other hand it is easy to see that denumerable unions of quasiopen sets are quasiopen.

The following proposition is a reformulation of Definition 6.4.8.

Proposition 6.4.9. *A set E is quasiopen if and only if for any given $\varepsilon > 0$ there is an open H such that $E \subset H$ and $C_{\alpha,p}(H \setminus E) < \varepsilon$.*

E is quasiclosed if and only if for any $\varepsilon > 0$ there is a closed $F \subset E$ such that $C_{\alpha,p}(E \setminus F) < \varepsilon$.

Proof. It is enough to prove the first half of the proposition, since the second half follows by taking complements.

Suppose that E is quasiopen and choose G as in Definition 6.4.8. Set $H = E \cup G$. Then H is open, and $C_{\alpha,p}(H \setminus E) < \varepsilon$.

Conversely, suppose that for any $\varepsilon > 0$ there is an open H as in the proposition. Then there is an open G such that $H \setminus E \subset G$ and $C_{\alpha,p}(G) < \varepsilon$. But then $E \setminus G = H \setminus G$, so $E \setminus G$ is open in G^c.

Note that if a set is quasiopen (or quasiclosed), then at the same time all sets equivalent to it are quasiopen (quasiclosed). Thus we can interpret "quasiopen" and "quasiclosed" as referring to equivalence classes of sets. The following proposition should be understood in this sense.

Proposition 6.4.10. *A real valued function f that is defined and finite quasi-everywhere is quasicontinuous if and only if the inverse image of every open interval is quasiopen.*

Proof. If f is quasicontinuous, then by the definitions the inverse image of any open set is quasiopen.

Let $\{I_n\}_1^\infty$ be an enumeration of all intervals (a, b) with rational a and b, and assume that all $f^{-1}(I_n)$ are quasiopen. Let $\varepsilon > 0$ and choose $\varepsilon_n > 0$ so that $\sum_1^\infty \varepsilon_n \leq \varepsilon$. Choose open G_n with $C_{\alpha,p}(G_n) < \varepsilon_n$ so that $f^{-1}(I_n) \setminus G_n$ are open in $(G_n)^c$, and set $G = \cup_1^\infty G_n$. Then $C_{\alpha,p}(G) < \varepsilon$, and $f^{-1}(I_n) \setminus G$ is open in G^c for all n. But any open $O \subset \mathbf{R}$ is a union of intervals I_n, and consequently $f^{-1}(O) \setminus G$ is a union of sets $f^{-1}(I_n) \setminus G$, and thus open in G^c. It follows that f restricted to G^c is continuous, and since ε is arbitrary f is quasicontinuous.

Proposition 6.4.11. *For any set E*

$$C_{\alpha,p}(E) = C_{\alpha,p}(\widetilde{E}) = C_{\alpha,p}(b(E)) \ .$$

Proof. Let $\mathcal{V}^\mu$ denote the capacitary potential for E. Then $\mathcal{V}^\mu(x) \geq 1$ everywhere on $b(E)$ by Corollary 6.3.15, and thus $\mathcal{V}^\mu(x) \geq 1$ q.e. on $\widetilde{E} = E \cup b(E)$ by Theorem 6.3.9. It follows that $C_{\alpha,p}(b(E)) \leq C_{\alpha,p}(\widetilde{E}) \leq C_{\alpha,p}(E)$. But on the other hand $E \subset \widetilde{E}$, and by the Kellogg property $C_{\alpha,p}(E \setminus b(E)) = 0$.

Proposition 6.4.12. *Any quasiopen set is equivalent to its fine interior. Any quasi-closed set is equivalent to its fine closure.*

Proof. It is enough to prove the second statement. Let E be quasiclosed, and let $\varepsilon > 0$. By Proposition 6.4.9 there is a closed F such that $F \subset E$ and $C_{\alpha,p}(E \setminus F) < \varepsilon$. Set $E \setminus F = A$. Then

$$\widetilde{E} = \widetilde{(F \cup A)} = \widetilde{F} \cup \widetilde{A} = F \cup \widetilde{A} \subset E \cup \widetilde{A} \ .$$

Thus $\widetilde{E} \setminus E \subset \widetilde{A}$ and by Proposition 6.4.11

$$C_{\alpha,p}(\widetilde{E} \setminus E) < C_{\alpha,p}(\widetilde{A}) = C_{\alpha,p}(A) < \varepsilon \ .$$

Proposition 6.4.13. *Any finely open set is quasiopen.*

Proof. Suppose that U is finely open. Then $U \subset e(U^c)$. By the Choquet property (Theorem 6.3.18) there is for given $\varepsilon > 0$ an open G such that $U \subset e(U^c) \subset G$ and $C_{\alpha,p}(G \setminus U) < \varepsilon$.

Second proof of Theorem 6.4.5. Suppose that f is quasicontinuous. As in the proof of Proposition 6.4.10 we let $\{I_n\}_1^\infty$ be an enumeration of all intervals (a, b) with rational a and b. Set $E_n = f^{-1}(I_n)$ and denote by A_n the fine interior of E_n. Then E_n is quasiopen by Proposition 6.4.10, and $C_{\alpha,p}(E_n \setminus A_n) = 0$ by Proposition 6.4.12. It follows that f is finely continuous everywhere outside $E = \cup_1^\infty(E_n \setminus A_n)$, and $C_{\alpha,p}(E) = 0$.

Proof of Theorem 6.4.6. Suppose that f is finely continuous quasieverywhere. Then the inverse image of any open set is equivalent to a finely open set. But any finely open set is quasiopen by Proposition 6.4.13, so f is quasicontinuous by Proposition 6.4.10.

For functions in $L^{\alpha,p}$ we can also give an explicit condition for fine continuity.

Theorem 6.4.14. *Let $1 < p \leq N/\alpha$ and let $f = G_\alpha * g$, where $g \in L^p_+(\mathbf{R}^N)$. Then f is (α, p)-finely continuous at all points a where $W^{g^p}_{\alpha,p}(a) < \infty$, and thus (α, p)-q.e.*

Proof. The proof is a simpler version of the proof of the sufficiency part of Theorem 6.3.11. Suppose that $f(a) < \infty$ and that there is a set E such that $a \in \overline{E}$ and

$$\liminf_{x\to a,\, x\in E\setminus\{a\}} f(x) - f(a) = \eta > 0 \ .$$

We claim that this implies that E is (α, p)-thin at a.

By multiplying g with a constant and restricting it to a sufficiently small ball $B(a, r)$ we can assume that $\eta = 1$ and that $f(a) < \varepsilon$ for some ε that remains to be chosen. Let $\rho < r$, denote by g_ρ the restriction of g to $B(a, \rho)$ and set $g = g_\rho + g'_\rho$. Let $|x - a| < \rho/2$. Then

$$\begin{aligned} G_\alpha * g'_\rho(x) &\leq A \int |y - x|^{\alpha-N} g'_\rho(y)\, dy \leq A \int |y - a|^{\alpha-N} g'_\rho(y)\, dy \\ &\leq A\, G_\alpha * g'_\rho(a) \leq A\,\varepsilon \leq 1/2 \ , \end{aligned}$$

if ε is small enough. But then

$$G_\alpha * g_\rho(x) = G_\alpha * g(x) - G_\alpha * g'_\rho(x) \geq 1/2$$

on $E \cap B(a, \rho/2)$. It follows from the definition of capacity that

$$C_{\alpha,p}(E \cap B(a, \rho/2)) \leq A \int_{B(a,\rho)} g^p\, dy \ .$$

The assumption now implies that E is (α, p)-thin at a.

6.5 Further Results

6.5.1. The idea of finding a representative with certain continuity properties in an equivalence class of almost everywhere equal functions is, of course, not limited to the spaces $L^{\alpha,p}$, $p > 1$. Another important example is given by the spaces $W^{\alpha,1}$ for integer α. For these the role of capacity is taken over by Hausdorff measure, or Hausdorff content of dimension $N - \alpha$ (see Section 5.1); cf. Section 5.6.3 above. For example, for a function in $W^{\alpha,1}$ all points are Lebesgue points, except for a set of $\Lambda_{N-\alpha}$-measure zero. This follows by standard methods from the results of D. R. Adams [13]; see in particular formula (14), p. 122. See also A. Carlsson and

V. G. Maz'ya [96]. These results depend, among other things, on an imbedding theorem of Maz'ya; see Section 7.6.10 below. See also Section 10.3 for a recent related result due to Yu. V. Netrusov.

$(N-1)$-dimensional Hausdorff measure is also the relevant measure in studying the important space BV of functions of bounded variation, which can be defined as the space of functions in $L^1(\mathbf{R}^N)$ whose weak (distribution) derivatives are measures. However, the situation with regard to Lebesgue points is more complicated than in $W^{1,1}$, and the set of exceptional points can have σ-finite $(N-1)$-dimensional measure. For the theory of these spaces, see e.g. the books by L. C. Evans and R. F. Gariepy [137], H. Federer [141], V. G. Maz'ya [308], and W. P. Ziemer [438], and concerning Lebesgue points especially Theorem 4.5.9, statements (16), (21), and (22) in [141], p. 483, and Theorem 5.14.4 in [438].

The following was communicated to the authors by Yu. V. Netrusov. Let BV^α, $\alpha = 1, 2, \ldots$, denote the space of functions in $L^1(\mathbf{R}^N)$, whose derivatives of order α are measures. Then again, for $\alpha \geq 2$, all points except for a set of zero $\Lambda_{N-\alpha}$-measure are Lebesgue points. This follows from the fact that $W^{\alpha-1,\infty} \cap \mathrm{BV}^\alpha$ is dense in BV^α. (But notice that $C^{\alpha-1} \cap \mathrm{BV}^\alpha$ is not dense in BV^α.)

6.5.2. J. H. Michael and W. P. Ziemer [324, 437] have proved the following extension of Proposition 6.1.2. Let $f \in W^{\alpha,p}(\Omega)$, where $\Omega \subset \mathbf{R}^N$, α is a positive integer, and $1 < p < \infty$. Assume that f is (α, p)-quasicontinuous. Let k be an integer such that $1 \leq k \leq \alpha$, and $(\alpha - k)p < N$. Then for any $\varepsilon > 0$ there is a function $f_\varepsilon \in C^k(\Omega)$ such that $\|f - f_\varepsilon\|_{\alpha,p} < \varepsilon$ and the set E where $f_\varepsilon(x) \neq f(x)$ has $C_{\alpha-k,p}(E) < \varepsilon$. See also Ziemer [438], Chapter 3, for this and other results. A related, Hölder type quasicontinuity result is due to J. Malý [285].

6.5.3. Yu. V. Netrusov [339] has announced the following result. Let $f \in F_\alpha^{p,q}$, where $\alpha > 0$, $1 \leq p < \infty$, and $1 \leq q \leq \infty$. (The results can be extended to $p, q > 0$.) Let $\beta > 0$ satisfy $0 < \alpha - \beta < N/p$. Then, for any $\varepsilon > 0$ there is a function f_ε, and an open set G_ε, such that $f_\varepsilon(x) = f(x)$ on G_ε^c, $\Lambda^{(1)}_{N-(\alpha-\beta)p}(G_\varepsilon) \leq \varepsilon$, and $f_\varepsilon \in B_{0,\beta}^{\infty,\infty}$, where the last space is defined as the closure of C_0^∞ in $B_\beta^{\infty,\infty}$.

In the same paper an analogous result for the space $B_\alpha^{p,q}$ was announced. Here the Hausdorff content $\Lambda^{(1)}_{N-(\alpha-\beta)p}$ has to be replaced by a set function $h_{N-(\alpha-\beta)p,q/p}$ defined in Netrusov [344]; cf. Section 5.6.7. In both cases the proof follows from the atomic representations of Section 4.6.

6.5.4. A. Nagel, W. Rudin, and Joel H. Shapiro [336] have investigated the tangential boundary behavior of Poisson integrals, and proved an extension of Corollary 6.2.3. Their proof uses the capacitary strong type inequality, Theorem 7.1.1, below.

6.5.5. There are many extensions of Theorem 6.3.3 (Wiener's Criterion). Here we try to give some brief hints to the literature.

W. Littman, G. Stampacchia, and H. Weinberger [275] proved that the necessary and sufficient condition of Theorem 6.3.3 extends to general linear uniformly elliptic second order equations in divergence form, whose coefficients are merely L^∞. See also E. B. Fabes, D. S. Jerison and C. E. Kenig [138], where the results of [275] are extended to degenerate elliptic equations in divergence form, and D. R. Adams [12], where the question of which elliptic equations have exactly the same regular points is considered for both divergence and non-divergence type operators. Other extensions are due to G. Dal Maso and U. Mosco [113].

The conclusion of [275] is no longer true for linear elliptic second order equation of non-divergence form, but a necessary and sufficient condition has been given by P. Bauman [48]. It was preceded by important examples of K. Miller [328]. See also M. Biroli [63].

Other notable results are Wiener type criteria for the heat equation due to E. M. Landis [262] (with the full proof belatedly published in [265]), and L. C. Evans and R. F. Gariepy [136]. The necessity of the condition of Evans and Gariepy was proved earlier by E. Lanconelli [260]. Their criterion has been extended to parabolic equations with variable coefficients by N. Garofalo and Lanconelli [178], and to the heat equation on the Heisenberg group by Garofalo and F. Segala [179]. The bibliographies of these papers contain additional references to related work. More information on parabolic equations, as well as on elliptic equations, both in divergence and non-divergence form, is found in the book by Landis [263], and in his more recent survey [264].

The boundary regularity of solutions is closely related to estimates of their modulus of continuity at the boundary. Sharp such estimates in terms of capacity for second order elliptic linear equations were first given by V. G. Maz′ya [297, 300]. In [302] Maz′ya obtained such estimates for a class of quasilinear equations including the p-Laplace equation,

$$\Delta_p u = \operatorname{div}\bigl(\nabla u |\nabla u|^{p-2}\bigr) = 0 \ ,$$

at the same time as he discovered the Wiener type condition for regularity given in Theorem 6.3.6.

A simpler proof of regularity in the case $p = N$ is given in S. Rickman [364], VII.4.12. In [177] R. F. Gariepy and W. P. Ziemer established results similar to those of Maz′ya for bounded solutions of a much larger class of quasilinear elliptic equations of the type $\operatorname{div} A(x, u, \nabla u) = B(x, u, \nabla u)$.

A problem related to the Dirichlet problem that has been much studied is the regularity of solutions of so called obstacle problems near an irregular obstacle. Both regularity criteria of the Wiener type, and estimates of the modulus of continuity of the Maz′ya type have been given. See e.g. G. Dal Maso, U. Mosco, and M. A. Vivaldi [114], and references found in that paper.

The question as to whether the sufficient condition in Theorem 6.3.6 is also necessary was open for a long time, even for the p-Laplace equation. Necessity was first proved for the case $p = N$ by P. Lindqvist and O. Martio [269], and their proof works also for $p > N - 1$. The problem was finally solved for all $p > 1$ by T. Kilpeläinen and J. Malý [250]. They also gave a characterization similar to Theorem 6.3.11 of $(1, p)$-thin sets in terms of so called p-superharmonic functions. These are essentially solutions of the inequality $-\Delta_p u \geq 0$, i.e., of $-\Delta_p u = \mu$, where $\mu \in \mathcal{M}^+(\mathbf{R}^N)$. The proofs are based on an interesting pointwise estimate for p-superharmonic functions in terms of a local version of the function $W^{\mu}_{1,p}$ (see Definition 4.5.1 and the remark following Theorem 4.5.4):

$$W^{\mu}_{1,p}(x,r) = \int_0^r \left(\frac{\mu(B(x,t))}{t^{N-p}}\right)^{p'-1} \frac{dt}{t} .$$

They proved that there are constants A_1, A_2, and A_3 such that if $-\Delta_p u = \mu$ in the ball $B(a,3r)$, then

$$A_1 W^{\mu}_{1,p}(a,r) \leq u(a) \leq A_2 \inf_{x\in B(a,r)} u(x) + A_3 W^{\mu}_{1,p}(a,2r) .$$

Thus, much of the classical connection between the Laplace equation and the Newton potential has now been extended to this nonlinear situation.

The results of Kilpeläinen and Malý, as well as those of Lindqvist and Martio, are given for equations of the type div $A(x,\nabla u) = 0$, where $A(x,\xi)\cdot\xi \approx |\xi|^p$. For operators of the general type studied by Gariepy and Ziemer much less is known; see D. R. Adams and A. Heard [19] for a result of this type in the semi-linear case. The subject is also treated in I. V. Skrypnik [382].

For equations of higher order very little is known, but Maz'ya [306] has given a sufficient condition of Wiener type for boundary regularity for the biharmonic equation, $\Delta^2 u = 0$, in dimensions $4 \leq N \leq 7$, and Maz'ya and T. Donchev [310] have extended this result to the equation $\Delta^k u = 0$, $k > 2$, for $2k \leq N \leq 2k+2$. See also Maz'ya [309].

6.5.6. The nonlinear potentials $V^{\mu}_{\alpha,p}$ do not satisfy Theorem 6.3.11 in general. They do, however, if $p > 2 - \alpha/N$.

We say that E is $\mathcal{F}_{\alpha,p}$-thin at a if there is a $\mu \in \mathcal{M}^+$ such that $V^{\mu}_{\alpha,p}$ is bounded, and

$$V^{\mu}_{\alpha,p}(a) < \liminf_{x\to a,\ x\in E\setminus\{a\}} V^{\mu}_{\alpha,p}(x) .$$

Then E is (α,p)-thin at a, $p > 1$, if E is $\mathcal{F}_{\alpha,p}$-thin there. If $p > 2 - \alpha/N$ the restriction to bounded potentials is superfluous, and then the converse is also true.

On the other hand E is $\mathcal{F}_{\alpha,p}$-thin at a for $1 < p < 2 - \alpha/N$, if

$$\int_0^1 \left(\frac{C_{\alpha,p}(E\cap B(a,r))}{r^{N-\alpha p}}\right)^{(N-\alpha)/(N-\alpha p)} \frac{dr}{r} < \infty ,$$

and for $p = 2 - \alpha/N$, if

$$\int_0^1 \left(\frac{C_{\alpha,p}(E\cap B(a,r))}{r^{N-\alpha p}} \log \frac{Ar^{N-\alpha p}}{C_{\alpha,p}(E\cap B(a,r))}\right)^{p'-1} \frac{dr}{r} < \infty .$$

The exponent $(N-\alpha)/(N-\alpha p)$ is best possible. These results are due to Adams and Meyers [22] and (in part) to Hedberg [207].

6.5.7. One can use the fine topologies to develop a theory of fine differentiability, i.e., $h \cdot \nabla f(x)$ is the fine differential of f at x if $f(x+h) - f(x) - h \cdot \nabla f(x) = o(h)$, as $h \to 0$ outside a thin set. Such an approach was taken by Y. Mizuta in [330]. The ideas were carried out further in Adams [10], Chapter 3, where the idea of a fine differential was connected to the existence of an L^p derivative. See also B. Fuglede [169, 170], and A. M. Davie and B. Øksendal [116].

6.5.8. There is no such natural ordering in the family of (α, p)-fine topologies as in the family of (α, p)-capacities (Theorem 5.5.1). Denote by $\tau_{\alpha,p}$ the class of (α, p)-finely open sets. Then $\tau_{\alpha,p} \subset \tau_{\beta,q}$ and $\tau_{\alpha,p} \neq \tau_{\beta,q}$ if either

(a) $0 < \beta q < \alpha p < N, \quad \beta/p' + N/p \leq \alpha/q' + N/q,$ or
(b) $0 < \beta q < \alpha p = N,$ or
(c) $\beta q = \alpha p = N, \quad q > p.$

Here the inclusions follow from Theorem 5.5.1. If $0 < \beta q \leq \alpha p < N$, and $\beta/p' + N/p > \alpha/q' + N/q$, then neither $\tau_{\alpha,p} \subset \tau_{\beta,q}$ nor $\tau_{\beta,q} \subset \tau_{\alpha,p}$ is true. See Adams and Hedberg [20], and also K. Hatano [198, 199].

6.5.9. If $G \subset \mathbf{R}^N$ is open and connected, then G is also connected in the (α, p)-fine topology for all (α, p) with $1 < p \leq N/\alpha$.

If $\alpha p > 1$ then G is also connected in the (α, p)-quasitopology, i.e. if $G = A \cup B$ where A and B are quasiopen and disjoint, then either A or B must have zero (α, p)-capacity.

The second result is difficult, and depends on the following inequality: If $\alpha p > 1$, then there is a constant A such that for any $E \subset \mathbf{R}^N$ and all open cubes Q

$$\min\{ C_{\alpha,p}(E \cap Q), C_{\alpha,p}(E^c \cap Q) \} \leq A C_{\alpha,p}(\widetilde{E} \cap \widetilde{E^c} \cap Q) \ .$$

Note that $\widetilde{E} \cap \widetilde{E^c}$ is the fine boundary of E.

If $\alpha p \leq 1$, then a hyperplane in $\mathbf{R}^N$ has (α, p)-capacity zero (Corollary 5.1.15). It follows that a closed halfspace is quasiopen, and thus $\mathbf{R}^N$ is disconnected in the (α, p)-quasitopology in this case.

These results are due to D. R. Adams and J. L. Lewis [21]. The proof of the main inequality also uses an argument of Burgess Davis. In the classical case the results are due to B. Fuglede [166, 167]. His proofs depend on balayage (sweeping), which is not available in the general situation.

6.5.10. If $E \subset \mathbf{R}^N$ is (α, p)-finely open and (α, p)-finely connected for (α, p) such that $\alpha p > 1$, then E is arcwise connected. The arcs can even be chosen as unions of line segments parallel to the coordinate axes, and their lengths can be bounded. The theorem is false for $\alpha p \leq 1$. See D. R. Adams and J. L. Lewis [21]. In the classical case the result is due to T. Lyons [282].

6.5.11. The (α, p)-fine topologies are locally connected if $\alpha p > 1$, but not if $\alpha p \leq 1$. The proof depends on the "quasi-Lindelöf property". This property, which holds for $\alpha p > 0$, says that any family of (α, p)-open sets has a denumerable subfamily whose union differs from the union of the whole family only by a set of (α, p)-capacity zero. See J. Heinonen, T. Kilpeläinen and J. Malý [220]. In the linear case these results are due to J. L. Doob [124] (see also [125]) and B. Fuglede [166, 167].

6.5.12. Potential theory related to weighted L^p-spaces, and a "weighted thinness", with applications to degenerate elliptic equations of second order, were investigated by D. R. Adams [12], and S. K. Vodop′yanov [420]. Many other results on weighted Sobolev spaces are found in J. Heinonen, T. Kilpeläinen and O. Martio [221], T. Kilpeläinen [248], and B. O. Turesson [408].

6.6 Notes

6.1. The existence of quasicontinuous and finely continuous representatives was proved in the classical case of $W^{1,2}$ by J. Deny in his thesis [119], Théorème IV.1, p. 171. The study was continued by J. Deny [120] (for $W^{1,p}$, $p \leq 2$), and by J. Deny and J. L. Lions [122]. See also N. Aronszajn [36], N. Aronszajn and K. T. Smith [39], and B. Fuglede [160].

Deny and Lions [122] also proved the uniqueness theorem, Theorem 6.1.4, in the same way as we do in Section 6.4 (Remark following Corollary 6.4.7). See Propositions II:3.2, II.3.5, and II.3.6, p. 353, and p. 355 in [122]. A different proof was given by H. Wallin [421]. The extension to the nonlinear case is due to V. P. Havin and V. G. Maz′ya [203], Lemma 5.8. This is the proof we have given here. See also T. Sjödin [380].

6.2. The method of proving Theorem 6.2.1 is essentially that of N. Wiener [434] in his proof of the Lebesgue differentiation theorem (Theorem III). See also E. M. Stein [389], Section I.1.5, p. 9. The theorem is due to H. Federer and W. P. Ziemer [142] for $\alpha = 1$, and to T. Bagby and W. P. Ziemer [46], C. P. Calderón, E. B. Fabes, and N. M. Riviere [90], and N. G. Meyers [320] in the general case. Lemma 6.2.2 was proved in Adams [3] for $q = 1$, and in Hedberg [212] (Lemma 4.2.a) for $q = p$. It is contained somewhat implicitly in [46], Theorem 3.1(c), p. 136. The subject is studied in much greater detail in W. P. Ziemer [438], Chapter 3.

6.3. The history of thin sets goes back at least to 1924, when N. Wiener in [433] published his criterion for the solvability of Dirichlet's problem, Theorem 6.3.3. Definition 6.3.1 and Theorem 6.3.2 are due to M. Brelot (see [73, 74, 77]). The Kellogg property, Theorem 6.3.4, was proved by G. C. Evans [133], after it had been conjectured by O. D. Kellogg (see [244], Chapter XI.20), and proved by him in the planar case [243]. See also O. Frostman's thesis [158], p. 79. The Choquet property, Theorem 6.3.5, is due to G. Choquet [104]. Much further information is found e.g. in Brelot [77], L. L. Helms [222], N. S. Landkof [266], and in J. L. Doob's monumental treatise [125].

The history of Theorem 6.3.6 was already told in 6.5.5 above.

Wiener criteria for potentials $V^{\mu}_{\alpha,p}$ were investigated by D. R. Adams and N. G. Meyers [22], and by Hedberg [206]. In particular, Adams and Meyers found that the properties here called (A) and (B) are not in general equivalent. See Section 6.5.6 above.

The definitions of an (α, p)-thin set (Definition 6.3.7) and of the (α, p)-fine topology were given by Meyers [321]. Theorem 6.3.11 is a variant of a result (Theorem 4) in Hedberg and Wolff [219]. In that paper the theorem was given for the nonlinear potential $W^{\mu}_{\alpha,p}$, and the proof depended on a theory for atomic nonlinear potentials similar to our $\mathfrak{V}^{\mu}_{\alpha,p}$ (see Section 4.7). The proof given here is similar to the proof of Theorems 5.1 and 5.3 in [22], and is related to Frostman's proof of the Wiener criterion in [159].

The estimate in Proposition 6.3.12 was proved for $V^{\mu}_{\alpha,p}$, $p \geq 2$, in [22] (Prop. 4.4), and extended to $p > 2-\alpha/N$ by Adams and Hedberg [20] (Lemma 4) (cf. the remark following Proposition 6.3.12 above). For $p = 2$ the result is classical and found e.g. in L. Carleson [92], Theorem III.5. The observation that $W^{\mu}_{\alpha,p}(x) < \infty$ (α, p)-q.e. is due to Meyers [321] (Theorem 2.1), and the proof of Proposition 6.3.12 given here (and in [20]) is a modification of his proof.

6.4. The first proof of Theorem 6.4.5 given here is essentially that given by Deny [119], p. 171. For Corollary 6.4.7, see Deny and Lions [122], Théorème II.3.2, p. 356. Fine continuity and quasicontinuity have been investigated in depth by B. Fuglede [163, 164, 166, 167]. See also M. Brelot [77], and the survey by J. Lukeš and J. Malý [280]. It was observed by Hedberg [207] that Fuglede's arguments apply in the nonlinear situation, and the extension to all (α, p) became possible after Th. H. Wolff had proved his inequality (Theorem 4.5.2). See Hedberg and Wolff [219]. The second proof of Theorem 6.4.5 and the proof of Theorem 6.4.6 are essentially taken from [164]. Theorem 6.4.14 is due to N. G. Meyers [321].

7. Trace and Imbedding Theorems

In Chapter 6 we investigated the continuity properties of distinguished representatives of the equivalence classes constituting elements of function spaces. In the present chapter we shall study the integrability properties of these representatives.

More precisely, our main interest is to characterize those measures μ in $\mathcal{M}^+(\mathbf{R}^N)$ for which $L^{\alpha,p}$, $\alpha > 0$, is continuously (or even compactly) imbedded in some $L^q(\mu)$. Equivalently, we want to characterize those μ such that the mapping $\mathcal{G}_\alpha : L^p \to L^q(\mu)$ defined by $f \mapsto G_\alpha * f$ is continuous, or compact. Such imbeddings are often referred to as *trace inequalities*, since we are restricting the potential $G_\alpha * f$ to the support of μ. For example, μ might be the surface measure on a smooth manifold in $\mathbf{R}^N$.

The results depend on an estimate of so called *capacitary strong type*, which we formulate for general potentials $\mathcal{G}f = g * f$, where g is a radially decreasing convolution kernel. Section 7.1 is mainly devoted to this inequality (Theorem 7.1.1), and then in Sections 7.2 and 7.3 we study the continuity and the compactness of the map $\mathcal{G} : L^p \to L^q(\mu)$.

In Section 7.4 we introduce a space $L^p(C_{\alpha,p})$ of (α, p)-quasicontinuous functions "integrable with respect to capacity", and we characterize its dual as the space of measures whose total variation belong to $L^{-\alpha,p'}$.

In Section 7.5 we prove a sharper capacitary strong type inequality, using an entirely different approach, which is based on the atomic representation of functions in Lizorkin–Triebel spaces presented in Section 4.6.

7.1 A Capacitary Strong Type Inequality

First of all, we notice that any (α, p)-quasicontinuous function u is measurable with respect to any $\mu \in \mathcal{M}$ that is absolutely continuous with respect to (α, p)-capacity, in the sense that $|\mu|(E) = 0$ for every (Borel) set E with $C_{\alpha,p}(E) = 0$. In fact, it follows from Definition 6.1.1 and the Tietze extension theorem that a quasicontinuous function is the pointwise limit of continuous functions outside some G_δ set of (α, p)-capacity zero.

Thus $\int_{\mathbf{R}^N} u \, d\mu$ is well defined as soon as $\int_{\mathbf{R}^N} |u| \, d|\mu| < \infty$. It follows from Theorem 6.1.4 that if u and v are quasicontinuous, then $\int_{\mathbf{R}^N} u \, d\mu = \int_{\mathbf{R}^N} v \, d\mu$, even if we only know that $u = v$ a.e.

In particular, the easy part of Theorem 2.5.1 gives that for any Borel set E and any $\mu \in \mathcal{M}^+$ we have

$$\mu(E) \le C_{\alpha,p}(E)^{1/p} \|G_\alpha * \mu\|_{p'} \ .$$

Thus all (α, p)-quasicontinuous u are measurable with respect to all μ such that $|\mu| \in L^{-\alpha,p'}$, and equality (α, p)-q.e. implies equality μ-a.e.

If $u \in L^{\alpha,p}$ for $p > 1$ and $\alpha > 0$, we can choose an (α, p)-quasicontinuous representative $G_\alpha * f$ for u with $f \in L^p$. Thus, for any $\mu \in \mathcal{M}(\mathbf{R}^N)$ such that $|\mu| \in L^{-\alpha,p'}$ we have

$$\begin{aligned} \int_{\mathbf{R}^N} |u|\, d|\mu| &\le \int_{\mathbf{R}^N} G_\alpha * |f|\, d|\mu| \\ &= \int_{\mathbf{R}^N} |f|(G_\alpha * |\mu|)\, dx \le \|f\|_p \|G_\alpha * |\mu|\|_{p'} < \infty \ . \end{aligned} \tag{7.1.1}$$

It follows that u belongs to $L^1(|\mu|)$ for such measures, and that $L^{\alpha,p}$ is continuously imbedded in $L^1(|\mu|)$. It follows, moreover, that in this case the duality between $L^{-\alpha,p'}$ and $L^{\alpha,p}$ is given by

$$\langle \mu, u \rangle = \int_{\mathbf{R}^N} u\, d\mu = \int_{\mathbf{R}^N} f(G_\alpha * \mu)\, dx \ . \tag{7.1.2}$$

This result is extended in Theorem 7.4.4 below.

In order to characterize those positive measures μ for which $L^{\alpha,p}$ is continuously imbedded in $L^q(\mu)$ for some q, we start from the elementary identity

$$\int_{\mathbf{R}^N} |u|^q\, d\mu = \int_0^\infty \mu(\{x : |u(x)| \ge \lambda\})\, d\lambda^q \ , \tag{7.1.3}$$

which is valid for all μ-measurable functions u, and $q > 0$, and is easily proved by writing

$$\mu(\{x : |u(x)| \ge \lambda\}) = \mu(E_\lambda) = \int_{\mathbf{R}^N} \chi_{E_\lambda}\, d\mu \ ,$$

and changing the order of integration. (Here $d\lambda^q$ means $q\lambda^{q-1} d\lambda$.)

In Chapter 2 we noted the elementary inequality (see Proposition 2.3.7)

$$C_{g,p}(\{x : \mathcal{G}f(x) \ge \lambda\}) \le \lambda^{-p} \int_{\mathbf{R}^N} f^p\, d\nu \ , \tag{7.1.4}$$

which is valid for a potential with respect to a general kernel g.

In this section we shall prove a stronger version of this estimate in the case when $\mathcal{G}f$ is given by a radially decreasing convolution kernel, $\mathcal{G}f = g * f$, where $g(x) = g_0(|x|)$, $x \in \mathbf{R}^N$, as considered in Section 2.6. When specialized to Riesz and Bessel kernels, this result will be our main tool in applying (7.1.3) to prove the desired imbedding theorems.

We always assume that g is a radially decreasing convolution kernel that satisfies the basic conditions that ensure that $C_{g,p}$ is neither trivial, nor different from zero on all nonempty sets. These conditions are, respectively,

$$\int_{|x|>1} g(x)^{p'}\,dx < \infty\ , \qquad \text{and} \qquad \int_{|x|<1} g(x)^{p'}\,dx = \infty\ ;$$

see Proposition 2.6.1.

Inequalities such as (7.1.4) are known as *weak type* estimates. (See e.g. Stein [389], Section I.4.) It is therefore natural that the following result has become known as a *capacitary strong type inequality.*

Theorem 7.1.1. *If g is a radially decreasing convolution kernel on $\mathbf{R}^N$, then there exists a constant A, depending only on N and p, $1 < p < \infty$, such that*

$$\int_0^\infty C_{g,p}(\{x : g * f(x) \geq \lambda\})\,d\lambda^p \leq A\,\|f\|_p^p\ , \tag{7.1.5}$$

for all $f \in L_+^p(\mathbf{R}^N)$.

Proof. Set

$$E_k = \{x : g * f(x) \geq 2^k\} \quad \text{for } k \in \mathbf{Z}\ .$$

As is easily seen it is equivalent to prove the inequality

$$\sum_{k=-\infty}^{\infty} 2^{kp} C_{g,p}(E_k) \leq A\,\|f\|_p^p\ . \tag{7.1.6}$$

We first prove (7.1.6) under the assumption that f is a nonnegative function in $C_0^\infty(\mathbf{R}^N)$. Note that the E_k are then compact subsets of $\mathbf{R}^N$, and empty for sufficiently large k. Let J be the sum on the left side of (7.1.6), and let μ_k be the extremal measure for the set E_k given in Theorem 2.5.3. Then

$$\begin{aligned} J &= \sum_{k=-\infty}^{\infty} 2^{kp} \int_{\mathbf{R}^N} d\mu_k \leq \sum_{k=-\infty}^{\infty} 2^{k(p-1)} \int_{\mathbf{R}^N} (g * f)\,d\mu_k \\ &= \sum_{k=-\infty}^{\infty} 2^{k(p-1)} \int_{\mathbf{R}^N} f(g * \mu_k)\,dx \leq \|f\|_p \Big\| \textstyle\sum_{k=-\infty}^{\infty} 2^{k(p-1)} g * \mu_k \Big\|_{p'}\ . \end{aligned}$$

Setting L equal to the p'-power of the second norm, we have

$$J \leq \|f\|_p L^{1/p'}.$$

We conclude the proof by showing that $L \leq A\,J$, for some constant A.

Case 1. $p \geq 2$. We begin by setting

$$\Lambda(x) = \sum_{k=-\infty}^{\infty} 2^{k(p-1)} g * \mu_k(x)\ ,$$

and

$$\Lambda_j(x) = \sum_{k=j}^{\infty} 2^{k(p-1)} g * \mu_k(x) \ .$$

Notice that due to our assumptions on f, we have $\Lambda_j \in L^{p'}$ and $\Lambda_j \nearrow \Lambda$, as $j \to -\infty$. Thus, in particular, $\Lambda_j(x) < \infty$ for a.e. x. Next we notice that

$$\begin{aligned}
\Lambda(x)^{p'} &= \lim_{n\to-\infty} \Lambda_n^{p'} = \lim_{n\to-\infty} \sum_{j=n}^{\infty} (\Lambda_j(x)^{p'} - \Lambda_{j+1}(x)^{p'}) \\
&\le p' \sum_{j=-\infty}^{\infty} \Lambda_j(x)^{p'-1} \big(\Lambda_j(x) - \Lambda_{j-1}(x)\big) \\
&= p' \sum_{j=-\infty}^{\infty} \Lambda_j(x)^{p'-1} 2^{j(p-1)} g * \mu_j(x)
\end{aligned}$$

for a.e. x, where the inequality follows from the mean value theorem. Thus, by the Hölder inequality

$$\begin{aligned}
L &\le p' \int_{\mathbf{R}^N} \sum_j 2^{jp(2-p')} (g * \mu_j)^{p'(2-p')} \cdot \Lambda_j^{p'-1} 2^{j(p'-1)} (g * \mu_j)^{(p'-1)^2} \, dx \\
&\le p' L_1^{2-p'} L_2^{p'-1} \ ,
\end{aligned}$$

where

$$\begin{aligned}
L_1 &= \int_{\mathbf{R}^N} \sum_j 2^{jp} (g * \mu_j)^{p'} \, dx = \sum_j 2^{jp} \int_{\mathbf{R}^N} V_{g,p}^{\mu_j} \, d\mu_j \\
&= \sum_j 2^{jp} C_{g,p}(E_j) = J \ ,
\end{aligned}$$

by Theorem 2.5.3, and

$$\begin{aligned}
L_2 &= \int_{\mathbf{R}^N} \sum_j \Lambda_j 2^j (g * \mu_j)^{p'-1} \, dx \\
&= \sum_j \sum_{k \ge j} 2^{k(p-1)} 2^j \int_{\mathbf{R}^N} (g * \mu_k)(g * \mu_j)^{p'-1} \, dx \\
&= \sum_j \sum_{k \ge j} 2^{k(p-1)} 2^j \int_{\mathbf{R}^N} V_{g,p}^{\mu_j} \, d\mu_k \\
&\le A \sum_j \sum_{k \ge j} 2^{k(p-1)} 2^j C_{g,p}(E_k) \\
&= 2A \sum_k 2^{kp} C_{g,p}(E_k) = 2A\, J \ ,
\end{aligned}$$

by the boundedness principle (Theorem 2.6.3).

Case 2. $1 < p < 2$. We define $\Lambda(x)$ as before, but this time we set

$$\Lambda_j(x) = \sum_{k=-\infty}^{j} 2^{k(p-1)} g * \mu_k(x) \ .$$

Analogously we get

$$\Lambda(x)^{p'} \le p' \sum_{j=-\infty}^{\infty} \Lambda_j(x)^{p'-1} 2^{j(p-1)} g * \mu_j(x)$$

for a.e. x. Hence

$$L \le p' \sum_{j=-\infty}^{\infty} 2^{j(p-1)} \|\Lambda_j\|_{p'-1;\sigma_j}^{p'-1} ,$$

where the last norm is taken with respect to the measure $d\sigma_j = (g * \mu_j)dx$. Thus we have

$$\begin{aligned} L &\le p' \sum_{j=-\infty}^{\infty} 2^{j(p-1)} \Big(\sum_{k=-\infty}^{j} 2^{k(p-1)} \|g * \mu_k\|_{p'-1;\sigma_j} \Big)^{p'-1} \\ &\le p' \sum_{j=-\infty}^{\infty} 2^{j(p-1)} \Big(\sum_{k=-\infty}^{j} 2^{k(p-1)} \Big(\int_{\mathbf{R}^N} V_{g,p}^{\mu_k} \, d\mu_j \Big)^{1/(p'-1)} \Big)^{p'-1} \\ &\le A \sum_{j=-\infty}^{\infty} 2^{j(p-1)} C_{g,p}(E_j) 2^j = A\, J \ , \end{aligned}$$

again by Theorem 2.6.3.

The result now follows by approximating $f \in L^p$ by elements of $C_0^\infty(\mathbf{R}^N)$. In fact, using Proposition 2.3.8 one easily shows that if $f_j \to f$ in L^p, then

$$C_{g,p}(\{x : g * f(x) \ge \lambda\}) \le \liminf_{j\to\infty} C_{g,p}(\{x : g * f_j(x) \ge \lambda\}) \ .$$

This finishes the proof.

7.2 Imbedding of Potentials

We now use Theorem 7.1.1 to investigate the continuity of the map

$$\mathcal{G} : L^p(\mathbf{R}^N) \to L^q(\mu, \mathbf{R}^N) \ ,$$

defined by $f \mapsto \mathcal{G}f = g * f$ for a $\mu \in \mathcal{M}^+(\mathbf{R}^N)$. As noted in the introduction to the chapter, when specialized to $g = G_\alpha$, $\alpha > 0$, this is equivalent to the continuity of the imbedding

$$L^{\alpha,p}(\mathbf{R}^N) \to L^q(\mu, \mathbf{R}^N) \ .$$

We denote by μ_E the restriction of μ to a set E.

Theorem 7.2.1. *Let g be a radially decreasing convolution kernel, and let $\mu \in \mathcal{M}^+(\mathbf{R}^N)$. Then for $1 < p \le q < \infty$ the following properties of μ are equivalent*:
(a) *There is a constant A_1 such that*

$$\Big(\int_{\mathbf{R}^N} |g * f|^q \, d\mu \Big)^{1/q} \le A_1 \|f\|_p \ , \tag{7.2.1}$$

for all $f \in L^p$.

(b) *There is a constant* A_2 *such that*

$$\|g * \mu_K\|_{p'} \leq A_2 \mu(K)^{1/q'} , \tag{7.2.2}$$

for all compact sets K.
(c) *There is a constant* A_3 *such that*

$$\sup_{t>0} t\, \mu(\{x : |g * f(x)| \geq t\})^{1/q} \leq A_3 \|f\|_p , \tag{7.2.3}$$

for all $f \in L^p$.
(d) *There is a constant* A_4 *such that*

$$\mu(K)^{1/q} \leq A_4 C_{g,p}(K)^{1/p} \tag{7.2.4}$$

for all compact sets K.
Moreover, the quantity

$$\Theta_q(\mu) \equiv \sup\{ \mu(K)^{1/q} C_{g,p}(K)^{-1/p} : K \text{ compact } \} \tag{7.2.5}$$

is comparable to the norm $\|\mathcal{G}\|$ *of the operator*

$$\mathcal{G} : L^p(\mathbf{R}^N) \to L^q(\mu, \mathbf{R}^N) ,$$

as are the least possible values of the constants A_1, A_2, *and* A_3.

Remark 1. Notice that in the language of harmonic analysis (c) says that the operator $\mathcal{G}$ satisfies an estimate of *weak type* (p, q), whereas (a) says that $\mathcal{G}$ is of *strong type* (p, q). See e.g. Stein [389], Section I.4. Hence, we are claiming here that $\mathcal{G}$ is of weak type (p, q) if and only if $\mathcal{G}$ is of strong type (p, q).

Remark 2. In Section 7.6.6 below we give a necessary and sufficient condition for the inequality (7.2.1) to hold in the more difficult case $0 < q < p$.

Proof. We show (a) $\Rightarrow$ (b) $\Rightarrow$ (c) $\Rightarrow$ (d) $\Rightarrow$ (a).

The first implication is immediate from

$$\begin{aligned}\int_{\mathbf{R}^N} f(g * \mu_K)\, dx &= \int_{\mathbf{R}^N} (g * f)\, d\mu_K \\ &\leq \|g * f\|_{L^q(\mu)}\, \mu(K)^{1/q'} \leq A_1 \|f\|_p\, \mu(K)^{1/q'} ,\end{aligned}$$

for all $f \in L^p$.

For the second one, let $f \in C_0^\infty(\mathbf{R}^N)$. Then, since $E_t = \{x : |g * f(x)| \geq t\}$ is compact, (b) implies

$$t\, \mu(E_t) \leq \int_{\mathbf{R}^N} |g * f|\, d\mu_{E_t} \leq \|f\|_p \|g * \mu_{E_t}\|_{p'} \leq A_2 \|f\|_p\, \mu(E_t)^{1/q'} .$$

The result follows upon approximating an arbitrary $f \in L^p$ by a sequence from $C_0^\infty(\mathbf{R}^N)$ in the L^p-norm.

For the third implication we merely choose f so that $g * f \geq 1$ on K. It follows from (c) that $\mu(K)^{1/q} \leq A_3\|f\|_p$, and then the definition of capacity gives $\mu(K)^{1/q} \leq A_3 C_{g,p}(K)^{1/p}$.

Finally, we apply Theorem 7.1.1 to $g * f$, with $f \in C_0^\infty(\mathbf{R}^N)$. Then, applying (7.1.3) and assuming (d),

$$\int_{\mathbf{R}^N} |f * g|^q \, d\mu = \int_0^\infty \mu(E_t)\, dt^q \leq \Theta_q(\mu)^q \int_0^\infty C_{g,p}(E_t)^{q/p}\, dt^q \ .$$

But $t\, C_{g,p}(E_t)^{1/p} \leq \|f\|_p$ by (7.1.4), and thus since $q \geq p$,

$$\int_0^\infty C_{g,p}(E_t)^{q/p}\, dt^q \leq (q/p)\|f\|_p^{q-p} \int_0^\infty C_{g,p}(E_t)\, dt^p \ .$$

Thus, by Theorem 7.1.1

$$\int_{\mathbf{R}^N} |f * g|^q \, d\mu \leq A\,(q/p)\Theta_q(\mu)^q \|f\|_p^q \ ,$$

and again we finish by approximating with C_0^∞-functions.

When $1 < p < q < \infty$ in Theorem 7.2.1, the condition in (d) can in many cases be replaced by a simpler one, in which the family of all compact sets is replaced by the family of all (closed) balls. We will demonstrate this for the Riesz and Bessel kernels.

Theorem 7.2.2. *If $1 < p < q < \infty$, then a necessary and sufficient condition for the map $f \mapsto I_\alpha * f$ to be continuous from $L^p(\mathbf{R}^N)$ into $L^q(\mu, \mathbf{R}^N)$ is that*

$$\sup_{x, r>0} \frac{\mu(B(x,r))}{\dot{C}_{\alpha,p}(B(x,r))^{q/p}} < \infty \ . \tag{7.2.6}$$

For the continuity of $f \mapsto G_\alpha * f$, replace (7.2.6) *by*

$$\sup_{x, 0<r\leq 1} \frac{\mu(B(x,r))}{C_{\alpha,p}(B(x,r))^{q/p}} < \infty \ . \tag{7.2.7}$$

Remark. It should be noted that this result cannot be extended to the case $p = q$. Indeed, if $g = G_\alpha$ and $h(r) = r^{N-\alpha p}$, $\alpha p < N$, and K is a compact set with $0 < \Lambda_h(K) < \infty$, then Proposition 5.1.5 and Frostman's theorem, Theorem 5.1.12, give a $\mu \in \mathcal{M}^+(K)$ satisfying $\mu(B(x,r)) \leq h(r)$ for all balls, and such that $0 < A\,\Lambda_h^{(\infty)}(K) \leq \mu(K)$. By Propositions 5.1.2 and 5.1.4 this measure satisfies Theorem 7.2.1(d) for all balls. If this implies that (d) holds for all compact sets we have $\mu(K) \leq A\, C_{\alpha,p}(K)$. But $C_{\alpha,p}(K) = 0$ by Theorem 5.1.9.

Proof. In order to show that (7.2.6) is sufficient, we show that (7.2.2) holds. Of course (7.2.6) is of interest only when $\alpha p < N$. We write (Lemma 3.1.1)

$$I_\alpha * \mu_K(x) = A \int_0^\infty \frac{\mu_K(B(x,r))}{r^{N-\alpha}} \frac{dr}{r} \ ,$$

and then by Minkowski's inequality (1.1.5)

$$\|I_\alpha * \mu_K\|_{p'} \le A \int_0^\infty \frac{\|\mu_K(B(\cdot\,, r))\|_{p'}}{r^{N-\alpha}} \frac{dr}{r} \ .$$

We conclude by estimating $\|\mu_K(B(\cdot\,, r))\|_{p'}$ in two ways. Without any assumption on μ we have for all r

$$\begin{aligned} \|\mu_K(B(\cdot\,, r))\|_{p'}^{p'} &= \int_{\mathbf{R}^N} \mu(K \cap B(x, r))^{p'}\, dx \\ &\le \mu(K)^{p'-1} \int_{\mathbf{R}^N} \int_{K \cap B(x,r)} d\mu(y)\, dx \\ &\le \mu(K)^{p'-1} A\, r^N \mu(K) \ . \end{aligned} \tag{7.2.8}$$

By (7.2.6) and Proposition 5.1.2 we have

$$\begin{aligned} \|\mu_K(B(\cdot\,, r))\|_{p'}^{p'} &\le A\, (r^{N-\alpha p})^{(p'-1)q/p} \int_{\mathbf{R}^N} \int_{K \cap B(x,r)} d\mu(y)\, dx \\ &= A\, r^{(N-\alpha p)(p'-1)q/p} r^N \mu(K) \ . \end{aligned} \tag{7.2.9}$$

Thus by (7.2.8)

$$\int_R^\infty \frac{\|\mu_K(B(\cdot\,, r))\|_{p'}}{r^{N-\alpha}} \frac{dr}{r} \le A\, \mu(K) R^{\alpha - N/p} \ ,$$

and by (7.2.9)

$$\int_0^R \frac{\|\mu_K(B(\cdot\,, r))\|_{p'}}{r^{N-\alpha}} \frac{dr}{r} \le A\, \mu(K)^{1/p'} R^{(\alpha - N/p)(1-q/p)} \ .$$

Now, choosing R so that $R^{\alpha - N/p} = \mu(K)^{-1/q}$ gives the result.

For the second half of the theorem, and $\alpha p < N$, we obtain (7.2.9) from (7.2.7) as before, but only when $r \le 1$. This gives the result if $\mu(K) \le 1$, so that $R \le 1$. But for $r > 1$ we have the estimate $\mu(B(x, r)) \le A\, r^N$. In fact, (7.2.7) gives a uniform estimate for μ on any ball of radius 1, and then the estimate for general r follows easily by a covering argument. Thus, applying (7.2.9) for $r \le 1$, and using the exponential decay of the Bessel kernel at infinity, we find in the same way

$$\begin{aligned} \|G_\alpha * \mu_K\|_{p'} &\le A \int_0^\infty \frac{\|\mu_K(B(\cdot\,, r))\|_{p'}}{r^{N-\alpha} e^{r/2}} \frac{dr}{r} \\ &\le A\, \mu(K)^{1/p'} + A\, \mu(K)^{1/p'} \int_1^\infty \frac{r^{Nq/p^2} r^{N/p'}}{r^{N-\alpha} e^{r/2}} \frac{dr}{r} \\ &\le A\, \mu(K)^{1/p'} \le A\, \mu(K)^{1/q'} \ , \end{aligned}$$

since $p < q$ and $\mu(K) \ge 1$. This gives (7.2.2).

Finally, we consider the case $\alpha p = N$. This case seems to lie somewhat deeper than the previous one. In fact, notice that if we use Proposition 5.1.3 to estimate $C_{\alpha,p}(B(x,r))$, then estimating the integral

$$\int_0^R \frac{\|\mu_K(B(\cdot\,,r))\|_{p'}}{r^{N-\alpha}} \frac{dr}{r}$$

as before does not give a finite result unless q is taken sufficiently large. But we are interested in all $q > p$. Consequently we are forced to make a subtler estimate of $G_\alpha * \mu_K$ than before. But applying Wolff's inequality, Theorem 4.5.2, we can write

$$\|G_\alpha * \mu_K\|_{p'}^{p'} \le A \int_K \int_0^1 \mu_K(B(x,r))^{p'-1} \frac{dr}{r}\, d\mu(x)\ .$$

By Proposition 5.1.3 and the estimate (7.2.9) we have

$$\mu_K(B(x,r)) \le A \left(\log \frac{e}{r}\right)^{(1-p)q/p}\ ,$$

and trivially $\mu_K(B(x,r)) \le \mu(K)$. Thus, for any $R \le 1$

$$\begin{aligned}\|G_\alpha * \mu_K\|_{p'}^{p'} &\le A\,\mu(K) \int_0^R \frac{dr}{r(\log e/r)^{q/p}} + A\,\mu(K)^{p'} \int_R^1 \frac{dr}{r} \\ &\le \frac{A\,\mu(K)}{(\log e/r)^{(q/p)-1}} + A\,\mu(K)^{p'} \log \frac{e}{r}\ ,\end{aligned}$$

since $q > p$. If $\mu(K) \le 1$ we can now choose $R \le 1$ so that $\mu(K) = (\log e/R)^{(1-p)q/p}$, and the result follows. If $\mu(K) > 1$ we choose $R = 1$, which gives

$$\|G_\alpha * \mu_K\|_{p'}^{p'} \le A\,\mu(K) \le A\,\mu(K)^{p'/q'}\ ,$$

since $p' > q'$.

The necessity of (7.2.6) and (7.2.7) follows immediately from Theorem 7.2.1.

7.3 Compactness of the Imbedding

We now turn our attention to the compactness of the map $\mathcal{G}f = g * f$, $\mathcal{G} : L^p(\mathbf{R}^N) \to L^q(\mu, \mathbf{R}^N)$. The following theorem is our main result.

Theorem 7.3.1. *Let g be a radially decreasing convolution kernel, and let $\mu \in \mathcal{M}^+(\mathbf{R}^N)$. For $1 < p \le q < \infty$ the map $\mathcal{G} : L^p(\mathbf{R}^N) \to L^q(\mu, \mathbf{R}^N)$ is compact if and only if the following two conditions are satisfied:*

$$\lim_{\delta\to 0} \sup_{\operatorname{diam} K \le \delta} \frac{\mu(K)}{C_{g,p}(K)^{q/p}} = 0\ ; \tag{7.3.1}$$

$$\lim_{\rho\to\infty} \sup_{K\subset \mathbf{R}^N\setminus B(0,\rho)} \frac{\mu(K)}{C_{g,p}(K)^{q/p}} = 0\ . \tag{7.3.2}$$

Proof. We prove the theorem in several steps.

Step 1. Suppose that μ has compact support. Then condition (7.3.2) is automatically satisfied. We claim that (7.3.1) implies that $\mathcal{G} : L^p(\mathbf{R}^N) \to L^q(\mu, \mathbf{R}^N)$ is compact.

Denote by g_δ the truncated kernel,

$$g_\delta(x) = \min\{g(x), g(\delta)\}, \quad \delta > 0 ,$$

and set $g^\delta = g - g_\delta$. Denote by $\mathcal{G}_\delta$ and $\mathcal{G}^\delta$ the corresponding operators. We first prove that

$$\lim_{\delta\to 0} \|\mathcal{G}^\delta\| = 0 , \tag{7.3.3}$$

where $\|\cdot\|$ denotes the operator norm.

For $\varepsilon > 0$ choose δ, so that if $\operatorname{diam} K < \delta$, then

$$\frac{\mu(K)^{1/q}}{C_{g,p}(K)^{1/p}} < \varepsilon . \tag{7.3.4}$$

Partition $\mathbf{R}^N$ into cubes $\{Q\}$ with disjoint interiors, each of side length s for an s such that $s\sqrt{N} < \delta/3$. Let $3Q$ be the cube concentric to Q with side $3s$. Denote by f_Q the restriction of f to Q. Then clearly $\operatorname{supp}(g^s * f_Q) \subset 3Q$.

Next observe that

$$|g^s * f(x)|^q = \Big|\sum_Q g^s * f_Q(x)\Big|^q \le 3^{N(q-1)} \sum_Q |g^s * f_Q(x)|^q ,$$

since at most 3^N cubes contribute to the sum for any fixed x. Thus

$$\int_{\mathbf{R}^N} |g^s * f(x)|^q \, d\mu \le 3^{N(q-1)} \sum_Q \int_{3Q} |g^s * f_Q(x)|^q \, d\mu .$$

We now use the assumption (7.3.4) in conjunction with the observation that $C_{g,p}(K) \le C_{g^s,p}(K)$ and the fact that $\operatorname{diam} 3Q < \delta$. It follows from the implication (d) $\Rightarrow$ (a) in Theorem 7.2.1 that

$$\int_{3Q} |g^s * f_Q(x)|^q \, d\mu \le A\varepsilon \|f_Q\|_p^q ,$$

and thus

$$\int_{\mathbf{R}^N} |g^s * f(x)|^q \, d\mu \le 3^{N(q-1)} A\varepsilon \sum_Q \|f_Q\|_p^q .$$

But, since $q \ge p$,

$$\sum_Q \Big(\int_Q |f|^p \, dx\Big)^{q/p} \le \Big(\sum_Q \int_Q |f|^p \, dx\Big)^{q/p} = \|f\|_p^q .$$

Thus

$$\|g^s * f\|_{L^q(\mu)} \le 3^{N/p} (A\varepsilon)^{1/q} \|f\|_p ,$$

whenever $s\sqrt{N} < \delta/3$, which proves (7.3.3).

The compactness of $\mathcal{G}$ now follows easily. In fact, $g_\delta \in L^{p'}(\mathbf{R}^N)$ for any $\delta > 0$. It follows that for any $\delta > 0$ the family $\{\mathcal{G}_\delta f : \|f\|_p \le 1\}$ is equicontinuous on $\mathbf{R}^N$, and thus by the Arzelà-Ascoli theorem (see e.g. Theorem 11.28 in Rudin [367]) $\mathcal{G}_\delta : L^p(\mathbf{R}^N) \to L^q(\mu, \mathbf{R}^N)$ is compact for any q, when $\operatorname{supp}\mu$ is compact. But the set of compact operators is norm closed (see e. g. Theorem 4.18 in Rudin [368]), and thus $\mathcal{G} : L^p(\mathbf{R}^N) \to L^q(\mu, \mathbf{R}^N)$ is compact.

Step 2. We now assume that $\mathcal{G} : L^p(\mathbf{R}^N) \to L^q(\mu, \mathbf{R}^N)$ is compact, and we shall prove that this implies (7.3.1), still under the assumption that $\operatorname{supp}\mu$ is compact.

The adjoint of $\mathcal{G}$ is the operator $\check{\mathcal{G}} : L^{q'}(\mu, \mathbf{R}^N) \to L^{p'}(\mathbf{R}^N)$, defined by $\check{\mathcal{G}}\varphi = g * (\varphi\mu)$. If $\mathcal{G}$ is compact, this implies that $\check{\mathcal{G}}$ is also compact (see e.g. Theorem 4.19 in Rudin [368]). Now, for any nonnegative f in L^p such that $g * f(x) \ge 1$ on K we have

$$\mu(K) \le \int_K g * f \, d\mu = \int_{\mathbf{R}^N} (g * \chi_K \mu) f \, dx \le \|f\|_p \|\check{\mathcal{G}}\chi_K\|_{p'} \ .$$

Thus, by the definition of capacity,

$$\mu(K) \le C_{g,p}(K)^{1/p} \|\check{\mathcal{G}}\chi_K\|_{p'} \ ,$$

or, if we choose $\varphi_K = \mu(K)^{-1/q'}\chi_K$,

$$\mu(K)^{1/q} C_{g,p}(K)^{-1/p} \le \|\check{\mathcal{G}}\varphi_K\|_{p'} \ .$$

Next, notice that if $\operatorname{diam} K_n \to 0$, as $n \to \infty$, then $C_{g,p}(K_n) \to 0$ by Proposition 5.1.1, and thus $\mu(K_n) \to 0$ by (7.2.4). Consequently, for any $F \in L^q(\mu, \mathbf{R}^N)$, by Hölder's inequality

$$\left| \int_{\mathbf{R}^N} \varphi_{K_n} F \, d\mu \right| \le \left(\int_{K_n} |F|^q \, d\mu \right)^{1/q} \to 0 \ .$$

Or, in other words, φ_{K_n} tends to zero, weakly in $L^{q'}(\mu, \mathbf{R}^N)$. But since the transformation $\check{\mathcal{G}} : L^{q'}(\mu, \mathbf{R}^N) \to L^{p'}(\mathbf{R}^N)$ is compact, it follows that

$$\|\check{\mathcal{G}}\varphi_{K_n}\|_{p'} \to 0 \ ,$$

and the result follows.

Step 3. We now free ourselves from the assumption that $\operatorname{supp}\mu$ is compact. Let χ_R be the characteristic function for the ball $B(0, R)$, and $\mu_R = \chi_R\mu$. Assume that μ satisfies (7.3.1). Then by the first step the mapping

$$\widetilde{\mathcal{G}}_R : L^p \to L^q(\mu_R, B(0, R)) \ ,$$

defined by $f \mapsto \mathcal{G} f \chi_R$, is compact for every $R > 0$.

On the other hand, choosing an arbitrary $\varepsilon > 0$ and assuming (7.3.2), we can find an R such that

$$\mu(K)^{1/q} C_{g,p}(K)^{-1/p} < \varepsilon$$

for any compact set $K \in \mathbf{R}^N \setminus B(0, R)$. Thus, by the implication (d) $\Rightarrow$ (a) in Theorem 7.2.1,

$$\int_{\mathbf{R}^N} |\mathcal{G}f - \widetilde{\mathcal{G}}_R f|^q \, d\mu = \int_{\mathbf{R}^N} |\mathcal{G}f|^q \, d\mu_R \le A\varepsilon \|f\|_p^q.$$

Thus $\|\mathcal{G} - \widetilde{\mathcal{G}}_R\| \to 0$, as $R \to \infty$, and it follows that $\mathcal{G}$ is compact.

To finish the proof, it remains to show the necessity of (7.3.2). But if $\{K_n\}_1^\infty$is a sequence of compact sets such that $K_n \subset \mathbf{R}^N \setminus B(0, R_n)$ for some $R_n \to \infty$, then again we have that φ_{K_n} tends to 0 weakly in $L^{q'}(\mu, \mathbf{R}^N)$. Hence, as before, $\|\check{\mathcal{G}}\varphi_{K_n}\|_{p'} \to 0$, which gives $\mu(K_n)^{1/q} C_{g,p}(K_n)^{-1/p} \to 0$, and the theorem is proved.

Just as in the case $1 < p < q < \infty$ of Section 7.2 we can simplify conditions (7.3.1) and (7.3.2) for Riesz and Bessel kernels by replacing the family of compact sets with the family of (closed) balls. We prove the result for Bessel kernels.

Theorem 7.3.2. *If $1 < p < q < \infty$, then a necessary and sufficient condition for the map $f \mapsto G_\alpha * f$ to be a compact mapping from $L^p(\mathbf{R}^N)$ into $L^q(\mu, \mathbf{R}^N)$ is that*

$$\lim_{\delta \to 0} \sup_{x, r \le \delta} \frac{\mu(B(x,r))}{C_{\alpha,p}(B(x,r))^{q/p}} = 0 \ , \tag{7.3.5}$$

and

$$\lim_{|x| \to \infty} \sup_{0<r<1} \frac{\mu(B(x,r))}{C_{\alpha,p}(B(x,r))^{q/p}} = 0 \ . \tag{7.3.6}$$

Proof. We first show that (7.3.5) implies (7.3.1) if $p < q$. We proceed as in the proof of Theorem 7.2.2 for Bessel kernels. An easy covering argument shows that (7.3.5) implies (7.2.7). Then, for $\varepsilon > 0$ we choose a $\delta > 0$ such that $\mu(B(x,r)) C_{\alpha,p}(B(x,r))^{-q/p} < \varepsilon$ if $r \le \delta$. This gives for an $R \le \delta$, and $\alpha p < N$,

$$\|G_\alpha * \mu_K\|_{p'} \le A\, \mu(K) R^{\alpha - N/p} + A\, \varepsilon^{1/p} \mu(K)^{1/p'} R^{(\alpha - N/p)(1-q/p)} \ . \tag{7.3.7}$$

We now assume that $\operatorname{diam} K \le \delta$. Then $\mu(K) C_{\alpha,p}(B(x,\delta))^{-q/p} < \varepsilon$. By Propositions 5.1.2 and 5.1.4 we can choose $R \le \delta$ so that

$$R^{\alpha - N/p} = A_1 \varepsilon^{1/p} \mu(K)^{-1/q} \ ,$$

with a constant A_1 depending only on N, p, and α.

If $\alpha p = N$ we again estimate as in the proof of Theorem 7.2.2, using Wolff's inequality. The choice of $R \le \delta$ is now given by $\mu(K) = A\varepsilon (\log e/R)^{(1-p)q/p}$, provided $\operatorname{diam} K \le \delta$.

In both cases we see that we can make A_2 in (7.2.2) arbitrarily small by choosing $\operatorname{diam} K$ small. Hence (7.3.1) follows.

To see that (7.3.6) implies (7.3.2) we return again to Wolff's inequality. Let $K \subset \mathbf{R}^N \setminus B(0, \rho)$. Since the centers of the balls to be dealt with in Wolff's inequality must lie in K, they must satisfy $|x| > \rho - 1$. Thus, for an $\varepsilon > 0$, choosing ρ large forces the estimate

$$\mu(B(x,r)) \leq \varepsilon\, C_{\alpha,p}(B(x,r))^{q/p}$$

for $0 < r < 1$, and $|x| > \rho - 1$. Hence, if $\alpha p < N$, we proceed to (7.3.7) for some $R < 1$, and then choose $R^{\alpha - N/p} = \mu(K)^{-1/q}$, when $\mu(K) < 1$. When $\mu(K) \geq 1$, we write

$$\|G_\alpha * \mu_K\|_{p'} \leq A\,\varepsilon^{1/p}\mu(K)^{1/p'} ,$$

as we have done earlier. Thus, in either case we get (7.3.2). The case $\alpha p = N$ follows the same lines.

7.4 A Space of Quasicontinuous Functions

From Section 7.1 it is natural to inquire as to whether or not the quantity

$$\left(\int_0^\infty C_{\alpha,p}(\{\,x : |u(x)| \geq t\,\})\,dt^p\right)^{1/p} \tag{7.4.1}$$

defines a norm on a linear space of functions u on $\mathbf{R}^N$. For general (α, p) the answer is not known; see however Sections 7.6.8 and 7.6.9 for cases when the answer is positive. But as we shall see, (7.4.1) can be replaced by an equivalent quantity which is a norm.

In fact, we define a new quantity $\Gamma_{\alpha,p}(u)$ by

$$\Gamma_{\alpha,p}(u) = \inf\{\,\|f\|_p^p : f \in \mathbf{K}_u\,\} , \tag{7.4.2}$$

where

$$\mathbf{K}_u = \{\,f \in L^p_+(\mathbf{R}^N) : G_\alpha * f(x) \geq |u(x)| \text{ for all } x \in \mathbf{R}^N\,\} . \tag{7.4.3}$$

We have the following result.

Proposition 7.4.1. *The function $\Gamma_{\alpha,p}(\,\cdot\,)^{1/p}$ defines a norm on $C_0(\mathbf{R}^N)$, and there is a constant A such that*

$$\tfrac{1}{4}\Gamma_{\alpha,p}(u) \leq \int_0^\infty C_{\alpha,p}(\{\,x : |u(x)| \geq t\,\})\,dt^p \leq A\,\Gamma_{\alpha,p}(u)$$

for all continuous functions u.

Proof. By Theorem 7.1.1 we have

$$\int_0^\infty C_{\alpha,p}(\{\,x : |u(x)| \geq t\,\})\,dt^p \leq A\,\|f\|_p^p$$

for any $f \in \mathbf{K}_u$. This proves the right hand inequality.

Moreover, if $u_i \in C_0$, and $f_i \in \mathbf{K}_{u_i}$ for $i = 1, 2$, then $f_1 + f_2 \in \mathbf{K}_{u_1+u_2}$, and thus

$$\Gamma_{\alpha,p}(u_1+u_2)^{1/p} = \inf\{\, \|f\|_p : f \in \mathbf{K}_{u_1+u_2} \,\} \le \|f_1+f_2\|_p \le \|f_1\|_p + \|f_2\|_p \ ,$$

whence

$$\Gamma_{\alpha,p}(u_1+u_2)^{1/p} \le \Gamma_{\alpha,p}(u_1)^{1/p} + \Gamma_{\alpha,p}(u_2)^{1/p} \ .$$

It is obvious that $\Gamma_{\alpha,p}(au) = |a|^p \Gamma_{\alpha,p}(u)$ for any constant a, and that $\Gamma_{\alpha,p}(u) = 0$ for $u \in C_0$ if and only if $u = 0$. This shows that $\Gamma_{\alpha,p}(\,\cdot\,)^{1/p}$ is a norm.

In addition $\Gamma_{\alpha,p}$ has the following properties:

$$\Gamma_{\alpha,p}(u_1) \le \Gamma_{\alpha,p}(u_2), \quad \text{if } |u_1| \le |u_2| \ ; \tag{7.4.4}$$

$$\Gamma_{\alpha,p}(\chi_E) = C_{\alpha,p}(E), \quad \text{for all } E \ ; \tag{7.4.5}$$

$$\Gamma_{\alpha,p}(u\chi(\cup_j K_j)) \le \sum_j \Gamma_{\alpha,p}(u\chi(K_j)) \ . \tag{7.4.6}$$

Of these, the first two are immediate from the definition. For (7.4.6) one proceeds as in proof of subadditivity of $C_{\alpha,p}$, Proposition 2.3.6.

We now set $K_j = \{\, x : \lambda^i \le |u(x)| \le \lambda^{i+1} \,\}$, and $E_j = \{\, x : |u(x)| \ge \lambda^i \,\}$ for $j \in \mathbf{Z}$ and $\lambda > 1$. It follows from (7.4.4)–(7.4.6) that

$$\Gamma_{\alpha,p}(u) \le \sum_{j=-\infty}^{\infty} \Gamma_{\alpha,p}(u\chi(K_j)) \le \sum_{j=-\infty}^{\infty} \lambda^{(j+1)p} C_{\alpha,p}(E_j) \ .$$

But

$$\int_0^\infty C_{\alpha,p}(\{\, x : |u(x)| \ge t \,\})\, dt^p \ge (1-\lambda^{-p}) \sum_{j=-\infty}^{\infty} \lambda^{jp} C_{\alpha,p}(E_j) \ .$$

We find

$$\Gamma_{\alpha,p}(u) \le \frac{\lambda^{2p}}{\lambda^p - 1} \int_0^\infty C_{\alpha,p}(\{\, x : |u(x)| \ge t \,\})\, dt^p \ ,$$

and the proposition follows if we choose $\lambda = 2^{1/p}$, the value that minimizes $\lambda^{2p}(\lambda^p - 1)^{-1}$.

We now define a new Banach space, $L^p(C_{\alpha,p})$, as the completion of $C_0^\infty(\mathbf{R}^N)$ in the norm $\Gamma_{\alpha,p}(\,\cdot\,)^{1/p}$. By the definition $\Gamma_{\alpha,p}(u)^{1/p} \le \|u\|_{\alpha,p}$ for $u \in C_0^\infty$, and C_0^∞ is dense in $L^{\alpha,p}$, so $L^{\alpha,p}$ is imbedded in $L^p(C_{\alpha,p})$.

This space has the following alternative characterization.

Theorem 7.4.2. *A function u on $\mathbf{R}^N$ belongs to $L^p(C_{\alpha,p})$ if and only if it is (α, p)-quasicontinuous, and*

$$\int_0^\infty C_{\alpha,p}(\{\, x : |u(x)| \ge t \,\})\, dt^p < \infty \ . \tag{7.4.7}$$

Proof. We first prove that if $u \in L^p(C_{\alpha,p})$, then u is (α, p)-quasicontinuous. In fact, if $u \in L^p(C_{\alpha,p})$, then there is a sequence $\{u_n\}_1^\infty$ of continuous, compactly supported functions such that

$$\int_0^\infty C_{\alpha,p}(\{\, x : |u_n(x) - u(x)| \ge t \,\})\, dt^p < 4^{-n} \ .$$

But then

$$C_{\alpha,p}(\{x : |u_n(x) - u(x)| \geq 2^{-n/p}\}) \int_0^{2^{-n/p}} dt^p \leq 4^{-n} \ ,$$

and thus

$$C_{\alpha,p}(\{x : |u_n(x) - u(x)| \geq 2^{-n/p}\}) \leq 2^{-n} \ .$$

Also, since $C_{\alpha,p}$ is an outer capacity, there is an open set G_n containing $\{x : |u_n(x) - u(x)| \geq 2^{-n/p}\}$ such that $C_{\alpha,p}(G_n) \leq 2^{-n+1}$. Thus, if $G = \cup_m^\infty G_n$, then $C_{\alpha,p}(G) \leq \sum_m^\infty C_{\alpha,p}(G_n) \leq 2^{-m+2}$. Moreover, $u_n(x) \to u(x)$ uniformly on G^c, and m is arbitrary, so u is (α, p)-quasicontinuous on $\mathbf{R}^N$.

We prove the converse in four steps. In the first step we prove that if $f \in L^p(\mathbf{R}^N)$, then $G_\alpha * f \in L^p(C_{\alpha,p})$.

Let $f \in L^p(\mathbf{R}^N)$ and choose a sequence $\{f_n\}_1^\infty$ in C_0^∞ such that $\|f - f_n\|_p \to 0$, as $n \to \infty$. Then $G_\alpha * f - G_\alpha * f_n \to 0$ in $L^p(C_{\alpha,p})$ by Theorem 7.1.1.

Define $G_{\alpha,R}(x)$ as $G_\alpha(x)$ for $|x| \leq R$ and as 0 for $|x| > R$. Then $G_{\alpha,R} * f_n \in C_0^\infty$, and

$$(G_\alpha - G_{\alpha,R}) * |f_n|(x) \leq \|f_n\|_\infty \int_{|y|>R} G_\alpha(y)\,dy \searrow 0, \qquad \text{as } R \to \infty \ .$$

This implies that $G_\alpha * f_n$, and thus $G_\alpha * f$ belong to $L^p(C_{\alpha,p})$.

Secondly, we observe that any continuous compactly supported function belongs to $L^p(C_{\alpha,p})$. In fact, any such function can be uniformly approximated by C_0^∞ functions supported in a fixed ball, and this implies convergence in $L^p(C_{\alpha,p})$ by Proposition 7.4.1.

The third step is to prove that any continuous function u that satisfies (7.4.7) belongs to $L^p(C_{\alpha,p})$. By Proposition 7.4.1 we have $\Gamma_{\alpha,p}(u) < \infty$, so there exists an $f \in L^p$ such that $|u| \leq G_\alpha * f$ on $\mathbf{R}^N$. Let $h \in C_0$ be arbitrary, and let $\alpha \in C_0$ be a cutoff function such that $\alpha(x) = 1$ on $\operatorname{supp} h$, and $0 \leq \alpha \leq 1$. Then $\alpha u \in C_0$, and $(1-\alpha)h = 0$, so $|u(x) - \alpha(x)u(x)| \leq (1 - \alpha(x))|G_\alpha * f(x) - h(x)|$, and the conclusion follows from the first two steps.

Finally, we let u be an arbitrary (α, p)-quasicontinuous function that satisfies (7.4.7). We can assume that u is real-valued. For any $M > 0$ we define u_M by $u_M(x) = u(x)$ when $|u(x)| \leq M$, $u_M(x) = M$ when $u(x) > M$, and $u_M(x) = -M$ when $u(x) < -M$. Then u_M is also quasicontinuous, and

$$\int_0^\infty C_{\alpha,p}(\{x : |u(x) - u_M(x)| \geq t\})\,dt^p$$
$$= \int_0^\infty C_{\alpha,p}(\{x : |u(x)| \geq M + t\})\,dt^p \leq \int_M^\infty C_{\alpha,p}(\{x : u(x)| \geq t\})\,dt^p \ ,$$

which is arbitrarily small.

Now, since u_M is quasicontinuous, we can for any $\varepsilon > 0$ find an open set G with $C_{\alpha,p}(G) < \varepsilon$ such that u_M restricted to G^c is continuous. By the Tietze extension theorem we can find a continuous function v such that $v|_{G^c} = u_M|_{G^c}$ and $\|v\|_\infty \leq M$. Then

$$\int_0^\infty C_{\alpha,p}(\{x : |v(x) - u_M(x)| \geq t\})\, dt^p \leq (2M)^p C_{\alpha,p}(G) < (2M)^p \varepsilon \ ,$$

and the theorem follows.

Theorem 7.4.2 has the following corollary.

Corollary 7.4.3. *If $u \in L^p(C_{\alpha,p})$, and if v is an (α, p)-quasicontinuous function such that $|v(x)| \leq |u(x)|$ a.e., then $v \in L^p(C_{\alpha,p})$, and $\|v\|_{L^p(C_{\alpha,p})} \leq \|u\|_{L^p(C_{\alpha,p})}$.*

Moreover, if $L^{\alpha,p}$ is imbedded in a Banach space B such that $\|\cdot\|_B$ is monotone in the sense that $\|v\|_B \leq \|u\|_B$ for all u and v in B such that $|v(x)| \leq |u(x)|$ everywhere, then B contains $L^p(C_{\alpha,p})$.

Proof. The first statement follows easily from the theorem. In fact, by Theorem 6.1.4 the assumptions imply that $|v(x)| \leq |u(x)|$ (α, p)-q.e. It follows that v satisfies (7.4.7), and thus $v \in L^p(C_{\alpha,p})$. An easy extension of (7.4.4) gives the norm inequality.

Suppose that $\|u\|_B \leq A\,\|u\|_{\alpha,p}$ for all $u \in L^{\alpha,p}$, and let $v \in C_0^\infty$. If B has the monotonicity property it follows that $\|v\|_B \leq \|G_\alpha * f\|_B \leq A\,\|f\|_p$ for all $f \in \mathbf{K}_v$, and thus $\|v\|_B \leq A\,\Gamma_{\alpha,p}(v)^{1/p}$. The conclusion follows from the definition of $L^p(C_{\alpha,p})$ as the closure of C_0^∞.

We also describe the dual space to $L^p(C_{\alpha,p})$. Cf. (7.1.2) above.

Theorem 7.4.4. *The dual space $L^p(C_{\alpha,p})^*$ can be identified with the space of all $\mu \in \mathcal{M}(\mathbf{R}^N)$ such that $G_\alpha * |\mu| \in L^{p'}$, i.e., $|\mu| \in L^{-\alpha,p'}$. If $u \in L^p(C_{\alpha,p})$ and $\mu \in L^p(C_{\alpha,p})^*$, then $u \in L^1(|\mu|)$, and the duality is given by*

$$\langle \mu, u \rangle = \int_{\mathbf{R}^N} u\, d\mu \ .$$

Moreover, the norm of μ in $L^p(C_{\alpha,p})^*$ is $\|\,|\mu|\,\|_{-\alpha,p'} = \|G_\alpha * |\mu|\,\|_{p'}$.

Proof. Suppose that $\mu \in \mathcal{M}(\mathbf{R}^N)$ is such that $\|G_\alpha * |\mu|\,\|_{p'} < \infty$, and let $u \in C_0$. Let $f \in \mathbf{K}_u$. Then

$$\int_{\mathbf{R}^N} |u|\, d|\mu| \leq \int_{\mathbf{R}^N} (G_\alpha * f)\, d|\mu| = \int_{\mathbf{R}^N} f(G_\alpha * |\mu|)\, dx \leq \|f\|_p \|G_\alpha * |\mu|\,\|_{p'} \ ,$$

and thus

$$\int_{\mathbf{R}^N} |u|\, d|\mu| \leq \Gamma_{\alpha,p}(u)^{1/p} \|G_\alpha * |\mu|\,\|_{p'} \ .$$

It follows that any Cauchy sequence in $L^p(C_{\alpha,p})$ is Cauchy in $L^1(|\mu|)$, and thus $L^p(C_{\alpha,p}) \subset L^1(|\mu|)$, since $L^p(C_{\alpha,p})$ is the completion of C_0. Moreover

$$\left|\int_{\mathbf{R}^N} u\,d\mu\right| \le \|f\|_p \|G_\alpha * |\mu|\,\|_{p'} \ ,$$

so μ defines a linear functional on $L^p(C_{\alpha,p})$ with norm $\|\mu\| \le \|\,|\mu|\,\|_{-\alpha,p'}$.

Conversely, let $\mu : u \mapsto \langle \mu, u\rangle$ be a bounded linear functional on $L^p(C_{\alpha,p})$ with norm $\|\mu\|$. Let K be a compact set, and let $\chi \in C_0$ with $\chi|_K = 1$, and $0 \le \chi \le 1$. Then for any $u \in C_0(K)$ we have $|u| \le \|u\|_\infty \chi$, and thus

$$|\langle \mu, u\rangle| \le \|\mu\|\, \Gamma_{\alpha,p}(u)^{1/p} \le \|\mu\|\, \Gamma_{\alpha,p}(\chi)^{1/p} \|u\|_\infty \ .$$

It follows that μ is a bounded linear functional on $C_0(K)$ for each compact K. By the Riesz representation theorem μ can be identified with a Radon measure, so that $\langle \mu, u\rangle = \int_{\mathbf{R}^N} u\,d\mu$ for all $u \in C_0$. Since C_0 is dense in $L^p(C_{\alpha,p})$, it only remains to prove that $\|\,|\mu|\,\|_{-\alpha,p'} \le \|\mu\|$.

We have

$$\int_{\mathbf{R}^N} u\,d|\mu| = \sup\left\{\int_{\mathbf{R}^N} v\,d\mu : v \in C_0,\ |v| \le u\right\} .$$

But by assumption

$$\left|\int_{\mathbf{R}^N} v\,d\mu\right| \le \|\mu\|\, \Gamma_{\alpha,p}(v)^{1/p} \ ,$$

and thus

$$\int_{\mathbf{R}^N} u\,d|\mu| \le \|\mu\|\, \Gamma_{\alpha,p}(u)^{1/p}$$

for all nonnegative u in C_0. By approximation from below it follows that

$$\int_{\mathbf{R}^N} (G_\alpha * f)\,d|\mu| \le \|\mu\|\, \Gamma_{\alpha,p}(G_\alpha * f)^{1/p} \le \|\mu\|\, \|f\|_p$$

for all nonnegative f in L^p. But $\int_{\mathbf{R}^N} (G_\alpha * f)\,d|\mu| = \int_{\mathbf{R}^N} f(G_\alpha * |\mu|)\,dx$, so by the converse of Hölder's inequality $\|G_\alpha * |\mu|\,\|_{p'} \le \|\mu\|$, and the theorem is proved.

7.5 A Capacitary Strong Type Inequality. Another Approach

There is an entirely different approach to the strong capacitary inequality, Theorem 7.1.1, based on some of the ideas exposed in Chapter 4. We recall that by Theorem 4.2.2 the space $L^{\alpha,p}$ can be identified with the Lizorkin–Triebel space $F_\alpha^{p,2}$, and that by Proposition 4.4.4 the capacities $C(\,\cdot\,; F_\alpha^{p,q})$, $1 < q < \infty$, are all equivalent to $C_{\alpha,p}(\,\cdot\,)$.

This approach has the advantage that it can be extended to general Lizorkin–Triebel spaces $F_\alpha^{p,q}$ for $p, q > 0$, and at the same time gives an extension of Proposition 4.4.4 to show the equivalence of the capacities $C(\,\cdot\,; F_\alpha^{p,q})$ in the full range $0 < q \le \infty$.

We limit ourselves to proving the following theorem, which clearly implies Theorem 7.1.1.

Theorem 7.5.1. *Let $\alpha > 0$, and $1 < p, q < \infty$. There is a constant A such that for any $f \in F_\alpha^{p,\infty}$*

$$\sum_{j=-\infty}^{\infty} 2^{jp} C\big(\{x : |f(x)| > 2^j\}; F_\alpha^{p,q}\big) \le A \, \|f\|_{F_\alpha^{p,\infty}} \ . \tag{7.5.1}$$

Proof. Let $f \in F_\alpha^{p,\infty}$. By Theorem 4.6.2, f can be represented as $f = \sum_{n=0}^{\infty} f_n$, where $f_n \in C^\infty$, and $\{2^{n\alpha} f_n\}_0^\infty \in L^p(l^\infty)$. Moreover, with the notation from Section 4.6, there is

$$g = \sum_{n=0}^{\infty} g_n \quad \text{with} \quad g_n(x) = \sum_{k \in \mathbf{Z}^N} \alpha_{nk} \chi_{nk}(x) \ ,$$

such that $|f_n| \le g_n$, and

$$\|\{2^{n\alpha} g_n\}_0^\infty\|_{L^p(l^\infty)} \le A \, \|f\|_{F_\alpha^{p,\infty}} \ .$$

Thus, it is enough to prove that

$$\sum_{j=-\infty}^{\infty} 2^{jp} C\big(\{x : g(x) \ge 2^j\}; F_\alpha^{p,q}\big) \le A \, \|\{2^{n\alpha} g_n\}_0^\infty\|_{L^p(l^\infty)}^p \ .$$

It follows from Theorem 4.6.2 and Proposition 4.7.2 that it is sufficient to prove the following claim:

Claim. Let g be defined by

$$g = \sum_{n=0}^{\infty} g_n \ , \qquad g_n = \sum_{k \in \mathbf{Z}^N} \alpha_{nk} \chi_{nk}(x) \ ,$$

where $\alpha_{nk} \ge 0$. Then there exist functions

$$\omega_j = \sum_{n=0}^{\infty} \omega_{jn} \ , \qquad j \in \mathbf{Z} \ ,$$

with

$$\omega_{jn} = \sum_{k \in \mathbf{Z}^N} \gamma_{jnk} \chi_{nk}(x) \ , \qquad \gamma_{jnk} \ge 0 \ ,$$

satisfying the conditions

$$\{x : \omega_j(x) \ge 2^j\} \supset \{x : g(x) \ge 2^{j+2}\} \ , \tag{7.5.2}$$

and

$$\sum_{j=-\infty}^{\infty} \Big\| \Big(\sum_{n=0}^{\infty} (2^{n\alpha} \omega_{jn})^q\Big)^{1/q} \Big\|_p^p \le A \, \| \sup_{n \ge 0} 2^{n\alpha} g_n \|_p^p \ . \tag{7.5.3}$$

We first prove the weaker result that there are ω_j as above, satisfying (7.5.2), such that

$$\sum_{j=-\infty}^{\infty}\sum_{n=0}^{\infty}\|2^{n\alpha}\omega_{jn}\|_p^p \leq A\sum_{n=0}^{\infty}\|2^{n\alpha}g_n\|_p^p \ . \tag{7.5.4}$$

This will imply the inequality

$$\sum_{n=-\infty}^{\infty} 2^{jp}C\big(\{x: |f(x)| > 2^j\}; F_\alpha^{p,p}\big) \leq A\,\|f\|_{F_\alpha^{p,p}} \ . \tag{7.5.5}$$

In the case $p \geq 2$, Theorem 7.1.1 will follow from this, since $C(\,\cdot\,; F_\alpha^{p,p})$ is equivalent to $C_{\alpha,p}(\,\cdot\,)$ by Proposition 4.4.4, and since $\|f\|_{F_\alpha^{p,p}} \leq \|f\|_{F_\alpha^{p,2}}$.

In order to prove (7.5.4) we set $\omega_j = \sum_{n=0}^{\infty}\omega_{jn}$, where

$$\omega_{jn}(x) = \Big(\min\big\{2^{j+1}, \textstyle\sum_{l=0}^{n} g_l(x)\big\} - \max\big\{2^j, \sum_{l=0}^{n-1} g_l(x)\big\}\Big)_+ \ .$$

It follows easily that $\omega_j(x) = 0$ if $g(x) \leq 2^j$, and $\omega_j(x) = 2^j$ if $g(x) \geq 2^{j+1}$. Moreover, $g_n = \sum_{j=-\infty}^{\infty}\omega_{jn}$, and thus $g = \sum_{j=-\infty}^{\infty}\omega_j$. Also, if $\omega_{rn}(x) \neq 0$, $\omega_{sn}(x) \neq 0$, and $r < j < s$, then $\omega_{jn}(x) = 2^j$. It follows that there is an A such that $\sum_{j=-\infty}^{\infty}\omega_{jn}(x)^p \leq A\,g_n(x)^p$, which gives (7.5.4).

For the proof of the full claim we have to refine the construction considerably. We choose $\varepsilon < 1$ so that $0 < 2\varepsilon < \alpha$, and set

$$h = \sum_{n=0}^{\infty} h_n \ , \qquad h_n = \sum_{k\in\mathbf{Z}^N} s_{nk}\chi_{nk}$$

with

$$s_{nk} = \sum_{m=n}^{\infty} 2^{(m-n)2\varepsilon+Nn}\|g_m\chi_{nk}\|_1 \ . \tag{7.5.6}$$

Then $g_n \leq h_n$, and

$$2^{n\alpha}h_n \leq \sum_{m=n}^{\infty} 2^{(m-n)2\varepsilon+n\alpha}Mg_m \leq A_0 \sup_{m\geq n} 2^{m\alpha}Mg_m \ .$$

Consequently, by the Fefferman–Stein theorem (Theorem 1.1.2),

$$\|\sup\nolimits_{n\geq 0} 2^{n\alpha}h_n\|_p \leq A_1\,\|\sup\nolimits_{n\geq 0} 2^{n\alpha}g_n\|_p \ . \tag{7.5.7}$$

Denote by Γ the subset of $\mathbf{N}\times\mathbf{Z}^N$ consisting of points (m,k), such that for any cube $Q_{nk'}$, such that $Q_{nk'} \supset Q_{mk}$, we have the inequality

$$s_{mk} > a_\varepsilon 2^{(n-m)\varepsilon}s_{nk'} \ , \tag{7.5.8}$$

where

$$a_\varepsilon = \tfrac{1}{2}(2^{-\varepsilon} + 2^{-2\varepsilon} + 2^{-3\varepsilon} + \ldots)^{-1} = \tfrac{1}{2}(2^\varepsilon - 1) < 1 \ .$$

It is easy to see that for any $(m,k) \notin \Gamma$ there is an $(n,k') \in \Gamma$ such that $Q_{mk} \subset Q_{nk'}$. We also define

$$E_n = \bigcup_{\{k:(n,k)\in\Gamma\}} Q_{nk} \ , \tag{7.5.9}$$

$$t_{nk} = \begin{cases} s_{nk} & \text{if } (n,k) \in \Gamma \ , \\ 0 & \text{if } (n,k) \notin \Gamma \ , \end{cases}$$

and

$$v = \sum_{n=0}^{\infty} v_n \ , \qquad v_n = \sum_{k\in\mathbf{Z}^N} t_{nk}\chi_{nk} \ .$$

It follows from the inequality

$$\sum_{n=0}^{\infty}\big(h_n(x) - v_n(x)\big) = \sum_{(n,k)\notin\Gamma} s_{nk}\chi_{nk}(x)$$
$$\leq \sum_{l=0}^{\infty} a_\varepsilon h_l(x)\big(2^{-\varepsilon} + 2^{-2\varepsilon} + \ldots\big) = \tfrac{1}{2}\sum_{l=0}^{\infty} h_l(x)$$

that

$$v(x) \geq \tfrac{1}{2}h(x) \ . \tag{7.5.10}$$

In addition, the definition of the functions h_n (see (7.5.6)), and the fact that $h_m \geq v_m$, give the inequalities

$$\|h_n\chi_{nk'}\|_1 \geq 2^{2(m-n)\varepsilon}\|h_m\chi_{nk'}\|_1 \geq 2^{2(m-n)\varepsilon}\|v_m\chi_{nk'}\|_1 \quad \text{for } m > n \ .$$

Hence, by means of the estimate (see (7.5.8))

$$v_m\chi_{nk'} \geq a_\varepsilon 2^{nN+(n-m)\varepsilon}\|h_n\chi_{nk'}\|_1\,\chi_{E_m}\chi_{nk'} \ ,$$

we obtain

$$|E_m \cap Q_{nk'}| \leq a_\varepsilon^{-1} 2^{-nN-\varepsilon(m-n)} \ . \tag{7.5.11}$$

We set $\omega_j = \sum_{n=0}^{\infty}\omega_{jn}$, where

$$\omega_{jn}(x) = \Big(\min\{2^{j+1}, \textstyle\sum_{l=0}^{n} v_l(x)\} - \max\{2^j, \textstyle\sum_{l=0}^{n-1} v_l(x)\}\Big)_+ \ .$$

It follows easily that $\omega_j(x) = 2^j$ if $v(x) \geq 2^{j+1}$. But $v(x) \geq \frac{1}{2}h(x)$ by (7.5.10), so (7.5.2) is satisfied, as claimed.

Moreover, $v_n = \sum_{j=-\infty}^{\infty}\omega_{jn}$, and thus

$$\|v_n\|_p^p = \Big\|\textstyle\sum_{j=-\infty}^{\infty}\omega_{jn}\Big\|_p^p$$
$$\geq \Big\|\big(\textstyle\sum_{j=-\infty}^{\infty}\omega_{jn}^p\big)^{1/p}\Big\|_p^p = \textstyle\sum_{j=-\infty}^{\infty}\|\omega_{jn}\|_p^p \ . \tag{7.5.12}$$

Using (7.5.10), the fact that $\operatorname{supp}\omega_{jn} \subset \operatorname{supp} v_n = E_n$, and the lemma given below, we obtain the inequalities

$$\Big\|\sup_{n\in\mathbf{N}} 2^{n\alpha} v_n\Big\|_p^p \geq A_2 \sum_{n=0}^{\infty}(2^{n\alpha}\|v_n\|_p)^p \ , \tag{7.5.13}$$

$$\Big\|\Big(\sum_{n=0}^{\infty}(2^{n\alpha}\omega_{jn})^q\Big)^{1/q}\Big\|_p^p \leq A_3 \sum_{n=0}^{\infty}\|2^{n\alpha}\omega_{jn}\|_p^p \ , \tag{7.5.14}$$

which by means of the estimates (7.5.7) and (7.5.12) proves (7.5.3), and the theorem.

Lemma 7.5.2. *Let* $1 < q < p < \infty$, $\Gamma \subset \mathbf{N} \times \mathbf{Z}^N$, *and suppose that* (7.5.11) *is satisfied, where the sets* E_n *are defined by* (7.5.9). *Then any function sequence* $\{b_n\}_{n=0}^{\infty}$, *where*

$$b_n(x) = \sum_{\{k\,:\,(n,k)\in\Gamma\}} \beta_{nk}\chi_{nk}(x) \ , \quad \beta_{nk} \geq 0 \ ,$$

satisfies the inequalities

$$A_1\Big\|\Big(\sum_{n=0}^{\infty} b_n^q\Big)^{1/q}\Big\|_p^p \leq \sum_{n=0}^{\infty}\|b_n\|_p^p \leq A_2\|\sup\nolimits_{n\geq 0} b_n\|_p^p \ .$$

Proof. Choose $S \in \mathbf{N}$ so large that $a_\varepsilon(2^{S\varepsilon}-1) \geq 2$, where as before $0 < 2\varepsilon < \alpha$ and $a_\varepsilon = \frac{1}{2}(2^\varepsilon - 1)$. We consider the sequence $\{b_{Sn}\}_{n=0}^{\infty} \subset \{b_n\}_{n=0}^{\infty}$, and prove the inequality

$$\Big\|\Big(\sum_{n=0}^{\infty} b_{Sn}^q\Big)^{1/q}\Big\|_p \leq A\,\|\sup\nolimits_{n\geq 0} b_{Sn}\|_p \ . \tag{7.5.15}$$

We define a set $E_{Sn,k}$, $(Sn, k) \in \Gamma$, by

$$E_{Sn,k} = Q_{Sn,k} \setminus \Big(\bigcup_{m=n+1}^{\infty} E_{Sm}\Big) \ .$$

It follows from (7.5.11) and the choice of S that

$$\begin{aligned} |E_{Sn,k}| &\geq |Q_{Sn,k}| - \sum_{m=n+1}^{\infty} |Q_{Sn,k} \cap E_{Sm}| \\ &\geq 2^{-SnN}\Big(1 - a_\varepsilon^{-1}\Big(\sum_{m=n+1}^{\infty} 2^{-(m-n)S\varepsilon}\Big)\Big) \\ &= 2^{-SnN}\Big(1 - a_\varepsilon^{-1}\big(2^{S\varepsilon}-1\big)^{-1}\Big) \geq 2^{-SnN-1} \ . \end{aligned} \tag{7.5.16}$$

We define functions d_{Sn} by

$$d_{Sn} = \sum_{\{k\,:\,(Sn,k)\in\Gamma\}} \beta_{Sn,k}\,\chi(E_{Sn,k}) \ .$$

Then (7.5.15) is a consequence of the Fefferman–Stein theorem (Theorem 1.1.2),

$$\Big\|\Big(\sum_{n=0}^{\infty}(Md_{Sn})^q\Big)^{1/q}\Big\|_p \leq A\,\Big\|\Big(\sum_{n=0}^{\infty} d_{Sn}^q\Big)^{1/q}\Big\|_p \ ,$$

and the relations

$$b_{Sn} \le A\, M d_{Sn} \ ,$$

$$\left(\sum_{n=0}^{\infty} d_{Sn}^q\right)^{1/q} \le \sup_{n\ge0} d_{Sn} \le \sup_{n\ge0} b_{Sn} \ ,$$

the first of which follows from (7.5.16), and the second one from the fact that the sets $E_{Sn,k}$, $(Sn, k) \in \Gamma$, are disjoint.

Similarly to (7.5.15) we prove the inequality

$$\left\| \left(\sum_{n=0}^{\infty} b_{Sn+r}^q\right)^{1/q} \right\|_p \le A \,\| \sup_{n\ge0} b_{Sn+r} \|_p$$

for $r = 1, 2, \ldots , S-1$. The desired inequalities follow from this and (7.5.15) by using the inequality

$$\sum_{n=0}^{\infty} \|b_n\|_p^p \le \left\| \left(\sum_{n=0}^{\infty} b_n^q\right)^{1/q} \right\|_p^p \ , \quad q > p \ .$$

7.6 Further Results

7.6.1. As was noticed in the beginning of the chapter, it follows from Theorem 2.5.1 that if $\mu \in \mathcal{M}^+(\mathbf{R}^N) \cap L^{-\alpha,p'}(\mathbf{R}^N)$, then $\mu(E) = 0$ for every μ-measurable set E such that $C_{\alpha,p}(E) = 0$. The following is an extension of this observation to signed measures, essentially due to M. Grun-Rehomme [187]:

Let $\mu \in \mathcal{M}(\mathbf{R}^N) \cap L^{-\alpha,p'}(\mathbf{R}^N)$, $\alpha > 0$, $1 < p \le N/\alpha$, and let $E \subset \mathbf{R}^N$ be an arbitrary set such that $C_{\alpha,p}(E) = 0$. Then $|\mu|(E) = 0$.

We give the proof. By the Hahn decomposition theorem (see e.g. Rudin [367]) there are μ_+ and μ_- in $\mathcal{M}^+$ such that $\mu = \mu_+ - \mu_-$, $|\mu| = \mu_+ + \mu_-$, and there are μ-measurable sets A and B such that $A \cap B = \emptyset$, $\mu_-(A) = 0$, and $\mu_+(B) = 0$. Note that the assumptions do not imply that $|\mu| \in L^{-\alpha,p'}$.

First let $K \in \mathbf{R}^N$ be compact such that $\mu_-(K) = 0$, and suppose that $C_{\alpha,p}(K) = 0$. Let $\varepsilon > 0$ and choose an open $G \supset K$ such that $|\mu|(G \setminus K) < \varepsilon$. Let $\chi \in C_0^\infty(G)$ satisfy $0 \le \chi \le 1$ and $\chi = 1$ on K. By Corollary 3.3.4 there are $\varphi_n \in C_0^\infty(\mathbf{R}^N)$ such that $0 \le \varphi_n \le 1$, $\varphi_n|_K = 1$, and $\|\varphi_n\|_{\alpha,p} \to 0$, as $n \to \infty$. Set $\psi_n = \chi\varphi_n$ Then also $\|\psi_n\|_{\alpha,p} \to 0$. If α is an integer this follows easily from the Leibniz rule. We omit the proof in the general case. (See R. S. Strichartz [392].) Since $\mu_-(K) = 0$ we have

$$\int_{\mathbf{R}^N} \psi_n \, d\mu = \int_K d\mu + \int_{K^c} \psi_n \, d\mu = \mu_+(K) + \int_{G\setminus K} \psi_n \, d\mu \ ,$$

whence

$$\mu_+(K) \le \|\psi_n\|_{\alpha,p} \|\mu\|_{-\alpha,p'} + |\mu|(G \setminus K) < 2\varepsilon \ ,$$

if n is large enough. This proves that $\mu_+(K) = 0$.

It follows that if $E \in \mathbf{R}^N$ is any μ-measurable set such that $\mu_-(E) = 0$ and $C_{\alpha,p}(E) = 0$, then $\mu_+(E) = 0$. In fact, $\mu_+(E) = \sup_{K\subset E} \mu_+(K) = 0$, since $C_{\alpha,p}(K) = 0$ for all compact $K \subset E$.

Now let K be any compact with $C_{\alpha,p}(K) = 0$. Then $K \cap A$ is μ-measurable, and $\mu_-(K \cap A) = 0$. Thus $\mu_+(K \cap A) = 0$, and similarly $\mu_-(K \cap B) = 0$, whence $|\mu|(K) = 0$. It follows that if E is μ-measurable and $C_{\alpha,p}(E) = 0$, then $|\mu|(E) = 0$.

If E is arbitrary with $C_{\alpha,p}(E) = 0$, then there is always a G_δ-set H (i.e. a denumerable intersection of open sets), such that $E \subset H$ and $C_{\alpha,p}(H) = 0$. But then $|\mu|(H) = 0$, and the result follows.

7.6.2. Let $\Omega \subset \mathbf{R}^N$ be open. For any compact $K \subset \Omega$ we define a so called *condenser capacity*

$$C_{1,p}(K;\Omega) = \inf\left\{\int_\Omega |\nabla\varphi|^p\,dx : \varphi \in C_0^\infty(\Omega),\ \varphi|_K \geq 1\right\}.$$

Then for any $p \geq 1$ and $u \in C_0^\infty(\Omega)$

$$\int_0^\infty C_{1,p}\big(\{x : u(x) \geq \lambda\};\Omega\big)\,d\lambda^p \leq \frac{p^p}{(p-1)^{p-1}}\int_\Omega |\nabla u|^p\,dx\,. \quad (7.6.1)$$

The constant on the right side is interpreted as 1 if $p = 1$. This result is due to V. G. Maz′ya [303], Theorem 3, and is also found in [308], Section 2.3.1. It can be seen from the proof that the constant is best possible. With a less precise constant the proof is simple.

Whether (7.6.1) can be extended to higher derivatives for arbitrary Ω seems to be an open problem. However, for second order derivatives and nonnegative functions there is a substitute. We define a capacity

$$C_{2,p}^+(K;\Omega) = \inf\left\{\int_\Omega \sum_{|\sigma|=2} |D^\sigma\varphi|^p\,dx : \varphi \in C_0^\infty(\Omega),\ \varphi \geq 0,\ \varphi|_K \geq 1\right\}.$$

Then for any $p > 1$ there is a constant A such that for any nonnegative $u \in C_0^\infty(\Omega)$

$$\int_0^\infty C_{2,p}^+\big(\{x : u(x) \geq \lambda\};\Omega\big)\,d\lambda^p \leq A\int_\Omega \sum_{|\sigma|=2} |D^\sigma u|^p\,dx\,.$$

The proof uses “smooth truncation”, see Section 3.7.5, and for details, Maz′ya [308], Section 8.2.1.

7.6.3. A related problem has recently been given a negative solution by V. G. Maz′ya and Yu. V. Netrusov [311]. Let $\widetilde{W}^{m,p}(\Omega)$ denote the space of functions $u \in L^p_{\mathrm{loc}}(\Omega)$ such that $D^\sigma u \in L^p(\Omega)$ for $|\sigma| = m$, and let the space be normed by

$$\|u\|_{\widetilde{W}^{m,p}(\Omega)} = \left(\int_K |u|^p\,dx + \int_\Omega \sum_{|\sigma|=m} |D^\sigma u|^p\,dx\right)^{1/p}$$

for some compact $K \subset \Omega$ of positive volume. Define the capacity for a relatively closed set $F \subset \Omega$ by

$$\widetilde{C}_{m,p}(F;\Omega) = \inf\left\{\|u\|^p_{\widetilde{W}^{m,p}(\Omega)} : \varphi \in C^\infty(\Omega),\ \varphi|_F \geq 1\right\}.$$

Then there is a domain $\Omega \subset \mathbf{R}^2$ and a measure $\mu \in \mathcal{M}^+(\Omega)$ satisfying

$$\mu(F) \leq \widetilde{C}_{2,2}(F; \Omega)$$

for all $F \subset \Omega$, such that $\widetilde{W}^{2,2}(\Omega)$ is not imbedded in $L^2(\mu, \Omega)$. It follows immediately that the corresponding strong type capacitary inequality must fail. See Maz'ya [308], Section 4.1.3, for the case $m = 1$.

7.6.4. For $\beta p = N$, $1 < p < \infty$, and a compactly supported $\mu \in \mathcal{M}^+(\mathbf{R}^N)$ there is a constant b such that

$$\sup_{\|f\|_p \leq 1} \int_{\mathbf{R}^N} \exp\big(b\,|G_\beta * f|^{p'}\big)\, d\mu < \infty \ ,$$

if and only if $\sup_{x\,,r>0} \mu(B(x,r))\, r^{-\alpha} < \infty$ for some $\alpha > 0$. See D. R. Adams [4], Theorem 3, for sufficiency, and Maz'ya [308], Corollary 8.6.2. See also V. I. Yudovich [436], and the notes to Chapter 3.

7.6.5. As we noted in the remark following Theorem 7.2.2 the equivalence of (a) and (d) in Theorem 7.2.1 for $p = q$ ceases to be true if (d) is required to hold only for e.g. all balls. However, R. Kerman and E. T. Sawyer [245] have proved that for (a) and (b) in the theorem to be equivalent it is enough that (b) is verified for all dyadic cubes. Moreover, (b) is equivalent to a similar inequality where $g * \mu_K$ is replaced by a fractional maximal function $M_g \mu_K$, which for $g = I_\alpha$ is the ordinary fractional maximal function $M_\alpha \mu_K$, again considered for dyadic cubes. A simplified proof of this result has been given (along with other results) by I. E. Verbitsky [412].

Other necessary and sufficient conditions for the trace inequality to be true have been found by V. G. Maz'ya and I. E. Verbitsky [314]. For example, if $p > 1$, and $\alpha > 0$, then the truth of

$$\left(\int_{\mathbf{R}^N} |G_\alpha * f|^p\, d\mu \right)^{1/p} \leq A\, \|f\|_p$$

for all $f \in L^p$ is equivalent to any of the following conditions:

$$G_\alpha * (G_\alpha * \mu)^{p'}(x) \leq A\, G_\alpha * \mu(x) < \infty \quad \text{a.e.;}$$

$$M_{\alpha,1}(M_{\alpha,1}\mu)^{p'}(x) \leq A\, M_{\alpha,1}\mu(x) < \infty \quad \text{a.e.;}$$

$$\int_K (G_\alpha * \mu)^{p'}\, dx \leq A\, C_{\alpha,p}(K) \quad \text{for all compact } K.$$

Here $M_{\alpha,1}\mu$ is the modified fractional maximal function defined in (1.1.3). In particular, the trace inequality holds for a positive measure μ if and only if it is valid for the absolutely continuous measure $(G_\alpha * \mu)^{p'}\, dx$. Their results strengthen a sufficient condition of C. Fefferman and D. H. Phong [143], Lemma C, p. 153. See also the discussion in Kerman and Sawyer [245].

More information on weighted inequalities is found in the notes to Section 3.6 at the end of Chapter 3.

7.6.6. In Theorem 7.2.1 and in Section 7.6.5 we have given conditions for $\mathcal{G}f$ to be an imbedding of $L^p(\mathbf{R}^N)$ in $L^q(\mu)$ in the case $q \geq p$. In order to formulate a condition for the "upper triangle case" $0 < q < p$ we let $\alpha > 0$ and define a quantity $\kappa(\mu; t)$ for $\mu \in \mathcal{M}^+$ and $t > 0$ by

$$\kappa(\mu; t) = \inf\{\, C_{\alpha,p}(F) : F \subset \mathbf{R}^N,\ F \text{ compact},\ \mu(F) \geq t\,\} \ .$$

Then it follows from Theorem 7.2.1(d) that for $1 < p \leq q$ there is a constant A such that

$$\left(\int_{\mathbf{R}^N} |G_\alpha * f|^q \, d\mu \right)^{1/q} \leq A \, \|f\|_p \tag{7.6.2}$$

for all $f \in L^p$, if and only if

$$\sup_{t>0} \frac{t^{p/q}}{\kappa(\mu; t)} < \infty \ .$$

Now let $0 < q < p$, and $p > 1$. Then (7.6.2) is true if and only if

$$\int_0^\infty \left(\frac{t^{p/q}}{\kappa(\mu; t)} \right)^{q/(p-q)} \frac{dt}{t} < \infty \ .$$

This result is due to V. G. Maz'ya and Yu. V. Netrusov [311] (Appendix, Theorem 2). A more unwieldy necessary and sufficient condition was proved earlier by Maz'ya; see [308], Section 8.4.2. A non-capacitary characterization of measures satisfying (7.6.2) for $q < p$ based on different ideas was given by I. E. Verbitsky [412].

7.6.7. The following covering theorem of V. G. Maz'ya and S. P. Preobrazhenskiĭ [312] (see also Maz'ya [308], Corollary 8.5.1) gives an alternative proof of Theorem 7.2.2. Let $1 < p \leq N/\alpha$, and let $K \subset \mathbf{R}^N$ be a compact set such that $C_{\alpha,p}(K) > 0$. Then for given $\varepsilon > 0$ there is a constant A, independent of K, such that K can be covered by balls $\{B(x_j, r_j)\}_1^\infty$, $r_j \leq 1$, satisfying

$$\sum_{j=1}^\infty C_{\alpha,p}(B(x_j, r_j))^{1+\varepsilon} \leq A \, C_{\alpha,p}(K)^{1+\varepsilon} \ .$$

7.6.8. A capacity C is termed *strongly subadditive* if

$$C(E_1 \cup E_2) + C(E_1 \cap E_2) \leq C(E_1) + C(E_2)$$

for all E_1 and E_2. If C is a capacity that has the properties of Propositions 2.3.4, 2.3.5, and 2.3.6, then the integral

$$\int_0^\infty C(\{\, x : |u(x)| \geq \lambda \,\}) \, d\lambda$$

is a sublinear functional of u if and only if C is strongly subadditive. See G. Choquet [102] or B. Anger [35]. It is classical that the capacities $C_{\alpha,2}$ are strongly subadditive for $\alpha \leq 1$, as are the capacities $C'_{1,p}$ of Definition 2.2.1. See e.g. N. S. Landkof [266], Section II:5, and T. Bagby [43]. See also the work of D. R. Adams and E. Nieminen contained in [16].

7.6.9. Another strongly subadditive capacity is the so called dyadic Hausdorff capacity, denoted $\widetilde{\Lambda}_{\alpha}^{(\infty)}$, and defined in the same way as the Hausdorff capacity $\Lambda_{\alpha}^{(\infty)}$ in Section 5.1, but using coverings with dyadic cubes instead of arbitrary balls. For this result, see R. Fefferman [145]. Of course, $\widetilde{\Lambda}_{\alpha}^{(\infty)}$ and $\Lambda_{\alpha}^{(\infty)}$ are comparable and, in particular, have the same nullsets. It then follows that for all $u \in C_0(\mathbf{R}^N)$ the integral

$$\int_0^\infty \Lambda_{\alpha}^{(\infty)}(\{x : |u(x)| \geq \lambda\})\, d\lambda$$

is comparable to

$$\sup_{\mu} \int_{\mathbf{R}^N} |u|\, d\mu\ ,$$

where $0 < \alpha \leq N$ and the supremum is taken over all $\mu \in \mathcal{M}^+$ for which $\mu(B(x,r))\, r^{-\alpha} \leq 1$ for all balls $B(x,r)$. See D. R. Adams [13].

7.6.10. The following result extends Theorems 7.2.1 and 7.2.2 to the case $p = 1$. Let α be an integer, $1 \leq \alpha \leq N$, let $q \geq 1$, and let $\mu \in \mathcal{M}^+(\mathbf{R}^N)$. Then $W^{\alpha,1}(\mathbf{R}^N)$ is continuously imbedded in $L^q(\mu)$ if and only if

$$\sup_{x,\, 0<r\leq 1} \frac{\mu(B(x,r))^{1/q}}{r^{N-\alpha}} < \infty\ .$$

It is remarkable that the condition $p < q$ in Theorem 7.2.2 here can be replaced by $1 = p \leq q$. This theorem is due to V. G. Maz′ya [307]. See also Maz′ya [308], Section 1.4.3, or Maz′ya and T. O. Shaposhnikova [313], Section 1.2.2 for the proof. An application is found in Section 6.5.1 above.

7.6.11. There is a space $L^1(\Lambda_{N-\alpha}^{(\infty)})$, which corresponds to $L^p(C_{\alpha,p})$ for $p = 1$. Theorem 7.4.4 has a counterpart, saying that the dual of $L^1(\Lambda_{N-\alpha}^{(\infty)})$ is the space of measures $\mu \in \mathcal{M}$ for which

$$\sup_{x,r} \frac{|\mu|(B(x,r))}{r^{N-\alpha}} < \infty\ .$$

See D. R. Adams [13]. Cf. Section 5.6.3 above.

7.6.12. The following surprising consequence of Lemma 7.5.2 and Theorems 4.6.2 and 4.6.3 is due to B. Jawerth [233] (with $E = \mathbf{R}^{N-1}$), and Yu. V. Netrusov [338, 340]. See also M. Frazier and B. Jawerth [155], Section 11.

Let $0 < p < \infty$, $0 < q \leq \infty$, $\alpha > 0$, and let $E \subset \mathbf{R}^N$ be porous in the sense that for some $\delta > 0$

$$\big|\{y \in B(x,r) : \operatorname{dist}(y,E) \leq \rho\}\big| \leq A r^{N-\delta} \rho^{\delta} \tag{7.6.3}$$

for all $x \in \mathbf{R}^N$, and all $0 < \rho \leq r < 1$. Then, for any function f_0 belonging to the Lizorkin–Triebel space $F_\alpha^{p,\infty}$ there exists a function $f \in F_\alpha^{p,q}$ with $\|f\|_{F_\alpha^{p,q}} \leq A\, \|f_0\|_{F_\alpha^{p,\infty}}$ that coincides with f_0 on E in the sense that for any $\varepsilon > 0$ there is a function $f_\varepsilon \in F_\alpha^{p,\infty}$ such that $\operatorname{supp} f_\varepsilon \cap E = \emptyset$, and $\|f_\varepsilon - (f - f_0)\|_{F_\alpha^{p,\infty}} < \varepsilon$. In other words, for a set E satisfying (7.6.3), the space of traces $f|_E$ on E of functions in $F_\alpha^{p,q}$ is independent of q for $0 < q \le \infty$.

If $\alpha p > N$ the above result has a converse, the proof of which is more complicated; see [340]. The result is the following:

Let $\alpha p > N$, $0 < q_1 < q_2 \le \infty$, and let $E \subset \mathbf{R}^N$. Suppose that for any function $f_1 \in F_\alpha^{p,q_1}$ there is a function $f_2 \in F_\alpha^{p,q_2}$, such that $f_1|_E = f_2|_E$. Then E satisfies (7.6.3).

7.7 Notes

7.1–7.3. The developments presented in this chapter all began with the 1962 note by V. G. Maz'ya [295], with details given in [298], although at the time these papers did not receive the attention they deserved. There Maz'ya proved the first versions of Theorems 7.2.1 and 7.3.1, while investigating the Schrödinger operator $L = -\Delta - \mu$ for a positive measure μ on a domain $\Omega \subset \mathbf{R}^N$, $N \ge 3$. (The measure μ can of course be absolutely continuous, so that $\mu = q\,m$ with a locally integrable function q, and m Lebesgue measure.)

A function $u \in W^{1,2}(\Omega)$ is a weak solution of $Lu = f$ for some $f \in L^2$ if

$$\int_\Omega \nabla u \cdot \nabla \varphi\, dx - \int_\Omega u\varphi\, d\mu = \int_\Omega f\varphi\, dx$$

for all $\varphi \in C_0^\infty(\Omega)$. In order to extend this relation to all $\varphi \in W_0^{1,2}(\Omega)$ one needs to prove an inequality

$$\int_\Omega \varphi^2\, d\mu \le A \int_\Omega |\nabla\varphi|^2\, dx \tag{7.7.1}$$

for all $\varphi \in C_0^\infty(\Omega)$. Maz'ya proved that a necessary and sufficient condition for (7.7.1) to be true is that

$$\mu(K) \le A\, C_{1,2}(K;\Omega) \tag{7.7.2}$$

for all compact $K \subset \Omega$, where $C_{1,2}(K;\Omega)$ is the condenser capacity defined in Section 7.6.2. When $\Omega = \mathbf{R}^N$ this is a special case of (d) in Theorem 7.2.1. The sufficiency of (7.7.2) follows immediately from (7.6.1), which is the earliest capacitary strong type inequality, but (7.6.1) was not proved until several years later in Maz'ya [303]. See also the books by Maz'ya [308], Chapter 2, and Maz'ya and T. O. Shaposhnikova [313] for further details.

The first order strong type inequality (7.6.1) is easily established by truncation in the class $W^{1,p}$, in fact, any Lipschitz function T with $T(0) = 0$ operates on $W^{1,p}$ for $1 \le p < \infty$, as pointed out in Theorem 3.3.1. In [303] (see Theorem 11) Maz'ya used "smooth truncation" to prove Theorem 7.1.1 in the case $g = I_2$, and to extend the equivalence of (a) and (d) in Theorem 7.2.1 to this case. However, by the result of B. E. J. Dahlberg, Theorem 3.3.2, it is not possible to extend this proof to higher derivatives. But by smoothly truncating potentials $I_\alpha * f$ for integer α and $f \in L^p_+$, and estimating by means of Proposition 3.1.2, D. R. Adams [6] proved Theorem 7.1.1 in this case. After several intermediate stages (see Maz'ya [308], Section 8.9 for details), the theorem was proved in the generality presented here by K. Hansson [192]. Hansson's proof was simplified by Maz'ya (see [308], Section 8.2.3), and the proof given here is a modification of that given by Maz'ya.

But it is clear that there are some cases where the smooth truncation method might be preferred, since the proof of Theorem 7.1.1 requires that we can represent the capacitary extremals as potentials that satisfy the boundedness principle, Theorem 2.6.3. Such cases arise, for example, with condenser capacities as in Section 7.6.2 above, and with parabolic (heat) potentials. In fact, a version of Theorem 5.5.1 was established for the parabolic analogue of the Riesz capacities $\dot{C}_{\alpha,p}$, using Morrey space estimates; see Adams [5] and compare with Section 5.6.6.

The results of Section 7.2 are due to Adams [6], (see also [10]), and Maz'ya, whose book [308] (Chapter 8) contains these and many related results. Theorem 7.2.2 goes back to Adams [2]. In the proof of this theorem we could have been a bit more precise by using Wolff's inequality (Theorem 4.5.2) to estimate $\|I_\alpha * \mu^E\|_{p'}$ or $\|G_\alpha * \mu^E\|_{p'}$. However, the point of our presentation is that this is not necessary here. If one wants to treat a general class of radially decreasing convolution kernels in the same way, it soon becomes clear, however, that something is lost in the estimates, unless some extra growth restrictions on the kernel are assumed. The ingredients for extending the Wolff inequality to more general kernels exist in the work by R. Kerman and E. T. Sawyer [245], where general Muckenhoupt–Wheeden estimates (see Theorem 3.6.1) are obtained. Further refinements of Theorem 7.2.2 are due to A. Jonsson and H. Wallin [238]. See also Yu. V. Netrusov [341].

Section 7.3 follows the exposition in K. Hansson [193]; see also Maz'ya [308], Sections 2.4.2 and 8.8.1. A forerunner of Theorem 7.3.2 was proved by M. Berger and M. Schechter [51]; see Theorem 2.8.

7.4. The results of this section are due mainly to K. Hansson [191, 192]; see also Adams [7].

7.5. Theorem 7.5.1 is due to Yu. V. Netrusov [341] (see Remark 2 to Theorem 1.1, and Theorem 1.1; the proof is sketched in Section 2.3). The details given here were provided by Netrusov. Lemma 7.5.2 is well known; see B. Jawerth and M. Frazier [154] (Lemma 3.1), [155] (Proposition 2.7), and Netrusov [338]. Our proof is adapted from [155].

8. Poincaré Type Inequalities

The term "Poincaré type inequality" is used, somewhat loosely, to describe a class of inequalities that generalize the classical Poincaré inequality,

$$\int_\Omega |f|^p\,dx \le A_\Omega \int_\Omega |\nabla f|^p\,dx \ ,$$

valid for $f \in W_0^{1,p}(\Omega)$ in a bounded open $\Omega \subset \mathbf{R}^N$. What the inequalities have in common is that an integral norm of a function is estimated in terms of integrals of its derivatives, and some information about the vanishing or the average of the function. Some such knowledge is clearly necessary, since estimates of this kind are false for non-zero constants.

The subject is treated systematically e.g. in the books by V. G. Maz'ya [308] and W. P. Ziemer [438]. For this reason we limit the treatment here to some aspects of the theory, mainly motivated by applications in Chapter 9.

Section 8.1 prepares the ground by proving a number of classical results. The central result of the chapter is Theorem 8.2.1, which gives an estimate of the size of a function in terms of capacities of the sets where the function and a number of its derivatives vanish. Section 8.2 is devoted to this theorem and its corollaries. In Section 8.3 these results are proved again by a different, abstract approach.

8.1 Some Basic Inequalities

We begin by proving a useful representation formula, due to S. L. Sobolev. Our starting point is the Taylor formula

$$\begin{aligned} f(x) &= \sum_{|\beta|\le m-1} \frac{(x-y)^\beta}{\beta!} D^\beta f(y) \\ &\quad + \sum_{|\beta|=m} \frac{(x-y)^\beta}{\beta!} \int_0^1 m(1-t)^{m-1} D^\beta f\bigl(y+t(x-y)\bigr)\,dt \\ &= P_y^{m-1} f(x) + R_y^{m-1} f(x) \ . \end{aligned} \tag{8.1.1}$$

Suppose that $f \in C^m(\Omega)$ for a bounded domain $\Omega \subset \mathbf{R}^N$, and let $G \subset \Omega$ be an open set such that Ω is starshaped with respect to all points in G. Let $\omega \in C_0^\infty(G)$, and $\int_G \omega\,dy = 1$. Then

$$f(x) = \int_G P_y^{m-1} f(x)\omega(y)\,dy + \int_G R_y^{m-1} f(x)\omega(y)\,dy \ . \tag{8.1.2}$$

Integrating the first term by parts, we get

$$\begin{aligned}\int_G P_y^{m-1} f(x)\omega(y)\,dy &= \sum_{|\beta|\le m-1} \frac{1}{\beta!}\int_G (x-y)^\beta D^\beta f(y)\omega(y)\,dy \\ &= \sum_{|\beta|\le m-1} \frac{(-1)^\beta}{\beta!}\int_G f(y) D^\beta((x-y)^\beta \omega(y))\,dy. \\ &= \sum_{|\beta|\le m-1} x^\beta \int_G f(y)\varphi_\beta(y)\,dy \ ,\end{aligned}$$

where $\varphi_\beta \in C_0^\infty(G)$, and

$$\|\varphi_\beta\|_\infty \le A \sum_{j=0}^{m-1} \|\nabla^j \omega\|_\infty \ . \tag{8.1.3}$$

In the second term, setting $x = 0$, introducing polar coordinates $y = r\sigma$, $r = |y|$, and defining the integrand to be 0 off the support of ω, we get

$$\begin{aligned}&\int_{\mathbf{R}^N} R_y^{m-1} f(0)\omega(y)\,dy \\ &= \int_0^\infty \int_{|\sigma|=1} \left(\sum_{|\beta|=m} \frac{r^m(-\sigma)^\beta}{\beta!} \int_0^1 m t^{m-1} D^\beta f(tr\sigma)\,dt \right) \omega(r\sigma)\, r^{N-1}\,dr\,d\sigma \\ &= \int_0^\infty \int_{|\sigma|=1} \left(\sum_{|\beta|=m} \frac{(-\sigma)^\beta}{\beta!} \int_0^r m t^{m-1} D^\beta f(t\sigma)\,dt \right) \omega(r\sigma)\, r^{N-1}\,dr\,d\sigma \\ &= \sum_{|\beta|=m} \frac{m}{\beta!} \int_0^\infty \int_{|\sigma|=1} \frac{(-\sigma)^\beta D^\beta f(t\sigma) \int_t^\infty \omega(r\sigma)\, r^{N-1}\,dr}{t^{N-m}}\, t^{N-1}\,dt\,d\sigma \\ &= \sum_{|\beta|=m} \frac{m}{\beta!} \int_{\mathbf{R}^N} \frac{(-\sigma)^\beta D^\beta f(\xi) \int_{|\xi|}^\infty \omega(r\sigma)\, r^{N-1}\,dr}{|\xi|^{N-m}}\, d\xi \ .\end{aligned}$$

Returning to an arbitrary $x \in \Omega$, this gives

$$\int_{\mathbf{R}^N} R_y^{m-1} f(x)\omega(y)\,dy = \sum_{|\beta|=m} \frac{1}{\beta!} \int_{\mathbf{R}^N} \frac{\psi(x,\xi)(-\sigma)^\beta D^\beta f(\xi)}{|x-\xi|^{N-m}}\, d\xi \ , \tag{8.1.4}$$

where

$$\psi(x,\xi) = m \int_{|\xi - x|}^\infty \omega(x + r\sigma) r^{N-1}\,dr, \quad \sigma = \frac{\xi - x}{|\xi - x|} \ . \tag{8.1.5}$$

We observe that the assumption that Ω is starshaped with respect to G implies that $\psi(x,\xi) = 0$ for all $\xi \notin \Omega$.

A passage to the limit proves the following theorem.

Theorem 8.1.1. *Let $\Omega \subset \mathbf{R}^N$ be a bounded domain, starshaped with respect to all points in an open $G \subset \Omega$, and suppose that $f \in W^{m,1}(\Omega)$. Then for a.e. $x \in \Omega$*

$$f(x) = P_{m-1}(x) + \sum_{|\beta|=m} \frac{1}{\beta!} \int_\Omega \frac{\psi(x,\xi)(x-\xi)^\beta D^\beta f(\xi)}{|x-\xi|^N}\, d\xi \ ,$$

where

$$P_{m-1}(x) = \sum_{|\beta|\le m-1} x^\beta \int_G f\varphi_\beta \, dy \ , \tag{8.1.6}$$

$\psi(x,\cdot) \in C_0^\infty(\Omega)$ for all x, $\varphi \in C_0^\infty(G)$, and there is a constant A such that $|\psi(x,\xi)| \le A$ for all x and ξ.

The theorem, and its proof have a number of consequences.

Corollary 8.1.2. *Suppose, in addition to the assumptions of Theorem* 8.1.1, *that $f \in W_0^{m,1}(\Omega)$. Then there is a constant A_N, only depending on N, such that*

$$f(x) = \sum_{|\beta|=m} \frac{mA_N}{\beta!} \int_\Omega \frac{(x-\xi)^\beta D^\beta f(\xi)}{|x-\xi|^N}\, d\xi \qquad \text{for a.e. } x \in \Omega \ .$$

Proof. Taylor's formula now gives, for any unit vector σ,

$$f(0) = \sum_{|\beta|=m} \frac{(-\sigma)^\beta}{\beta!} \int_0^\infty mt^{m-1} D^\beta f(t\sigma)\, dt \ .$$

Then, as before,

$$f(0) \int_{|\sigma|=1} d\sigma = \sum_{|\beta|=m} \frac{m}{\beta!} \int_{\mathbf{R}^N} \frac{(-\sigma)^\beta D^\beta f(\xi)}{|\xi|^{N-m}}\, d\xi \ .$$

Corollary 8.1.3. *Suppose, in addition to the assumptions of Theorem* 8.1.1, *that G contains a ball B_δ of radius δ. Then there is a constant A, independent of δ, such that*

$$|f(x) - P_{m-1}(x)| \le \frac{A}{\delta^{N-1}} \int_\Omega \frac{|\nabla^m f(\xi)|}{|x-\xi|^{N-m}}\, d\xi$$

for a.e. $x \in \Omega$, and

$$\sup_{x\in\Omega} |P_{m-1}(x)| \le \frac{A}{\delta^{N+m-1}} \int_{B_\delta} |f|\, d\xi \ .$$

Proof. We can clearly assume that the function ω in the proof of Theorem 8.1.1 satisfies $\|\omega\|_\infty \le A\delta^{-N}$, and $\|\nabla^j \omega\|_\infty \le A\delta^{-N-j}$ for $j \le m-1$. It follows from (8.1.3) that $\|\varphi_\beta\|_\infty \le A\delta^{-N-m+1}$, and then the estimate of P_{m-1} follows from (8.1.6). The definition (8.1.5) of ψ implies that $\|\psi\|_\infty \le A\delta^{-N+1}$. The corollary follows.

Corollary 8.1.4. *Suppose that Ω satisfies the assumptions of Corollary 8.1.3, and let $f \in W^{m,p}(\Omega)$ for some $p \geq 1$. Then there is a constant A, independent of δ, such that*

$$\|\nabla^k (f - P_{m-1})\|_{L^p(\Omega)} \leq \frac{A}{\delta^{N-1}} \|\nabla^m f\|_{L^p(\Omega)}, \quad k = 0, 1, \dots, m-1 .$$

If $p > 1$ and $mp < N$ there is a constant A such that

$$\|f - P_{m-1}\|_{L^{p^*}(\Omega)} \leq \frac{A}{\delta^{N-1}} \|\nabla^m f\|_{L^p(\Omega)}, \quad p^* = \frac{Np}{N - mp} .$$

Proof. The first inequality follows from Corollary 8.1.3 applied to $\nabla^k f$, and from the fact that the Riesz kernels I_{m-k}, $k \leq m-1$, belong to L^1_{loc}, and that convolution with such kernels is bounded on $L^p(\Omega)$. The second inequality follows from Theorem 3.1.4.

We state the next corollary in a form that will be useful to us later, although more general formulations are possible.

Corollary 8.1.5. *Let B_δ be a ball of radius δ, let $G \subset B_\delta$ be a ball of radius $\eta\delta$, and let $f \in W^{m,p}(B_\delta)$. Then there is a constant A independent of G and of δ such that for $k = 0, 1, 2, \dots, m-1$*

$$\delta^{kp} \int_{B_\delta} |\nabla^k f|^p \, dx \leq \frac{A}{\eta^{N+(m-1)p}} \int_G |f|^p \, dx + \frac{A\,\delta^{mp}}{\eta^{(N-1)p}} \int_{B_\delta} |\nabla^m f|^p \, dx .$$

Proof. By homogeneity it is enough to take $\delta = 1$. The result follows directly from Corollaries 8.1.3 and 8.1.4.

For later reference we also give the following closely related proposition.

Proposition 8.1.6. *Let B_δ be a ball of radius δ, and let $f \in W^{1,p}(B_\delta)$. Then there is a constant A such that for any ball $B_{\delta/2} \subset B_\delta$,*

$$\left| \left(\frac{1}{|B_\delta|} \int_{B_\delta} |f|^p \, dx \right)^{1/p} - \left(\frac{1}{|B_{\delta/2}|} \int_{B_{\delta/2}} |f|^p \, dx \right)^{1/p} \right| \leq A\delta \left(\frac{1}{|B_\delta|} \int_{B_\delta} |\nabla f|^p \, dx \right)^{1/p} .$$

Proof. As before we can assume that $\delta = 1$. By Theorem 1.2.2 we can assume that $f \in W^{1,p}(\mathbf{R}^N)$. The left hand side in the inequality equals a constant times

$$\left| \left(\int_{B_{1/2}} \int_{B_1} |f(x)|^p \, dx\, dy \right)^{1/p} - \left(\int_{B_1} \int_{B_{1/2}} |f(y)|^p \, dy\, dx \right)^{1/p} \right| \leq \left(\int_{B_{1/2}} \int_{B_1} |f(x) - f(y)|^p \, dx\, dy \right)^{1/p} .$$

We first assume that $f \in C^1$. By Taylor's formula (8.1.1) for $m = 1$, using polar coordinates as in proving (8.1.4), we obtain for any $y \in B_{1/2}$

$$\int_{B_1} |f(x) - f(y)|\,dx \le \int_{|y-x|\le 2} |f(x) - f(y)|\,dx \le A \int_{|\xi - y|\le 2} \frac{|\nabla f(\xi)|}{|\xi - y|^{N-1}}\,d\xi \ ,$$

whence

$$\int_{B_{1/2}} \int_{B_1} |f(x) - f(y)|^p\,dx\,dy \le A \int_{|\xi|\le 5/2} |\nabla f(\xi)|^p\,d\xi \ ,$$

as in the proof of Corollary 8.1.4. The proposition follows by a passage to the limit.

8.2 Inequalities Depending on Capacities

In Corollary 8.1.3 we estimated the size of a function f in terms of $\nabla^m f$, given that the function is zero on a ball of given radius. We shall now give such estimates of a more precise form, if we are given that the function and a number of its derivatives vanish on some set that in general is not open.

We recall the notion of the trace of a function on a set E. (See Definition 6.1.6.) If $f \in W^{m,p}(\mathbf{R}^N)$, and $E \subset \mathbf{R}^N$ is an arbitrary set, then $f|_E = 0$ means that f has an (m, p)-quasicontinuous representative $\widetilde{f}$ that satisfies $\widetilde{f}(x) = 0$ for (m, p)-q.e. $x \in E$. Moreover, by Theorem 6.2.1,

$$\lim_{r\to 0} \frac{1}{r^N} \int_{B(x,r)} |f(y)|\,dy = 0 \qquad (m, p)\text{-q.e. on } E \ .$$

Also, if β is a multiindex with $|\beta| < m$, then $D^\beta f|_E = 0$ means that $D^\beta f$, defined as the distribution derivative of f and considered as an element in $W^{m-|\beta|,p}(\mathbf{R}^N)$, has an $(m - |\beta|, p)$-quasicontinuous representative $\widetilde{D^\beta f}$ that satisfies

$$\widetilde{D^\beta f}(x) = 0 \qquad (m - |\beta|, p)\text{-q.e. on } E.$$

Theorem 8.2.1. *Let $\Omega \subset \mathbf{R}^N$ be a bounded domain, starshaped with respect to all points in an open $G \subset \Omega$, and suppose that $f \in W^{m,p}(\Omega)$, $1 < p < \infty$. Let $K \subset \Omega$ be closed, and $C_{m-k,p}(K) > 0$ for some $k = 0, 1, \dots, m-1$. Suppose that*

$$D^\beta f|_K = 0 \quad \textit{for all } \beta, \quad 0 \le |\beta| \le k \ .$$

Then

$$f(x) = P_{m-1}(x) + R_{m-1}(x) \quad (m, p)\textit{-q.e.} \ ,$$

where $P_{m-1}(x)$ is a polynomial of degree $\le m-1$, and

$$|R_{m-1}(x)| \le A \int_\Omega \frac{|\nabla^m f(\xi)|}{|x - \xi|^{N-m}}\,d\xi + A\,\frac{\|\nabla^m f\|_{L^p(\Omega)}}{C_{m-k,p}(K)^{1/p}} \ .$$

If $k = m-1$ the polynomial P_{m-1} is the zero polynomial, and in general

$$P_{m-1}(x) = \sum_{0\le|\gamma|\le m-1} a_\gamma x^\gamma \ ,$$

where the coefficients a_γ are given by

$$a_\gamma = \sum_{|\beta|=k+1} \int_G D^\beta f(y)\varphi_{\beta,\gamma}(y)\,dy$$

for functions $\varphi_{\beta,\gamma} \in C_0^\infty(G)$.

Proof. The proof is a variation of the proof of Theorem 8.1.1. Suppose first that $f \in C^m(\Omega)$. We shall derive a representation formula for f under the assumption that $f(z)$ and $D^\beta f(z)$ are given on K for $|\beta| \le k$, $k = 0, 1, \dots, m-1$. Thus, by (8.1.1),

$$f = P_z^k f + R_z^k f \ ,$$

where $P_z^k f$ is a known polynomial when $z \in K$.

In order to bring out the idea of the proof clearly we first treat the case $k = 0$. Then for any x and y in Ω, and $z \in K$,

$$\begin{aligned} f(x) - f(z) &= P_y^{m-1}f(x) - P_y^{m-1}f(z) + R_y^{m-1}f(x) - R_y^{m-1}f(z) \\ &= \sum_{1\le|\beta|\le m-1} \frac{1}{\beta!}\big((x-y)^\beta - (z-y)^\beta\big)D^\beta f(y) \\ &\quad + R_y^{m-1}f(x) - R_y^{m-1}f(z) \ . \end{aligned}$$

Let $\omega \in C_0^\infty(G)$, and $\int_G \omega\,dy = 1$. Multiplying by $\omega(y)$ and integrating, we obtain

$$\begin{aligned} f(x) - f(z) &= \int_\Omega \big(P_y^{m-1}f(x) - P_y^{m-1}f(z)\big)\omega(y)\,dy \\ &\quad + \int_\Omega R_y^{m-1}f(x)\omega(y)\,dy - \int_\Omega R_y^{m-1}f(z)\omega(y)\,dy \ , \end{aligned}$$

where by (8.1.4)

$$\left|\int_\Omega R_y^{m-1}f(x)\omega(y)\,dy\right| \le A\, I_m * (\chi_\Omega|\nabla^m f|)(x) \ , \tag{8.2.1}$$

and

$$\left|\int_\Omega R_y^{m-1}f(z)\omega(y)\,dy\right| \le A\, I_m * (\chi_\Omega|\nabla^m f|)(z) \ ,$$

χ_Ω denoting the characteristic function for Ω. Now let μ be a positive measure on K, such that $\mu(K) = 1$, and

$$\|I_m * \mu\|_{L^{p'}(\Omega)} \le A\, C_{m,p}(K)^{-1/p} \ .$$

That such measures exist follows from the dual definition of capacity, Theorem 2.5.1. (If $mp > N$ we choose μ as the Dirac measure at any $z \in K$.) Because of the relation between the Bessel and Riesz kernels the constant A depends on the diameter of Ω. Integrating, we obtain

$$f(x) - \int_K f(z)\,d\mu(z) = P_{m-1}(x) + R_{m-1}(x) \ ,$$

where

$$P_{m-1}(x) = \sum_{1\le|\beta|\le m-1} \frac{1}{\beta!}\int_\Omega\left((x-y)^\beta - \int_K (z-y)^\beta\, d\mu(z)\right) D^\beta f(y)\omega(y)\, dy \ ,$$

and, using Fubini's theorem and Hölder's inequality,

$$\begin{aligned}|R_{m-1}(x)| &\le A\, I_m * (\chi_\Omega|\nabla^m f|)(x) + A\int_K I_m * (\chi_\Omega|\nabla^m f|)\, d\mu \\ &= A\, I_m * (\chi_\Omega|\nabla^m f|)(x) + A\int_\Omega (I_m * \mu)|\nabla^m f|\, dy \\ &\le A\, I_m * (\chi_\Omega|\nabla^m f|)(x) + A\, C_{m,p}(K)^{-1/p}\|\nabla^m f\|_{L^p(\Omega)} \ .\end{aligned}$$

After integration by parts,

$$P_{m-1}(x) = \sum_{0\le|\beta|\le m-1} a_\beta x^\beta \ ,$$

where

$$a_\beta = \sum_{|\gamma|=1}\int_\Omega D^\gamma f(y)\varphi_{\beta,\gamma}(y)\, dy \ ,$$

and $\varphi_{\beta,\gamma} \in C_0^\infty(\Omega)$. The important thing to observe here is that the coefficients a_β depend on ∇f only, and not on f.

We now return to the general case, $0 \le k \le m-1$. We want to approximate $f(x)$ and $P_z^k f(x)$ by their Taylor expansions at y. We have

$$f(x) = \sum_{|\alpha|\le m-1} \frac{1}{\alpha!}(x-y)^\alpha D^\alpha f(y) + R_y^{m-1} f(x) \ ;$$

$$P_z^k f(x) = \sum_{|\alpha|\le k} \frac{1}{\alpha!}(x-z)^\alpha D^\alpha f(z) \ .$$

Substituting

$$D^\alpha f(z) = \sum_{|\beta|\le m-|\alpha|-1} \frac{1}{\beta!}(z-y)^\beta D^{\alpha+\beta} f(y) + R_y^{m-1}(D^\alpha f)(z) \ ,$$

we obtain for $k < m-1$ the expansion

$$\begin{aligned}P_z^k f(x) &= \sum_{|\alpha|\le k}\ \sum_{|\beta|\le m-|\alpha|-1} \frac{1}{\alpha!\beta!}(x-z)^\alpha (z-y)^\beta D^{\alpha+\beta} f(y) \\ &\quad + \sum_{|\alpha|\le k}\frac{1}{\alpha!}(x-z)^\alpha R_y^{m-|\alpha|-1}(D^\alpha f)(z) \\ &= \sum\sum_{|\alpha+\beta|\le k} + \sum\sum_{\substack{|\alpha|\le k \\ k+1\le|\alpha+\beta|\le m-1}} \frac{1}{\alpha!\beta!}(x-z)^\alpha(z-y)^\beta D^{\alpha+\beta}f(y) + R(x,y,z) \\ &= \sum_{|\gamma|\le k}\frac{1}{\gamma!}(x-y)^\gamma D^\gamma f(y) + Q(x,y,z) + R(x,y,z) \ . \qquad (8.2.2)\end{aligned}$$

In fact

$$(x-y)^\gamma = \sum_{\alpha+\beta=\gamma} \frac{\gamma!}{\alpha!\beta!}(x-z)^\alpha (z-y)^\beta .$$

Thus

$$\begin{aligned} f(x) - P_z^k f(x) &= \sum_{k+1\le|\alpha|\le m-1} \frac{1}{\alpha!}(x-y)^\alpha D^\alpha f(y) - Q(x,y,z) \\ &\quad + R_y^{m-1} f(x) - R(x,y,z) . \end{aligned} \tag{8.2.3}$$

If $k = m-1$ the result is the same, except for the term $Q(x,y,z)$, which is missing. If we multiply by $\omega(y)$ and integrate (setting $Q(x,y,z) = 0$ in the case $k = m-1$), we obtain

$$f(x) - P_z^k f(x) = \int_\Omega (P_y^{m-1} f(x) - P_y^k f(x) - Q(x,y,z))\omega(y)\,dy + S(x,z) . \tag{8.2.4}$$

Here

$$S(x,z) = \int_\Omega R_y^{m-1}(x)\omega(y)\,dy - \sum_{|\alpha|\le k} \frac{1}{\alpha!}(x-z)^\alpha \int_\Omega R_y^{m-|\alpha|-1}(D^\alpha f)(z)\omega(y)\,dy , \tag{8.2.5}$$

By (8.1.4) or (8.2.1) applied to f and to $D^\alpha f$ we have

$$\begin{aligned} |S(x,z)| &\le A\, I_m * (\chi_\Omega|\nabla^m f|)(x) \\ &\quad + A \sum_{|\alpha|\le k} A\, I_{m-|\alpha|} * (\chi_\Omega |\nabla^{m-|\alpha|} D^\alpha f|)(z) \\ &\le A\, I_m * (\chi_\Omega|\nabla^m f|)(x) + A\, I_{m-k} * (\chi_\Omega|\nabla^m f|)(z) . \end{aligned}$$

Now let μ be a positive measure on K, such that $\mu(K) = 1$ and

$$\| I_{m-k} * \mu \|_{L^{p'}(\Omega)} \le A\, C_{m-k,p}(K)^{-1/p} ,$$

and integrate. We obtain the representation formula

$$f(x) - \int_K P_z^k f(x)\,d\mu(z) = P_{m-1}(x) + R_{m-1}(x) , \tag{8.2.6}$$

where

$$|R_{m-1}(x)| \le A \int_\Omega \frac{|\nabla^m f(\xi)|}{|x-\xi|^{N-m}}\,d\xi + A \frac{\|\nabla^m f\|_{L^p(\Omega)}}{C_{m-k,p}(K)^{1/p}} ,$$

and

$$P_{m-1}(x) = \int_K \int_\Omega (P_y^{m-1} f(x) - P_y^k f(x) - Q(x,y,z))\omega(y)\,dy\,d\mu(z)$$

is a polynomial of degree $m-1$, which is 0 in the case $k=m-1$. If $k<m-1$, we write

$$P_{m-1}(x) = \sum_{0\le|\gamma|\le m-1} a_\gamma x^\gamma \ .$$

It is easily seen from (8.2.2) that the coefficients a_γ depend linearly on the derivatives $D^\alpha f$ for $k+1\le|\alpha|\le m$, but not on any derivatives of f of order $\le k$. Integration by parts shows, as before, that there are $\varphi_{\beta,\gamma}\in C_0^\infty(\Omega)$ such that

$$a_\gamma = \sum_{|\beta|=k+1}\int_\Omega D^\beta f(y)\varphi_{\beta,\gamma}(y)\,dy \ .$$

Now suppose only that $f\in W^{m,p}(\Omega)$, and that f is (m,p)-quasicontinuous. It follows from Theorem 1.2.1 and Proposition 2.3.8 (see also Proposition 6.1.2) that there is a sequence $\{f_n\}_1^\infty$ of functions in $C^\infty(\Omega)$, such that $f_n\to f$ in $W^{m,p}(\Omega)$ and $D^\beta f_n(x)$, $|\beta|=k$, converges to an $(m-k,p)$-quasicontinuous representative of $D^\beta f(x)$ uniformly outside an open set of arbitrarily small $(m-k,p)$-capacity. If $C_{m-k,p}(K)>0$ we can choose the measure μ above so that the convergence is uniform to zero on supp μ. It follows that (8.2.6) is true for (m,p)-q.e. x in Ω. The theorem follows.

Corollary 8.2.2. *Let Ω, K, and f satisfy the assumptions of Theorem* 8.2.1. *Then there is a polynomial P_{m-1} of degree $\le m-1$, which is the zero polynomial if $k=m-1$, and otherwise satisfies*

$$\sup_{x\in\Omega}|P_{m-1}(x)|\le A\,\|\nabla^{k+1}f\|_{L^1(\Omega)} \ ,$$

such that

$$\|f-P_{m-1}\|_{L^p(\Omega)}\le A\,\frac{\|\nabla^m f\|_{L^p(\Omega)}}{C_{m-k,p}(K)^{1/p}} \ .$$

Proof. This follows immediately from the theorem, and the fact that the kernel I_m belongs to L^1_{loc}, which for bounded Ω implies that $\|I_m*\nabla^m f\|_{L^p(\Omega)}\le A\,\|\nabla^m f\|_{L^p(\Omega)}$.

In order to give the results a form that is invariant under change of scale we introduce some new notation. In Section 3.6 we defined a modified Riesz kernel $I_{k,\delta}$ by

$$I_{k,\delta}(x)=I_k(x)\quad\text{for } |x|<\delta\,,\qquad I_{k,\delta}=0\quad\text{for } |x|\ge\delta \ .$$

We denote capacity with respect to this kernel by $C_{k,p;\delta}(\,\cdot\,)$.

Definition 8.2.3. Let $K\subset\mathbf{R}^N$, let k be a positive integer, and $1<p<\infty$. Then the *relative capacity* of K is

$$c_{k,p}(x_0,K,\delta)=\frac{C_{k,p;3\delta}(K\cap B(x_0,\delta))}{\delta^{N-kp}} \ .$$

Proposition 8.2.4. *Let $\delta K = \{\delta x : x \in K\}$. Then for any $\delta > 0$, and all k and p*

$$c_{k,p}(0, \delta K, \delta) = c_{k,p}(0, K, 1) \ .$$

In other words, the relative capacity is invariant under a change of scale.

Proof. For any measure μ we define a measure μ_δ by setting $\mu_\delta(\delta E) = \mu(E)$ for all sets E. Then, by a change of variable,

$$\int_{\mathbf{R}^N} (I_{k,\delta} * \mu_\delta)^{p'} \, dx = \frac{\delta^N}{\delta^{(N-k)p'}} \int_{\mathbf{R}^N} (I_{k,1} * \mu)^{p'} \, dx \ ,$$

and the result follows from Theorem 2.5.1.

Proposition 8.2.5. *There are constants $A > 0$ and $c > 0$, such that for any compact $K \subset \mathbf{R}^N$ and $0 < \delta \le 1$,*
(a) *for $kp < N$*

$$A^{-1} c_{k,p}(x_0, K, \delta) \le \frac{C_{k,p}(K \cap B(x_0, \delta))}{\delta^{N-kp}} \le A\, c_{k,p}(x_0, K, \delta) \ ;$$

(b) *for $kp = N$*

$$A^{-1} c_{k,p}(x_0, K, \delta) \le C_{k,p}(K \cap B(x_0, \delta)) \le A\, c_{k,p}(x_0, K, \delta) \ ,$$

if $C_{k,p}(K \cap B(x_0, \delta)) < c\, C_{k,p}(B(x_0, \delta))$;
(c) *for $kp > N$*

$$c_{k,p}(x_0, K, \delta) \ge A^{-1} > 0 \ ,$$

if $K \cap B(x_0, \delta) \ne \emptyset$.

Proof. (a) The right hand inequality is evident. We prove that $C_{k,p;3\delta}(K) \le A\, C_{k,p}(K)$.

Suppose that $K \subset B(0, \delta)$, and suppose first that $C_{k,p}(K) < c\, \delta^{N-kp}$, where c is a number to be fixed. Let $\mu \in \mathcal{M}^+(K)$ be any measure with unit mass. Then

$$\int_{\mathbf{R}^N} (I_k * \mu)^{p'} \, dx \ge C_{k,p}(K)^{1-p'} \ .$$

On the other hand

$$\int_{|x|>2\delta} (I_k * \mu)^{p'} \, dx \le 2^{(N-k)p'} \int_{|x|>2\delta} (I_k)^{p'} \, dx = A_1 \delta^{(kp-N)(p'-1)} \ ,$$

since $|x - y| \ge |x| - \delta \ge |x|/2$ if $|x| \ge 2\delta$ and $|y| \le \delta$. Thus, if c is chosen suitably,

$$\int_{|x|\le 2\delta} (I_k * \mu)^{p'} \, dx \ge \tfrac{1}{2} C_{k,p}(K)^{1-p'} \ .$$

But

$$\int_{\mathbf{R}^N} (I_{k,3\delta} * \mu)^{p'} \, dx \ge \int_{|x|\le 2\delta} (I_k * \mu)^{p'} \, dx \ ,$$

which gives $C_{k,p;3\delta}(K) \le A\, C_{k,p}(K)$.

If $C_{k,p}(K) \geq c\,\delta^{N-kp}$, we estimate $C_{k,p;3\delta}(K)$ by $C_{k,p;3\delta}(B(0,\delta))$. But $C_{k,p;3\delta}(B(0,\delta)) = A\,\dot{C}_{k,p}(B(0,\delta)) = A\,\delta^{N-kp}$, both capacities having the same homogeneity, and the result follows from Proposition 5.1.4.

(b) Now suppose that $C_{k,p}(K) < c(\log(1/\delta))^{1-p}$. The argument in (a) gives (cf. the proof of Proposition 5.1.3)

$$\int_{|x|>2\delta} (G_k * \mu)^{p'}\,dx \leq A \log\frac{2}{\delta}\ ,$$

which again gives the required inequality if c is small enough.

(c) In this case $I_{k,\delta} \in L^{p'}(\mathbf{R}^N)$, which implies that

$$\|I_{k,\delta} * \mu\|_{p'} \leq \mu(K)\|I_{k,\delta}\|_{p'} < \infty \quad \text{for all } \mu \in \mathcal{M}^+(K)\ .$$

This finishes the proof.

Corollary 8.2.6. *If $kp \leq N$ a set $E \subset \mathbf{R}^N$ is (k,p)-thin at x_0 if and only if*

$$\int_0^1 \left(c_{k,p}(x_0, K, \delta)\right)^{p'-1} \frac{d\delta}{\delta} < \infty\ .$$

Proof. Only the case $kp = N$ needs proof, and in this case only the claim that thinness implies convergence of the above integral.

Suppose that E is (k,p)-thin at x_0, i.e.,

$$\int_0^1 \left(C_{k,p}(E \cap B(x_0,\delta))\right)^{p'-1} \frac{d\delta}{\delta} < \infty\ .$$

Then $\lim_{\delta\to 0}(C_{k,p}(E \cap B(x_0,\delta)))^{p'-1}\log(1/\delta) = 0$, since for an increasing function $\varphi(\delta)$

$$\int_r^1 \varphi(\delta)\,\frac{d\delta}{\delta} = \varphi(r)\log\frac{1}{r} + \int_r^1 \log\frac{1}{\delta}\,d\varphi(\delta) \geq \varphi(r)\log\frac{1}{r}\ .$$

Thus $\lim_{r\to 0}\varphi(r)\log(1/r)$ exists if the integral converges, but if the limit is not 0 this is impossible. The claim now follows from Proposition 8.2.5.

Because of homogeneity the following corollaries now follow from Theorem 8.2.1 or Corollary 8.2.2 in the same way as in the proof of Corollary 8.1.4.

Corollary 8.2.7. *Let B_δ be a ball of radius δ, and let $f \in W^{m,p}(B_\delta)$, $1 < p < \infty$. Let $K \subset B_\delta$ be a closed set such that $C_{1,p}(K) > 0$, and suppose that*

$$D^\beta f|_K = 0 \quad \textit{for all } \beta,\quad 0 \leq |\beta| \leq m-1\ .$$

Then there is a constant A, independent of δ, such that

$$\int_{B_\delta} |f|^p\,dx \leq \frac{A\,\delta^{mp}}{c_{1,p}(0,K,\delta)} \int_{B_\delta} |\nabla^m f|^p\,dx\ ,$$

and

$$\int_{B_\delta} |\nabla^i f|^p\,dx \leq \frac{A\,\delta^{(m-i)p}}{c_{1,p}(0,K,\delta)} \int_{B_\delta} |\nabla^m f|^p\,dx\ , \quad i = 1, 2, \ldots, m-1\ .$$

Corollary 8.2.8. *Let B_δ be a ball of radius δ, and let $f \in W^{m,p}(B_\delta)$, $1 < p < \infty$. Let $K \subset B_\delta$ be a closed set such that $C_{m-k,p}(K) > 0$, for some $k = 1, 2, \ldots, m-2$, and suppose that*

$$D^\beta f|_K = 0 \quad \textit{for all } \beta, \quad 0 \le |\beta| \le k.$$

Then there is a constant A, independent of δ, such that

$$\int_{B_\delta} |f|^p\,dx \le A\,\delta^{(k+1)p} \int_{B_\delta} |\nabla^{k+1} f|^p\,dx + \frac{A\,\delta^{mp}}{C_{m-k,p}(0, K, \delta)} \int_{B_\delta} |\nabla^m f|^p\,dx\,,$$

and

$$\int_{B_\delta} |\nabla^i f|^p\,dx \le A\,\delta^{(k+1-i)p} \int_{B_\delta} |\nabla^{k+1} f|^p\,dx + \frac{A\,\delta^{(m-i)p}}{C_{m-k,p}(0, K, \delta)} \int_{B_\delta} |\nabla^m f|^p\,dx\,,$$

for $i = 1, 2, \ldots, k$.

These two corollaries will play a crucial role in the next chapter.

Remark. Let K belong to the unit ball B, let $f \in W^{m,p}(B)$, and suppose that

$$D^\beta f|_K = 0 \quad \text{for all } \beta, \quad 0 \le |\beta| \le k\ ,$$

for some $k \le m-1$. It is a consequence of the closed graph theorem that there exists a constant A_K, independent of f, and a polynomial P of degree $\le m-1$ such that $D^\beta P|_K = 0$ for $0 \le |\beta| \le k$, and

$$\int_B |f - P|^p\,dx \le A_K \int_B |\nabla^m f|^p\,dx\ .$$

In particular

$$\int_B |f|^p\,dx \le A_K \int_B |\nabla^m f|^p\,dx\ .$$

if there are no such polynomials. See Maz'ya [308], Section 10.3.1, and also Ziemer [438], Section 4.6.

However, it is difficult to say anything about A_K that is useful for our purposes, because A_K depends on algebraic properties of K. For example, as pointed out by V. G. Maz'ya in conversation with one of the authors, if $m = p = N = 2$, then there is a constant A_K such that $\int_B |f|^2\,dx \le A_K \int_B |\nabla^2 f|^2\,dx$, if K consists of three non collinear points, but not if K consists of two points. The idea of Theorem 8.2.1 comes from this observation. It is by allowing a polynomial P_{m-1} which does not vanish on K, that we can obtain an estimate in terms of capacity.

Maz'ya [305] has given estimates for the best value of the constant A_K in terms of a "polynomial capacity". See also Maz'ya [308], Section 10.3. A lower estimate of this capacity that depends on certain algebraic properties of the set, was recently given by K. Nyström [351]. See also Section 11.5 below, and the notes to that section.

8.3 An Abstract Approach

In this section we give alternative, non-constructive proofs of the results of the previous section. We start by the following lemma.

Lemma 8.3.1. *Let X_0 be a Banach space with norm $\|\cdot\|_0$, and let $X \subset X_0$ be a Banach space with norm $\|\cdot\|_X = \|\cdot\|_0 + \|\cdot\|_1$, where $\|\cdot\|_1$ is a seminorm with null space $Y \neq \{0\}$. If the imbedding $X \subset X_0$ is compact, there is a constant A such that*

$$\|x - Lx\|_0 \leq A\,\|L\|\,\|x\|_1$$

for all $x \in X$ and all projections L of X onto Y.

Proof. First let $L' : X \to Y$ be a fixed projection. We will show that

$$\|x - L'x\|_0 \leq A'\,\|x\|_1 \ ,$$

where A' depends on L' but not on x. Suppose that this were not true. By normalizing we may then assume that there is a sequence $\{x_n\}_1^\infty$ such that $\|x_n\|_1 \to 0$ and $\|x_n - L'x_n\|_0 = 1$. Setting $z_n = x_n - L'x_n$ we obtain

$$\|z_n\|_1 \leq \|x_n\|_1 + \|L'x_n\|_1 = \|x_n\|_1 \ ,$$

since $L'x_n \in Y$. By the compactness assumption there is a subsequence, again denoted by $\{z_n\}_1^\infty$, that converges in X_0 to some element z. But $\|z_n\|_1 \to 0$, so $\{z_n\}_1^\infty$ is Cauchy in X. It follows that $z \in X$, that $\|z_n - z\|_X \to 0$, and that $L'z = z$, since $\|z\|_1 = 0$. On the other hand $L'z_n = 0$, since L' is a projection. But L' is continuous on X, so $L'z_n \to L'z = z$, which gives a contradiction, since $\|z\|_0 = 1$.

Now we let $L : K \to Y$ be an arbitrary projection. Then $LL' = L'$, and thus

$$\begin{aligned}\|x - Lx\|_0 &= \|x - L'x - L(x - L'x)\|_0 \leq \|x - L'x\|_0 + \|L\|\,\|x - L'x\|_X \\ &\leq \big(A' + \|L\|(A'+1)\big)\|x\|_1 \leq (2A'+1)\|L\|\,\|x\|_1 \ .\end{aligned}$$

Here the last inequality follows from the fact that $\|L\| \geq 1$, since $L^2 = L$.

Recall that a domain $\Omega \subset \mathbf{R}^N$ is called an extension domain if for any domain Ω' such that $\overline{\Omega} \subset \Omega'$ there is for every positive integer m a bounded linear extension operator $E_m : W^{m,p}(\Omega) \to W_0^{m,p}(\Omega')$. Thus, for any $f \in W^{m,p}(\Omega)$ there is $E_m f \in W_0^{m,p}(\Omega')$ such that $E_m f|_\Omega = f$. See Section 1.2, in particular Theorem 1.2.2.

In Section 7.1 (see (7.1.2)) we observed that if $f \in W^{k,p}(\mathbf{R}^N)$ is (k,p)-quasicontinuous, and if $\mu \in \mathcal{M}^+(\mathbf{R}^N)$ satisfies $G_k * \mu \in L^{p'}(\mathbf{R}^N)$, then $f \in L^1(\mu)$, and

$$\left|\int_{\mathbf{R}^N} f\,d\mu\right| = |\langle \mu, f\rangle| \leq \|f\|_{k,p}\|G_k * \mu\|_{p'} \ . \tag{8.3.1}$$

Lemma 8.3.2. *Let $\Omega \subset \mathbf{R}^N$ be a bounded extension domain. Let $1 < p < \infty$, let m be a positive integer, and let $\mathfrak{P}_{m-1}$ denote the space of polynomials of degree $\le m-1$. For all multiindices σ, $|\sigma| \le m-1$, let μ_σ be a positive measure supported in Ω such that $\mu_\sigma(\Omega) = 1$ and $G_{m-|\sigma|} * \mu_\sigma \in L^{p'}(\mathbf{R}^N)$.*

Then there is a unique linear operator $L : W^{m,p}(\Omega) \to \mathfrak{P}_{m-1}$ such that for all $|\sigma| \le m-1$, and all $f \in W^{m,p}(\Omega)$

$$\int_\Omega D^\sigma f \, d\mu_\sigma = \int_\Omega D^\sigma (Lf) \, d\mu_\sigma \; . \tag{8.3.2}$$

Moreover, L is a projection, if $\mathfrak{P}_{m-1}$ is identified with a subspace of $W^{m,p}(\Omega)$, and

$$\|L\| \le A \sum_{|\sigma| \le m-1} \|G_{m-|\sigma|} * \mu_\sigma\|_{p'} \; . \tag{8.3.3}$$

Proof. It should be noticed that if the measures μ_σ are all equal to the Dirac measure at a point x, then Lf is just the Taylor expansion of f. Now assume that the projection L exists and set $Lf(x) = \sum_{|\gamma| \le m-1} a_\gamma x^\gamma$. We have $D^\sigma x^\gamma = \frac{\gamma!}{(\gamma-\sigma)!} x^{\gamma-\sigma}$ for $\sigma \le \gamma$, and $D^\sigma x^\gamma = 0$ for $\sigma > \gamma$. Thus

$$D^\sigma (Lf(x)) = \sum_{0 \le |\beta| < m-|\sigma|} a_{\sigma+\beta} \frac{(\sigma+\beta)!}{\beta!} x^\beta \; .$$

The condition (8.3.2) then becomes

$$\int_\Omega D^\sigma f \, d\mu_\sigma = \sum_{0 \le |\beta| < m-|\sigma|} a_{\sigma+\beta} \frac{(\sigma+\beta)!}{\beta!} \int_\Omega x^\beta \, d\mu_\sigma \, , \quad |\sigma| \le m-1 \; .$$

For $|\sigma| = m-1$ this gives

$$a_\sigma = \frac{1}{\sigma!} \int_\Omega D^\sigma f \, d\mu_\sigma \; , \tag{8.3.4}$$

and for $0 \le |\sigma| < m-1$ we obtain the recursion formula

$$a_\sigma = \frac{1}{\sigma!} \int_\Omega D^\sigma f \, d\mu_\sigma - \sum_{1 \le |\beta| < m-|\sigma|} a_{\sigma+\beta} \binom{\sigma+\beta}{\sigma} \int_\Omega x^\beta \, d\mu_\sigma \; . \tag{8.3.5}$$

The formula proves the uniqueness and also the existence of an operator L that satisfies (8.3.2). Because of the uniqueness we also conclude that $Lp = p$ if $p \in \mathfrak{P}_{m-1}$, in other words L is a projection onto $\mathfrak{P}_{m-1}$.

Now let $\|f\|_{W^{m,p}(\Omega)} = 1$. Because of the boundedness of Ω there is a constant A such that

$$\|Lf\|_{W^{m,p}(\Omega)} \le A \sum_{0 \le |\sigma| \le m-1} |a_\sigma| \; .$$

By (8.3.4), (8.3.1), and the extension property we obtain for $|\sigma| = m-1$

$$|a_\sigma| \le A \, \|E_1(D^\sigma f)\|_{1,p} \|G_1 * \mu_\sigma\|_{p'} \le A \, \|G_1 * \mu_\sigma\|_{p'} \; .$$

Similarly, (8.3.5) gives for $0 \le |\sigma| < m-1$

$$|a_\sigma| \le A \, \|G_{m-|\sigma|} * \mu_\sigma\|_{p'} + A \sum_{1 \le |\beta| < m-|\sigma|} |a_{\sigma+\beta}| \ ,$$

since $\left|\int_\Omega x^\beta \, d\mu_\sigma\right|$ is bounded by a constant depending on Ω. Induction gives the formula (8.3.3).

Theorem 8.3.3. *Let $\Omega \subset \mathbf{R}^N$ be a bounded extension domain, and suppose that $f \in W^{m,p}(\Omega)$, $1 < p < \infty$. Let $K \subset \Omega$ be closed, and $C_{m-k,p}(K) > 0$ for some $k = 0, 1, \ldots, m-1$. Suppose that*

$$D^\beta f|_K = 0 \quad \text{for all } \beta, \quad 0 \le |\beta| \le k \ .$$

Then there is a polynomial $P_{m-1} \in \mathfrak{P}_{m-1}$, which is zero if $k = m-1$, and otherwise satisfies

$$\sup_{x \in \Omega} |P_{m-1}(x)| \le A \, \|\nabla^{k+1} f\|_{L^1(\Omega)} \ , \tag{8.3.6}$$

such that

$$\|f - P_{m-1}\|_{L^p(\Omega)} \le A \, \frac{\|\nabla^m f\|_{L^p(\Omega)}}{C_{m-k,p}(K)^{1/p}} \ . \tag{8.3.7}$$

Remark. Corollaries 8.2.7 and 8.2.8 follow from Theorem 8.3.3 in the same way as from Theorem 8.2.1 or Corollary 8.2.2.

Proof. We apply Lemma 8.3.1 to $X_0 = L^p(\Omega)$, $X_1 = W^{m,p}(\Omega)$ with the seminorm $\|f\|_1 = \|\nabla^m f\|_{L^p(\Omega)}$, and a suitable projection L onto the nullspace $Y = \mathfrak{P}_{m-1}$, as constructed in Lemma 8.3.2.

To this end we let μ_σ, $0 \le |\sigma| \le k$, be the $(m-|\sigma|, p)$-capacitary measure for K, normalized so that $\mu_\sigma(K) = 1$, and consequently

$$\|G_{m-|\sigma|} * \mu_\sigma\|_{p'} = C_{m-|\sigma|,p}(K)^{-1/p} \ .$$

The existence of this measure was established in Theorem 2.2.7 and Theorem 2.5.3.

If $k < m-1$, we also choose a function $\omega \in C_0^\infty(\Omega)$ such that $\int_\Omega \omega \, dx = 1$, and we set $d\mu_\sigma = \omega \, dx$ for all σ with $k+1 \le |\sigma| \le m-1$. Let L be the projection associated by Lemma 8.3.2 to the measures $\{\mu_\sigma\}$, and set $P_{m-1} = Lf$.

The estimate (8.3.7) now follows from Lemma 8.3.2. Indeed, for $k+1 \le |\sigma| \le m-1$ we have

$$\|G_{m-|\sigma|} * \mu_\sigma\|_{p'} \le \|G_1 * |\omega|\|_{p'} \le A \ .$$

Thus, by (8.3.3),

$$\|L\| \le A \left(\sum_{j=0}^{k} C_{m-j,p}(K)^{-1/p} + 1 \right) \le A \, C_{m-k,p}(K)^{-1/p} \ ,$$

and (8.3.7) now follows from Lemma 8.3.1.

In order to prove (8.3.6) we observe that if $k < m - 1$, and $0 \le \gamma \le \sigma$, then for $k + 1 \le |\sigma| \le m - 1$ we have by integration by parts

$$\int_\Omega D^\sigma f \, d\mu_\sigma = \int_\Omega D^\sigma f \, \omega \, dx = (-1)^{|\gamma|} \int_\Omega D^\gamma f \, D^{\sigma-\gamma} \omega \, dx \ .$$

Thus, choosing γ with $|\gamma| = k + 1$, we find by (8.3.4)

$$|a_\sigma| \le A \, \|\nabla^{k+1} f\|_{L^1(\Omega)}$$

for $|\sigma| = m - 1$. By using (8.3.5) recursively we obtain similarly

$$|a_\sigma| \le A \, \|\nabla^{k+1} f\|_{L^1(\Omega)} + A \sum_{1 \le |\beta| < m - |\sigma|} |a_{\sigma+\beta}| \le A \, \|\nabla^{k+1} f\|_{L^1(\Omega)}$$

for $k + 1 \le |\sigma| < m - 1$.

For $0 \le |\sigma| \le k \le m - 1$ we have by assumption

$$\int_\Omega D^\sigma f \, d\mu_\sigma = 0 \ .$$

Thus, by (8.3.4) and (8.3.5)

$$|a_\sigma| \le A \sum_{1 \le |\beta| < m - |\sigma|} |a_{\sigma+\beta}| \ ,$$

which gives $P_{m-1} = 0$ in the case $k = m - 1$, and (8.3.6) in the other cases.

Remark. If no assumption about the vanishing of f and its derivatives on K is made, the proof gives

$$|a_\sigma| \le A \sum_{0 \le |\beta+\sigma| \le k} \int_K |D^{\sigma+\beta} f| \, d\mu_{\sigma+\beta} + A \, \|\nabla^{k+1} f\|_{L^1(\Omega)} \ ,$$

if $k < m - 1$, and

$$|a_\sigma| \le A \sum_{0 \le |\beta+\sigma| \le m-1} \int_K |D^{\sigma+\beta} f| \, d\mu_{\sigma+\beta} \ ,$$

if $k = m - 1$. Rewriting the integrals by means of (7.1.3), and using the fact that for all Borel sets E the capacitary measures μ_σ satisfy

$$\mu_\sigma(E) \le \frac{C_{m-|\sigma|,p}(E)}{C_{m-|\sigma|,p}(K)}$$

(see Theorem 2.5.5), we obtain the estimate

$$\begin{aligned} \int_K |D^\sigma f| \, d\mu_\sigma &= \int_0^\infty \mu_\sigma(\{x : |D^\sigma f(x)| > \lambda|\}) \, d\lambda \\ &\le \frac{1}{C_{m-|\sigma|,p}(K)} \int_0^\infty C_{m-|\sigma|,p}(\{x \in K : |D^\sigma f(x)| > \lambda|\}) \, d\lambda \ . \end{aligned}$$

It follows that the conclusion of Theorem 8.3.3 is still true with P_{m-1} satisfying

$$\sup_{x\in\Omega} |P_{m-1}(x)| \le A \sum_{j=0}^{k} \frac{1}{C_{m-j,p}(K)} \int_0^\infty C_{m-j,p}(\{x \in K : |\nabla^j f(x)| > \lambda|\})\, d\lambda$$
$$+ A \int_\Omega |\nabla^{k+1} f|\, dx \ ,$$

if $k < m - 1$, and

$$\sup_{x\in\Omega} |P_{m-1}(x)|$$
$$\le A \sum_{j=0}^{m-1} \frac{1}{C_{m-j,p}(K)} \int_0^\infty C_{m-j,p}(\{x \in K : |\nabla^j f(x)| > \lambda|\})\, d\lambda \ ,$$

if $k = m - 1$. Cf. Corollary 8.1.3.

8.4 Notes

8.1. The results of this section are due to S. L. Sobolev [384] and [385]. See also the books by R. A. Adams [26], and V. G. Maz'ya [308], Chapter 1, where many more results and references are found.

8.2. Theorem 8.2.1 was proved by Hedberg in [213], after a version which contained Corollary 8.2.7 had been proved in [212]. Maz'ya has given a somewhat different proof of Corollary 8.2.8, which is also valid for $p = 1$, in [308], Section 10.1.3. A variant of this proof is found in [216]. The history of this kind of capacitary estimates, which contains many rediscoveries, goes back at least to 1953, when A. M. Molchanov [331] proved a theorem similar to Corollary 8.2.7 for $m = 1$ and $p = 2$. Maz'ya began a systematic study of such inequalities in 1963, see [296, 305], and he gave a detailed exposition of the whole subject, including estimates of the best constants in the inequalities, in [308], Chapter 10. See also the remark at the end of Section 8.2 above.

8.3. The abstract approach is due to N. G. Meyers [322]. Lemma 8.3.1 is a version of one of the standard proofs of the Poincaré inequality; see e.g. W. P. Ziemer [438], Theorem 4.1.1. Meyers' original method did not, however, give the correct power of the capacity, as it now appears in Theorem 8.3.3, and it did not give the crucial estimate (8.3.6). A missing step was provided by D. R. Adams, and appears as Corollary 4.1.5 in Ziemer [438]. Theorem 4.5.1 in [438] is a version of Theorem 8.3.3, but again without (8.3.6). The method was finally extended by Anders Carlsson [94] to give (a slightly more general result than) Theorem 8.3.3.

9. An Approximation Theorem

This chapter, and part of the next, are devoted to a result which can be viewed either as an approximation theorem, or as a theorem characterizing the kernel of a trace operator for arbitrary sets. In the present chapter we treat the case of Sobolev spaces $W^{m,p}(\mathbf{R}^N)$ for integer m and $1 < p < \infty$. The main result, Theorem 9.1.3, and a number of corollaries are stated and discussed at some length in Section 9.1. The proof, which uses much of the nonlinear potential theory developed previously in the book, occupies the rest of the chapter. The contents are outlined at the end of Section 9.1.

In Chapter 10 we prove, by quite different means, a more general theorem (Theorem 10.1.1), due to Yu. V. Netrusov, which contains Theorem 9.1.3 as a special case. The beautiful proof depends on some of the theory for Lizorkin–Triebel spaces developed in Chapter 4, and is considerably shorter. In spite of this, the original proof is probably still the simplest one for sets with some regularity, and it also has some other features which make us think it worth including here in its entirety.

9.1 Statement of Results

Before formulating the main theorem we recall a result for the space C^m of m times continuously differentiable functions.

Theorem 9.1.1. *Let $f \in C^m(\mathbf{R}^N)$, and let $K \in \mathbf{R}^N$ be compact. Suppose that $D^\beta f(x) = 0$ for all $x \in K$, and for all multiindices β with $|\beta| \leq m$. Then, for any neighborhood V of K and any $\varepsilon > 0$ there is a function $\eta \in C^\infty(\mathbf{R}^N)$ such that $0 \leq \eta \leq 1$, $\operatorname{supp}(1-\eta) \subset V$, $\eta = 0$ on a neighborhood of K and*

$$\sum_{|\beta| \leq m} \|D^\beta(f - \eta f)\|_\infty < \varepsilon \ .$$

We give the proof in Section 9.3 below. The definition of support of a distribution immediately gives the following corollary.

Corollary 9.1.2. *Let T be a distribution of order m with compact support, i.e., T is a bounded linear functional on $C^m(\mathbf{R}^N)$. Then $\langle T, f\rangle = 0$ for every $f \in C^m(\mathbf{R}^N)$ such that $D^\beta f(x) = 0$ for all $x \in \operatorname{supp} T$, and for all $|\beta| \leq m$.*

Our purpose is to prove an analogous result in $W^{m,p}(\mathbf{R}^N)$ for integer m, and $1 < p < \infty$. In order to formulate the theorem more easily we use the notion of the trace of a function on a set E. (See Definition 6.1.6.)

Thus, $f|_E = 0$ means that there is an (m, p)-quasicontinuous representative $\widetilde{f}$ of f that satisfies

$$\widetilde{f}(x) = 0 \qquad (m, p)\text{-q.e. on } E.$$

Moreover, by Theorem 6.2.1, $\widetilde{f}(x)$ can always be defined in the Lebesgue sense, i.e., we can interpret the statement $f|_E = 0$ as meaning simply that

$$\lim_{r\to 0} \frac{1}{r^N} \int_{B(x,r)} |f(y)|\, dy = 0 \qquad \text{for } (m, p)\text{-q.e. } x \in E.$$

Similarly, if $f \in W^{m,p}(\mathbf{R}^N)$, and β is a multiindex with $|\beta| < m$, then $D^\beta f|_E = 0$ means that $D^\beta f$, defined as the distribution derivative of f, has an $(m-|\beta|, p)$-quasicontinuous representative $\widetilde{D^\beta f}$, defined in the sense of Lebesgue, that satisfies

$$\widetilde{D^\beta f}(x) = 0 \qquad (m - |\beta|, p)\text{-q.e. on } E.$$

If $C_{m,p}(E) = 0$, then we interpret $f|_E = 0$ as being true for all f in $W^{m,p}(\mathbf{R}^N)$.

We also define a vector valued trace operator Tr_E by setting

$$\mathrm{Tr}_E\, f = \{D^\beta f|_E\}_{|\beta|\le m-1}. \tag{9.1.1}$$

We can now formulate our main result.

Theorem 9.1.3. *Let m be a positive integer, let $1 < p < \infty$, and let $f \in W^{m,p}(\mathbf{R}^N)$. Let $\Omega \subset \mathbf{R}^N$ be an arbitrary open set, and denote its complement by K. Then the following statements are equivalent*:

(a) $D^\beta f|_K = 0$ *for all multiindices* β, $0 \le |\beta| \le m - 1$;
(b) $f \in W_0^{m,p}(\Omega)$;
(c) *for any* $\varepsilon > 0$ *and any compact* $F \subset \Omega$ *there is a function* $\eta \in C_0^\infty(\Omega)$ *such that* $\eta = 1$ *on* F, $0 \le \eta \le 1$, *and* $\|f - \eta f\|_{m,p} < \varepsilon$.

By means of the kernel of operator Tr_K the main result can be formulated concisely as follows.

Corollary 9.1.4. *Under the assumptions of Theorem* 9.1.3

$$\mathrm{Ker}\, \mathrm{Tr}_K = W_0^{m,p}(\Omega)\ .$$

A few comments are in order. The implications (b)⇒(a), and (c)⇒(b) are easy, and the proof of the theorem will consist in proving the implication (a)⇒(c).

In fact, if $f \in W_0^{m,p}(\Omega)$, then by definition there is a sequence $\{\varphi_n\}_1^\infty$, $\varphi_n \in C_0^\infty(\Omega)$ such that $\lim_{n\to\infty} \|f - \varphi_n\|_{m,p} = 0$. But then, by Proposition 2.3.8, there is a subsequence that converges pointwise (m, p)-q.e. to a quasicontinuous representative $\widetilde{f}$ of f. But $\varphi_n(x) = 0$ everywhere on K, so $\widetilde{f}(x) = 0$ (m, p)-q.e. on K. Similarly $\widetilde{D^\beta f}(x) = 0$ $(m - |\beta|, p)$-q.e. on K for $|\beta| \le m - 1$, so (b)⇒(a).

The implication (c)⇒(b) is clear, because if η has the required properties, then the convolution of ηf with a suitable test function with support in a small enough ball will be a function in $C_0^\infty(\Omega)$ that approximates f.

We do not know any proof of either of the implications (a)⇒(b), or (b)⇒(c), that does not also give the implication (a)⇒(c).

Before turning to the proof of the theorem, we give a few more corollaries.

Corollary 9.1.5. *Let $\Omega \subset \mathbf{R}^N$ be an arbitrary open set, m a positive integer, and $1 < p < \infty$. Suppose that $f \in W_0^{m,p}(\Omega)$ and that $f(x) \geq 0$ a.e. Then there is a sequence of functions $\{\varphi_n\}_1^\infty$, such that $\varphi_n \in C_0^\infty(\Omega)$, $\varphi_n \geq 0$, and* $\lim_{n\to\infty} \|f - \varphi_n\|_{m,p} = 0$.

Proof. Just regularize the functions ηf in (c) by suitable convolutions.

Corollary 9.1.6. *Let $T \in W^{-m,p'}(\mathbf{R}^N)$, where m is a positive integer, and $1 < p < \infty$. Denote* $\operatorname{supp} T$ *by K. Then $\langle T, f\rangle = 0$ for all $f \in W^{m,p}(\mathbf{R}^N)$ such that $D^\beta f|_K = 0$ for all β, $0 \leq |\beta| \leq m-1$.*

Proof. By assumption $\langle T, \varphi\rangle = 0$ for all $\varphi \in C_0^\infty(K^c)$. By the theorem, functions f that satisfy the assumptions of the corollary are in $W_0^{m,p}(K^c)$. But then there is a sequence of functions $\varphi_n \in C_0^\infty(K^c)$ such that $\varphi_n \to f$ in $W^{m,p}(\mathbf{R}^N)$. The corollary follows.

The following corollary is a dual formulation of the equivalence between (a) and (b).

Corollary 9.1.7. *Let $T \in W^{-m,p'}(\mathbf{R}^N)$ for a positive integer m and $1 < p' < \infty$. Then, for any $\varepsilon > 0$ there are compactly supported (signed) measures*

$$\mu_\beta \in W^{|\beta|-m,p'}(\mathbf{R}^N), \quad 0 \leq |\beta| \leq m-1\ ,$$

with $\operatorname{supp} \mu_\beta \subset \operatorname{supp} T$, *such that*

$$\left\| T - \sum\nolimits_{|\beta|\leq m-1} \mu_\beta \right\|_{-m,p'} < \varepsilon.$$

Proof. Denote $\operatorname{supp} T$ by K. Then the annihilator of $W_0^{m,p}(K^c)$ is $W^{-m,p'}(K)$, the elements in $W^{-m,p'}(\mathbf{R}^N)$ with support in K. It is easily seen that if $f \in W^{m,p}(\mathbf{R}^N)$ satisfies $\langle D^\beta \mu, f\rangle = 0$ for all compactly supported $\mu \in W^{|\beta|-m,p'}(K)$, then $D^\beta f|_K = 0$. In fact, by (7.1.2)

$$\langle D^\beta \mu, f\rangle = (-1)^{|\beta|}\langle \mu, D^\beta f\rangle = \int_K \widetilde{D^\beta f}\, d\mu\ .$$

Thus, by the theorem, if $\langle D^\beta \mu, f\rangle = 0$ for all $\mu \in W^{|\beta|-m,p'}(K)$, and for all $|\beta| \leq m-1$, then $f \in W_0^{m,p}(K^c)$, i.e., f belongs to the annihilator of $W_0^{m,p}(K^c)$. It follows by reflexivity that the linear span of $\{D^\beta \mu_\beta\}_{|\beta|\leq m-1}$, where the μ_β are compactly supported measures in $W^{|\beta|-m,p'}(K)$, is dense in $W^{|\beta|-m,p'}(K)$.

Another corollary is a uniqueness theorem for the Dirichlet problem. We take the operator Δ^m as a model and consider the equation $\Delta^m u = 0$ in an arbitrary bounded domain Ω. Then it is well known that the Dirichlet problem has a unique solution in the sense that for any given $g \in W^{m,2}(\Omega)$ there exists a unique $u \in W^{m,2}(\Omega)$ such that $\Delta^m u = 0$ in Ω (in the weak sense or in the distribution sense), and such that $u-g \in W_0^{m,2}(\Omega)$. See e.g. G. B. Folland [150]. An alternative way of formulating the Dirichlet problem is by means of the notion of trace on the boundary, and then Theorem 9.1.3 has the following consequence. See also Section 9.12.1 below.

Corollary 9.1.8. *Let $\Omega \subset \mathbf{R}^N$ be a bounded open set, and let $g \in W^{m,2}(\mathbf{R}^N)$ be given. Then there exists a unique $u \in W^{m,2}(\mathbf{R}^N)$ such that $\Delta^m u = 0$ in Ω, and $D^\beta(u-g)|_{\Omega^c} = 0$ for all β with $|\beta| \le m-1$.*

Proof. There exists a unique solution u to the equation such that $u-g \in W_0^{m,2}(\Omega)$. Then $D^\beta(u-g)|_{\Omega^c} = 0$ for all β with $|\beta| \le m-1$, so the only problem is to show uniqueness under this hypothesis. But suppose that v is another solution such that $D^\beta(v-g)|_{\Omega^c} = 0$ for all β with $|\beta| \le m-1$. Then also $D^\beta(v-u)|_{\Omega^c} = 0$, and thus $v-u \in W_0^{m,2}(\Omega)$ by the theorem. But then $v = u$.

Remark. In Corollary 9.1.8 one can replace the requirement $D^\beta(u-g)|_{\Omega^c} = 0$ by $D^\beta(u-g)|_{\partial\Omega} = 0$ for all β with $|\beta| \le m-1$. In fact, by the theorem $u-g$ can be approximated in $W^{m,2}(\mathbf{R}^N)$ by functions in $C_0^\infty(\mathbf{R}^N \setminus \partial\Omega)$. It follows that the continuation of $u-g$ to $\mathbf{R}^N \setminus \overline{\Omega}$ by 0 is in $W_0^{m,2}(\Omega)$, and thus the restriction of u to Ω is uniquely determined by the traces $D^\beta u|_{\partial\Omega}$, $|\beta| \le m-1$.

In order to illustrate Corollary 9.1.8 we discuss its relation to a classic theorem of S. L. Sobolev [383] (also in Sobolev [385]), concerning the case when Ω is bounded by a finite union of manifolds of arbitrary codimension.

Let $f \in W^{m,2}(\Omega)$, and let M be a compact piece of an $(N-k)$-dimensional smooth manifold in Ω, where $1 \le k \le N$. Let f_δ, $\delta > 0$, be the averaged function, obtained by taking the convolution with a smooth approximate identity with support in a ball with radius δ. Then Sobolev proved the following trace theorem:

If $k < 2m$, then $\lim_{\delta\to 0} f_\delta(x)$, denoted $f|_M(x)$, exists a.e. on M in the $(N-k)$-dimensional sense, and $\lim_{\delta\to 0} \|f_\delta - f|_M\|_{L^2(M)} = 0$. The condition $k < 2m$ is sharp. (If $N < 2m$, then f and $f|_M$ are continuous, and $f_\delta(x)$ converges pointwise.)

Now let $\Omega \subset \mathbf{R}^N$ be a bounded domain, whose boundary consists of a finite union of smooth manifolds, not forming cusps. We write

$$\partial\Omega = \bigcup_{k=1}^{N} M_k \ ,$$

where each M_k is a finite union of $(N-k)$-manifolds. Then Sobolev proved, by letting the above M approach the boundary, that f also has well-defined traces, denoted $f|_{M_k}$, on the boundary components for which $k < 2m$.

It follows that $D^\beta f|_{M_k}$ is also well defined, provided $k < 2(m-|\beta|)$, i.e., $|\beta| < m - \frac{1}{2}k$.

Sobolev then formulated and solved the Dirichlet problem for the polyharmonic equation $\Delta^m u = 0$ in the following form.

Theorem 9.1.9 (S. L. Sobolev). *Let $\Omega \subset \mathbf{R}^N$ be open and bounded, and let $\partial\Omega = \bigcup_{k=1}^N M_k$ satisfy the above assumptions. Let $g \in W^{m,2}(\Omega)$ be given. Then there exists a unique $u \in W^{m,2}(\Omega)$ such that $\Delta^m u = 0$ in Ω, and $D^\beta(u-g)|_{M_k} = 0$ for $1 \le k < 2(m-|\beta|)$, and $0 \le |\beta| \le m-1$.*

Example. Consider the biharmonic equation $\Delta^2 u = 0$ in $\Omega \subset \mathbf{R}^3$, where $\partial\Omega = M_1 \cup M_2 \cup M_3$, and M_1 is the two-dimensional part of the boundary. Let $u \in W^{2,2}(\Omega)$. Then, by the above discussion, $u|_{M_k}$ is defined for $k = 1, 2, 3$, and $\nabla u|_{M_1}$ is defined. If, for example, Ω is the unit ball with the center removed, then M_1 is the unit sphere, $M_2 = \emptyset$, and $M_3 = \{0\}$. It is easily seen that $u(x) = (1-|x|)^2$ satisfies $\Delta^2 u = 0$ in Ω, and that $u \in W^{2,2}(\mathbf{R}^3)$. By the theorem this is the only solution such that $u|_{M_1} = 0$, $\nabla u|_{M_1} = 0$, and $u|_{M_3} = u(0) = 1$. Note that u is not differentiable at 0.

Proof of Theorem 9.1.9. It is clear from the discussion preceding Corollary 9.1.8 that solutions always exist. In fact, it is easily seen that the boundary conditions are satisfied if $u - g \in W_0^{m,2}(\Omega)$. What is not evident is that this solution is the only one, but this was proved by Sobolev. We shall deduce this result from Theorem 9.1.3.

It is enough to prove that if $f \in W^{m,2}(\Omega)$, and $D^\beta f|_{M_k} = 0$ in the sense of Sobolev for $1 \le k < 2(m-|\beta|)$, $0 \le |\beta| \le m-1$, then $f \in W_0^{m,2}(\Omega)$.

Suppose first that f is defined in all of $\mathbf{R}^N$ so that $f \in W^{m,2}(\mathbf{R}^N)$. We can assume that $D^\beta f$ is $(m-|\beta|, 2)$-quasicontinuous, and we claim that $D^\beta f(x) = 0$ $(m-|\beta|, 2)$-q.e. on each M_k, $1 \le k \le N$, for $0 \le |\beta| \le m-1$. For $k \ge 2(m-|\beta|)$ this is clear, since then $C_{m-|\beta|,2}(M_k) = 0$ by Theorem 5.1.9 or Corollary 5.1.15. For $k < 2(m-|\beta|)$ the assumption implies that $D^\beta f(x) = 0$ a.e. on M_k in the $(N-k)$-dimensional sense. This implies that $D^\beta f|_{M_k} = 0$ $(m-|\beta|, 2)$-q.e. on M_k. A quick way of seeing this is to refer to the fact that $D^\beta f$ is $(m-|\beta|, 2)$-finely continuous $(m-|\beta|, 2)$-q.e. according to Theorem 6.4.5. As in the proof of Corollary 6.4.7 it follows that $D^\beta f(x) = 0$ at all points of fine continuity belonging to M_k. The claim follows, and thus $f \in W_0^{m,2}(\Omega)$ by Theorem 9.1.3.

The condition that $f \in W^{m,2}(\mathbf{R}^N)$ is a restriction only in appearance. We need the following lemma.

Lemma 9.1.10. *Let $f \in W^{m,p}(\Omega \setminus F)$, where $\Omega \subset \mathbf{R}^N$ is open, and $F \subset \Omega$ is relatively closed with $(N-1)$-dimensional Hausdorff measure $\Lambda_{N-1}(F) = 0$. Then $f \in W^{m,p}(\Omega)$.*

Proof. By a partition of unity we can assume that f has compact support in Ω, i.e., we can assume that $f \in W^{m,p}(\mathbf{R}^N \setminus F)$. Then f and $D^\beta f$, $|\beta| \le m$, are defined almost everywhere in $\mathbf{R}^N$, so we can consider them as elements in $L^p(\mathbf{R}^N)$. It is enough to prove that

$$\int_{\mathbf{R}^N} \varphi\, D^\beta f\, dx = (-1)^{|\beta|} \int_{\mathbf{R}^N} f\, D^\beta \varphi\, dx \quad \text{for all } \varphi \in C_0^\infty(\mathbf{R}^N).$$

First let $|\beta| = 1$. By Fubini's theorem

$$\int_{\mathbf{R}^N} \varphi\, \partial_j f\, dx = \int_{\mathbf{R}^{N-1}} dx' \int_{-\infty}^{\infty} \varphi\, \partial_j f\, dx_j$$

($dx' = dx_1 \dots dx_{j-1} dx_{j+1} \dots dx_N$). By assumption, for almost every $a' \in \mathbf{R}^{N-1}$ the line $x' = a'$ is disjoint from F. In fact, the orthogonal projection of F on the hyperplane $x_j = 0$ has $(N-1)$-dimensional measure zero. Thus, by a well-known property of Sobolev spaces, f is absolutely continuous on almost all such lines. (See e.g. W. P. Ziemer [438], Theorem 2.1.4, p. 44, or L. C. Evans and R. F. Gariepy [137], Section 4.9.2, p. 164.) By integration by parts

$$\int_{\mathbf{R}^{N-1}} dx' \int_{-\infty}^{\infty} \varphi\, \partial_j f\, dx_j = -\int_{\mathbf{R}^{N-1}} dx' \int_{-\infty}^{\infty} f\, \partial_j \varphi\, dx_j = -\int_{\mathbf{R}^N} f\, \partial_j \varphi\, dx \ .$$

The lemma follows by induction.

Now let $f \in W^{m,2}(\Omega)$, where $\partial\Omega$ satisfies the assumptions of Theorem 9.1.9. It follows from the lemma that $\bigcup_2^N M_k$ is removable for $W^{m,2}$, or in other words, that f can be extended to a domain bounded by M_1, i.e., a finite union of smooth hypersurfaces not forming cusps. This means that M_1 locally satisfies a Lipschitz condition, and thus, by the extension theorem of A. P. Calderón [88] (see also E. M. Stein [389], Theorem VI.5, p. 181), f can be extended to $W^{m,2}(\mathbf{R}^N)$. The theorem follows, by the remark following Corollary 9.1.8.

The following corollary of H. Brezis and F. E. Browder [82] extends Theorem 3.4.2.

Corollary 9.1.11. *Let $\Omega \subset \mathbf{R}^N$ be an arbitrary open set, and let $u \in W_0^{m,p}(\Omega)$, where m is a positive integer, and $1 < p < \infty$. Let $S \in W^{-m,p'}(\Omega) \cap L^1_{\text{loc}}(\Omega)$, and suppose that $S(x)u(x) \geq -|f(x)|$ a.e. in Ω for some $f \in L^1(\Omega)$. Then $Su \in L^1(\Omega)$ and*

$$\langle S, u\rangle = \int_\Omega S(x)u(x)\, dx \ .$$

Proof. By Theorem 9.1.3 applied to u there is a sequence $\{u_n\}_1^\infty$ such that $\operatorname{supp} u_n \subset \Omega$, $u_n u \geq 0$, $|u_n| \leq |u|$, and $u_n \to u$ pointwise and in $W^{m,p}$. By Theorem 3.4.1 we can assume that each $u_n \in L^\infty$. We know that $\langle S, \varphi\rangle = \int_\Omega S(x)\varphi(x)\, dx$ if $\varphi \in C_0^\infty(\Omega)$. There are $\varphi_k \in C_0^\infty(\Omega)$ that converge to u_n boundedly and in $W^{\alpha,p}$, and thus $\langle S, u_n\rangle = \int_\Omega S(x)u_n(x)\, dx$.

But $\lim_{n\to\infty}\langle S, u_n\rangle = \langle S, u\rangle$, since $u_n \to u$ in $W_0^{m,p}(\Omega)$. On the other hand $S(x)u_n(x) \geq -|f(x)|$, so by Fatou's lemma $Su \in L^1(\mathbf{R}^N)$. But $|S(x)u_n(x)| \leq |S(x)u(x)|$, so by dominated convergence $\lim_{n\to\infty} \int_\Omega Su_n\, dx = \int_\Omega Su\, dx$, which proves the theorem.

The proof of Theorem 9.1.3 presented here is in its full generality quite long and complicated. It depends on results from several earlier chapters. In particular, the Kellogg property (Corollary 6.3.17) and the Poincaré type inequalities in Chapter 8 play an important part in the proof.

In the next section we prove the theorem in the special case $m = 1$. In Section 9.3 we then prove Theorem 9.1.1 and give an outline of the proof of Theorem 9.1.3 for $m \geq 2$. In Section 9.4 we prove the theorem in a special case which includes all regular domains, and the rest of the chapter is then devoted to the proof in the general case. Sections 9.6 – 9.11 are very technical, and are recommended only for highly motivated readers.

9.2 The Case $m = 1$

Because of the fact that $W^{1,p}(\mathbf{R}^N)$ is closed under truncation (see Theorem 3.3.1), the proof of (most of) Theorem 9.1.3 is much easier for $m = 1$ than in the case of higher derivatives. For this reason we give that proof separately here.

We assume that $f \in W^{1,p}(\mathbf{R}^N)$, and that $f|_K = 0$. Then $f_+ = \max\{f, 0\}$ satisfies the same conditions. This is an immediate consequence of Theorem 6.2.1, i.e., the fact that quasiall points are Lebesgue points. It follows that we can assume that f is nonnegative.

We can also assume that f is bounded. Otherwise, again using Theorem 3.3.1, we can approximate f by $f_n = \min\{f, n\}$ for some sufficiently large n. By multiplying f with a suitable cut-off function we can also assume that $\operatorname{supp} f$ is bounded. It is clear that after these modifications f still satisfies $f|_K = 0$.

First let $p > N$. Then f is continuous, and $f(x) = 0$ everywhere on K. Thus, the function f_ε, defined for $\varepsilon > 0$ by

$$f_\varepsilon(x) = \max\{f(x) - \varepsilon, 0\} ,$$

vanishes on a neighborhood of K. Moreover, by Theorem 3.3.1,

$$\begin{aligned} \nabla f_\varepsilon(x) &= \nabla f(x) \quad && \text{a.e. on } \{x : f(x) > \varepsilon\} , \\ \nabla f_\varepsilon(x) &= 0 && \text{a.e. on } \{x : f(x) \leq \varepsilon\} , \\ \nabla f(x) &= 0 && \text{a.e. on } \{x : f(x) = 0\} . \end{aligned}$$

It follows that

$$\begin{aligned} \|f - f_\varepsilon\|_{1,p}^p &= \int_{\mathbf{R}^N} |f - f_\varepsilon|^p \, dx + \int_{\mathbf{R}^N} |\nabla f - \nabla f_\varepsilon|^p \, dx \\ &\leq \varepsilon^p |\operatorname{supp} f| + \int_{\{x : 0 < f(x) \leq \varepsilon\}} |\nabla f|^p \, dx , \end{aligned}$$

which can be made arbitrarily small by choosing ε small. Regularization of f_ε by convolution with a smooth function with sufficiently small support shows that $f \in W_0^{1,p}(\Omega)$.

If $1 < p \leq N$, we only know that f is quasicontinuous. By Definition 6.1.1 there is, corresponding to any $\delta > 0$, an open G such that $C_{1,p}(G) < \delta$ and $f|_{G^c} \in C(G^c)$.

Let $\omega \in W^{1,p}$ be such that $\omega(x) = 1$ on G, $0 \le \omega \le 1$, and $\|\omega\|_{1,p}^p < \delta$. If f_ε is defined as before, it follows that $(1-\omega)f_\varepsilon$ vanishes on a neighborhood of K. Moreover

$$\|f - (1-\omega)f_\varepsilon\|_{1,p} \le \|f - f_\varepsilon\|_{1,p} + \|\omega f_\varepsilon\|_{1,p} .$$

Here $\|f - f_\varepsilon\|_{1,p}$ tends to zero with ε as before. For any $\varepsilon > 0$

$$\begin{aligned}\|\omega f_\varepsilon\|_{1,p}^p &\le \int_{\mathbf{R}^N} |\omega f|^p \, dx + A \int_{\mathbf{R}^N} \omega^p |\nabla f|^p \, dx + A \int_{\mathbf{R}^N} |\nabla \omega|^p |f|^p \, dx \\ &\le A\,\delta (\max f)^p + A \int_{\mathbf{R}^N} \omega^p |\nabla f|^p \, dx .\end{aligned}$$

As $\delta \to 0$, $\omega = \omega_\delta \to 0$ in L^p. We can thus choose a subsequence $\{\delta_i\}_1^\infty$ such that $\omega_{\delta_i}(x) \to 0$ a.e. Then $\int \omega_{\delta_i}^p |\nabla f|^p \, dx \to 0$ by dominated convergence, and it follows as before that $f \in W_0^{1,p}(\Omega)$.

If we define a function η by

$$\begin{aligned}\eta(x) &= \frac{\big(1-\omega(x)\big) f_\varepsilon(x)}{f(x)}, &&\text{if } f(x) > 0 , \\ \eta(x) &= 0, &&\text{if } f(x) = 0 ,\end{aligned}$$

we obtain Theorem 9.1.3(c), except for the statement that $\eta \in C^\infty$. This fact seems to require the more complicated construction given later.

It is tempting to try to prove Theorem 9.1.3 for higher derivatives by means of a "smooth truncation", but so far all such attempts have failed.

Remark. If $E \subset \mathbf{R}^N$ is an arbitrary set, the same proof shows that if $f \in W^{1,p}(\mathbf{R}^N)$, and $f|_E = 0$, then f can be approximated by functions in $W^{1,p}$ with compact support contained in E^c. Cf. Theorem 10.1.1 below.

9.3 The General Case. Outline

We first prove the easy Theorem 9.1.1. Let f satisfy the assumptions of the theorem, and let an open $V \supset K$ and $\varepsilon > 0$ be given. For $\delta > 0$ we set $G_\delta = \{\, x : \operatorname{dist}(x, K) < \delta \,\}$, and choose δ so small that $G_{3\delta} \subset V$. By convolving the characteristic function for $G_{2\delta}$ with a suitable function in $C_0^\infty(B(0,\delta))$ we can construct an $\omega_\delta \in C_0^\infty(G_{3\delta})$ such that $\omega_\delta = 1$ on G_δ, and

$$|D^\beta \omega_\delta| \le A\,\delta^{-|\beta|}, \quad |\beta| \le m .$$

By uniform continuity we can choose δ so small that

$$\sup_{x \in G_{3\delta}} |D^\beta f(x)| < \varepsilon, \quad |\beta| = m .$$

It follows from Taylor's theorem, applied to $D^\beta f$ for $|\beta| \le m-1$, that there is a constant A, independent of δ, such that

$$\sup_{x\in G_{3\delta}} |D^\beta f(x)| < A\varepsilon\delta^{m-|\beta|}, \quad 0 \le |\beta| \le m \ .$$

By the Leibniz formula there is A such that

$$\sum_{|\beta|\le m} \left\| D^\beta(\omega f) \right\|_\infty < A\varepsilon \ ,$$

which proves the theorem with $\eta = 1 - \omega$.

Theorem 9.1.3 is proved by constructing a multiplier ω with similar properties, but there are many difficulties. First of all, a replacement has to be found for Taylor's formula.

Recall the Poincaré type inequality, Corollary 8.2.7: If $f \in W^{m,p}(\mathbf{R}^N)$, and

$$D^\beta f|_K = 0 , \qquad 0 \le |\beta| \le m-1 \ , \tag{9.3.1}$$

then

$$\int_{B_n(x_0)} |f|^p \, dx \le A \frac{2^{-nmp}}{c_{1,p}(x_0, K, 2^{-n})} \int_{B_n(x_0)} |\nabla^m f|^p \, dx \ , \tag{9.3.2}$$

where $c_{1,p}$ is the relative capacity of Definition 8.2.3. Thus

$$c_{1,p}(x_0, K, 2^{-n}) = C_{1,p}\big(K \cap B_n(x_0)\big) 2^{n(N-p)}, \quad 1 < p < N \ ;$$
$$c_{1,N}(x_0, K, 2^{-n}) = C_{1,N}\big(K \cap B_n(x_0); B_{n-1}(x_0)\big) \ ;$$
$$c_{1,p}(x_0, K, 2^{-n}) \ge c > 0, \quad p > N, \quad \text{if } K \cap B_n(x_0) \ne \emptyset \ .$$

If $c_{1,p}(x_0, K, 2^{-n})$ is bounded below as $n \to \infty$ for all $x_0 \in K$, as it is e.g. if $p > N$, the proof is again easy. The main idea of the proof is that a suitable ω can be constructed, even if $c_{1,p}(x_0, K, 2^{-n}) \to 0$, as $n \to \infty$, provided the rate of decay is not too rapid. It turns out that the allowed rate is exactly that given by the requirement that K is $(1, p)$-thick (Definition 6.3.7) at all $x_0 \in K$.

Now recall the Kellogg property (Corollary 6.3.17): For any E and any $k = 1, \dots, m$, we have $C_{k,p}\big(e_{k,p}(E) \cap E\big) = 0$. Thus, the subset of K where K is $(1, p)$-thin has zero $(1, p)$-capacity, and it remains to deal with such sets.

But if it is assumed that $C_{1,p}(K) = 0$, then (9.3.2) gives no information about the behavior of f near K. It has to be replaced by the inequality in Corollary 8.2.8. This inequality enables us to split K into two sets, one where, roughly speaking, $f(x) = O(|x - x_0|^{m-1})$, as $x \to x_0 \in K$, and another one where an inequality similar to (9.3.2) holds. On the latter set the first construction can be applied, except for a subset of zero $(2, p)$-capacity, where K is $(2, p)$-thin, etc.

To be a little more precise, we can always assume that f has compact support. In fact, we can always approximate f by χf for a C^∞ cut-off function χ that vanishes outside a large ball. By the Leibniz rule (Theorem 6.1.5) the function χf also satisfies (9.3.1). Thus it is no restriction to assume that K is compact.

The following proposition is an immediate consequence of Corollary 6.3.17.

Proposition 9.3.1. *K can be split into $m+1$ disjoint Borel sets, $K = E_0 \cup E_1 \cup \ldots \cup E_m$, with the following properties:*

(a) *K is $(1, p)$-thick everywhere on E_0;*
(b) *$C_{k,p}(E_k) = 0$, and E_k (and thus K) is $(k+1, p)$-thick everywhere on E_k, $k = 1, \ldots, m-1$;*
(c) *$C_{m,p}(E_m) = 0$.*

By assumption f satisfies (9.3.1).

Let us for the sake of this outline assume that the sets E_k are compact. The idea of the proof is to construct the multiplier $\eta = 1 - \omega$ inductively as follows.

Let $\varepsilon > 0$ be given. First construct a function $\eta_0 \in C_0^\infty(E_0^c)$ such that $0 \leq \eta_0 \leq 1$ and $\|f - \eta_0 f\|_{m,p} < \varepsilon$. By the Leibniz rule $f_0 = \eta_0 f$ satisfies (9.3.1).

The next step is to construct $\eta_1 \in C_0^\infty(E_1^c)$ so that $0 \leq \eta_1 \leq 1$, and $\|f_0 - \eta_1 f_0\|_{m,p} < \varepsilon/2$. Then $f_1 = \eta_1 f_0$ satisfies (9.3.1).

By induction one constructs $\eta_k \in C_0^\infty(E_k^c)$ so that $0 \leq \eta_k \leq 1$,

$$\|f_{k-1} - \eta_k f_{k-1}\|_{m,p} < \varepsilon/2^k \ ,$$

and $f_k = \eta_k f_{k-1}$ for $k = 1, \ldots, m$. Then

$$f_m = \eta_m f_{m-1} = \eta_m \eta_{m-1} \cdots \eta_0 f = \eta f \ ,$$

$\|f - \eta f\|_{m,p} < 2\varepsilon$, and $\eta \in C_0^\infty(K^c)$.

In general the sets E_k are not compact, and the argument has to be modified.

In order to carry out the programme we shall prove the following theorem. To complete the proof of Theorem 9.1.3 is then relatively easy.

Theorem 9.3.2. *Let $f \in W^{m,p}(\mathbf{R}^N)$, $1 < p < \infty$. Let $K \subset \mathbf{R}^N$ be compact, and suppose that $D^\beta f|_K = 0$, for all β, $0 \leq |\beta| \leq m-1$. If K satisfies one of the following conditions:*

(a) *K is $(1, p)$-thick at all of its points;*
(b) *K is $(k+1, p)$-thick at all of its points and $C_{k,p}(K) = 0$ for some k, $k = 1, \ldots, m-1$;*
(c) *$C_{m,p}(K) = 0$,*

then, for any $\varepsilon > 0$ and any compact set F disjoint from K, there is a function $\eta \in C_0^\infty(K^c)$ such that $0 \leq \eta \leq 1$, $\eta = 1$ on F, and $\|f - \eta f\|_{m,p} < \varepsilon$.

The next four sections are devoted to the proof of the first part of this theorem. The construction is complicated, so in order to bring out the idea clearly, we first treat a special case in Section 9.4, the case when K is "uniformly $(1, p)$-thick". In Section 9.5 we then treat the general case, with the exception of the proof of a crucial lemma, which is given in Sections 9.6 and 9.7.

In the following Section 9.8 we prove a number of properties of nonlinear potentials that are needed for the proofs of the second and third parts of the theorem.

The third part is then proved in Section 9.9. The proof of the second part of the theorem consists in combining the method of Section 9.9 with the construction used in proving the first part. This is carried out in Section 9.10, and then the proof of Theorem 9.1.3 is concluded in Section 9.11.

9.4 The Uniformly $(1, p)$-Thick Case

We now assume that K is uniformly $(1, p)$-thick. By this we mean that there exists a sequence $\{a_n\}_1^\infty$ of positive numbers such that

$$\sum_{n=1}^{\infty} a_n^{p'} = \infty , \tag{9.4.1}$$

and such that for all $x_0 \in K$

$$c_{1,p}(x_0, K, 2^{-n}) \geq a_n^p, \quad n = 1, 2, \ldots . \tag{9.4.2}$$

This is a weak condition. For example, if K satisfies an interior cone condition, then a_n can be chosen equal to a constant. Here the cone does not have to be a solid, circular cone — any conical set with positive $(1, p)$-capacity will do. Note also that any K is uniformly $(1, p)$-thick at all of its points if $p > N$.

We shall now construct a function $\omega = 1 - \eta$ such that

(a) $\omega \in C_0^\infty(V)$ for some arbitrarily chosen neighborhood V of K,
(b) $\omega(x) = 1$ on a neighborhood of K included in V, and
(c) $\|\omega f\|_{m,p}^p \leq A \int_V |\nabla^m f|^p \, dx$, where A is a constant independent of f and of K.

The last integral is arbitrarily small. In fact, the assumption (9.3.1) implies that $\nabla^{m-1} f(x) = 0$ a.e. on K. But $\nabla^{m-1} f \in W^{1,p}(\mathbf{R}^N)$, so by the truncation theorem, Theorem 3.3.1, $\nabla^m f(x) = 0$ a.e. on K. Thus $\int_V |\nabla^m f|^p \, dx = \int_{V\setminus K} |\nabla^m f|^p \, dx$, which can be made arbitrarily small.

For any integer n we set

$$G_n = \left\{ x : \operatorname{dist}(x, K) < 2^{-n-1} \right\} .$$

Then we can easily construct a function $\omega_n \in C^\infty$ such that

(a) $\omega_n(x) = 1$ on G_{n+1},
(b) $0 \leq \omega_n \leq 1$, and
(c) $\max_x |\nabla^k \omega_n(x)| \leq A\, 2^{nk}$, for $1 \leq k \leq m$.

Since $\{a_n\}_1^\infty$ is bounded, and $\sum_1^\infty a_n^{p'} = \infty$, we can for a suitably chosen A and arbitrarily large M find $P > M$ so that

$$1 \leq \sum_M^P a_n^{p'} \leq A .$$

We set

$$\omega(x) = \frac{\sum_M^P a_n^{p'} \omega_n(x)}{\sum_M^P a_n^{p'}} .$$

Then $\omega \in C_0^\infty(G_M)$, $0 \le \omega(x) \le 1$, $\omega(x) = 1$ on G_{P+1}, and

$$|\nabla^k \omega(x)| = a_n^{p'} |\nabla^k \omega_n(x)| \le A\, 2^{nk} a_n^{p'}, \quad 1 \le k \le m , \tag{9.4.3}$$

on $G_n \setminus G_{n+1}$ for $M \le n \le P$.

We have to estimate

$$\|\omega f\|_{m,p}^p = \sum_{|\beta| \le m} \int_{\mathbf{R}^N} |D^\beta(\omega f)|^p \, dx .$$

It is enough to estimate the terms of order m, so by the Leibniz formula it is enough to estimate

$$\int_{\mathbf{R}^N} |\nabla^k \omega|^p |\nabla^{m-k} f|^p \, dx$$

for $k = 0, 1, 2, \ldots, m$.

The case $k = 0$ is clear, since

$$\int_{\mathbf{R}^N} \omega^p |\nabla^m f|^p \, dx \le \int_{G_M} |\nabla^m f|^p \, dx .$$

For $k = 1, \ldots, m$ we apply (9.3.2) to $\nabla^{m-k} f$. By (9.4.2) this gives that for all $x \in K$, and $n = 1, 2, \ldots,$

$$\int_{B_n(x)} |\nabla^{m-k} f|^p \, dx \le A\, 2^{-nkp} a_n^{-p} \int_{B_n(x)} |\nabla^m f|^p \, dx .$$

For any $x \in G_n \setminus G_{n+1}$ there is a point $\overline{x} \in K$, such that

$$B_{n+2}(x) \subset B_n(\overline{x}) \subset B_{n-1}(x) ,$$

and thus, for any $x \in G_n \setminus G_{n+1}$,

$$\int_{B_{n+2}(x)} |\nabla^{m-k} f|^p \, dx \le A\, 2^{-nkp} a_n^{-p} \int_{B_n(\overline{x})} |\nabla^m f|^p \, dx .$$

Now $\mathbf{R}^N$, and thus $G_n \setminus G_{n+1}$, can be covered by balls $B_{n+2}(x_i)$ in such a way that no point in $\mathbf{R}^N$ belongs to more than a fixed number (depending on N) of the balls $B_{n-1}(x_i)$. It follows that

$$\int_{G_n \setminus G_{n+1}} |\nabla^{m-k} f|^p \, dx \le A\, 2^{-nkp} a_n^{-p} \int_{G_{n-3}} |\nabla^m f|^p \, dx ,$$

and thus, by (9.4.3),

$$\int_{G_n \setminus G_{n+1}} |\nabla^k \omega|^p |\nabla^{m-k} f|^p \, dx \le A\, a_n^{pp'-p} \int_{G_{n-3}} |\nabla^m f|^p \, dx$$
$$\le A\, a_n^{p'} \int_{G_{M-3}} |\nabla^m f|^p \, dx .$$

Summing over n we find

$$\int_{\mathbf{R}^N} |\nabla^k \omega|^p |\nabla^{m-k} f|^p \, dx = \sum_{n=M}^{P} \int_{G_n \setminus G_{n+1}} |\nabla^k \omega|^p |\nabla^{m-k} f|^p \, dx$$
$$\leq A \sum_{n=M}^{P} a_n^{p'} \int_{G_{M-3}} |\nabla^m f|^p \, dx \leq A \int_V |\nabla^m f|^p \, dx \ ,$$

if M is large enough. This finishes the proof.

Remark. Notice that, as a subset of the above, we have now proved Theorem 9.1.3 for all domains Ω in the case $p > N$, and for all Ω with "nice" boundary, e.g. domains satisfying an exterior cone condition, in the case $1 < p \leq N$. Thus, in what follows we can always assume that $p \leq N$.

9.5 The General Thick Case

In the general case the construction of the function ω is much more complicated than in the previous section, because of the lack of uniformity, but the guiding idea is the same. The construction is contained in the proof of Lemma 9.5.2 below, the proof of which is postponed to the following sections.

We shall use the following notation for the L^p-average of a function over a ball $B_n(x)$:

$$[f]_n(x) = \left(\frac{1}{|B_n(x)|} \int_{B_n(x)} |f|^p \, dy \right)^{1/p} . \tag{9.5.1}$$

Suppose now that f satisfies the assumption (9.3.1), and that K is $(1, p)$-thick at a point x. By the Poincaré type inequality (9.3.2),

$$[f]_n(x) \leq A \, 2^{-nm} c_{1,p}(x, K, 2^{-n})^{-1/p} [\nabla^m f]_n(x) \ .$$

Thus, if $[f]_n(x) \neq 0$,

$$\left(\frac{2^{-nm} [\nabla^m f]_n(x)}{[f]_n(x)} \right)^{p'} \geq A^{-1} c_{1,p}(x, K, 2^{-n})^{p'-1} ,$$

so by the definition of $(1, p)$-thickness, Definition 6.3.7, either

$$\sum_{n=1}^{\infty} \left(\frac{2^{-nm} [\nabla^m f]_n(x)}{[f]_n(x)} \right)^{p'} = \infty \ , \tag{9.5.2}$$

or $[f]_n(x) = 0$ for some n. But in the latter case $f \equiv 0$ in a neighborhood of x, i.e. $x \notin \operatorname{supp} f$.

We shall prove the following theorem.

Theorem 9.5.1. *Let K be compact, and let f be a function in $W^{m,p}(\mathbf{R}^N)$, $1 < p \le N$, that satisfies (9.5.2) for all $x \in K \cap \operatorname{supp} f$. Then, given any $\varepsilon > 0$, and any sufficiently small neighborhood V of K, there is a function ω in $C_0^\infty(V)$ such that $0 \le \omega \le 1$, $\omega = 1$ on a neighborhood of $K \cap \operatorname{supp} f$, and*

$$\|\omega f\|_{m,p}^p \le A \int_V |\nabla^m f|^p \, dx < \varepsilon \, .$$

Remark 1. The conclusion is the same if instead of assuming that $f \in W^{m,p}(\mathbf{R}^N)$ we assume that $f \in W^{m,p}(K^c)$ and extend f to $\mathbf{R}^N$ by setting $f(x) = 0$ on K^c. It follows that the extended function belongs to $W_0^{m,p}(K^c)$, and thus to $W^{m,p}(\mathbf{R}^N)$. (The integrals in (9.5.2) should be interpreted as integrals over $B_n(x) \setminus K$.)

Remark 2. It is easily seen that if $f \in W^{m,p}(\mathbf{R}^N)$ and satisfies (9.5.2) (m,p)-q.e. on K, then $D^\beta f|_K = 0$ for $0 \le |\beta| \le m-1$. In fact, by Proposition 6.3.12

$$W_{m,p}^{\mu}(x) < \infty \qquad (m,p)\text{-q.e.}$$

for any positive finite measure μ, thus in particular if $d\mu = |\nabla^m f|^p \, dx$. It follows that

$$\sum_{n=1}^{\infty} \left(2^{-nm} [\nabla^m f]_n(x)\right)^{p'} < \infty \qquad (m,p)\text{-q.e.} \, .$$

Thus by (9.5.2)

$$\liminf_{n\to\infty} [f]_n(x)^{p'} = 0 \quad (m,p)\text{-q.e. on } K \, .$$

But for (m,p)-q.e. x we know that

$$f(x) = \lim_{n\to\infty} \frac{1}{|B_n(x)|} \int_{B_n(x)} f(y) \, dy \, ,$$

so by Hölder's inequality $f(x) = 0$ (m,p)-q.e. on K, i.e. $f|_K = 0$.

By using the inequality (Corollary 8.1.5)

$$2^{-nkp} \int_{B_n} |\nabla^k f|^p \, dx \le A \int_{B_n} |f|^p \, dx + A \, 2^{-nmp} \int_{B_n} |\nabla^m f|^p \, dx,$$

the conclusion for $D^\beta f$ follows easily in the same way.

This remark will be amplified in the proof of Theorem 11.5.10 below.

In proving the theorem we can clearly assume that $K \subset \operatorname{supp} f$.

In order to construct the function ω we decompose $\mathbf{R}^N$ into a mesh of closed unit cubes with disjoint interiors, and we denote this mesh by $\mathcal{Q}_0$. By successively decomposing each cube into 2^N equal cubes, we obtain meshes $\mathcal{Q}_1, \mathcal{Q}_2, \ldots$, so that $\mathcal{Q}_n$ is a mesh of cubes with side 2^{-n}.

The cubes in $\mathcal{Q}_n$ will now usually be denoted $\{Q_{ni}\}_{i=1}^\infty$ with some arbitrary enumeration. If Q is a cube with side $l(Q)$, we denote by rQ, $r > 0$, the concentric cube with side $r\,l(Q)$. We set $\widetilde{Q}_{ni} = 7Q_{ni}$, and we denote

$$Q(f) = \left(\frac{1}{|Q|} \int_Q |f|^p \, dx \right)^{1/p} .$$

For any Q_{ni} such that $3Q_{ni}$ intersects K we set

$$\lambda_{ni} = \min\left\{ 1, \left(\frac{2^{-nm} \widetilde{Q}_{ni}(\nabla^m f)}{\widetilde{Q}_{ni}(f)} \right)^{p'} \right\} . \tag{9.5.3}$$

If $3Q_{ni}$ does not intersect K, we set $\lambda_{ni} = 0$.

We can now formulate a lemma which gives a function ω with the desired properties.

Lemma 9.5.2. *Under the assumptions of Theorem* 9.5.1 *there exists a function* $\omega \in C_0^\infty(V)$, *such that*

(a) $0 \le \omega \le 1$;
(b) $\omega(x) = 1$ *on a neighborhood of* K;
(c) *for every* $x \in \operatorname{supp} \omega$ *there are* n *and* i *such that* $x \in Q_{ni} \subset \widetilde{Q}_{ni} \subset V$, *and there are constants* A_k *such that*

$$|\nabla^k \omega(x)| \le A_k \lambda_{ni} 2^{nk}, \quad k = 1, 2, \ldots \ ; \tag{9.5.4}$$

(d) *there is a constant* A, *only depending on* N, *such that for all* x

$$\sum_{n=0}^{\infty} \sum_i \lambda_{ni} \chi(x; \widetilde{Q}_{ni}) \le A \ , \tag{9.5.5}$$

where $\chi(\cdot\,; E)$ *denotes the characteristic function for* E, *and the sum is extended only over those indices* i *for which* $\nabla\omega(x) \not\equiv 0$ *on* Q_{ni}.

We postpone the proof of the lemma, and proceed with the proof of the theorem.

We have to estimate $\|\omega f\|_{m,p}$. It is enough to estimate

$$\int_{\mathbf{R}^N} |\nabla^k \omega|^p |\nabla^{m-k} f|^p \, dx \quad \text{for } k = 0, 1, 2, \ldots, m \ .$$

By (9.5.3) we have

$$\int_{Q_{ni}} |f|^p \, dx \le \int_{\widetilde{Q}_{ni}} |f|^p \, dx \le A \frac{2^{-nmp}}{\lambda_{ni}^{p-1}} \int_{\widetilde{Q}_{ni}} |\nabla^m f|^p \, dx \ , \tag{9.5.6}$$

and by combining this with the inequality (Corollary 8.1.5 with balls replaced by cubes)

$$\int_{Q_{ni}} |\nabla^{m-k} f|^p \, dx \le A \, 2^{n(m-k)p} \int_{Q_{ni}} |f|^p \, dx + A \, 2^{-nkp} \int_{Q_{ni}} |\nabla^m f|^p \, dx \ ,$$

we obtain

$$\int_{Q_{ni}} |\nabla^{m-k} f|^p \, dx \le A \frac{2^{-nkp}}{\lambda_{ni}^{p-1}} \int_{\widetilde{Q}_{ni}} |\nabla^m f|^p \, dx \ , \tag{9.5.7}$$

for $k = 1, 2, \ldots, m-1$.

By (c) in Lemma 9.5.2 we can decompose $\operatorname{supp}\nabla\omega$ into a union of disjoint sets Q'_{ni}, the indices (n,i) belonging to a set I, such that $Q'_{ni} \subset Q_{ni}$ and

$$|\nabla^k \omega(x)| \leq A_k \lambda_{ni} 2^{nk} \quad \text{on } Q'_{ni} .$$

Note that I is finite, since $\operatorname{supp}\nabla\omega \cap K = \emptyset$, and $\lambda_{ni} = 0$ if $3Q_{ni} \cap K = \emptyset$.

Thus, for $k = 1, 2, \ldots, m$, we obtain, using (9.5.7), (9.5.6), and (9.5.5)

$$\begin{aligned}\int_{\mathbf{R}^N} |\nabla^k\omega|^p |\nabla^{m-k} f|^p \, dx &\leq A \sum_{(n,i)\in I} \lambda_{ni}^p 2^{nkp} \int_{Q'_{ni}} |\nabla^{m-k} f|^p \, dx \\ &\leq A \sum_{(n,i)\in I} \lambda_{ni} \int_{\widetilde{Q}_{ni}} |\nabla^m f|^p \, dx \\ &= A \int_{\mathbf{R}^N} \Bigg(\sum_{(n,i)\in I} \lambda_{ni}\, \chi(x; \widetilde{Q}_{ni}) \Bigg) |\nabla^m f|^p \, dx \\ &\leq A \int_V |\nabla^m f|^p \, dx .\end{aligned}$$

For $k = 0$ we have

$$\int_{\mathbf{R}^N} |\omega|^p |\nabla^m f|^p \, dx \leq \int_V |\nabla^m f|^p \, dx .$$

But $\int_V |\nabla^m f|^p \, dx$ is arbitrarily small as in Section 9.4, and Theorem 9.5.1 follows.

9.6 Proof of Lemma 9.5.2 for $m = 1$

In order to bring out the idea clearly we first prove a version of Lemma 9.5.2 where the estimate (9.5.4) is proved only for $k = 1$. This is enough to prove Theorem 9.5.1 for $m = 1$.

For each n we define a function ρ_n by setting (see (9.5.3))

$$\rho_n(x) = \lambda_{ni} 2^n \quad \text{if } x \in Q_{ni} . \tag{9.6.1}$$

The value of $\rho_n(x)$ on the sides of Q_{ni} is unimportant. We choose n_0 so large that

$$G_n = \operatorname{supp} \rho_n \subset V \quad \text{for } n \geq n_0 .$$

It is no loss of generality to assume that $n_0 = 0$. Then, we set

$$\rho(x) = \max_{n\geq 0} \rho_n(x) .$$

Note that for any $x \notin K$ there is only a finite number of $n \geq 0$ with $\rho_n(x) \neq 0$. Clearly,

$$G_{n+1} \subset G_n , \qquad \text{and} \qquad K \in \bigcap_{n=0}^{\infty} G_n .$$

Now, for any $x \in G_0$ we let $\Gamma(x)$ denote the family of all paths that join ∂G_0 to x, and set

$$\omega(x) = \min\left\{1, \inf_{\gamma\in\Gamma(x)} \int_\gamma \rho(t)\,|dt|\right\} .$$

For $x \notin G_0$ we set $\omega(x) = 0$. Clearly, for any h,

$$\omega(x+h) - \omega(x) \le \int_0^1 \rho(x+sh)\,|h|\,ds \le |h| \max_{0\le s\le 1} \rho(x+sh) .$$

It follows that ω is locally Lipschitz on K^c, and

$$|\nabla\omega(x)| \le \rho(x) \quad \text{a.e.} ,$$

which proves (9.5.4) for $k = 1$.

We now prove that $\omega(x) = 1$ on a neighborhood of K. Fix a point $x_0 \in K$ and let

$$\{Q_{n0}\}_{n=0}^{\infty}, \qquad Q_{n0} \in \mathcal{Q}_n ,$$

be a sequence of nested cubes that contain x_0. Consider the sequence of expanded cubes $\{3Q_{n0}\}_{n=0}^{\infty}$. Let

$$\overline{\lambda}_n(x_0) = \max_i\{\lambda_{ni} : Q_{ni} \subset 3Q_{n0}\} , \tag{9.6.2}$$

and

$$\underline{\lambda}_n(x_0) = \min_i\{\lambda_{ni} : Q_{ni} \subset 3Q_{n0}\} . \tag{9.6.3}$$

We claim that if (9.5.2) is satisfied at x_0, then

$$\sum_{n=0}^{\infty} \underline{\lambda}_n(x_0) = \infty . \tag{9.6.4}$$

Recall that $7Q_{ni}$ is denoted $\widetilde{Q}_{ni}$. Then, for any $Q_{ni} \subset 3Q_{n0}$, we have $\widetilde{Q}_{ni} \supset 3Q_{n0} \supset B_n(x_0)$ (but not $B_n(x_0) \supset Q_{ni}$). By Corollary 8.1.5 we have

$$\int_{\widetilde{Q}_{ni}} |f|^p\,dx \le A \int_{B_n(x_0)} |f|^p\,dx + A\,2^{-nmp} \int_{\widetilde{Q}_{ni}} |\nabla^m f|^p\,dx .$$

Set

$$\frac{2^{-nm}[\nabla^m f]_n(x_0)}{[f]_n(x_0)} = c_n(x_0) ,$$

so that

$$\sum_{n=0}^{\infty} c_n(x_0)^{p'} = \infty ,$$

by (9.5.2). It follows that

$$\int_{\widetilde{Q}_{ni}} |f|^p\,dx \le A\big(c_n(x_0)^{-p} + 1\big)2^{-nmp} \int_{\widetilde{Q}_{ni}} |\nabla^m f|^p\,dx ,$$

so that

$$\lambda_{ni} \geq A^{-1}\left(\frac{c_n(x_0)^p}{1+c_n(x_0)^p}\right)^{p'/p} ,$$

whence

$$\underline{\lambda}_n(x_0) \geq A^{-1}2^{-p'/p}\min\{c_n(x_0)^{p'},1\} ,$$

which proves the claim.

Now, on $3Q_{n0}$

$$\rho(x) \geq \rho_n(x) \geq \underline{\lambda}_n(x_0)2^n .$$

The distance from the boundary of $3Q_{n0}$ to $3Q_{n+1,0}$ is 2^{-n-1}, so for any x in $3Q_{n+1,0}$, and any $\gamma \in \Gamma(x)$, the contribution to $\int_\gamma \rho(t)\,|dt|$ from the part of γ that lies in $3Q_{n0} \setminus 3Q_{n+1,0}$ is at least $\frac{1}{2}\underline{\lambda}_n(x_0)$. Thus

$$\omega(x) \geq \tfrac{1}{2}\sum_{k=0}^{n}\underline{\lambda}_k(x_0) ,$$

so it follows from (9.6.4) that $\omega(x) = 1$ on $3Q_{n+1,0}$ if n is large enough.

It remains to prove (9.5.5). Consider a point $x \in V$. Then

$$x \in \bigcup_{n=0}^{n_1}\bigcup_i \widetilde{Q}_{ni} , \qquad n_1 = n_1(x) ,$$

where the union is taken over all indices such that $\lambda_{ni} \neq 0$ and such that $\nabla\omega(x) \not\equiv 0$ on Q_{ni}. Fix one of the cubes $\widetilde{Q}_{n_1 i}$. There are only a certain number A_N of other $\widetilde{Q}_{n_1 j}$ in the union, and they are all contained in $3\widetilde{Q}_{n_1 i}$. Moreover, $3Q_{n_1 i}$ intersects K. Let $x_0 \in 3Q_{n_1 i} \cap K$, and let again $\{Q_{n0}\}$ be a sequence of nested cubes containing x_0. We can assume that $Q_{n_1 0} \subset 3Q_{n_1 i}$. A moment of thought shows that if $Q_{n_1-3,k} \subset 3Q_{n_1-3,0}$ then $3\widetilde{Q}_{n_1 i} \subset 3\widetilde{Q}_{n_1-3,k}$, so that $\widetilde{Q}_{nj} \subset \widetilde{Q}_{n_1-3,k}$ for all the $\widetilde{Q}_{nj}$ above.

Clearly, for all $n \leq n_1$, all $\widetilde{Q}_{nj}$ in the union above are contained in some $\widetilde{Q}_{n-3,k}$, where $Q_{n-3,k} \subset 3Q_{n-3,0}$.

It follows from Corollary 8.1.5 that

$$\int_{\widetilde{Q}_{n-3,k}} |f|^p\,dx \leq A\int_{\widetilde{Q}_{nj}} |f|^p\,dx + A\,2^{-nmp}\int_{\widetilde{Q}_{n-3,k}} |\nabla^m f|^p\,dx .$$

So by (9.5.3)

$$\begin{aligned}\widetilde{Q}_{n-3,k}(f) &\leq A\big(\lambda_{nj}^{-1/p'}+1\big)2^{-nm}\widetilde{Q}_{n-3,k}(\nabla^m f)\\ &\leq A\lambda_{nj}^{-1/p'}2^{-nm}\widetilde{Q}_{n-3,k}(\nabla^m f) ,\end{aligned}$$

and thus

$$\lambda_{nj} \leq A\left(\frac{2^{-nm}\widetilde{Q}_{n-3,k}(\nabla^m f)}{\widetilde{Q}_{n-3,k}(f)}\right)^{p'} = A\,\lambda_{n-3,k} ,$$

unless $\lambda_{n-3,k} = 1$.

It follows that $\lambda_{nj} \le A\,\underline{\lambda}_{n-3}(x_0)$ for all λ_{nj} that appear in the sum (9.5.5). On the other hand we assumed that $\nabla\omega \not\equiv 0$ on $Q_{n_1 i}$, so that $\omega(y) < 1$ for some $y \in Q_{n_1 i}$. Now $Q_{n_1 i} \subset 3Q_{n_1 0}$, so

$$\omega(y) \ge \tfrac{1}{2} \sum_{k=0}^{n_1-1} \underline{\lambda}_k(x_0) \quad \text{on } Q_{n_1 i}$$

by the argument above, and thus

$$\sum_{k=0}^{n_1-1} \underline{\lambda}_k(x_0) \le 2 \ .$$

Thus, in (9.5.5)

$$\sum_{n=0}^{\infty} \sum_{i} \lambda_{ni}\,\chi(x;\widetilde{Q}_{ni}) \le A A_N \sum_{n=0}^{n_1} \underline{\lambda}_{n-3}(x_0) \le A \ ,$$

since we can replace λ_{ni} by 1 for $n = 0, 1, 2$.

The function ω can easily be regularized by convolution with a smooth approximate identity with sufficiently small support, and Theorem 9.5.1 follows for $m = 1$.

However, it is clear that we cannot by regularizing in this way control the higher derivatives of ω. In the next section we shall modify the construction accordingly.

9.7 Proof of Lemma 9.5.2

The simple regularization of ω by convolution at the end of the previous section is much too crude to give the desired control (9.5.4) of the higher derivatives of ω.

We can use the same basic idea—to build the function using integrals over paths—but now we have to proceed step by step, carefully regularizing the function obtained before taking the next step.

We start by modifying the construction of the function ρ. We define

$$\rho_n(x) = \lambda_{ni} 2^n \ ,$$

as in (9.6.1), and we assume that

$$\operatorname{supp} \rho_n \subset V \quad \text{for } n \ge 0 \ .$$

We now define λ^*_{ni} and a function ρ^*_n by setting

$$\rho^*_n(x) = \lambda^*_{ni} 2^n = \max_{0 \le k \le n} \rho_k(x) \ , \tag{9.7.1}$$

and we cut this function down slightly by setting

$$\rho'_n(x) = \min\{\, \rho^*_n(y) : |y - x| \le 2^{-n-3} \,\} \ . \tag{9.7.2}$$

This is in order to have

$$\rho'_n * \varphi(x) \le \rho^*_n(x) \tag{9.7.3}$$

for all nonnegative $\varphi \in C_0^\infty(B(0, 2^{-n-3}))$ such that $\int \varphi\, dx = 1$.

Next, we observe that if $x_0 \in K$ and if $\overline{\lambda}_n(x_0)$ and $\underline{\lambda}_n(x_0)$ are defined by (9.6.2) and (9.6.3), then there is a constant M, independent of x_0, f, and K, such that

$$\overline{\lambda}_{n+1}(x_0) \le M\, \underline{\lambda}_n(x_0) \ . \tag{9.7.4}$$

In fact, if $Q_{n+1,j} \subset 3Q_{n+1,0}$ and $Q_{ni} \subset 3Q_{n0}$ then $\widetilde{Q}_{n+1,j} \subset \widetilde{Q}_{ni}$. Thus, by Corollary 8.1.5

$$\int_{\widetilde{Q}_{ni}} |f|^p\, dx \le A \int_{\widetilde{Q}_{n+1,i}} |f|^p\, dx + A\, 2^{-nmp} \int_{\widetilde{Q}_{ni}} |\nabla^m f|^p\, dx \ .$$

So by (9.5.3)

$$\widetilde{Q}_{ni}(f) \le A\big(\lambda_{n+1,j}^{-1/p'} + 1\big) 2^{-nm}\, \widetilde{Q}_{ni}(\nabla^m f) \le A\, \lambda_{n+1,j}^{-1/p'} 2^{-nm}\, \widetilde{Q}_{ni}(\nabla^m f) \ ,$$

and thus, unless $\lambda_{ni} = 1$,

$$\lambda_{n+1,j} \le A \left(\frac{2^{-nm}\, \widetilde{Q}_{ni}(\nabla^m f)}{\widetilde{Q}_{ni}(f)} \right)^{p'} = A\, \lambda_{ni} \ ,$$

which proves (9.7.4).

We have to be careful in order not to lose property (b) in Lemma 9.5.2. With this in mind we set

$$G_0 = \{\, x : \rho_0(x) > 0 \,\} \ , \tag{9.7.5}$$

and

$$G_n = \left\{ x : \rho_n(x) \ge \frac{\rho^*_n(x)}{2M} \right\}, \quad n = 1, 2, \ldots \ . \tag{9.7.6}$$

Thus, G_n is the union of certain Q_{ni}. As we shall see later, $K \subset G_n$ for each n, but $\{G_n\}_0^\infty$ is no longer a decreasing sequence.

Then we denote

$$G'_n = \left\{ x \in G_n : \operatorname{dist}(x, \partial G_n) \ge 2^{-n-3} \right\}, \quad n = 0, 1, 2, \ldots \ . \tag{9.7.7}$$

After these preparations we can proceed to construct our function ω. We shall follow an inductive procedure.

If $x \in G'_0$ we denote by $\Gamma_0(x)$ the family of all paths that join x to $\partial G'_0$, and we set

$$\omega_0(x) = 0\,, \quad x \notin G_0'\,,$$
$$\omega_0(x) = \inf_{\gamma \in \Gamma_0(x)} \int_\gamma \rho_0'(t)\,|dt|\,, \quad x \in G_0'\,.$$

Clearly,

$$|\nabla \omega_0(x)| \le \rho_0'(x)\,.$$

Let $\varphi \ge 0$ be a fixed function in $C_0^\infty(B_0(0))$ such that $\int \varphi\,dx = 1$. Set

$$\varphi_n(x) = 2^{nN}\varphi(2^n x), \quad n = 0, 1, 2, \dots\,.$$

We observe that

$$\operatorname{supp}(\varphi_n * \varphi_{n+1} * \cdots * \varphi_{n+m}) \subset B_{n-1}(0) \quad \text{for all } m\,. \tag{9.7.8}$$

We adjust and regularize ω_0 by setting

$$\widetilde{\omega}_0 = \big(\min\{1, \omega_0\}\big) * \varphi_4\,.$$

It follows that

$$|\nabla \widetilde{\omega}_0| \le |\nabla \omega_0| * \varphi_4 \le \rho_0' * \varphi_4\,.$$

Thus, by the definition (9.7.2) of ρ_0',

$$|\nabla \widetilde{\omega}_0(y)| \le \rho_0(x) \quad \text{for } |y - x| \le 2^{-3} - 2^{-4} = 2^{-4}.$$

Consequently, for any k and l,

$$\big|\nabla^k(\widetilde{\omega}_0 * \varphi_5 * \cdots * \varphi_l)(x)\big| \le |\nabla \omega_0| * |\nabla^{k-1}\varphi_4| * \varphi_5 * \cdots * \varphi_l(x) \le A_k \rho_0(x)\,.$$

We now assume that ω_m has been constructed for all $m = 0, 1, \dots, n-1$, as well as

$$\widetilde{\omega}_m = \big(\min\{1, \omega_m\}\big) * \varphi_{m+4}\,.$$

We assume that $\widetilde{\omega}_{n-1}$ satisfies

$$|\nabla \widetilde{\omega}_{n-1}(y)| \le \rho_{n-1}^*(x) \quad \text{for } |y - x| \le 2^{-n-3}\,, \tag{9.7.9}$$

and that for arbitrary k and l

$$\big|\nabla^k(\widetilde{\omega}_{n-1} * \varphi_{n+4} * \cdots * \varphi_{n+l})(x)\big| \le A_k 2^{mk} \rho_m^*(x) \tag{9.7.10}$$

for all $x \in G_m \setminus \big(\bigcup_{m+1}^{n-1} G_j\big)$, if $m = 0, 1, \dots, n-2$, and for all $x \in G_{n-1}$, if $m = n-1$.

We denote by $\Gamma(y, x)$ the set of all paths that join two points x and y. We define ω_n by setting

$$\omega_n(x) = \widetilde{\omega}_{n-1}(x)\,, \quad x \notin G_n'\,;$$
$$\omega_n(x) = \inf_{\substack{y \in \partial G_n' \\ \gamma \in \Gamma(y,x)}} \left\{\widetilde{\omega}_{n-1}(y) + \int_\gamma \max\big\{|\nabla \widetilde{\omega}_{n-1}(t)|, \rho_n'(t)\big\}\,|dt|\right\}, \quad x \in G_n'\,.$$

Then we again adjust and regularize by setting

$$\widetilde{\omega}_n = \big(\min\{1, \omega_n\}\big) * \varphi_{n+4} \ .$$

We claim that

$$|\nabla \widetilde{\omega}_n(y)| \le \rho_n^*(x) \quad \text{for } |y - x| \le 2^{-n-4} \ , \tag{9.7.11}$$

and that for arbitrary k and l

$$\big|\nabla^k(\widetilde{\omega}_n * \varphi_{n+5} * \cdots * \varphi_{n+l})(x)\big| \le A_k 2^{mk} \rho_m^*(x) \tag{9.7.12}$$

for all $x \in G_m \setminus \big(\bigcup_{m+1}^n G_j\big)$, if $m = 0, 1, \ldots, n-1$, and for all $x \in G_n$, if $m = n$.

First we observe hat the appearance of the term $|\nabla \widetilde{\omega}_{n-1}(t)|$ in the integral guarantees that $\omega_n(x)$ is uniquely defined at points on $\partial G_n'$, or in other words, that ω_n is continuous.

Then, clearly,

$$\nabla \omega_n(x) = \nabla \widetilde{\omega}_{n-1}(x) \ , \quad x \notin G_n' \ ,$$

and

$$|\nabla \omega_n(x)| \le \max\big\{|\nabla \widetilde{\omega}_{n-1}(x)|, \rho_n'(x)\big\} \ , \quad x \in G_n' \ .$$

By the induction hypothesis (9.7.9)

$$|\nabla \widetilde{\omega}_{n-1}(y)| \le \rho_{n-1}^*(x) \le \rho_n^*(x) \quad \text{for } |y - x| \le 2^{-n-3} \ ,$$

and by (9.7.2)

$$\rho_n'(y) \le \rho_n^*(x) \quad \text{for } |y - x| \le 2^{-n-3} \ .$$

It follows that for $x \in G_n$

$$|\nabla \widetilde{\omega}_n(y)| \le \big(|\nabla \omega_n| * \varphi_{n+3}\big)(y) \le \rho_n^*(x) \quad \text{for } |y - x| \le 2^{-n-4} \ .$$

Thus

$$\big(|\nabla \widetilde{\omega}_n| * \varphi_{n+5} * \cdots * \varphi_{n+l}\big)(x) \le \rho_n^*(x) \ ,$$

and

$$\big(|\nabla^k \widetilde{\omega}_n| * \varphi_{n+5} * \cdots * \varphi_{n+l}\big)(x) \le A_k 2^{mk} \rho_m^*(x)$$

for all $x \in G_n$ and all k and l. For $x \notin G_n$, the distance from x to $\partial G_n'$ is at least 2^{-n-3}, so

$$\widetilde{\omega}_n(x) = \big(\widetilde{\omega}_n * \varphi_{n+4}\big)(x) \ ,$$

and (9.7.12) is clearly satisfied.

We claim that for n large enough, $\widetilde{\omega}_n(x) = 1$ on a neighborhood of K. Let $x_0 \in K$ and let $\{Q_{n0}\}_0^\infty$ be a sequence of nested cubes containing x_0. Then $3Q_{00} \in G_0$. Consider the sequence $\big\{\overline{\lambda}_n(x_0)2^n\big\}_0^\infty$. Clearly $\limsup_{n\to\infty} \overline{\lambda}_n(x_0)2^n = \infty$, since otherwise we would have $\sum_0^\infty \overline{\lambda}_n(x_0) < \infty$, which contradicts (9.6.4). We denote by

$$\left\{\overline{\lambda}_{n_\nu+1}(x_0)2^{n_\nu+1}\right\}_{\nu=0}^{\infty}$$

the subsequence of successive maxima, i.e.

$$\begin{aligned}\overline{\lambda}_n(x_0)2^n &< \overline{\lambda}_{n_\nu+1}(x_0)2^{n_\nu+1} \qquad \text{for } n \le n_\nu \ ,\\ \overline{\lambda}_n(x_0)2^n &\le \overline{\lambda}_{n_\nu+1}(x_0)2^{n_\nu+1} \qquad \text{for } n_\nu < n \le n_{\nu+1} \ ,\\ \overline{\lambda}_{n_\nu+1}(x_0)2^{n_\nu+1} &< \overline{\lambda}_{n_{\nu+1}+1}(x_0)2^{n_{\nu+1}+1} \ .\end{aligned}$$

Then

$$\sum_{n=n_\nu+1}^{n_{\nu+1}} \overline{\lambda}_n(x_0) \le \overline{\lambda}_{n_\nu+1}(x_0)2^{n_\nu+1} \sum_{n=n_\nu+1}^{n_{\nu+1}} 2^{-n} \le 2\,\overline{\lambda}_{n_\nu+1}(x_0) \ ,$$

so that

$$\sum_{n=0}^{\infty} \overline{\lambda}_n(x_0) \le 2 \sum_{\nu=0}^{\infty} \overline{\lambda}_{n_\nu+1}(x_0) \ . \tag{9.7.13}$$

Thus, by (9.6.4),

$$\sum_{\nu=0}^{\infty} \overline{\lambda}_{n_\nu+1}(x_0) = \infty \ ,$$

and by (9.7.4)

$$\sum_{\nu=0}^{\infty} \underline{\lambda}_{n_\nu}(x_0) = \infty \ . \tag{9.7.14}$$

Moreover, we claim that (9.7.4) implies that $3Q_{n_\nu,0} \subset G_{n_\nu}$. In fact, $\overline{\lambda}_{n_\nu+1}(x_0) \le M\underline{\lambda}_{n_\nu}(x_0)$ by (9.7.4), and for all i such that $Q_{n_\nu,i} \subset 3Q_{n_\nu,0}$ we have

$$\lambda^*_{n_\nu,i}2^{n_\nu} < \overline{\lambda}_{n_\nu+1}(x_0)2^{n_\nu+1} \ .$$

In fact, if $Q_{n_\nu,i} \subset Q_{kj}$ for some $k \le n_\nu$, then clearly $Q_{kj} \subset 3Q_{k0}$, so

$$\lambda_{kj}2^k \le \overline{\lambda}_k(x_0)2^k < \overline{\lambda}_{n_\nu+1}(x_0)2^{n_\nu+1} \ .$$

Thus, by (9.7.1)

$$\lambda^*_{n_\nu,i} < 2M\underline{\lambda}_{n_\nu}(x_0) \le 2M\lambda_{n_\nu,i} \ ,$$

which proves the claim.

Thus, by (9.7.7),

$$\tfrac{11}{4}Q_{n_\nu,0} \subset G'_{n_\nu} \ .$$

It follows that

$$\begin{aligned}\operatorname{dist}\big(\partial G'_{n_\nu}, 3Q_{n_\nu+1,0}\big) &\ge \operatorname{dist}\big((3Q_{n_\nu,0})^c, 3Q_{n_\nu+1,0}\big) - 2^{-n_\nu-3}\\ &= 2^{-n_\nu-1} - 2^{-n_\nu-3} = \tfrac{3}{8}2^{-n_\nu} \ .\end{aligned}$$

Thus $\omega_{n_\nu}(x)$ increases by at least $\frac{3}{8}2^{-n_\nu}\underline{\lambda}_{n_\nu}(x_0)2^{n_\nu} = \frac{3}{8}\underline{\lambda}_{n_\nu}(x_0)$, as x moves from $\partial G'_{n_\nu}$ to $3Q_{n_\nu+1,0}$.

Suppose that $\widetilde{\omega}_{n_\nu-1}(x) \geq L$ on $\frac{11}{4}Q_{n_\nu,0}$. It follows that

$$\omega_{n_\nu}(x) \geq L + \tfrac{3}{8}\underline{\lambda}_{n_\nu}(x_0) \quad \text{on } 3Q_{n_\nu+1,0} \ .$$

Unless this number is ≥ 1,

$$\widetilde{\omega}_{n_\nu}(x) = \omega_{n_\nu} * \varphi_{n_\nu+4}(x) \geq L + \tfrac{3}{8}\underline{\lambda}_{n_\nu}(x_0) \ ,$$

and

$$\widetilde{\omega}_{n_\nu} * \varphi_{n_\nu+5} * \cdots * \varphi_{n_\nu+l}(x) \geq L + \tfrac{3}{8}\underline{\lambda}_{n_\nu}(x_0) \quad \text{on } \tfrac{11}{4}Q_{n_\nu+1,0} \ ,$$

since by (9.7.8)

$$\operatorname{supp}(\varphi_{n_\nu+4} * \cdots * \varphi_{n_\nu+l}) \subset B_{n_\nu+3}(0) \ ,$$

and

$$\operatorname{dist}\bigl(\partial(3Q_{n_\nu+1,0}), \tfrac{11}{4}Q_{n_\nu+1,0}\bigr) = 2^{-n_\nu-3} \ .$$

But $n_{\nu+1} \geq n_\nu + 1$, so by induction

$$\widetilde{\omega}_{n_{\nu+1}} \geq \tfrac{3}{8}\sum_{\mu=0}^{\nu} \underline{\lambda}_{n_\mu}(x_0) \quad \text{on } \tfrac{11}{4}Q_{n_\nu+1,0} \ .$$

By (9.7.14) the series $\sum_{\mu=0}^{\nu} \underline{\lambda}_{n_\mu}(x_0)$ diverges. This implies that $\widetilde{\omega}_n(x) = 1$ on a neighborhood of x_0 for all sufficiently large n. By the compactness of K there is a finite number of such neighborhoods that cover K. It follows that if n is sufficiently large, then $\widetilde{\omega}_n(x) = 1$ on a neighborhood of K. We finally fix n so that this is the case, and define $\omega = \widetilde{\omega}_n$.

In order to finish the proof of the lemma we have to prove (9.5.5). But this is proved exactly as in the previous section, once we observe that by (9.7.13) and (9.7.4)

$$\sum_{\mu=0}^{n_\nu} \overline{\lambda}_n(x_0) \leq 2\sum_{\mu=0}^{\nu} \overline{\lambda}_{n_\mu+1}(x_0) \leq 2M \sum_{\mu=0}^{\nu} \underline{\lambda}_{n_\mu}(x_0) \ .$$

This completes the proof of Lemma 9.5.2. Thus Theorems 9.5.1 and 9.3.2(a) have been proved.

Remark. For later use (Section 9.10) we make an observation concerning the situation when (9.5.2) is true only on a noncompact subset of K. For any $P \geq 0$ we set

$$E_P = \left\{ x \in K \cap \operatorname{supp} f : \sum_{n=0}^{P} \left(\frac{2^{-nm}[\nabla^m f]_n(x)}{[f]_n(x)} \right)^{p'} > L \right\} ,$$

where L is a constant. Then, if L is chosen large enough, the above construction gives that for any P there is n_P such that $\widetilde{\omega}_{n_P} = 1$ on a neighborhood of $\overline{E}_P$. Moreover, it is easy to see that there are neighborhoods G_P of $\overline{E}_P$ such that $\widetilde{\omega}_{n_P} = 1$ on G_P, and such that $G_P \subset G_{P+1}$ for all P.

9.8 Estimates for Nonlinear Potentials

In this section we collect some estimates for nonlinear potentials, $V^{\nu}_{\alpha,p} = G_\alpha * (G_\alpha * \nu)^{p'-1}$, that will be used in the proofs of parts (b) and (c) of Theorem 9.3.2. The proofs can of course be simplified in the linear case $p = 2$. We first prove a "Harnack property".

Lemma 9.8.1. *Let ν be a positive measure, and let x_0 be a point off* $\operatorname{supp}\nu$. *Suppose that* $\operatorname{dist}(x_0, \operatorname{supp}\nu) = \delta \le 1$. *Then there is a constant A independent of ν such that*

$$V^{\nu}_{\alpha,p}(x) \le A\, V^{\nu}_{\alpha,p}(x_0) \quad \textit{for all } x \in B(x_0, \delta/4)\ .$$

Proof. Without loss of generality we set $x_0 = 0$. We split $V^{\nu}_{\alpha,p}(x)$ into two parts,

$$V^{\nu}_{\alpha,p}(x) = \int_{|y|\le\delta/2} + \int_{|y|>\delta/2} G_\alpha(x-y)\varphi(y)\,dy\ ,$$

where $\varphi(y) = \left(\int G_\alpha(y-z)\,d\nu(z)\right)^{p'-1}$. If $|x| \le \frac{\delta}{4}$ and $|y| > \frac{\delta}{2}$, we have $|x-y| \ge |y| - \frac{\delta}{4} \ge \frac{1}{2}|y|$, and $|x-y| \ge |y| - \frac{1}{4}$, so by (1.2.14) and (1.2.16), $G_\alpha(x-y) \le A\,G_\alpha(x)$, and

$$\int_{|y|>\delta/2} G_\alpha(x-y)\varphi(y)\,dy \le A \int_{\mathbf{R}^N} G_\alpha(y)\varphi(y)\,dy = A\,V^{\nu}_{\alpha,p}(0)\ .$$

For the remaining part of $V^{\nu}_{\alpha,p}(x)$ we write

$$\int_{|y|\le\delta/2} G_\alpha(x-y)\varphi(y)\,dy = \int_{|x-y|\le\delta/2} G_\alpha(y)\varphi(x-y)\,dy\ ,$$

and $\varphi(x-y) = \left(\int G_\alpha(x-y-z)\,d\nu(z)\right)^{p'-1}$. Here $|z| \ge \delta$, so $|x-y-z| \ge |z| - |x-y| \ge |z| - \frac{\delta}{2} \ge \max\left\{\frac{1}{2}|z|, |z| - \frac{1}{2}\right\}$. But $|y-z| \ge \frac{\delta}{2} + |z| \le \min\left\{\frac{3}{2}|z|, |z| + \frac{1}{2}\right\}$, so $|x-y-z| \ge \frac{1}{3}|y-z|$, or $|x-y-z| \ge |y-z| - 1$. Thus $\varphi(x-y) \le A\,\varphi(y)$, and it follows that

$$\int_{|x-y|\le\delta/2} G_\alpha(x-y)\varphi(y)\,dy \le A \int_{|x-y|\le\delta/2} G_\alpha(y)\varphi(y)\,dy \le A\,V^{\nu}_{\alpha,p}(0)\ ,$$

which proves the lemma.

Next, we prove a lower estimate for nonlinear potentials.

Lemma 9.8.2. *Let $F \subset \mathbf{R}^N$ be compact, and let ν be a positive measure such that $V^{\nu}_{\alpha,p}$ satisfies $V^{\nu}_{\alpha,p}(x) \le M$ everywhere, and $V^{\nu}_{\alpha,p}(x) \ge 1$ (α, p)-q.e. on F. Suppose that F contains a cube Q with diameter ≤ 1. Then there is a constant $c > 0$, independent of ν, F, and Q, such that $V^{\nu}_{\alpha,p}(x) \ge c$ everywhere on $9Q$.*

The lemma is an immediate consequence of the following somewhat more general lemma.

Lemma 9.8.3. *Let ν satisfy the assumptions in Lemma 9.8.2. Suppose that for some δ, $0 < \delta \le 1$, and some $c > 0$*

$$\frac{C_{\alpha,p}(F \cap B(x_0, \delta))}{\delta^{N-\alpha p}} \ge c \quad \text{if } N > \alpha p ,$$

or

$$C_{\alpha,p}(F \cap B(x_0, \delta))(\log \tfrac{2}{\delta})^{p-1} \ge c \quad \text{if } N = \alpha p .$$

Then

$$V^{\nu}_{\alpha,p}(x_0) \ge A\, c^{p'-1} ,$$

where A is a constant, independent of ν, F, x_0, δ, and c.

Lemma 9.8.2 follows if we set $\delta = 5 \operatorname{diam} Q$. In fact, if $x_0 \in 9Q$, where $Q \subset F$, it follows that $Q \subset F \cap B(x_0, \delta)$. Thus $C_{\alpha,p}(F \cap B(x_0, \delta)) \ge C_{\alpha,p}(Q)$, and there is $c > 0$ so that $C_{\alpha,p}(Q) \ge c\delta^{N-\alpha p}$ for $N > \alpha p$, and $C_{\alpha,p}(Q) \ge c(\log \frac{2}{\delta})^{1-p}$ for $N = \alpha p$.

Proof of Lemma 9.8.3. The proof is based on the following observation:

If $V^{\nu}_{\alpha,p}(x) \le M$ everywhere, and if $V^{\nu}_{\alpha,p}(x) \ge 1$ (α, p)-q.e. on E, then $C_{\alpha,p}(E) \le M\, \nu(\mathbf{R}^N)$.

In fact, if $V^{\nu}_{\alpha,p}(x) = G_\alpha * (G_\alpha * \nu)^{p'-1}(x) \ge 1$ (α, p)-q.e. on E, then by the definition of capacity

$$C_{\alpha,p}(E) \le \int_{\mathbf{R}^N} (G_\alpha * \nu)^{p'}\, dx = \int_{\mathbf{R}^N} V^{\nu}_{\alpha,p}\, d\nu \le M\, \nu(\mathbf{R}^N) .$$

Now assume that the assumptions of the lemma are satisfied. Let ν' be the restriction of ν to the ball $B(x_0, 2\delta)$, and set $\nu = \nu' + \nu''$. Then, for any $y \in F \cap B(x_0, \delta)$

$$V^{\nu}_{\alpha,p}(y) \le A\, V^{\nu'}_{\alpha,p}(y) + A\, V^{\nu''}_{\alpha,p}(y) \le A\, V^{\nu'}_{\alpha,p}(y) + A_1 V^{\nu}_{\alpha,p}(x_0) ,$$

by Lemma 9.8.1. We can assume that $A_1 V^{\nu}_{\alpha,p}(x_0) \le \frac{1}{2}$, since otherwise there is nothing to prove. It follows that $V^{\nu'}_{\alpha,p}(y) \ge 1/(2A)$ (α, p)-q.e. on $F \cap B(x_0, \delta)$. Thus, by the above observation,

$$C_{\alpha,p}(F \cap B(x_0, \delta)) \le AM\, \nu'(\mathbf{R}^N) = AM\, \nu(B(x_0, 2\delta)) .$$

But by Proposition 2.6.9 we have the lower estimate

$$V^{\nu}_{\alpha,p}(x_0) \ge A \int_0^4 \left(\frac{\nu(B(x_0, t))}{t^{N-\alpha p}} \right)^{p'-1} \frac{dt}{t} \quad \text{for } N \ge \alpha p ,$$

whence

$$V^{\nu}_{\alpha,p}(x_0) \ge A\, \nu(B(x_0, 2\delta))^{p'-1} \int_{2\delta}^4 \frac{1}{t^{(N-\alpha p)(p'-1)}} \frac{dt}{t} .$$

For $N > \alpha p$ we obtain

$$V_{\alpha,p}^{\nu}(x_0) \geq A\left(\frac{C_{\alpha,p}(F \cap B(x_0,\delta))}{M\delta^{N-\alpha p}}\right)^{p'-1} \geq A\left(\frac{c}{M}\right)^{p'-1} ,$$

and for $N = \alpha p$

$$V_{\alpha,p}^{\nu}(x_0) \geq A\left(\frac{C_{\alpha,p}(F \cap B(x_0,\delta))}{M}\right)^{p'-1} \log\frac{2}{\delta} \geq A\left(\frac{c}{M}\right)^{p'-1} .$$

We now extend Lemma 9.8.1.

Lemma 9.8.4. *Let ν be a positive measure, and suppose that $0 < \delta(x) = \operatorname{dist}(x, \operatorname{supp}\nu) \leq 1$. Set $V_{\alpha,p}^{\nu} = G_\alpha * \varphi$.*

(a) *For all j*

$$|\nabla^j V_{\alpha,p}^{\nu}(x)| \leq A\,\delta(x)^{-j} V_{\alpha,p}^{\nu}(x) . \tag{9.8.1}$$

(b) *There is a function $\eta \geq 0$ such that*

$$\|\eta\|_p \leq A\,\|\varphi\|_p , \tag{9.8.2}$$

and

$$|\nabla^j V_{\alpha,p}^{\nu}(x)| \leq A\,\delta(x)^{\alpha-j}\eta(x), \quad j \geq \alpha . \tag{9.8.3}$$

Moreover, η has the Harnack property, i.e.

$$A^{-1}\eta(x) \leq \eta(y) \leq A\,\eta(x), \quad y \in B(x, \tfrac{1}{2}\delta(x)) . \tag{9.8.4}$$

Proof. Without loss of generality we assume that $x = 0$, and $\delta(0) = \delta > 0$. We split the kernel by setting

$$G_\alpha = \chi G_\alpha + (1-\chi)G_\alpha = G'_\alpha + G''_\alpha ,$$

where χ is a C^∞ cut-off function such that $0 \leq \chi \leq 1$, and

$$\begin{aligned} \chi(y) &= 1, \quad |y| \leq \tfrac{\delta}{4} , \\ \chi(y) &= 0, \quad |y| \geq \tfrac{\delta}{2} , \\ |\nabla^j \chi(y)| &\leq A_j \delta^{-j}, \quad j = 1, 2, \ldots . \end{aligned}$$

Then

$$|\nabla^j G''_\alpha(y)| \leq A\,\delta^{\alpha-N-j} \quad \text{for } \tfrac{\delta}{4} \leq |y| \leq \tfrac{\delta}{2} ,$$

by the Leibniz formula, (1.2.27), and (1.2.14). Thus

$$|\nabla^j G''_\alpha(y)| \leq A\,\delta^{-j} G_\alpha(y) \quad \text{for all } y , \tag{9.8.5}$$

by (1.2.27) and (1.2.28), and also

$$|\nabla^j G''_\alpha(y)| \leq A\,|y|^{\alpha-N-j} \quad \text{for all } y, \text{ and all } j . \tag{9.8.6}$$

Now, φ is C^∞ off $\operatorname{supp}\nu$, so we can write

$$\nabla^j(G'_\alpha * \varphi)(y) = (G'_\alpha * \nabla^j\varphi)(y) = \int_{\mathbf{R}^N} G'_\alpha(t)\nabla^j\varphi(y-t)\,dt$$

$$= \int_{\mathbf{R}^N} G'_\alpha(t)\nabla^j_y\left(\int_{\mathbf{R}^N} G_\alpha(y-t-z)\,d\nu(z)\right)^{p'-1} dt \ .$$

In this integral $|t| \le \frac{\delta}{2}$, $|z| \ge \delta$, and $y = 0$, so $|y-t-z| \ge \frac{\delta}{2}$, and thus for all j

$$\left|\nabla^j_y \int_{\mathbf{R}^N} G_\alpha(y-t-z)\,d\nu(z)\right| = \left|\int_{\mathbf{R}^N} \nabla^j G_\alpha(y-t-z)\,d\nu(z)\right|$$

$$\le A\,\delta^{-j} \int_{\mathbf{R}^N} G_\alpha(y-t-z)\,d\nu(z) \ .$$

As in the proof of Theorem 3.3.3 we obtain by the chain rule

$$|\nabla^j\varphi| = \left|\nabla^j(G_\alpha * \nu)^{p'-1}\right| \le A\sum_{k=1}^{j}(G_\alpha * \nu)^{p'-1-k}\sum\prod_{l=1}^{k}\left|\nabla^{\beta_l}(G_\alpha * \nu)\right| \ ,$$

where the last sum is taken over all k-tiples $\beta_1, \dots, \beta_k$, such that $\sum_1^k \beta_l = j$ and all $\beta_l \ge 1$. Hence

$$|\nabla^j\varphi| \le A\sum_{k=1}^{j}(G_\alpha * \nu)^{p'-1-k}\delta^{-j}(G_\alpha * \nu)^k \le A\,\delta^{-j}\varphi \ ,$$

so

$$|\nabla^j(G'_\alpha * \varphi)(0)| \le A\,\delta^{-j}(G'_\alpha * \varphi)(0) \ . \tag{9.8.7}$$

On the other hand, by (9.8.5) above

$$|\nabla^j(G''_\alpha * \varphi)(0)| = \left|\int_{\mathbf{R}^N} \nabla^j G''_\alpha(-t)\varphi(t)\,dt\right| \le A\,\delta^{-j}(G''_\alpha * \varphi)(0) \ , \tag{9.8.8}$$

which proves (a).

In order to prove (b) we first observe that (9.8.7) and Lemma 3.1.1 give that

$$|\nabla^j(G'_\alpha * \varphi)(0)| \le A\,\delta^{\alpha-j}M\varphi(0) \quad \text{for all } j \ . \tag{9.8.9}$$

For $j > \alpha$ we also have

$$|\nabla^j(G''_\alpha * \varphi)(0)| \le A\int_{|t|\ge\delta/4} |t|^{\alpha-N-j}\varphi(t)\,dt \le A\,\delta^{\alpha-j}M\varphi(0) \ , \tag{9.8.10}$$

by (9.8.6) and Lemma 3.1.1. This proves (b) for $j > \alpha$ with $\eta = M\varphi$. For $j = \alpha$ we have

$$\|\nabla^\alpha(G_\alpha * \varphi)\|_p \le A\,\|\varphi\|_p$$

by Theorem 1.2.3, so (9.8.3) is true with

$$\eta = M\varphi + |\nabla^\alpha(G_\alpha * \varphi)| \ .$$

It remains to prove the Harnack property (9.8.4). The maximal function $M\varphi$ is easily seen to have this property. Let in fact $|y| \leq \frac{\delta}{4}$. It is clear that $\varphi = (G_\alpha * \nu)^{p'-1}$ has the Harnack property, so for $r \leq \frac{\delta}{4}$

$$\frac{1}{|B(y,r)|} \int_{|B(y,r)|} \varphi(t)\, dt \leq A\, \varphi(0) \leq A\, M\varphi(0) \ .$$

For $r \geq \frac{\delta}{4}$ we have

$$\frac{1}{|B(y,r)|} \int_{|B(y,r)|} \varphi(t)\, dt \leq \frac{2^N}{|B(0,2r)|} \int_{|B(0,2r)|} \varphi(t)\, dt \leq 2^N M\varphi(0) \ ,$$

so $M\varphi(y) \leq A\, M\varphi(0)$.

We then consider $|\nabla^\alpha (G_\alpha * \varphi)|$. For $|y| \leq \frac{\delta}{8}$ we have

$$\begin{aligned} |\nabla^\alpha (G_\alpha * \varphi)(y)| &\leq |\nabla^\alpha (G_\alpha * \varphi)(y) - \nabla^\alpha (G_\alpha * \varphi)(0)| + |\nabla^\alpha (G_\alpha * \varphi)(0)| \\ &\leq |\nabla^\alpha (G'_\alpha * \varphi)(y)| + |\nabla^\alpha (G'_\alpha * \varphi)(0)| \\ &\quad + |\nabla^\alpha (G''_\alpha * \varphi)(y) - \nabla^\alpha (G''_\alpha * \varphi)(0)| + |\nabla^\alpha (G_\alpha * \varphi)(0)| \\ &\leq A\, M\varphi(y) + A\, M\varphi(0) \\ &\quad + |\nabla^\alpha (G''_\alpha * \varphi)(y) - \nabla^\alpha (G''_\alpha * \varphi)(0)| + |\nabla^\alpha (G_\alpha * \varphi)(0)| \ , \end{aligned}$$

where the last inequality follows from (9.8.9) above. Now observe that by (9.8.6)

$$|\nabla^{\alpha+1} G''_\alpha(y)| \leq A\, |y|^{-N-1} \quad \text{for } |y| \leq \tfrac{\delta}{4} \ ,$$

so that

$$\begin{aligned} &\left|\nabla^\alpha (G''_\alpha * \varphi)(y) - \nabla^\alpha (G''_\alpha * \varphi)(0)\right| \\ &\qquad = \left| \int_{\mathbf{R}^N} \big(\nabla^\alpha G''_\alpha(y-t) - \nabla^\alpha G''_\alpha(-t)\big)\varphi(t)\, dt \right| \\ &\qquad \leq \int_{|t| \geq \delta/4} \frac{|y|}{(|t| - \frac{\delta}{8})^{N+1}} \varphi(t)\, dt \\ &\qquad \leq A\, |y| M\varphi(0) \delta^{-1} \leq A\, M\varphi(0) \end{aligned}$$

by Lemma 3.1.1. Thus

$$|\nabla^\alpha (G''_\alpha * \varphi)(y)| \leq A\, M\varphi(0) + A\, |\nabla^\alpha (G''_\alpha * \varphi)(0)| \ ,$$

so that if

$$\eta = M\varphi + |\nabla^\alpha (G''_\alpha * \varphi)| \ ,$$

there is a constant A such that

$$\eta(y) \leq A\, \eta(0) \quad \text{for } |y| \leq \tfrac{\delta}{8} \ .$$

Iteration finishes the proof of Lemma 9.8.4.

Finally we prove the following lemma.

Lemma 9.8.5. *Let ν, $V^{\nu}_{\alpha,p} = G_\alpha * \varphi$, and $\delta(x)$ be as in Lemma 9.8.4, and let Φ be a C^∞ function defined on $[0,\infty)$ such that $0 \le \Phi(r) \le 1$ for all r, and $\Phi(r) = 1$ for $r \ge 1$. Set $\omega = \Phi \circ V^{\nu}_{\alpha,p}$. Then there is a function $\eta \ge 0$, with the Harnack property* (9.8.4), *such that*

(a) $\|\eta\|_p \le A\,\|\varphi\|_p$;
(b) $|\nabla^j \omega(x)| \le A\,\eta(x)^{j/\alpha}, \quad 1 \le j \le \alpha$;
(c) $|\nabla^j \omega(x)| \le A\,\delta(x)^{\alpha-j}\eta(x), \quad j \ge \alpha$.

Proof. We set $V^{\nu}_{\alpha,p} = \psi$. Then as in the proof of Theorem 3.3.3 we find by the chain rule

$$|\nabla^j \omega| \le A \sum_{k=1}^{j} |\Phi^{(k)} \circ \psi| \sum_{\beta} \prod_{l=1}^{k} |\nabla^{\beta_l}\psi| ,$$

where the last sum is taken over all k-tiples $\beta = (\beta_1, \dots, \beta_k)$ such that $\sum_1^k \beta_l = j$, and all $\beta_l \ge 1$. We also observe that $\nabla^j \omega(x) = 0$ wherever $\psi(x) > 1$. In fact, $\omega(x) = 1$ in a neighborhood of every such point, since ψ is lower semicontinuous.

By Proposition 3.1.8 we have for $\beta_l < \alpha$

$$|\nabla^{\beta_l}\psi| \le A\,(M\varphi)^{\beta_l/\alpha}\psi^{1-\beta_l/\alpha} .$$

Thus, for $j < \alpha$

$$|\nabla^j \omega| \le A \sum_{k=1}^{j} \sum_{\beta} \prod_{l=1}^{k} (M\varphi)^{\beta_l/\alpha}\psi^{1-\beta_l/\alpha} = A \sum_{k=1}^{j} \sum_{\beta} (M\varphi)^{j/\alpha}\psi^{k-j/\alpha}$$

$$\le A \sum_{\beta} (M\varphi)^{j/\alpha}\psi^{-j/\alpha} \sum_{k=1}^{j} \psi^k \le A\,(M\varphi)^{j/m} .$$

If $j = \alpha$ we also use (9.8.3) in Lemma 9.8.4:

$$|\nabla^\alpha \psi| \le A\,\eta ,$$

and obtain similarly

$$|\nabla^j \omega| \le A(M\varphi + \eta) .$$

If $j > \alpha$ we combine Proposition 3.1.8 with (9.8.1):

$$|\nabla^{\beta_l}\psi| \le A\,\delta^{-\beta_l}\psi ,$$

and (9.8.3):

$$|\nabla^{\beta_l}\psi| \le A\,\delta^{\alpha-\beta_l}\eta, \quad \beta_l \ge \alpha .$$

In this way we obtain for $\sum_{l=1}^k \beta_l = j > \alpha$ that

$$\prod_{l=1}^{k} |\nabla^{\beta_l}\psi| \le A\,\delta^{\alpha-j} M\varphi ,$$

if all $\beta_l < \alpha$, and

$$\prod_{l=1}^{k} |\nabla^{\beta_l} \psi| \leq A\, \delta^{\alpha - j} \eta \ ,$$

if $\beta_l \geq \alpha$ for at least some l. Thus for $j > \alpha$

$$|\nabla^j \omega| \leq A\, \delta^{\alpha - j} (M\varphi + \eta) \ .$$

But η has the Harnack property by Lemma 9.8.4. It follows as in the proof of that lemma that $M\varphi + \eta$ has the same property.

9.9 The Case $C_{m,p}(K) = 0$

In this section we shall prove (c) in Theorem 9.3.2. The same construction will then be used in the following section in order to prove the more complicated case (b).

The assumption (9.3.1) is vacuous if $C_{m,p}(K) = 0$, so the theorem can be reformulated in the following way. We assume that $mp \leq N$, since otherwise only the empty set has zero (m, p)-capacity.

Theorem 9.9.1. *Let $f \in W^{m,p}(\mathbf{R}^N)$, $1 < p < \infty$, $m \in \mathbf{Z}^+$, $mp \leq N$, and let $K \subset \mathbf{R}^N$ be a compact set with $C_{m,p}(K) = 0$. Then, for any $\varepsilon > 0$ and any neighborhood V of K there is a function $\omega \in C_0^\infty(V)$ such that $0 \leq \omega \leq 1$, $\omega = 1$ on a neighborhood of K, and $\|\omega f\|_{m,p} < \varepsilon$.*

The function ω will be constructed as a modification of a nonlinear potential. In Section 9.5 we defined meshes $\mathcal{Q}_n$ of closed cubes Q with sidelength 2^{-n}. According to the Whitney covering lemma (Theorem 1.4.2) the complement of the given compact set K is a union of cubes with disjoint interiors, such that each Q belongs to some $\mathcal{Q}_n$, and such that for each Q

$$\operatorname{diam} Q \leq \operatorname{dist}(Q, K) \leq 4 \operatorname{diam} Q \ . \tag{9.9.1}$$

Moreover, the covering can be made to satisfy the following requirement: If Q_1 and Q_2 belong to the same covering and touch one another, i.e. $\partial Q_1 \cap \partial Q_2 \neq \emptyset$, then

$$\tfrac{1}{4} \operatorname{diam} Q_1 \leq \operatorname{diam} Q_2 \leq 4 \operatorname{diam} Q_1 \ . \tag{9.9.2}$$

See e.g. Stein [389], Chapter VI.1.3.

We choose such a covering of $K^c = \Omega$:

$$\Omega = \bigcup Q_{ni}, \quad Q_{ni} \in \mathcal{Q}_n \ .$$

We represent the given function f as $f = G_m * g$, and let $\lambda > 0$ be a large number. We denote by U'_λ the union of all Q_{ni} such that

$$\frac{1}{|Q_{ni}|} \int_{Q_{ni}} (G_m * |g|)^p \, dx > \lambda^p \ .$$

If x belongs to such a Q_{ni} it is clear that

$$\frac{A}{|B(x,r)|}\int_{B(x,r)}(G_m * |g|)^p\,dx > \lambda^p$$

for some r and some $A = A(N)$. Thus, by Lemma 6.2.2 there is a constant A such that

$$C_{m,p}(U'_\lambda) < \frac{A}{\lambda^p}\|g\|_p^p \ .$$

We now define U_λ as the union of U'_λ and a compact neighborhood U''_λ of K such that

$$C_{m,p}(U''_\lambda) < \frac{1}{\lambda^p}\|g\|_p^p \ .$$

We also assume, as we may, that $U''_\lambda \setminus K$ is a union of Q_{ni}. Thus there is A such that

$$C_{m,p}(U_\lambda) < \frac{A}{\lambda^p}\|g\|_p^p \ , \tag{9.9.3}$$

and for all Q_{ni} that do not belong to U_λ

$$\frac{1}{|Q_{ni}|}\int_{Q_{ni}}(G_m * |g|)^p\,dx \le \lambda^p \ . \tag{9.9.4}$$

We note that if λ is chosen large enough, then (9.9.1) ensures that U_λ is contained in an arbitrarily given neighborhood of K. In fact, (9.9.3) puts a bound on the size of the Q_{ni} constituting $U_\lambda \setminus K$.

We shall now construct ω by modifying the capacitary potential for U_λ. Let ν be the (m,p)-capacitary measure for U_λ, and set $V^\nu_{m,p} = G_m * \varphi$, $\varphi = (G_m * \nu)^{p'-1}$. Then

$$\begin{aligned} &\operatorname{supp}\nu \subset U_\lambda \ ; \\ &\|\varphi\|_p^p = C_{m,p}(U_\lambda) \le A\,\lambda^{-p}\|g\|_p^p \ ; \\ &V^\nu_{m,p}(x) \le M < \infty \quad \text{ for all } x \ . \end{aligned} \tag{9.9.5}$$

The last inequality is the boundedness principle, Theorem 2.6.3.

By Lemma 9.8.2 there is a constant $c > 0$ such that $V^\nu_{m,p}(x) \ge c$ everywhere on the set $\widetilde{U}_\lambda$ defined as $\widetilde{U}_\lambda = \cup(9Q_{ni})$, the union being taken over all Q_{ni} in U_λ.

We let Φ be a non-decreasing C^∞ function on $\mathbf{R}$ such that

$$\Phi(r) = 0, \quad r \le \tfrac{c}{2}, \qquad \text{and} \qquad \Phi(r) = 1, \quad r \ge c\,.$$

We define ω by

$$\omega = \Phi \circ V^\nu_{m,p} \ .$$

Clearly $\omega \in C^\infty$, $\omega(x) = 1$ on $\widetilde{U}_\lambda$, and $\operatorname{supp}\omega$ is compact. In fact, we claim that $\operatorname{supp}\omega$ can be made to be contained in any prescribed neighborhood V of K. The function $\omega(x) = \Phi \circ V^\nu_{m,p}(x)$ vanishes whenever $V^\nu_{m,p}(x) \le \frac{c}{2}$. Suppose that $x_0 \in \Omega$ is such that $V^\nu_{m,p}(x_0) > \frac{c}{2}$. The Harnack property, Lemma 9.8.1, implies that there is a constant A such that $V^{\nu}_{m,p}(x) > c/(2A)$ for all x in any Whitney cube Q_{ni} containing x_0. But ν is the capacitary measure for U_λ, which is a set with arbitrarily small capacity if λ is made large. It follows from the definition of capacity that the capacity of the set where $V^{\nu}_{m,p}(x) > c/(2A)$ is also arbitrarily small. Thus n has to be large if $x_0 \in Q_{ni}$, and then (9.9.1) gives that $\operatorname{dist}(x_0, K)$ has to be small. The claim follows.

Next, we need the following observation about Whitney cubes: If Q is a cube in the Whitney covering of Ω, and if Q is not contained in $\widetilde{U}_\lambda$, then there is a constant A such that

$$\operatorname{dist}(Q, U_\lambda) \geq A \operatorname{diam} Q .$$

To prove this, we note that according to property (9.9.2) of Whitney cubes, if Q' is a Whitney cube that touches Q, then the ratio of the sidelengths of Q and Q' is between 4 and $\frac{1}{4}$. Thus, if Q is not contained in $\widetilde{U}_\lambda$, it cannot touch a Q' contained in U_λ, since any Whitney cube touching Q' is contained in $9Q'$. Moreover, $\frac{3}{2}Q$ is contained in the union of all Whitney cubes touching Q, since they have sides at least $\frac{1}{4}$ of the side of Q. Thus $\frac{3}{2}Q$ does not intersect U_λ, so

$$\operatorname{dist}(Q, U_\lambda) \geq \frac{1}{4\sqrt{N}} \operatorname{diam} Q \ .$$

(Cf. Stein [389], Chapter VI.1.3.)

It follows after iteration that Lemma 9.8.5 can be applied to ω in all those cubes Q_{ni} where $\nabla\omega$ is not identically zero.

We now set $f = G_m * g$ and estimate

$$\int_{Q_{ni}} |\nabla^j \omega|^p |\nabla^{m-j} f|^p \, dx, \quad j = 1, 2, \ldots, m$$

for all Q_{ni} not contained in $\widetilde{U}_\lambda$.

If we let x_{ni} be the center of Q_{ni}, we obtain from Lemma 9.8.5, Proposition 3.1.8, and Hölder's inequality that for $j < m$

$$\begin{aligned}
\int_{Q_{ni}} |\nabla^j \omega|^p |\nabla^{m-j} f|^p \, dx &\leq A\, \eta(x_{ni})^{pj/m} \int_{Q_{ni}} |\nabla^{m-j} f|^p \, dx \\
&\leq A\, \eta(x_{ni})^{pj/m} \int_{Q_{ni}} (Mg)^{(m-j)p/m} (G_m * |g|)^{pj/m} \, dx \\
&\leq A \left(\eta(x_{ni})^p |Q_{ni}|\right)^{j/m} \left(\int_{Q_{ni}} (Mg)^p \, dx \right)^{(m-j)/m} \\
&\quad \times \left(\frac{1}{|Q_{ni}|} \int_{Q_{ni}} (G_m * |g|)^p \, dx \right)^{j/m} \\
&\leq A \left(\int_{Q_{ni}} \eta^p \, dx \right)^{j/m} \left(\int_{Q_{ni}} (Mg)^p \, dx \right)^{(m-j)/m} \lambda^{pj/m} \ .
\end{aligned}$$

Here the last inequality follows from (9.9.4). By summing over cubes and applying Hölder's inequality for sums we now obtain for $1 \leq j < m$

$$\int_{\mathbf{R}^N} |\nabla^j \omega|^p |\nabla^{m-j} f|^p \, dx \le A\Big(\int_{\mathbf{R}^N} \eta^p \, dx\Big)^{j/m} \Big(\int_V (Mg)^p \, dx\Big)^{(m-j)/m} \lambda^{pj/m}$$
$$\le A\|g\|_p^{j/m} \Big(\int_V (Mg)^p \, dx\Big)^{(m-j)/m} \lambda^{pj/m} ,$$

where the last inequality follows from Lemma 9.8.5 and formula (9.9.5). But V has arbitrarily small measure, so the integral to be estimated is arbitrarily small.

For $j = m$ we obtain similarly

$$\int_{Q_{ni}} |\nabla^m \omega|^p |f|^p \, dx \le A\eta(x_{ni})^p |Q_{ni}| \frac{1}{|Q_{ni}|} \int_{Q_{ni}} |f|^p \, dx \le A \int_{Q_{ni}} \eta^p \, dx \, \lambda^p ,$$

so that by Lemma 9.8.1 and (9.9.5)

$$\int_V |\nabla^m \omega|^p |f|^p \, dx \le A \int_V \eta^p \, dx \, \lambda^p \le A \int_{\mathbf{R}^N} |g|^p \, dx .$$

For $j = 0$ we have

$$\int_{\mathbf{R}^N} |\omega|^p |\nabla^m f|^p \, dx \le \int_V |\nabla^m| f|^p \, dx ,$$

which is arbitrarily small.

Thus, if ω now is denoted ω_λ, we have proved that

$$\|\omega_\lambda f\|_{m,p} \le A \, \|f\|_{m,p} ,$$

independently of λ, as $\lambda \to \infty$. On the other hand, $\operatorname{supp} \omega_\lambda f \subset V = V_\lambda$, where $|V_\lambda| \to 0$, as $\lambda \to \infty$. It follows from the weak compactness of the unit ball in $W^{m,p}$ that there is a subsequence $\{\lambda_i\}_1^\infty$ such that $\omega_\lambda f \to 0$ in the weak topology, as $\lambda_i \to \infty$. But then, by the Mazur lemma (Theorem 1.3.5), there are convex combinations $\omega = \sum_i \alpha_i \omega_{\lambda_i}$, such that $\|\omega f\|_{m,p} \to 0$. This proves Theorem 9.9.1, since $\omega(x) = 1$ on a neighborhood of K.

9.10 The Case $C_{k,p}(K) = 0$, $1 \le k < m$

We shall now combine the constructions in Sections 9.7 and 9.9 in order to prove Theorem 9.3.2(b). The theorem can be reformulated in the following way, analogous to Theorem 9.9.1.

Theorem 9.10.1. *Let $f \in W^{m,p}(\mathbf{R}^N)$, $1 < p < \infty$, and let $K \subset \mathbf{R}^N$ be compact. Suppose that for some $k = 1, 2, \dots, m-1$ the following conditions are satisfied*:

(a) $C_{k,p}(K) = 0$;
(b) *K is $(k+1, p)$-thick at all of its points.*

Suppose that $D^\beta f|_K = 0$ for all multiindices β, $0 \le |\beta| < m - k$. Then, for any given $\varepsilon > 0$ and any neighborhood V of K, there is a function $\omega \in C_0^\infty(V)$, such that $0 \le \omega \le 1$, $\omega(x) = 1$ on a neighborhood of K, and $\|\omega f\|_{m,p} < \varepsilon$.

We recall the notation (see (9.5.1))

$$[f]_n(x) = \left(\frac{1}{|B_n(x)|} \int_{B_n(x)} |f|^p \, dy \right)^{1/p} .$$

Instead of the Poincaré type inequality in Corollary 8.2.7 used in the proof of Theorem 9.5.1 we now have to use Corollary 8.2.8. Under the assumptions of Theorem 9.10.1 it gives for all n

$$[f]_n(x) \le A\, 2^{-n(m-k)} [\nabla^{m-k} f]_n(x) + A \frac{2^{-nm}}{c_{k+1,p}(K, B_n(x))^{1/p}} [\nabla^m f]_n(x) .$$

Thus, denoting $2A$ by A, for each n either

$$[f]_n(x) \le A \frac{2^{-nm}}{c_{k+1,p}(K, B_n(x))^{1/p}} [\nabla^m f]_n(x) ,$$

or

$$[f]_n(x) \le A\, 2^{-n(m-k)} [\nabla^{m-k} f]_n(x) .$$

It follows that for a fixed $x \in K$, either there is an $n(x)$ such that

$$[f]_n(x) \le A \frac{2^{-nm}}{c_{k+1,p}(K, B_n(x))^{1/p}} [\nabla^m f]_n(x), \quad n \ge n(x) , \tag{9.10.1}$$

or else there is a sequence $\{n_j\}_{j=0}^{\infty}$ tending to infinity, such that

$$[f]_{n_j}(x) \le A\, 2^{-n_j(m-k)} [\nabla^{m-k} f]_{n_j}(x) . \tag{9.10.2}$$

In the former case

$$A \frac{2^{-nm} [\nabla^m f]_n(x)}{[f]_n(x)} \ge c_{k+1,p}(K, B_n(x))^{1/p} \quad \text{for } n \ge n(x) ,$$

so if K is $(k+1, p)$-thick at x, then

$$\sum_{n=1}^{\infty} \left(\frac{2^{-nm} [\nabla^m f]_n(x)}{[f]_n(x)} \right)^{p'} = \infty ,$$

i.e., (9.5.2) is satisfied. Thus, Theorem 9.5.1 can be applied to every compact set where (9.10.1) holds.

We shall need the following lemma in order to deal with the case (9.10.2).

Lemma 9.10.2. *Let $f \in W^{m,p}(\mathbf{R}^N)$, and suppose that $D^\beta f|_K = 0$ for all multi-indices β, $0 \le |\beta| < m-k$. Suppose that K is $(k+1, p)$-thick at a point x. Then, either* (9.5.2) *holds at x, or there is an $n_0 = n_0(x)$, and a constant A independent of x, such that*

$$[f]_n(x) \le A\, 2^{-n(m-k)} \big[I_k * |\nabla^m f| \big]_n(x) \tag{9.10.3}$$

for all $n \ge n_0$.

Proof. Suppose that (9.5.2) does not hold at x. Then

$$\lim_{n\to\infty} \frac{2^{-nm}[\nabla^m f]_n(x)}{[f]_n(x)} = 0 .$$

This implies that also

$$\lim_{n\to\infty} \frac{2^{-nm}[\nabla^m f]_n(x)}{[f]_{n+r}(x)} = 0 \tag{9.10.4}$$

for any fixed positive r. In fact, by Corollary 8.1.5 there is A such that

$$[f]_n(x) \le A\,[f]_{n+r}(x) + A\,2^{-nm}[\nabla^m f]_n(x) ,$$

and thus

$$\frac{[f]_{n+r}(x)}{2^{-nm}[\nabla^m f]_n(x)} \ge A^{-1}\frac{[f]_n(x)}{2^{-nm}[\nabla^m f]_n(x)} - 1 ,$$

which tends to ∞ as $n \to \infty$.

Now let $\{n_j\}_{j=0}^{\infty}$ be a sequence tending to infinity, such that (9.10.2) holds. Such a sequence exists according to the argument above. By (9.10.4) we can assume that n_0 is so large that

$$2^{-nm}[\nabla^m f]_{n-r}(x) < \varepsilon\,[f]_n(x) \tag{9.10.5}$$

for all $n \ge n_0$, and for certain $\varepsilon > 0$ and $r > 0$ that will be chosen later.

The lemma follows if we can prove that there is A such that (9.10.3) holds for $n_0 \le n \le n_j$ for arbitrarily large j. We know by the assumption (9.10.2) that (9.10.3) is true for $n = n_j$ for a certain A, since there is A such that

$$|\nabla^{m-k} f(x)| \le A\,I_k * |\nabla^m f(x)| . \tag{9.10.6}$$

(See the proof of Lemma 3.1.7.)

We prove (9.10.3) for $n_0 \le n < n_j$ by induction. We assume that

$$[f]_n(x) \le M\,2^{-n(m-k)}\big[I_k * |\nabla^m f|\big]_n(x) \tag{9.10.7}$$

for $n = n_j,\ n_j - 1,\ \ldots,\ n_j - \nu + 1$, where M is a constant which is still to be determined. We shall prove that if M is chosen large enough the inequality will be true with the same constant for $n = n_j - \nu$, if $n_j - \nu \ge n_0$. We denote constants independent of the choice of M by A and A_i.

By Corollary 8.1.5 we have for all n

$$[\nabla^{m-k-i}f]_n(x)\le A\,2^{n(m-k-i)}[f]_n(x)+A\,2^{-ni}[\nabla^{m-k}f]_n(x)\ .$$

By (9.10.2) this gives

$$[\nabla^{m-k-i}f]_{n_j}(x)\le A\,2^{-n_j i}[\nabla^{m-k}f]_{n_j}(x) \tag{9.10.8}$$

for $i=1,\dots,m-k$. We claim that (9.10.5) and the induction hypothesis (9.10.7) imply

$$\big[I_k*|\nabla^m f|\big]_n(x)\le \tfrac{3}{2}\big[I_k*|\nabla^m f|\big]_{n-1}(x)\ , \tag{9.10.9}$$

for $n=n_j, n_j-1,\dots,n_j-\nu+1$, if ε and r have been chosen suitably (depending on M).

Assuming this for the moment, we find by (9.10.8) and Proposition 8.1.6 for $n=n_j-1,\dots,n_j-\nu$

$$\begin{aligned}[\nabla^{m-k-1}f]_n(x)&=[\nabla^{m-k-1}f]_{n_j}(x)\\ &\quad+\sum_{\mu=1}^{n_j-n}\big([\nabla^{m-k-1}f]_{n_j-\mu}(x)-[\nabla^{m-k-1}f]_{n_j-\mu+1}(x)\big)\\ &\le A\,2^{-n_j}[\nabla^{m-k}f]_{n_j}(x)+A\sum_{\mu=1}^{n_j-n}2^{-n_j+\mu}[\nabla^{m-k}f]_{n_j-\mu}(x)\ .\end{aligned}$$

By (9.10.6) and (9.10.9) this gives for $n=n_j, n_j-1,\dots,n_j-\nu$

$$\begin{aligned}[\nabla^{m-k-1}f]_n(x)&\le A\,2^{-n_j}\big[I_k*|\nabla^m f|\big]_n(x)\sum_{\mu=0}^{n_j-n}2^{\mu}\big(\tfrac{3}{2}\big)^{n_j-n-\mu}\\ &\le A_1\,2^{-n}\big[I_k*|\nabla^m f|\big]_n(x)\ .\end{aligned} \tag{9.10.10}$$

But then in the same way, by (9.10.8) and Proposition 8.1.6

$$\begin{aligned}[\nabla^{m-k-2}f]_n(x)&=[\nabla^{m-k-2}f]_{n_j}(x)\\ &\quad+\sum_{\mu=1}^{n_j-n}\big([\nabla^{m-k-2}f]_{n_j-\mu}(x)-[\nabla^{m-k-2}f]_{n_j-\mu+1}(x)\big)\\ &\le A\,2^{-2n_j}[\nabla^{m-k}f]_{n_j}(x)\\ &\quad+A\sum_{\mu=1}^{n_j-n}2^{-n_j+\mu}[\nabla^{m-k-1}f]_{n_j-\mu}(x)\ .\end{aligned}$$

By (9.10.8) and (9.10.10)

$$\begin{aligned}[\nabla^{m-k-2}f]_n(x)&\le A\,2^{-2n_j}\big[I_k*|\nabla^m f|\big]_{n_j}(x)\\ &\quad+AA_1\sum_{\mu=1}^{n_j-n}2^{-2(n_j-\mu)}\big[I_k*|\nabla^m f|\big]_{n_j-\mu}(x)\ ,\end{aligned}$$

and by (9.10.9)

$$[\nabla^{m-k-2} f]_n(x) \le AA_1\, 2^{-2n_j}\big[I_k * |\nabla^m f|\big]_n(x) \sum_{\mu=0}^{n_j-n} 2^{2\mu}(\tfrac{3}{2})^{n_j-n-\mu}$$
$$\le A_2 2^{-2n}\big[I_k * |\nabla^m f|\big]_n(x) \ ,$$

and so on for $i = 1, 2, \ldots, m-k$. Thus, after $m-k$ steps

$$[f]_n(x) \le A_k 2^{-(m-k)n}\big[I_k * |\nabla^m f|\big]_n(x) \ ,$$

for $n = n_j, n_j - 1, \ldots, n_j - \nu$ with a constant A_k that does not depend on ν. If we choose $M = A_k$, this proves (9.10.7) for all $n \ge n_0$.

Finally we prove (9.10.9). We assume without loss of generality that $x = 0$, we let $n_j \ge n \ge n_j - \nu + 1$, and we temporarily fix r. We define

$$g_1(y) = |\nabla^m f(y)|, \quad y \in B_{n-r}(0) = B_{n-r} \ ;$$
$$g_1(y) = 0, \qquad y \notin B_{n-r} \ ;$$
$$g_2(y) = |\nabla^m f(y)| - g_1(y) \ .$$

Then

$$\big[I_k * |\nabla^m f|\big]_n(0) \le [I_k * g_1]_n(0) + [I_k * g_2]_n(0) \ .$$

In estimating the first term we can replace $I_k(y)$ by 0 for $|y| > 2^{-n+r+1}$, and extend the integral over all of $\mathbf{R}^N$. Then, using the inequality $\|h * g\|_p \le \|h\|_1 \|g\|_p$ (see (1.1.6)), we find

$$\left(\int_{B_n} (I_k * g_1)^p\, dy\right)^{1/p} \le \|g_1\|_p \int_{B_{n-r-1}} I_k(y)\, dy$$
$$\le A\, 2^{-nk}\left(\int_{B_{n-r}} |\nabla^m f|^p\, dy\right)^{1/p} .$$

Thus

$$\big[I_k * |\nabla^m f|\big]_n(0) \le A_1 2^{-nk}[\nabla^m f]_{n-r}(0) + [I_k * g_2]_n(0) \ , \qquad (9.10.11)$$

where A_1 is a constant depending on r. Similarly

$$\big[I_k * |\nabla^m f|\big]_{n-1}(0) \ge [I_k * g_2]_{n-1}(0) - [I_k * g_1]_{n-1}(0)$$
$$\ge [I_k * g_2]_{n-1}(0) - A_2 2^{-nk}[\nabla^m f]_{n-r}(0) \ . \qquad (9.10.12)$$

But by (9.10.5)

$$2^{-nm}[\nabla^m f]_{n-r}(0) < \varepsilon [f]_n(0) \ .$$

Thus, by the induction hypothesis (9.10.7), i.e.

$$[f]_n(0) \le M\, 2^{-n(m-k)}\big[I_k * |\nabla^m f|\big]_n(0) \ ,$$

we have

$$2^{-nk}[\nabla^m f|_{n-r}(0) \le M\,\varepsilon\left[I_k * |\nabla^m f|\right]_n(0)\ .$$

Hence, by (9.10.11), if ε is small enough,

$$(1 - A_1 M\,\varepsilon)\left[I_k * |\nabla^m f|\right]_n(0) \le [I_k * g_2]_n(0)\ . \tag{9.10.13}$$

On the other hand, by (9.10.12),

$$\left[I_k * |\nabla^m f|\right]_{n-1}(0) \ge [I_k * g_2]_{n-1}(0) - A_2 M\,\varepsilon\left[I_k * |\nabla^m f|\right]_n(0)\ . \tag{9.10.14}$$

We now observe that if $x \in B_{n-1}$ and $y \in B^c_{n-r}$, then

$$(1 - 2^{-r+1})^{N-k} \le \frac{I_k(y)}{I_k(x-y)} \le (1 + 2^{-r+1})^{N-k}\ .$$

Thus, for $x \in B_{n-1}$

$$(1 - 2^{-r+1})^{N-k} I_k * g_2(x) \le I_k * g_2(0) \le (1 + 2^{-r+1})^{N-k} I_k * g_2(x)\ ,$$

so that

$$[I_k * g_2]_n(0) \le \left(\frac{1 + 2^{-r+1}}{1 - 2^{-r+1}}\right)^{N-k} [I_k * g_2]_{n-1}(0)\ .$$

We now choose r so large that

$$\left(\frac{1 + 2^{-r+1}}{1 - 2^{-r+1}}\right)^{N-k} < \frac{4}{3}\ .$$

Then, by (9.10.13) and (9.10.14)

$$\begin{aligned}(1 - A_1 M\,\varepsilon)\left[I_k * |\nabla^m f|\right]_n(0) &\le \tfrac{4}{3}[I_k * g_2]_{n-1}(0)\\ &\le \tfrac{4}{3}\left[I_k * |\nabla^m f|\right]_{n-1}(0) + \tfrac{4}{3} A_2 M\,\varepsilon\left[I_k * |\nabla^m f|\right]_n(0)\ ,\end{aligned}$$

and thus

$$\left[I_k * |\nabla^m f|\right]_n(0) \le \frac{\frac{4}{3}}{1 - A_1 M\,\varepsilon - \frac{4}{3} A_2 M\,\varepsilon}\left[I_k * |\nabla^m f|\right]_{n-1}(0)\ .$$

Choosing ε so small that $1 - A_1 M\,\varepsilon - \frac{4}{3} A_2 M\,\varepsilon > \frac{8}{9}$ we are done. The proof of Lemma 9.10.2 is complete.

Theorem 9.10.1 is now essentially a consequence of Lemma 9.10.2 and the following theorem.

Theorem 9.10.3. *Let $f \in W^{m,p}(\mathbf{R}^N)$, $1 < p < \infty$, and let $K \subset \mathbf{R}^N$ be compact. Suppose that $C_{k,p}(K) = 0$ for some $k = 1, 2, \ldots, m-1$, and suppose that for any $x \in K$ there is an $n_0 = n_0(x)$ such that (9.10.3) holds for all $n \ge n_0$. Then, for any given $\varepsilon > 0$ and any neighborhood V of K, there is a function $\omega \in C_0^\infty(V)$, such that $0 \le \omega \le 1$, $\omega(x) = 1$ on a neighborhood of K, and $\|\omega f\|_{m,p} < \varepsilon$.*

Proof. Part of the proof is similar to the proof of Theorem 9.9.1.

We first assume that n_0 is independent of x, so that (9.10.3) holds uniformly for all $n \geq n_0$.

As in the proof of Theorem 9.9.1 we cover $K^c = \Omega$ by Whitney cubes, $\Omega = \cup Q_{ni}$, where the side of a Q_{ni} is 2^{-n}.

We now assume, without loss of generality, that the support of f is contained in the ball $B_1 = \{x : |x| \leq \frac{1}{2}\}$.

We define I'_j by setting $I'_j(x) = I_j(x)$ on the unit ball and zero outside. It then follows that for $|x| \leq \frac{1}{2}$ and $j = 1, \ldots, m$

$$\begin{aligned} |\nabla^{m-j} f(x)| &\leq A\,(I_j * |\nabla^m f|)(x) \\ &= A\,(I'_j * |\nabla^m f|)(x) \leq A\,(G_j * |\nabla^m f|)(x) \ . \end{aligned} \tag{9.10.15}$$

Denote by U'_λ the union of all Q_{ni} such that

$$Q_{ni} \subset \{x : M((G_k * |\nabla^m f|)^p)(x) > \lambda^p\} \ .$$

By Lemma 6.2.2 there is A such that

$$C_{k,p}(U'_\lambda) \leq \frac{A}{\lambda^p} \|f\|^p_{m,p} \ .$$

We again define U_λ as the union of U'_λ and a (closed) neighborhood U''_λ of K such that

$$C_{k,p}(U''_\lambda) \leq \frac{1}{\lambda^p} \|f\|^p_{m,p} \ ,$$

and we assume that $U''_\lambda \setminus K$ is a union of Whitney cubes.

Then, for any Q_{ni} not in U_λ, we have for any $x_0 \in Q_{ni}$, and a constant A depending only on N,

$$\begin{aligned} \frac{1}{|Q_{ni}|} \int_{Q_{ni}} |\nabla^{m-j} f(x)|^p \, dx &\leq \frac{A}{|Q_{ni}|} \int_{Q_{ni}} (G_k * |\nabla^m f|)^p \, dx \\ &\leq A\, M((G_k * |\nabla^m f|)^p)(x_0) \leq A\, \lambda^p \ . \end{aligned} \tag{9.10.16}$$

Now ω is constructed exactly as in the proof of Theorem 9.9.1 by modifying the (k,p)-capacitary potential for U_λ,

$$\omega = \Phi \circ V^\nu_{k,p} \ ,$$

where Φ is chosen so that so that $\omega(x) = 1$ on $\widetilde{U}_\lambda = \cup(9Q_{ni})$, $Q_{ni} \subset U_\lambda$. Here $V^\nu_{k,p} = G_k * \varphi$, $\varphi = (G_k * \nu)^{p'-1}$, and

$$\|\varphi\|^p_p = C_{k,p}(\widetilde{U}_\lambda) \leq \frac{A}{\lambda^p} \|f\|^p_{m,p} \ . \tag{9.10.17}$$

We estimate $\int_{\mathbf{R}^N} |\nabla^j \omega|^p |\nabla^{m-j} f|^p \, dx$ for $0 \leq j \leq m$.

For $1 \leq j \leq k$ we proceed exactly as in the proof of Theorem 9.9.1. Note that because of the assumption that $C_{k,p}(K) = 0$ we do not have any information about $\nabla^{m-j} f(x)$ near K for $j \le k$. But it is exactly this assumption that will give us the desired estimate.

Let Q_{ni} be a Whitney cube where $\nabla \omega$ is not identically zero, and estimate $\int_{Q_{ni}} |\nabla^j \omega|^p |\nabla^{m-j} f|^p \, dx$. If x_{ni} is the center of Q_{ni}, we obtain for $1 \le j < k$, using Lemma 9.8.5, (9.10.15), Proposition 3.1.2(b), Hölder's inequality, and (9.10.16)

$$\begin{aligned}
&\int_{Q_{ni}} |\nabla^j \omega|^p |\nabla^{m-j} f|^p \, dx \\
&\quad \le A\, \eta(x_{ni})^{pj/k} \int_{Q_{ni}} (I_j * |\nabla^m f|)^p \, dx \\
&\quad \le A\, \eta(x_{ni})^{pj/k} \int_{Q_{ni}} (M(\nabla^m f))^{(k-j)p/k} (G_k * |\nabla^m f|)^{pj/k} \, dx \\
&\quad \le A\, (\eta(x_{ni})^p |Q_{ni}|)^{j/k} \left(\int_{Q_{ni}} (M(\nabla^m f))^p \, dx \right)^{(k-j)/k} \\
&\qquad \times \left(\frac{1}{|Q_{ni}|} \int_{Q_{ni}} (G_k * |\nabla^m f|)^p \, dx \right)^{j/k} \\
&\quad \le A \left(\int_{Q_{ni}} \eta^p \, dx \right)^{j/k} \left(\int_{Q_{ni}} (M(\nabla^m f))^p \, dx \right)^{(k-j)/k} \lambda^{pj/k} \, .
\end{aligned}$$

By summing over cubes, and by using Hölder's inequality for sums, Lemma 9.8.5, and (9.10.17) we find

$$\begin{aligned}
&\int_{\mathbf{R}^N} |\nabla^j \omega|^p |\nabla^{m-j} f|^p \, dx \\
&\qquad \le A \left(\int_{\mathbf{R}^N} \eta^p \, dx \right)^{j/k} \left(\int_{V_\lambda} (M(\nabla^m f))^p \, dx \right)^{(k-j)/k} \lambda^{pj/k} \\
&\qquad \le A\, \|f\|_{m,p}^{pj/k} \left(\int_{V_\lambda} (M(\nabla^m f))^p \, dx \right)^{(k-j)/k} .
\end{aligned}$$

But again

$$\int_{\mathbf{R}^N} (M(\nabla^m f))^p \, dx \le A \int_{\mathbf{R}^N} |\nabla^m f|^p \, dx \le A\, \|f\|_{m,p}^p \, ,$$

and V_λ has arbitrarily small measure, so for $j < k$ the integral to be estimated tends to zero as $\lambda \to \infty$.

For $j = k$

$$\begin{aligned}
&\int_{Q_{ni}} |\nabla^k \omega|^p |\nabla^{m-k} f|^p \, dx \\
&\qquad \le A\, \eta(x_{ni})^p |Q_{ni}| \frac{1}{|Q_{ni}|} \int_{Q_{ni}} |\nabla^{m-k} f|^p \, dx \le A\, \lambda^p \int_{Q_{ni}} \eta^p \, dx
\end{aligned}$$

by (9.10.16), so by (9.10.17)

$$\int_{V_\lambda} |\nabla^k \omega|^p |\nabla^{m-k} f|^p \, dx \le A\,\lambda^p \int_{\mathbf{R}^N} \eta^p \, dx \le A \, \|f\|_{m,p}^p \ .$$

Finally, for $k < j \le m$ we have to use (9.10.3). Consider a Whitney cube Q_{ni}, and let $x_0 \in K$ be a point with $\text{dist}(x_0, Q_{ni}) \le 4 \,\text{diam}\, Q_{ni}$. Then for any $x_0 \in Q_{ni}$

$$Q_{ni} \subset B(x_0, 5N^{1/2}2^{-n}) \subset B(x, 10N^{1/2}2^{-n}) \ ,$$

so if r is chosen so that $5N^{1/2} \le 2^r$, by (9.10.3)

$$\int_{Q_{ni}} |f|^p \, dy \le \int_{B_{n-r}(x_0)} |f|^p \, dy$$
$$\le A\,2^{-n(m-k)p} \int_{B_{n-r-1}(x)} (I_k * |\nabla^m f|)^p \, dy \ ,$$

if $n - r \ge n_0$.

Thus, if $Q_{ni} \not\subset U_\lambda$, so that x can be chosen in $Q_{ni} \setminus U_\lambda$, by (9.10.16)

$$\frac{1}{|Q_{ni}|} \int_{Q_{ni}} |f|^p \, dy \le A\,2^{-n(m-k)p} \lambda^p \ .$$

As before, using Corollary 8.1.5, it follows that

$$\int_{Q_{ni}} |\nabla^{m-j} f|^p \, dy$$
$$\le A\,2^{n(m-j)p} \left(\int_{Q_{ni}} |f|^p \, dy + A\,2^{-n(m-k)p} \int_{Q_{ni}} |\nabla^{m-k} f|^p \, dy \right)$$
$$\le A\,2^{n(k-j)p} \lambda^p |Q_{ni}| , \quad k < j \le m \ . \tag{9.10.18}$$

Applying this and Lemma 9.8.5(c), we obtain for $k < j \le m$

$$\int_{Q_{ni}} |\nabla^j \omega|^p |\nabla^{m-j} f|^p \, dy \le A\,\eta(x_{ni})^p 2^{-n(k-j)p} \int_{Q_{ni}} |\nabla^{m-j} f|^p \, dy$$
$$\le A\,\eta(x_{ni})^p |Q_{ni}| \lambda^p \ ,$$

whence

$$\int_{\mathbf{R}^N} |\nabla^j \omega|^p |\nabla^{m-j} f|^p \, dy \le A\,\lambda^p \int_{\mathbf{R}^N} \eta^p \, dy \le A \, \|f\|_{m,p}^p \ .$$

Thus,

$$\|\omega f\|_{m,p} \le A \, \|f\|_{m,p} \ ,$$

independently of λ, as $\lambda \to \infty$.

As in the proof of Theorem 9.9.1, it follows from weak compactness and the Mazur lemma that there are ω such that $\|\omega f\|_{m,p}$ is arbitrarily small, which finishes the proof under the assumption that (9.10.3) holds uniformly for $n \le n_0$.

In order to complete the proof of Theorem 9.10.1 we have to remove this hypothesis. We set

$$E_M = \{x \in K : \text{ (9.10.3) holds for all } n \ge M\}.$$

Then E_M is clearly closed, $E_M \subset E_{M+1}$, and by assumption $K = \bigcup_{M=M_0}^{\infty} E_M$ for any M_0.

If M is large enough, we can, by applying the above construction to E_M, construct a function ω_M, such that

(a) $\omega_M \in C_0^\infty(V)$,
(b) $\omega_M(x) = 1$ on a neighborhood G_M of E_M, and
(c) $\|\omega_M f\|_{m,p} < \frac{\varepsilon}{2}$,

where we assume that $0 < \varepsilon < \|f\|_{m,p}$.

Set $f_M = f(1 - \omega_M)$, so that $|f_M(x)| \le |f(x)|$, and $\|f - f_M\|_{m,p} < \frac{\varepsilon}{2}$. Then $\|f_M\|_{m,p} \le 2\|f\|_{m,p}$, $f_M(x) = 0$ on G_r, and for $x \in E_{M+1}$ we have by (9.10.3)

$$[f_M]_n(x) \le [f]_n(x) \le A\, 2^{-n(m-k)} \big[I_k * |\nabla^m f|\big]_n(x) \ , \qquad (9.10.19)$$

for $n \ge r + 1$. We want to construct ω_{M+1} as in the first part of the proof, so that

$$\|\omega_{M+1} f_M\|_{m,p} < \varepsilon 2^{-2} \ ,$$

but we have to make a small modification of the proof because of the fact that (9.10.3) now is replaced by (9.10.19).

Thus, instead of (9.10.18) we now obtain for an $x \in Q_{ni} \setminus U_\lambda$, using Corollary 8.1.5 and (9.10.19)

$$\begin{aligned} &\int_{Q_{ni}} |\nabla^{m-j} f_M|^p \, dy \\ &\quad \le A\, 2^{n(m-j)p} \left(\int_{Q_{ni}} |f_M|^p \, dy + A\, 2^{-n(m-k)p} \int_{Q_{ni}} |\nabla^{m-k} f_M|^p \, dy \right) \\ &\quad \le A\, 2^{-n(j-k)p} \left(\int_{B_{n-r-1}(x)} |I_k * |\nabla^m f|^p \, dy + \int_{Q_{ni}} |\nabla^{m-k} f_M|^p \, dy \right) , \end{aligned} \qquad (9.10.20)$$

for $k < j \le m$. We modify the definition of U_λ by defining U'_λ as the union of all Q_{ni} such that

$$Q_{ni} \subset \left\{ x : M((G_k * |\nabla^m f|)^p)(x) + M((G_k * |\nabla^m f_M|)^p)(x) > \lambda^p \right\} .$$

Then we still have (9.10.17), and (9.10.20) gives

$$\int_{Q_{ni}} |\nabla^{m-j} f_M|^p \, dy \le A\, 2^{-n(j-k)p} \lambda^p |Q_{ni}| , \quad k < j \le m \ .$$

After this change the proof works as before, and we construct $\omega_{M+1} \in C_0^\infty(V)$ so that $\omega_{M+1}(x) = 1$ on a neighborhood G_{M+1} of E_{M+1}, and

$$\|\omega_{M+1} f_M\|_{m,p} < \varepsilon 2^{-2} \ .$$

We set $f_{M+1} = f_M(1 - \omega_{M+1})$, so that $f_{M+1}(x) = 0$ on $G_M \cup G_{M+1}$, and

$$\|f_M - f_{M+1}\|_{m,p} < \varepsilon 2^{-2} \ .$$

Proceeding inductively, we construct f_{M+j} so that

$$\|f_M - f_{M+j}\|_{m,p} < \varepsilon \sum_{i=1}^{j+1} 2^{-i} < \varepsilon \ ,$$

and $f_{M+j}(x) = 0$ on $\bigcup_{i=0}^{j} G_{M+j}$, $G_{M+j} \supset E_{M+j}$. But K is compact, so $K \subset \bigcup_{i=0}^{j} G_{M+j}$, if j is large enough. This proves Theorem 9.10.3 if we set $\omega = 1 - (1-\omega_M)\cdots(1-\omega_{M+j})$.

We shall now by means of Lemma 9.10.2 combine Theorem 9.10.3 and Theorem 9.5.1 in order to prove Theorem 9.10.1.

Fix a large number M, and set for any $P \geq M$

$$E_P = \left\{ x \in K : \sum_{n=M}^{P} \left(\frac{2^{-nm}[\nabla^m f]_n(x)}{[f]_n(x)} \right)^{p'} > 1 \right\} .$$

Then, if M is large enough, there is by Theorem 9.5.1, and the remark following the proof of Lemma 9.5.2 (Section 9.7), for each $P \geq M$ a function $\omega_P \in C_0^\infty(V)$ such that $\omega_P(x) = 1$ on a neighborhood G_P of $\overline{E}_P$, and

$$\|\omega_P f\|_{m,p}^p \leq A \int_V |\nabla^m f|^p \, dx < \varepsilon^p \ .$$

Moreover, we can assume that $G_P \subset G_{P+1}$.

We observe that E_P is open in the relative topology on K, since $[f]_n(x)$ and $[\nabla^m f]_n(x)$ are continuous, and $[f]_n(x) \neq 0$, $M \leq n \leq P$. Thus $\bigcup_M^\infty E_P$ is also open, so that $K \setminus \left(\bigcup_M^\infty E_P\right) = K'$ is compact.

By weak compactness a subsequence of $\{(1-\omega_P)f\}_{P=M}^\infty$ converges weakly in $W^{m,p}$ to a function f_0 with $\|f - f_0\|_{m,p} < \varepsilon$.

By the Mazur lemma f_0 is the strong limit of a sequence of averages of $(1-\omega_P)f$. We can write these averages as $(1-\widetilde{\omega}_{P_i})f$, where $\widetilde{\omega}_{P_i} = \sum_{j=i} \alpha_j \omega_{P_j}$. It follows from the monotonicity of $\{G_P\}$ that $\widetilde{\omega}_{P_i}(x) = 1$ on G_{P_i}.

Assumption (b) in the theorem, and Lemma 9.10.2 imply that f satisfies (9.10.3) for all $x \in K'$. Moreover, $[f_0]_n(x) \leq [f]_n(x)$, so (9.10.20) follows for f_0. Thus, as before we can construct a function $\omega' \in C_0^\infty(V)$ such that $\omega'(x) = 1$ on a neighborhood G' of K', and $\|\omega' f_0\|_{m,p} < \varepsilon$. Thus $(1-\omega')f_0 = 0$ on G'. But

$$\lim_{i\to\infty} \left\| f_0 - (1-\widetilde{\omega}_{P_i})f \right\|_{m,p} = 0 \ ,$$

so

$$\left\|(1-\omega')f_0 - (1-\omega')(1-\widetilde{\omega}_{P_i})f\right\|_{m,p} \leq A \max_{|\beta|\leq m} |D^\beta(1-\omega')| \left\| f_0 - (1-\widetilde{\omega}_{P_i})f \right\|_{m,p} < \varepsilon \ ,$$

if i is large enough. But $(1-\omega')(1-\widetilde{\omega}_{P_i})=0$ on $G'\cup G_{P_i}$, $G_{P_i}\subset G_{P_{i+1}}$, and $K\subset G'\bigcup(\bigcup_i G_{P_i})$. By compactness $K\subset G'\cup G_{P_i}$ if i is large enough.

Now set $\omega=1-(1-\omega')(1-\widetilde{\omega}_{P_i})$. Then

$$\begin{aligned}\|\omega f\|_{m,p} &= \left\|f-(1-\omega')(1-\widetilde{\omega}_{P_i})f\right\|_{m,p}\\ &\le \|f-f_0\|_{m,p}+\|f_0-(1-\omega')f_0\|_{m,p}\\ &\quad+\left\|(1-\omega')f_0-(1-\omega')(1-\widetilde{\omega}_{P_i})f\right\|_{m,p}<3\varepsilon\ ,\end{aligned}$$

so ω satisfies all the requirements of the theorem. Theorem 9.10.1 is proved.

9.11 Conclusion of the Proof

In order to prove Theorem 9.1.3 we now only need to modify the proof of Theorem 9.10.1 slightly.

Suppose that $f\in W^{m,p}(\mathbf{R}^N)$ and that $D^\beta f|_K=0$ for all β, $0\le|\beta|\le m-1$. Let $\varepsilon\ge 0$, and fix a large number M. We now set

$$E_{0P}=\left\{x\in K:1<\sum_{n=M}^{P}\left(\frac{2^{-nm}[\nabla^m f]_n(x)}{[f]_n(x)}\right)^{p'}<\infty\right\}\ ,$$

and

$$E_0=\bigcup_{P=M}^{\infty}E_{0P}\ .$$

Again E_0 is open, and $K\setminus E_0$ is compact.

By repeated applications of the Kellogg property (Corollary 6.3.17) and Theorem 9.10.3 we find that there is a set $E_m\subset K$ with $C_{m,p}(E_m)=0$ such that $K\setminus(E_0\cup E_m)$ is the union of compact sets E_{kP}, $k=1,\ldots,m-1$; $P=M$, $M+1,\ldots$, with the properties

$$\begin{gathered}C_{k,p}(E_{kP})=0\quad\text{for all } P;\\ [f]_n(x)\le A\,2^{-n(m-k)}\left[I_k*|\nabla^m f|\right]_n(x)\end{gathered}$$

for all $x\in E_{kP}$ and all $n\ge P$, i.e. (9.10.3) is satisfied on E_{kP}.

Let $\varepsilon>0$. By the previous results, for every P there is $\omega_P\in C_0^\infty(V)$ such that $0\le\omega_P\le 1$, $\omega_P(x)=1$ on a neighborhood G_P of $\bigcup_{k=0}^{m-1}E_{kP}$, and $\|(1-\omega_P)f\|_{m,p}<\varepsilon$. Again the G_P can be chosen so that they form an increasing sequence. There is a sequence of averages of $(1-\omega_P)f$, again denoted $(1-\widetilde{\omega}_{P_i})f$, that converges strongly to a function f_0. Then $(1-\widetilde{\omega}_{P_i})f=0$ on G_{P_i}, and thus f_0 vanishes on $\bigcup_1^\infty G_{P_i}$. But $K\setminus(\bigcup_1^\infty G_{P_i})=K'$ is compact, and $C_{m,p}(K')=0$. By Theorem 9.9.1 there is $\omega'\in C_0^\infty(V)$ such that $\omega'(x)=1$ on a neighborhood G' of K', and $\|\omega' f_0\|_{M,p}<\varepsilon$. By the compactness of $K\setminus G'$, $(1-\widetilde{\omega}_{P_i})(1-\omega')f$ vanishes on a neighborhood of K if i is large enough.

As in the proof of Theorem 9.10.1, $\omega=1-(1-\widetilde{\omega}_{P_i})(1-\omega')$ satisfies all the requirements of Theorem 9.1.3 if i is large enough.

9.12 Further Results

9.12.1. It is a restriction in the formulation of Corollary 9.1.8 that the boundary data are given by a function g that has to belong to $W^{m,2}$ in all of $\mathbf{R}^N$, or at least across $\partial\Omega$. This can be avoided by formulating a "fine Dirichlet problem", and the result is the following (T. Kolsrud [256]): Suppose that $\Omega \subset \mathbf{R}^N$ is a bounded open set, and let $g \in W^{m,2}(\Omega)$. Then there is a unique $u \in W^{m,2}(\Omega)$ such that $\Delta^m u = 0$, and such that for $(m - |\beta|, 2)$-q.e. $x_0 \in \partial\Omega$

$$\lim_{x \to x_0, x \in \Omega} D^\beta(u(x) - g(x)) = 0 \quad \text{for all } \beta,\ 0 \le |\beta| \le m - 1 \ ,$$

where the limit is understood in the sense of the $(m - |\beta|, 2)$-fine topology. (See Definition 6.4.1.)

In fact, a function f in $W^{m,p}(\Omega)$, $p > 1$, belongs to $W_0^{m,p}(\Omega)$ if and only if for $(m - |\beta|, p)$-q.e. $x_0 \in \partial\Omega$

$$\lim_{x \to x_0, x \in \Omega} D^\beta f(x) = 0 \quad \text{for all } \beta,\ 0 \le |\beta| \le m - 1 \ ,$$

with the limit in the $(m - |\beta|, p)$-fine sense.

This characterization of $W_0^{1,2}(\Omega)$ is due to J. Deny and J. L. Lions [122], Théorème 5.1, who deduced it from a uniqueness theorem for harmonic functions of M. Brelot [75]. See also B. Fuglede [168], Theorem 9.1. Here the procedure is the opposite one; the uniqueness theorem is proved using a characterization of $W_0^{m,p}(\Omega)$.

9.13 Notes

Theorem 9.1.1 and its corollary are well known. They are found, together with various extensions, in Hörmander's book [230], Theorem 2.3.3.

The origin of Theorem 9.1.3 goes back at least to S. L. Sobolev's fundamental paper [383], which appeared in 1937, but was translated into English only in 1963. See also his book [385] from 1950 (also translated in 1963), §§14–15. Generalizing earlier results of K. Friedrichs [157], Sobolev studied this approximation problem in connection with the uniqueness problem for solutions of the Dirichlet problem for the polyharmonic equation in the case of a domain Ω which is bounded by a finite union of smooth manifolds of arbitrary codimension. See Theorem 9.1.9 and the discussion in Section 9.1 above. The example following Theorem 9.1.9 also comes from [385].

The simple Lemma 9.1.10 is probably well known. It appears in print in L. I. Hedberg [214] and T. Kolsrud [257]. Kolsrud gave a characterization of the sets removable for $W^{1,p}$ in the sense of the lemma (references to earlier work are given in his paper), but it seems to be an open problem to characterize the removable sets for $W^{m,p}$ for $m > 1$.

Theorem 9.1.3 (and Corollary 9.1.4) is easy and well known in the case when Ω is bounded by a hypersurface satisfying the so called segment property. See e.g.

the books by G. B. Folland [150], Proposition 6.46, p. 278, and F. Treves [401], Theorem 26.9, p. 244.

Generalizing a theorem of A. Beurling [59] (Theorem 2, p. 28), J. Deny [119] proved Theorem 9.1.3 (in the form of the equivalent Corollary 9.1.7) for a class of Hilbert spaces containing $W^{1,2}$ and $L^{\alpha,2}(\mathbf{R}^N)$, $0 < \alpha \le 1$. See [119], Théorème II:2, p. 143, and also pp. 144–145. Deny's proof depends on the maximum principle, which is satisfied for the Riesz and Bessel kernels $I_{2\alpha}$ and $G_{2\alpha}$ for $\alpha \le 1$. See also the notes to Section 2.7 at the end of Chapter 2.

The result was extended to abstract "Dirichlet spaces" by Beurling and Deny in [62]. See also Deny [121], p. 168 and p. 172. For $W^{1,p}$, $1 < p < \infty$, Theorem 9.1.3 is due to V. P. Havin [200] and T. Bagby [43]. The proof given in Section 9.2 comes from L. I. Hedberg [207]. All these proofs depend on the fact that the spaces are closed under truncation. See also Beurling [60].

Approximation results such as those of Beurling and Deny are called "spectral synthesis" theorems. The reason for this is briefly the following. The spectrum of a locally integrable function, or of a distribution f in $\mathcal{S}'$, is defined as the support of its Fourier transform $\widehat{f}$. (Beurling's definition was of course different.) The problem of spectral synthesis is whether a given f, belonging to some space, can be approximated in the sense of this space by elements that are Fourier transforms of measures, whose support is contained in the spectrum of f. The above result for $L^{\alpha,2}$ shows that this is true for all elements in the space $L^2(w, \mathbf{R}^N)$, where w is the weight $w(\xi) = \widehat{G}_{2\alpha}(\xi) = (1 + |\xi|^2)^{-\alpha}$, $0 < \alpha \le 1$.

The spectral synthesis problem for weak* L^∞ can be formulated in the following way: Let $A(\mathbf{R}^N)$ be the Banach space (Banach algebra) of functions f normed by the L^1-norm of their Fourier transforms, $\|f\|_A = \|\widehat{f}\|_1$. Can every $f \in A(\mathbf{R}^N)$ that vanishes on a set K be approximated by functions that vanish on neighborhoods of K? As is well known, counterexamples were given by L. Schwartz [372] for $N \ge 3$ and by P. Malliavin [286] in the general case. See also e.g. Y. Katznelson [241].

Corollary 9.1.5 is also well known for $m = 1$. See D. Kinderlehrer and G. Stampacchia [251], Chapter II, Exercise 17, p. 81. In the general case we know of no direct proof of this natural-looking result.

The general formulation of the uniqueness problem given in Corollary 9.1.8 is due to B. Fuglede (unpublished, but see B.-W. Schulze and G. Wildenhain [371], Ch. IX, §5.1), who also pointed out the equivalence of this problem with the approximation problem in $W^{m,2}$. See also L. I. Hedberg [214], and G. Wildenhain [435] for discussions of different formulations of the Dirichlet problem for higher order equations.

Fuglede used the term "$2m$-spectral synthesis", and he was followed by Hedberg [213] and others. Here we have not used this terminology, because experience has shown that it can lead to confusion for somebody unfamiliar with the spectral synthesis problem in harmonic analysis.

Theorem 9.1.3 was proved in special cases by V. I. Burenkov [86], J. C. Polking [359], and L. I. Hedberg [210]. Polking and Hedberg were led to this problem by an interest in an approximation problem for harmonic functions (see

Chapter 11). The theorem was proved by Hedberg in [212] and [213] under the restriction $p > 2 - \frac{1}{N}$. As was pointed out in [212] (p. 74) and [213] (p. 242), the reason for this restriction was that the Kellogg property was only known to be true for (α, p)-capacities for $p > 2 - \frac{\alpha}{N}$. P. W. Jones drew the attention of Th. H. Wolff to this problem, and Wolff's solution of it was published in the joint paper by Hedberg and Wolff [219].

Corollary 9.1.11 is due to H. Brezis and F. E. Browder [82]. See the notes to Chapter 3.

The idea of constructing a weight function like the function ω used in the proof of Theorem 9.1.3 goes back to L. V. Ahlfors [27] (the "Ahlfors mollifier"). See also L. Bers [52] and [53]. It was developed and used by Hedberg in [206] and [210].

The Harnack property, Lemma 9.8.1, was proved by D. R. Adams and N. G. Meyers [22].

10. Two Theorems of Netrusov

In this chapter we apply the powerful "smooth atomic" method of representing elements in function spaces that was exposed in Chapter 4. In Section 10.1 we give the generalization of Theorem 9.1.3 that was announced in the introduction to Chapter 9. In Section 10.2 similar methods are used to extend H. Whitney's classic characterization of closed ideals of differentiable functions.

10.1 An Approximation Theorem, Another Approach

Theorem 9.1.3 will be extended to the spaces $L^{\alpha,p}(\mathbf{R}^N) = L^{\alpha,p}$ for $\alpha > 0$, and $1 < p < \infty$ (see also the notes at the end of the chapter), and to arbitrary sets E instead of closed sets K. For any set E we denote by $L_0^{\alpha,p}(E)$ the closure in $L^{\alpha,p}$ of the functions in $L^{\alpha,p}$ with compact support contained in E. If G is open, then clearly $L_0^{\alpha,p}(G)$ is the closure of $C_0^\infty(G)$.

As before, functions f in $L^{\alpha,p}$, and their derivatives $D^\beta f$ for $|\beta| < \alpha$, can be assumed to be defined pointwise $(\alpha - |\beta|, p)$-q.e., and to be $(\alpha - |\beta|, p)$-quasicontinuous. We can then formulate the result.

Theorem 10.1.1 (Yu. V. Netrusov). *Let $\alpha > 0$, let $1 < p < \infty$, and let $f \in L^{\alpha,p}(\mathbf{R}^N)$. Let $E \subset \mathbf{R}^N$ be an arbitrary set. Then the following statements are equivalent*:

(a) $D^\beta f(x) = 0$ *$(\alpha - |\beta|, p)$-q.e. on E for all multiindices β, $0 \le |\beta| < \alpha$;*
(b) $f \in L_0^{\alpha,p}(E^c)$;
(c) *for any $\varepsilon > 0$ there is a function η such that $\eta = 0$ on a neighborhood of E, $0 \le \eta \le 1$, and $\|f - \eta f\|_{\alpha,p} < \varepsilon$.*

With the notation defined in (9.1.1) we have the following corollary.

Corollary 10.1.2. *Under the assumptions of the theorem*

$$\operatorname{Ker} \operatorname{Tr}_E = L_0^{\alpha,p}(E^c)\ .$$

The implications (c) ⇒ (b) ⇒ (a) are easy, and proved as in Chapter 9. The proof of the implication (a) ⇒ (c) will occupy the rest of this section.

The first step in the proof is the identification of $L^{\alpha,p}$ with the Lizorkin–Triebel space $F_\alpha^{p,2}$ (Theorem 4.2.2). In addition, the proof depends on the representation of functions by means of smooth atoms in Section 4.6. As usual when working in the Lizorkin–Triebel spaces, the Fefferman–Stein maximal theorem, Theorem 1.1.2, is a basic tool.

As in the proof of Theorem 9.1.3 the main difficulty is the construction of a multiplier η. The main difference between the proof given below and the proof of the corresponding result in Chapter 9 is that in this chapter we use properties of the function f more efficiently, whereas in Chapter 9 the proof depended more on properties of the set K. This makes for a considerable simplification of the proof. In particular, the elementary Lemma 10.1.8 below takes the place of the Wiener Criterion and the Kellogg property.

We proved in Theorem 6.2.1 that for a function f in $L^{\alpha,p}$ (α, p)-quasievery point is a Lebesgue point. We shall now reprove and extend this result, using the atomic representation.

Definition 10.1.3. If f is an integrable function we say that a polynomial π_s of degree at most s is a *differential of order* s to f at a point x if

$$\lim_{r\to 0}\frac{1}{r^{N+s}}\int_{B(x,r)}|f(y)-\pi_s(y)|\,dy=0\,.$$

It is easily seen that such a polynomial is unique, and thus, if $f\in C^s$ it has to be the Taylor polynomial,

$$P_x^s f(y)=\sum_{|\beta|\le s}D^\beta f(x)\frac{(y-x)^\beta}{\beta!}\,.$$

To say that f has a differential of order 0 at x is just another way of saying that x is a Lebesgue point for f.

Dyadic cubes with side 2^{-n} are denoted Q_{nk}, $k\in\mathbf{Z}^N$, and we write $\widetilde{Q}_{nk}$ for $3Q_{nk}$. Their characteristic functions are denoted χ_{nk} and $\widetilde{\chi}_{nk}$, respectively.

By Theorems 4.2.2 and 4.6.2, a function f in $L^{\alpha,p}$, $1<p<\infty$, $\alpha>0$, can be represented as

$$f=\sum_{n=0}^{\infty}f_n\ ,\ \text{ where } f_n\in C^\infty,\ \text{ and } \{2^{n\alpha}f_n\}_0^\infty\in L^p(l^2)\,. \tag{10.1.1}$$

Moreover, the f_n are such that if g_n is defined by

$$g_n(x)=\sum_{k\in\mathbf{Z}^N}s_{nk}\chi_{nk}(x)\ , \tag{10.1.2}$$

where

$$s_{nk}=\max\left\{2^{-n|\beta|}|D^\beta f_n(x)|:x\in\widetilde{Q}_{nk}\,,\ |\beta|\le S\right\} \tag{10.1.3}$$

with $S>\alpha$, then

$$\left\|\{2^{n\alpha}g_n\}_0^\infty\right\|_{L^p(l^2)}\le A\,\|f\|_{\alpha,p}\ , \tag{10.1.4}$$

where A is independent of f. Conversely, by Theorem 4.6.3, every f that can be so represented belongs to $L^{\alpha,p}$.

Proposition 4.7.2 gives that

$$\sum_{n=0}^{\infty} |f_n(x)| < \infty \quad (\alpha, p)\text{-q.e.} \tag{10.1.5}$$

It follows that if the value of f at the points of absolute convergence is defined by

$$f(x) = \sum_{n=0}^{\infty} f_n(x) \ , \tag{10.1.6}$$

then f is (α, p)-quasicontinuous. In fact,

$$C_{\alpha,p}\big(\{x : |f(x) - \textstyle\sum_{m=0}^{n} f_m(x)| > \lambda\}\big) \leq A\,\lambda^{-p} \big\| \{2^{n\alpha} f_m\}_{n+1}^{\infty} \big\|_{L^p(l^2)} \ ,$$

which easily implies quasicontinuity. Theorem 6.2.1, and the uniqueness of quasicontinuous representatives (Theorem 6.1.4) imply that for (α, p)-q.e. x such that (10.1.5) and (10.1.6) are satisfied, $f(x)$ is the value of f at x in the sense of Lebesgue, i.e.,

$$\lim_{r \to 0} \frac{1}{r^N} \int_{B(x,r)} |f(y) - f(x)|\, dy = 0 \ . \tag{10.1.7}$$

Similarly, for any multiindex β with $|\beta| < \alpha$, the sum

$$D^\beta f(x) = \sum_{n=0}^{\infty} D^\beta f_n(x)$$

is absolutely convergent $(\alpha - |\beta|, p)$-q.e., and defines an $(\alpha - |\beta|, p)$-quasicontinuous representative of the derivative $D^\beta f$. It follows that for any $s < \alpha$, the Taylor polynomial $P_x^s f$ is well defined for x off a set with zero $(\alpha - s, p)$-capacity. If $\mathrm{Tr}_E\, f = 0$ then $P_x^s f = 0$ for $(\alpha - s, p)$-q.e. $x \in E$ for all s, $0 \leq s < \alpha$.

The following theorem makes these ideas much more precise.

Theorem 10.1.4. *Let $f \in L^{\alpha,p}$, $1 < p < \infty$, $\alpha > 0$, and suppose that $\{f_n\}_0^\infty$ and $\{g_n\}_0^\infty$ are functions such that the conditions* (10.1.1) – (10.1.4) *are satisfied. Let s be an integer, $s < \alpha$, and let $P_x^s f_n$ be the Taylor polynomial of degree s of f_n at x. Then there is a set F_s with $C_{\alpha-s,p}(F_s) = 0$, such that for every $x \in F_s^c$:*

(a) $\sum_{n=0}^{\infty} |D^\beta f_n(x)| < \infty$ *for* $|\beta| \leq s$;
(b) *f has a differential of order s at x;*
(c) $D^\beta f(x) = \sum_{n=0}^{\infty} D^\beta f_n(x)$ *for* $|\beta| \leq s$;
(d) *The differential is* $P_x^s f(y) = \sum_{|\beta| \leq s} D^\beta f(x)(y - x)^\beta / \beta!$.

In what follows, when we write $f(x)$ for an $f \in L^{\alpha,p}$, we will always tacitly assume that x is a Lebesgue point for f, and that $f(x)$ is given by (10.1.7).

The proof of the theorem depends on the following basic lemma.

Lemma 10.1.5. *Let* $g_n(x) = \sum_{k\in\mathbf{Z}^N} s_{nk}\chi_{nk}(x)$, $s_{nk} \geq 0$, *for* $n \in \mathbf{N}$, *and suppose that* $0 < \left\|\{2^{n\alpha}g_n\}_0^\infty\right\|_{L^p(l^2)} < \infty$. *Then there exist* h_n, $n \in \mathbf{N}$, *with the following properties*:

(a) $h_n(x) = \sum_{k\in\mathbf{Z}^N} t_{nk}\chi_{nk}(x)\,, \quad t_{nk} \geq 0\,;$
(b) $g_n \leq h_n\,;$
(c) *For all* Q_{nk}

$$2^{nN}\int_{\widetilde{Q}_{nk}}\left(\sum_{m=n}^{\infty} g_m\right)dx \leq t_{nk}\ ;$$

(d) *There is a constant* $A > 0$ *independent of* $\{g_n\}_0^\infty$ *such that*

$$\left\|\{2^{n\alpha}h_n\}_0^\infty\right\|_{L^p(l^2)} \leq A\left\|\{2^{n\alpha}g_n\}_0^\infty\right\|_{L^p(l^2)}\ ;$$

(e) $\sum_{n=0}^\infty 2^{n\alpha}h_n(x) = \infty$ *for all* $x\,;$
(f) $t_{nk} \geq 2^{-N-2}t_{nk'}$ *if* $|k-k'| = 1$, $n \in \mathbf{N}$.

Proof. Set

$$t_{nk} = 2^{nN}\int_{\widetilde{Q}_{nk}}\left(\sum_{m=n}^{\infty} g_m\right)dx\ .$$

Then (b) and (c) are trivially satisfied.

In order to prove (d) we observe that for $x \in Q_{nk}$, by the definition of the Hardy–Littlewood maximal operator M,

$$h_n(x) = t_{nk} \leq A\sum_{m=n}^{\infty} Mg_m(x)\ .$$

We want to estimate

$$\sum_{n=0}^{\infty}\left(2^{n\alpha}\sum_{m=n}^{\infty} Mg_m(x)\right)^2\,,$$

under the assumption that

$$\sum_{n=0}^{\infty}\left(2^{n\alpha}Mg_n(x)\right)^2 < \infty\ .$$

We proceed as in the proof of Theorem 4.1.4, and choose an ι with $0 < \iota < \alpha$. Denote $Mg_m(x)$ by c_m. By Cauchy's inequality

$$\left(\sum_{m=n}^{\infty} c_m\right)^2 \leq \sum_{m=n}^{\infty} 2^{2m\iota}c_m^2\sum_{m=n}^{\infty} 2^{-2m\iota} \leq A\,2^{-2n\iota}\sum_{m=n}^{\infty} 2^{2m\iota}c_m^2\ ,$$

and it follows that

$$\sum_{n=0}^{\infty}\left(2^{n\alpha}\sum_{m=n}^{\infty}c_m\right)^2 \le A\sum_{n=0}^{\infty}2^{2n(\alpha-\iota)}\sum_{m=n}^{\infty}2^{2m\iota}c_m^2$$

$$= A\sum_{m=n}^{\infty}2^{2m\iota}c_m^2\sum_{n=0}^{m}2^{2n(\alpha-\iota)}$$

$$\le A\sum_{m=0}^{\infty}2^{2m\iota}c_m^2 2^{2m(\alpha-\iota)} = A\left\|\{2^{m\alpha}c_m\}_0^{\infty}\right\|_{l^2}^2 .$$

Thus, by the Fefferman–Stein theorem (Theorem 1.1.2)

$$\left\|\{2^{n\alpha}h_n\}_0^{\infty}\right\|_{L^p(l^2)} \le A\left\|\{2^{n\alpha}Mg_n\}_0^{\infty}\right\|_{L^p(l^2)} \le A\left\|\{2^{n\alpha}g_n\}_0^{\infty}\right\|_{L^p(l^2)} .$$

This proves that (d) is satisfied.

If now (e) is not satisfied, we can modify the sequence t_{nk} by adding suitable functions to h_n. We choose $\varepsilon_n \ge 0$ so that $\sum_0^{\infty}\varepsilon_n = \infty$, and $\sum_0^{\infty}\varepsilon_n^2 < \infty$, and we define

$$h_n'(x) = h_n(x) + \varepsilon_n 2^{-n\alpha}\sum_{k\in\mathbf{Z}^N}\frac{\chi_{0k}(x-k)}{1+|k|^{N+1}} .$$

If the ε_n are chosen sufficiently small, the modified sequence $\{h_n'\}_0^{\infty}$ satisfies conditions (a) – (e).

In order to satisfy (f), we write the function already constructed as

$$h_n' = \sum_{k\in\mathbf{Z}^N}t_{nk}'\chi_{nk} ,$$

and we define a modified function h_n'' by setting

$$h_n'' = \sum_{k\in\mathbf{Z}^N}t_{nk}'u_{nk} = \sum_{k\in\mathbf{Z}^N}t_{nk}''\chi_{nk} ,$$

where

$$u_{nk}(x) = u_0(2^n x - k) , \qquad \text{and} \qquad u_0(x) = \sum_{k\in\mathbf{Z}^N}\frac{\chi_{0k}(x-k)}{1+|k|^{N+1}} .$$

It is then easy to see that h_n'' satisfies (a) – (c), and (e). That (f) is satisfied follows from the construction, and the trivial inequality

$$\frac{1+|k|^{N+1}}{1+(|k|+1)^{N+1}} \ge 2^{-N-2} ,$$

valid for $|k| \ge 1$.

Finally, (d) follows from the theorem of Fefferman–Stein and the estimate

$$h_n'' = \sum_{k\in\mathbf{Z}^N}t_{nk}'u_{nk} \le 2^{nN}4^{-N-1}h_n' * u_{n0}$$

$$\le 2^{nN}(12)^{-N-1}h_n' * \frac{1}{1+(2^n|x|)^{N+1}} \le A\,Mh_n' ,$$

where the last inequality follows from Lemma 4.3.7. (Cf. Lemma 4.6.5.)

Proof of Theorem 10.1.4. By assumption $|D^\beta f_n| \le A\,2^{n|\beta|} g_n$. Let $\{h_n\}_{n=0}^\infty$ be the functions constructed in Lemma 10.1.5. Set

$$F_s = \left\{ x : \textstyle\sum_{n=0}^\infty 2^{ns} h_n(x) = \infty \right\}, \qquad s = 0, 1, \dots \quad . \tag{10.1.8}$$

We note that $F_s = \mathbf{R}^N$ for $s \ge \alpha$ by (e) in the lemma. Let $s < \alpha$. Then $C_{\alpha - s, p}(F_s) = 0$ by Proposition 4.7.2, and

$$\sum_{n=0}^\infty |D^\beta f_n(x)| \le \sum_{n=0}^\infty 2^{ns} g_n(x) < \infty$$

for all $x \in F_s^c$ and $|\beta| \le s$. Thus, we can define $P_x^s f = \sum_{n=0}^\infty P_x^s f_n$. We claim that

$$\lim_{j\to\infty} 2^{j(N+s)} \int_{B_j(x)} |f(y) - P_x^s f(y)|\, dy = 0, \quad x \in F_s^c \ .$$

In fact, for $n \le j$

$$2^{j(N+s)} \int_{B_j(x)} |f_n(y) - P_x^s f_n(y)|\, dy$$
$$\le A\,2^{-j} \sum_{|\beta| = s+1} \max_{y \in B_j(x)} |D^\beta f_n(y)| \le A\,2^{n(s+1)-j} g_n(x) \ ,$$

and thus, by Lemma 10.1.5

$$\begin{aligned} 2^{j(N+s)} & \int_{B_j(x)} \left| \sum_{n=0}^\infty f_n - \sum_{n=0}^\infty P_x^s f_n \right| dy \\ & \le \sum_{n=0}^j 2^{j(N+s)} \int_{B_j(x)} |f_n - P_x^s f_n|\, dy \\ & \quad + 2^{j(N+s)} \int_{B_j(x)} \sum_{n=j+1}^\infty (|f_n| + |P_x^s f_n|)\, dy \\ & \le A \left(\sum_{n=0}^j 2^{n(s+1)-j} h_n(x) + 2^{(j+1)s} h_{j+1}(x) + \sum_{n=j+1}^\infty 2^{ns} h_n(x) \right) . \end{aligned}$$

The theorem now follows from the elementary fact that

$$\lim_{j\to\infty} \left(\sum_{n=0}^j a_n 2^{n-j} + \sum_{n=j+1}^\infty a_n \right) = 0$$

if $a_n \ge 0$ and $\sum_{n=0}^\infty a_n < \infty$.

In proving Lemma 10.1.5 and Theorem 10.1.4 we have actually established the following corollary, which will be used in Section 10.2.

Corollary 10.1.6. *Let* $f \in L^{\alpha,p}$, $1 < p < \infty$, $\alpha > 0$, $f \not\equiv 0$. *Then there exist functions* $\{f_n\}_0^\infty$ *and* $\{g_n\}_0^\infty$ *satisfying* (10.1.1) – (10.1.4), *and functions* $\{h_n\}_0^\infty$ *satisfying* (a), (b), *and* (d) – (f) *of Lemma* 10.1.5, *such that conditions* (a) – (d) *of Theorem* 10.1.4 *are satisfied for* $s \in \mathbf{N}$, $s < \alpha$, *for all* $x \in \mathbf{R}^N$ *such that*

$$\sum_{n=0}^{\infty} 2^{ns} h_n(x) < \infty \ . \tag{10.1.9}$$

The only way the assumption $\mathrm{Tr}_E f = 0$ will be used in proving Theorem 10.1.1 is through the following corollary.

Corollary 10.1.7. *Let* $f \in L^{\alpha,p}$, $1 < p < \infty$, $\alpha > 0$, $f \not\equiv 0$. *Let* $E \subset \mathbf{R}^N$, *and suppose that* $D^\beta f(x) = 0$ *for* $(\alpha - |\beta|, p)$*-q.e.* $x \in E$ *for all multiindices* β *with* $|\beta| < \alpha$. *Then there exist functions* $\{f_n\}_0^\infty$ *and* $\{g_n\}_0^\infty$ *satisfying* (10.1.1) – (10.1.4), *and functions* $\{h_n\}_0^\infty$ *satisfying* (a), (b), *and* (d) – (f) *of Lemma* 10.1.5, *such that if* $x \in E$, *and* $\sum_{n=0}^{\infty} 2^{ns} h_n(x) < \infty$, $0 \le s < \alpha$, *then* $\sum_{n=0}^{\infty} |D^\beta f_n(x)| < \infty$, *and* $\sum_{n=0}^{\infty} D^\beta f_n(x) = 0$ *for all* β, $|\beta| \le s$.

Proof. Let $\{f_n\}_0^\infty$ and $\{h_n\}_0^\infty$ satisfy the conditions in Lemma 10.1.5. Let $E_s = \{x \in E : P_x^s f \ne 0\}$, $0 \le s < \alpha$. Then, by Proposition 4.7.2, for any $\varepsilon > 0$ there exist functions $h'_{sn}(x)$ such that

$$h'_{sn} = \sum_{k \in \mathbf{Z}^N} t'_{snk} \chi_{nk} \,, \quad t'_{snk} \ge 0 \ ,$$
$$\|\{2^{n\alpha} h'_{sn}\}_{n=0}^\infty\|_{L^p(l^2)} \le \varepsilon \, \|f\|_{\alpha,p} \ ,$$

and

$$\sum_{n=0}^{\infty} 2^{ns} h'_{sn}(x) = \infty \qquad \text{for all } x \in E_s \ .$$

Set

$$h'_n = \sum_{0 \le s < \alpha} h'_{sn} = \sum_{k \in \mathbf{Z}^N} t'_{nk} \chi_{nk} \ .$$

We define functions h''_n by setting

$$h''_n = h_n + \sum_{k \in \mathbf{Z}^N} t'_{nk} u_{nk} = \sum_{k \in \mathbf{Z}^N} t''_{nk} \chi_{nk} \ ,$$

where u_{nk} is defined as in the proof of Lemma 10.1.5. Then h''_n satisfies all required conditions.

The following elementary lemma will play an important role in the construction of a multiplier.

Lemma 10.1.8. *Let $a_n > 0$, $n = 0, 1, 2, \dots$, and let $s \geq 0$.*

(a) *If $\sum_{n=0}^{\infty} a_n = \infty$, then*

$$\sum_{n=0}^{\infty} \frac{a_n}{A_n} = \infty \ ,$$

where $A_n = \sum_{m=0}^{n} a_m$.

(b) *If $\sum_{n=0}^{\infty} a_n < \infty$, then*

$$\sum_{n=0}^{\infty} \frac{a_n}{B_n} = \infty \ ,$$

where $B_n = \sum_{m=n}^{\infty} a_m$.

(c) *If $\sum_{n=0}^{\infty} a_n 2^{ns} < \infty$, but $\sum_{n=0}^{\infty} a_n 2^{n(s+1)} = \infty$, then*

$$\sum_{n=0}^{\infty} \frac{a_n 2^{ns}}{2^{-n} A_n + B_n} = \infty \ ,$$

where $A_n = \sum_{m=0}^{n} a_m 2^{m(s+1)}$, and $B_n = \sum_{m=n+1}^{\infty} a_m 2^{ms}$.

Proof. Parts (a) and (b) are well-known (see the notes at the end of the chapter). To prove (a) we observe that if $\lim_{n\to\infty} A_n = \infty$, then for any k

$$\liminf_{n\to\infty} \sum_{m=k}^{n} \frac{a_m}{A_m} \geq \liminf_{n\to\infty} \frac{1}{A_n} \sum_{m=k}^{n} a_m = \liminf_{n\to\infty} \frac{A_n - A_{k-1}}{A_n} = 1 \ .$$

In (b) similarly, if all $B_n > 0$ and $\lim_{n\to\infty} B_n = 0$, then for any k

$$\liminf_{n\to\infty} \sum_{m=k}^{n} \frac{a_m}{B_m} \geq \liminf_{n\to\infty} \frac{1}{B_k} \sum_{m=k}^{n} a_m = \liminf_{n\to\infty} \frac{B_k - B_{n+1}}{B_k} = 1 \ .$$

In proving (c) we consider two cases: we either have

$$\liminf_{n\to\infty} 2^{-n} A_n / B_n \geq 1 \ ,$$

or

$$\liminf_{n\to\infty} 2^{-n} A_n / B_n < 1 \ .$$

In the first case we can choose k so large that $B_m \leq 2 \cdot 2^{-m} A_m$ for $m \geq k$. It follows that for $m \geq k$

$$\frac{a_m 2^{ms}}{2^{-m} A_m + B_m} \geq \frac{a_m 2^{ms}}{3\, 2^{-m} A_m} = \frac{a_m 2^{m(s+1)}}{3 A_m} \ ,$$

and thus

$$\liminf_{n\to\infty} \sum_{m=k}^{n} \frac{a_m 2^{ms}}{2^{-m} A_m + B_m} \geq \liminf_{n\to\infty} \sum_{m=k}^{n} \frac{a_m 2^{m(s+1)}}{3 A_m}$$

$$\geq \liminf_{n\to\infty} \frac{1}{3A_n} \sum_{m=k}^{n} a_m 2^{m(s+1)} = \liminf_{n\to\infty} \frac{A_n - A_{k-1}}{3 A_n} = \tfrac{1}{3} \ .$$

In the second case there are arbitrarily large k such that $2^{-k}A_k/B_k \leq 2$, and then for $m \geq k$ we have

$$2^{-m}A_m + B_m = 2^{-m}A_k + \sum_{j=k+1}^{m} 2^{j-m}a_j 2^{js} + B_m$$

$$\leq 2^{-k}A_k + \sum_{j=k+1}^{m} a_j 2^{js} + B_m = 2^{-k}A_k + B_k \leq 3B_k \ .$$

It follows that

$$\liminf_{n\to\infty} \sum_{m=k}^{n} \frac{a_m 2^{ms}}{2^{-m}A_m + B_m} \geq \liminf_{n\to\infty} \frac{1}{3B_k} \sum_{m=k}^{n} a_m 2^{ms}$$

$$= \liminf_{n\to\infty} \frac{B_k - B_n}{3B_k} = \tfrac{1}{3} \ ,$$

which proves the lemma.

We now denote the partial sums of $\sum_{n=0}^{\infty} f_n$ by

$$\Phi_n = \sum_{m=0}^{n} f_m \ , \tag{10.1.10}$$

and we set

$$d_{nk} = \max\left\{ 2^{-n|\beta|} |D^\beta \Phi_n(x)| : x \in \widetilde{Q}_{nk} \ , \ |\beta| \leq S \right\} \ . \tag{10.1.11}$$

Lemma 10.1.9. *Let the functions f, $\{f_n\}_0^\infty$, $\{g_n\}_0^\infty$, and $\{h_n\}_0^\infty$ satisfy the conditions of Corollary* 10.1.7, *let F_s be defined by*

$$F_s = \left\{ x : \textstyle\sum_{n=0}^{\infty} 2^{ns} h_n(x) = \infty \right\} \ , \qquad s = 0,\ 1,\ \ldots \ ,$$

and let $x \in E \setminus F_s$, $s < \alpha$, so that $\sum_{n=0}^{\infty} D^\beta f_n(x) = 0$ for all β, $|\beta| \leq s$. Then

$$\sum_{n=0}^{\infty} \sum_{k\in \mathbf{Z}^N} \frac{t_{nk}}{d_{nk}} \chi_{nk}(x) = \infty \ . \tag{10.1.12}$$

Proof. We first observe that (10.1.12) holds if $x \in F_0$. In fact, if $x \in Q_{nk}$, then clearly

$$2^{-n|\beta|} |D^\beta \Phi_n(x)| \leq \sum_{m=0}^{n} 2^{(m-n)|\beta|} g_m(x) \ ,$$

and thus

$$d_{nk} \leq \sum_{Q_{mj} \supset Q_{nk}} s_{mj} \ .$$

The conclusion follows directly from Lemma 10.1.8(a).

We have $E \setminus F_0 = \bigcup_{s<\alpha}\big(E \cap (F_{s+1} \setminus F_s)\big)$, since $F_s = \mathbf{R}^N$ for $s \geq \alpha$ by condition (e) in Lemma 10.1.5. Let $x \in E \cap (F_{s+1} \setminus F_s)$, $0 \leq s < \alpha$, so that $\sum_{n=0}^{\infty} 2^{ns} h_n(x) < \infty$, and $\sum_{n=0}^{\infty} 2^{n(s+1)} h_n(x) = \infty$.

We claim that if $x \in Q_{nk}$, then

$$d_{nk} \leq A \sum_{m=0}^{n} 2^{(m-n)(s+1)} g_m(x) + A \sum_{m=n+1}^{\infty} 2^{(m-n)s} g_m(x) \ . \qquad (10.1.13)$$

Then the result follows from (c) in Lemma 10.1.8, since $g_m \leq h_m$.

First let β be a multiindex with $|\beta| \geq s+1$. Then, for any $y \in Q_{nk}$, we have

$$\begin{aligned} 2^{-n|\beta|}|D^\beta \Phi_n(y)| &\leq \sum_{m=0}^{n} 2^{(m-n)|\beta|} g_m(y) \\ &\leq \sum_{m=0}^{n} 2^{(m-n)(s+1)} g_m(y) \ . \end{aligned} \qquad (10.1.14)$$

Now let $|\beta| \leq s$. Then, the crucial observation is that

$$\begin{aligned} |D^\beta \Phi_n(x)| = \left|\sum_{m=0}^{n} D^\beta f_m(x)\right| &= \left|\sum_{m=n+1}^{\infty} D^\beta f_m(x)\right| \\ &\leq \sum_{m=n+1}^{\infty} 2^{m|\beta|} g_m(x) \ . \end{aligned} \qquad (10.1.15)$$

This follows from Corollary 10.1.7. As we already noted, this is the only place in the entire proof of Theorem 10.1.1 where the assumption $\mathrm{Tr}_E\, f = 0$ is used.

By Taylor's formula, applied to $D^\beta \Phi_n(y)$ for $y \in Q_{nk}$ (recall that we assume $x \in Q_{nk}$),

$$\begin{aligned} &2^{-n|\beta|}|D^\beta \Phi_n(y)| \\ &\quad \leq A \sum_{|\beta|\leq|\gamma|\leq s} 2^{-n|\gamma|}|D^\gamma \Phi_n(x)| + A \sum_{|\gamma|=s+1} \max_{z \in Q_{nk}} 2^{-n|\gamma|}|D^\gamma \Phi_n(z)| \ , \end{aligned}$$

and by (10.1.15) and (10.1.14)

$$\begin{aligned} &2^{-n|\beta|}|D^\beta \Phi_n(y)| \\ &\quad \leq A \sum_{|\beta|\leq|\gamma|\leq s} 2^{-n|\gamma|} \sum_{m=n+1}^{\infty} 2^{m|\gamma|} g_m(x) + A \sum_{m=0}^{n} 2^{(m-n)(s+1)} g_m(x) \\ &\quad \leq A \sum_{m=n+1}^{\infty} 2^{(m-n)s} g_m(x) + A \sum_{m=0}^{n} 2^{(m-n)(s+1)} g_m(x) \ . \end{aligned}$$

The inequality (10.1.13) follows from this and (10.1.14), and Lemma 10.1.9 is proved.

Proof of Theorem 10.1.1. We only have to prove that (a) in the theorem implies (c). Let f, $\{f_n\}$, $\{g_n\}$, $\{h_n\}$ be functions satisfying the conditions in Corollary 10.1.7, and let $\{d_{nk}\}$ be the numbers defined by (10.1.11). We use (10.1.12) to construct a multiplier ω. Let $\varphi \geq 0$ be a function in $C_0^\infty(\mathbf{R}^N)$ such that $\operatorname{supp}\varphi \subset 2Q_{00}$ and $\varphi(x) \geq 1$ on $\frac{3}{2}Q_{00}$. Define $\varphi_{nk}(x) = \varphi(2^n x - k)$ for $n \in \mathbf{N}$ and $k \in \mathbf{Z}^N$, and set

$$\kappa_n(x) = \sum_{k \in \mathbf{Z}^N} \min\left\{1, \frac{t_{nk}}{d_{nk}}\right\} \varphi_{nk}(x) \ .$$

Then fix a large number n_0, and denote

$$K_n = \sum_{m=n_0}^{n} \kappa_m \quad \text{for } n \geq n_0 \ .$$

Let $\Psi \in C^\infty(\mathbf{R})$ be a function such that $\Psi(0) = 0$, $0 \leq \Psi(t) \leq 1$, $\Psi(t) = 1$ for all $t \geq 1$, and set

$$\omega = \lim_{n \to \infty} \Psi(K_n) = \Psi\left(\sum_{n=n_0}^{\infty} \kappa_n\right) ,$$

Setting

$$\begin{aligned} \omega_{n_0} &= \Psi(\kappa_{n_0}) \ ; \\ \omega_n &= \Psi(K_n) - \Psi(K_{n-1}) , \quad n \geq n_0 + 1 \ , \end{aligned}$$

we have $\omega = \sum_{n=n_0}^{\infty} \omega_n$. Then Lemma 10.1.9 implies that $\omega(x) = 1$ in a neighborhood of E, and by choosing n_0 we can make this neighborhood arbitrarily small.

We shall finish the proof of Theorem 10.1.1 by showing that ωf can be made arbitrarily small by choosing n_0 sufficiently large. By Theorem 4.6.3 it is enough to prove that there are functions u_n, $n \geq n_0$, and a constant A, such that

$$\omega f = \sum_{n=n_0}^{\infty} u_n \ ,$$

and

$$|D^\beta u_n| \leq A\, 2^{n|\beta|} \sum_{k \in \mathbf{Z}^\mathbf{N}} t_{nk} \chi_{nk}$$

for all $|\beta| \leq S$.

We have

$$\begin{aligned} \omega f = \sum_{m=0}^{\infty} f_m \sum_{n=n_0}^{\infty} \omega_n &= \sum_{n=n_0}^{\infty} \omega_n \left(\sum_{m=0}^{n} f_m + \sum_{m=n+1}^{\infty} f_m\right) \\ &= \sum_{n=n_0}^{\infty} \omega_n \Phi_n + \sum_{m=n_0+1}^{\infty} f_m \sum_{n=n_0}^{m-1} \omega_n \\ &= \sum_{n=n_0}^{\infty} \omega_n \Phi_n + \sum_{n=n_0+1}^{\infty} f_n \Psi(K_{n-1}) \ . \end{aligned}$$

We set

$$u_n = \omega_n \Phi_n + \Psi(K_{n-1}) f_n \ ,$$

and we shall estimate $D^\beta u_n$ for $|\beta| \le S$. We first observe that

$$|D^\beta \kappa_n(x)| \le A\, 2^{n|\beta|} \sum_{k\in\mathbf{Z}^N} \min\Big\{1, \frac{t_{nk}}{d_{nk}}\Big\} \widetilde{\chi}_{nk}(x) \ , \tag{10.1.16}$$

and that

$$|D^\beta K_n(x)| \le \sum_{m=n_0}^{n} |D^\beta \kappa_m(x)| \le A \sum_{m=n_0}^{n} 2^{m|\beta|} \le A\, 2^{n|\beta|} \ . \tag{10.1.17}$$

We write

$$\omega_n = \Psi(K_n) - \Psi(K_{n-1}) = \kappa_n \int_0^1 \Psi'(K_{n-1} + \kappa_n t)\, dt = \kappa_n Y_n \ . \tag{10.1.18}$$

By repeatedly using the chain rule, (10.1.16) and (10.1.17) (cf. also the proof of Theorem 3.3.3) we find

$$|D^\beta Y_n(x)| \le A\, 2^{n|\beta|} \ .$$

It follows from (10.1.18), using the Leibniz formula and (10.1.16), that

$$2^{-n|\beta|} |D^\beta \omega_n(x)| \le A \sum_{k\in\mathbf{Z}^N} \frac{t_{nk}}{d_{nk}} \widetilde{\chi}_{nk}(x) \ .$$

For an $x \in Q_{nl}$ the last sum contains only those 3^N terms $\big(t_{nk}/d_{nk}\big)\widetilde{\chi}_{nk}(x)$ for which $Q_{nk} \subset \widetilde{Q}_{nl}$. Thus, by (10.1.11)

$$|D^\beta \Phi_n(x)| \le 2^{n|\beta|} \min_{Q_{nk}\subset \widetilde{Q}_{nl}} d_{nl} \ ,$$

and thus, again by the Leibniz formula,

$$|D^\beta(\omega_n \Phi_n)| \le A\, 2^{n|\beta|} \sum_{k\in\mathbf{Z}^N} t_{nk} \widetilde{\chi}_{nk} \le A\, 2^{n|\beta|} \sum_{k\in\mathbf{Z}^N} t_{nk} \chi_{nk}$$

for $|\beta| \le S$.

Finally, by the chain rule and (10.1.17)

$$|D^\beta \Psi(K_{n-1})| \le A\, 2^{n|\beta|} \ ,$$

and by (10.1.3)

$$|D^\beta f_n| \le 2^{n|\beta|} \sum_{k\in\mathbf{Z}^N} s_{nk} \chi_{nk} \ ,$$

so again by the Leibniz formula

$$|D^\beta(\Psi(K_{n-1}) f_n)| \le 2^{n|\beta|} \sum_{k\in\mathbf{Z}^N} s_{nk} \chi_{nk} \ .$$

This proves the claim and finishes the proof of Theorem 10.1.1.

10.2 A Generalization of a Theorem of Whitney

The space $C^m(\mathbf{R}^N)$ of m times continuously differentiable functions is also an algebra. A subspace which is closed under multiplication by elements in $C^m(\mathbf{R}^N)$ is called an ideal. The closed ideals (in the topology of uniform convergence on compact sets) were characterized by H. Whitney [431], and our purpose is to give an extension of his theorem to $L^{\alpha,p}(\mathbf{R}^N)$, $\alpha > 0$, $1 < p < \infty$. Before doing so we will describe the classical results.

The situation is particularly simple if $m = 0$.

Theorem 10.2.1. *The closed ideals in* $C(\mathbf{R}^N)$ *are of the form* $\mathfrak{M} = \mathfrak{M}_E = \{ f \in C(\mathbf{R}^N) : f|_E = 0 \}$, *where* E *is any closed subset of* $\mathbf{R}^N$.

We denote by $\mathfrak{P}$ the ring of all polynomials in N real variables, by $\mathfrak{P}_m$ the subspace of polynomials of degree at most m, and by $\mathfrak{I}_m$, $m \in \mathbf{N}$, the ideal

$$\mathfrak{I}_m = \{ p \in \mathfrak{P} : p(x) = \textstyle\sum_{|\sigma|>m} a_\sigma x^\sigma \} \ .$$

The ring of residue classes (the quotient ring) mod $\mathfrak{I}_m$ is denoted $\mathfrak{P}/\mathfrak{I}_m$. We denote elements in $\mathfrak{P}_m$ by the same letters as their images in $\mathfrak{P}/\mathfrak{I}_m$ under the natural mapping.

For each $y \in \mathbf{R}^N$ the Taylor formula defines a ring homomorphism

$$P_y^m : C^m \to \mathfrak{P}/\mathfrak{I}_m$$

by

$$P_y^m f(x) = \sum_{|\sigma|\leq m} D^\sigma f(y) \frac{x^\sigma}{\sigma!} \ ,$$

and the image of a closed ideal $\mathfrak{M}$ in C^m is an ideal in $\mathfrak{P}/\mathfrak{I}_m$, called the local ideal, $\mathfrak{M}_y$, of $\mathfrak{M}$ at y.

Theorem 10.2.2 (H. Whitney). *Let* $\mathfrak{M}$ *be a closed ideal in* $C^m(\mathbf{R}^N)$, *and let* $g \in C^m(\mathbf{R}^N)$. *Then* $g \in \mathfrak{M}$ *if and only if* $P_y^m g \in \mathfrak{M}_y$ *for all* $y \in \mathbf{R}^N$.

In $L^{\alpha,p}(\mathbf{R}^N)$ the problem cannot be posed in exactly the same way, since $L^{\alpha,p}$ is not an algebra if $\alpha p \leq N$. It is, however, always closed under multiplication by functions in C_0^∞, i.e., it is a C_0^∞-module. We shall characterize the closed C_0^∞-submodules of $L^{\alpha,p}(\mathbf{R}^N)$.

If $m < \alpha$, the mapping $P_y^m : f \mapsto P_y^m f \in \mathfrak{P}/\mathfrak{I}_m$ is well defined for $(\alpha - m, p)$-q.e. $y \in \mathbf{R}^N$ if $f \in L^{\alpha,p}$ is given, and $P_y^m f$ is the differential of order m at y of f. It is easily seen that P_y^m is a homomorphism in the sense that if $f \in L^{\alpha,p}$, and f has an m-differential at y, then φf has an m-differential at y for all $\varphi \in C_0^\infty$, and

$$P_y^m \varphi \, P_y^m f = P_y^m(\varphi f) \pmod{\mathfrak{I}_m} \ . \tag{10.2.1}$$

We shall prove the following theorem.

Theorem 10.2.3 (Yu. V. Netrusov). *Let $\mathfrak{M}$ be a closed C_0^∞-submodule of $L^{\alpha,p}(\mathbf{R}^N)$, $\alpha > 0$, and $1 < p < \infty$. Let S be any integer greater than or equal to the integer part $[\alpha]$ of α. Then there is a family of ideals $\{\mathfrak{E}_y \subset \mathfrak{P}/\mathfrak{I}_S : y \in \mathbf{R}^N\}$ with the property that a function f_0 in $L^{\alpha,p}$ belongs to $\mathfrak{M}$ if and only if for all $m \in \mathbf{N}$, $m < \alpha$,*

$$P_y^m f_0 \in \mathfrak{E}_y \pmod{\mathfrak{I}_m} \quad \textit{for } (\alpha - m, p)\textit{-q.e. } y \in \mathbf{R}^N \ . \qquad (10.2.2)$$

Theorem 10.2.3 is an easy consequence of the following result, which is of independent interest.

We denote by $\mathcal{L}(V)$ the linear hull of a subset V of a vector space, i.e. $\mathcal{L}(V)$ is the set of finite linear combinations of elements in V.

Theorem 10.2.4. *Let $f_i \in L^{\alpha,p}$, $i \in \mathbf{N}$. Then the following conditions are equivalent:*

(a) *For all $m \in \mathbf{N}$, $m < \alpha$, and for $(\alpha - m, p)$-q.e. $y \in \mathbf{R}^N$, the polynomial $P_y^m f_0$ belongs $\pmod{\mathfrak{I}_m}$ to the ideal in $\mathfrak{P}/\mathfrak{I}_m$ generated by the polynomials $\{P_y^m f_i : i = 1, 2, \dots\}$.*
(b) *The function f_0 belongs to the closed C_0^∞-module generated by $\{f_i\}_1^\infty$, i.e., for any $\varepsilon > 0$ there is a finite subset Q of $\mathbf{N} \setminus \{0\}$, and functions $\varphi_i \in C_0^\infty$, $i \in Q$, such that*

$$\| f_0 - \textstyle\sum_{i \in Q} \varphi_i f_i \|_{\alpha,p} < \varepsilon \ .$$

Proof of Theorem 10.2.3. Let $\mathfrak{M}$ be a C_0^∞-submodule of $L^{\alpha,p}$, let $\{f_i\}_1^\infty$ span $\mathfrak{M}$, and let B_m, $0 \le m < \alpha$, be the common domain of definition of the differentials of order m of the functions f_i, $i = 1, 2, \dots$. Then $C_{\alpha-m,p}(B_m^c) = 0$ by Theorem 10.1.4. For $y \in \mathbf{R}^N$ we denote by $m(y)$ the largest integer m such that $y \in B_m$, and we define $\mathfrak{E}_y$ as the ideal in $\mathfrak{P}/\mathfrak{I}_S$ generated by the polynomials $P_y^{m(y)} f_i$, $i = 1$, $2, \dots$. If $y \notin B_0$ we set $\mathfrak{E}_y = \mathfrak{P}/\mathfrak{I}_S$.

It follows from the definition of the ideals $\mathfrak{E}_y$ that (10.2.2) is equivalent to (a) in Theorem 10.2.4. This implies, by Theorem 10.2.4 and the choice of the functions f_i, that $f_0 \in \mathfrak{M}$ if and only if (10.2.2) is satisfied.

Proof of Theorem 10.2.4. We first prove that (b) implies (a). By (10.2.1) it is clear that (a) is satisfied by all f_0 which have a representation as a finite sum, $f_0 = \sum_{i \in Q} \varphi_i f_i$, $\varphi_i \in C_0^\infty$. But the set of functions satisfying (a) is easily seen to be closed in $L^{\alpha,p}$. In fact, if a sequence $\{g_i\}_{i=1}^\infty$ converges to f_0 in $L^{\alpha,p}$, then by Proposition 2.3.8 there is a subsequence $\{g_{n_i}\}_{i=1}^\infty$ such that $\{D^\sigma g_{n_i}(y)\}_{i=1}^\infty$ converges to $D^\sigma f_0(y)$ for $(\alpha - |\sigma|, p)$-q.e. $y \in \mathbf{R}^N$, $0 \le |\sigma| < \alpha$. Theorem 10.1.4, which says that $P_y^m f_0(x) = \sum_{|\sigma| \le m} D^\sigma f_0(y)(x - y)^\sigma / \sigma!$ for $(\alpha - m, p)$-q.e. y, now implies that $\{P_y^m g_{n_i}(y)\}_{i=1}^\infty$ converges to $P_y^m f_0(y)$ for $(\alpha - m, p)$-q.e. y, $0 \le m < \alpha$.

To prove the converse statement is considerably harder. Suppose that condition (a) holds. It is no loss of generality to assume that supp f_0 is compact. Indeed, it is easy to see that for any $\varepsilon > 0$ there is $\omega \in C_0^\infty$ such that $\| f_0 - \omega f_0 \|_{\alpha,p} < \varepsilon$, and by (10.2.1) the function ωf_0 also satisfies (a).

Let the function $\varphi \in C_0^\infty$ be equal to 1 on a neighborhood of supp f_0. The conclusion follows from the following theorem, applied to the set of functions $\{\varphi x^\sigma f_i : 0 \le |\sigma| < S,\ i = 1, 2, \dots\}$.

Theorem 10.2.5. *Let $f_i \in L^{\alpha,p}$, $i \in \mathbf{N}$, be such that for all $m \in \mathbf{N}$, $m < \alpha$,*

$$P_y^m f_0 \in \mathcal{L}\big(\{P_y^m f_i\}_{i=1}^\infty\big) \quad \textit{for } (\alpha - m, p)\textit{-q.e. } y \in \mathbf{R}^N . \tag{10.2.3}$$

Then f_0 belongs to the closed C_0^∞-module generated by $\{f_i\}_1^\infty$.

The rest of this section will be devoted to the proof of Theorem 10.2.5. We will use the same technique as in Section 10.1, and we first prove three technical lemmas.

Lemma 10.2.6. *Let Γ be a finite subset of $\mathbf{N}\times\mathbf{Z}^N$, let $K \subset \mathbf{R}^N$, and let $\{a_{nk}\}_{(n,k)\in\Gamma}$ be positive numbers such that $\sum_{(n,k)\in\Gamma} a_{nk}\chi(\frac{4}{3}Q_{nk}) \ge 1$ on K. Then there are nonnegative functions ω_{nk}, $(n,k) \in \Gamma$, and a constant A, such that*

$$\operatorname{supp}\omega_{nk} \subset \tfrac{3}{2}Q_{nk} \ ; \tag{10.2.4}$$

$$\|D^\beta \omega_{nk}\|_\infty \le A \min\{1, a_{nk}\} 2^{i|\beta|} , \quad |\beta| \le S \ ; \tag{10.2.5}$$

$$\sum_{(n,k)\in\Gamma} \omega_{nk} \le 1 \ ; \tag{10.2.6}$$

$$\sum_{(n,k)\in\Gamma} \omega_{nk}(x) = 1 \quad \textit{for all } x \in K \ . \tag{10.2.7}$$

Proof. It is no loss of generality to assume that $a_{nk} \le 1$. We enumerate $\Gamma = \{(i_1, k_1), \dots, (i_r, k_r)\}$, so that the sequence $\{i_s\}_1^r$ is non-decreasing. Let $\varphi \in C_0^\infty(\mathbf{R}^N)$ and $\eta \in C^\infty(\mathbf{R})$ be functions satisfying

$$0 \le \varphi \le 1 , \quad \operatorname{supp}\varphi \subset \tfrac{3}{2}Q_{00} , \quad \varphi = 1 \text{ on } \tfrac{4}{3}Q_{00} ,$$

and

$$0 \le \eta \le 1 , \quad \eta(0) = 0 , \quad \eta(t) = 1 \text{ for } t \ge 1 .$$

We set

$$\varphi_s = a_{i_s k_s}\varphi(x2^{i_s} - k_s) \quad \text{for } s = 1, 2, \dots, r ,$$

and define

$$\omega_{i_1 k_1} = \eta(\varphi_1) ,$$
$$\omega_{i_u k_u} = \eta\big(\textstyle\sum_{s=1}^u \varphi_s\big) - \eta\big(\sum_{s=1}^{u-1}\varphi_s\big), \quad u = 2, \dots, r .$$

Then (10.2.4), (10.2.6), and (10.2.7) follow directly from the construction, and the estimate (10.2.5) is proved in the same way as the corresponding estimate in the proof of Theorem 10.1.1.

Lemma 10.2.7. *Let Γ be a finite subset of $\mathbf{N} \times \mathbf{Z}^N$, and let $\{a_{ik}\}_{(i,k)\in\Gamma}$ and $\{\omega_{ik}\}_{(i,k)\in\Gamma}$ have the same meaning as in Lemma* 10.2.6. *Let*

$$h_n = \sum_{k\in\mathbf{Z}^N} t_{nk}\chi_{nk} \ , \quad n \in \mathbf{N} \ ,$$

and let f_{ik}, $(i,k) \in \Gamma$, be functions that can be split as $f_{ik} = f'_{ik} + f''_{ik}$, where f'_{ik} satisfies

$$\sup_{x\in 2Q_{ik}} \sup_{|\beta|\le S} 2^{-i|\beta|}|D^\beta f'_{ik}(x)| \le A\frac{t_{ik}}{a_{ik}} \ , \tag{10.2.8}$$

and f''_{ik} can be written $f''_{ik} = \sum_{n=i}^{\infty} f''_{ikn}$ with

$$2^{-n|\beta|}|D^\beta f''_{ikn}| \le A\,h_n \ , \quad n \ge i \ , \quad |\beta| \le S \ . \tag{10.2.9}$$

Suppose that $t_{nk} \ge 2^{-N-2}t_{nk'}$ if $|k-k'| = 1$, $n \in \mathbf{N}$. Then the function $\sum_{(i,k)\in\Gamma}\omega_{ik}f_{ik}$ admits the representation

$$\sum_{(i,k)\in\Gamma}\omega_{ik}f_{ik} = \sum_{n=s}^{\infty} v_n \ , \quad s = \min\{\, i : (i,k) \in \Gamma \,\} \ , \tag{10.2.10}$$

where

$$2^{-n|\beta|}|D^\beta v_n| \le A\,h_n \quad \text{for all } n \in \mathbf{N}, \text{ and } |\beta| \le S \ . \tag{10.2.11}$$

Proof. It is enough to prove that each of the functions $\sum_{(i,k)\in\Gamma}\omega_{ik}f'_{ik}$ and $\sum_{(i,k)\in\Gamma}\omega_{ik}f''_{ik}$ admit the desired representation. For the first one of these, this follows directly from the Leibniz formula, (10.2.5), and (10.2.8). We write $\sum_{(i,k)\in\Gamma}\omega_{ik}f''_{ik} = \sum_{i=s}^{\infty} v_i$, where $v_i = \sum_{(n,k)\in\Gamma(i)}\omega_{nk}f''_{nki}$, and $\Gamma(i) = \{\,(n,k) \in \Gamma : n \le i\,\}$. By the Leibniz formula, and (10.2.9) we have

$$|D^\beta v_i| \le A \sum_{0\le\sigma\le\beta}\Bigg(\sum_{(n,k)\in\Gamma(i)} |D^\sigma\omega_{nk}|\Bigg)2^{i|\beta-\sigma|}h_i \ .$$

This implies (10.2.11), because $\sum_{(n,k)\in\Gamma(i)}|D^\sigma\omega_{nk}| \le A\,2^{i|\sigma|}$ for $|\sigma| \le S$ by (10.2.5) and (10.2.6).

Lemma 10.2.8. *Let $\{f_y\}_{y\in\mathbf{R}^N}$ be functions on $\mathbf{R}^N$ and suppose that*

$$f_y = \sum_{n=0}^{\infty} f_{y,n} \ , \quad f_{y,n} \in C^\infty \ . \tag{10.2.12}$$

Suppose that there are functions $h_{y,n}$ such that

$$h_{y,n} = \sum_{k\in\mathbf{Z}^N} t_{y,nk}\chi_{nk}\ , \quad t_{y,nk} \geq 0\ , \tag{10.2.13}$$

$$|D^\beta f_{y,n}| \leq A_\beta 2^{n|\beta|} h_{y,n} \quad \textit{for } |\beta| \leq S, \tag{10.2.14}$$

$$\frac{t_{y,nk}}{t_{y,nk'}} \geq 2^{-N-2} \quad \textit{if} \quad |k-k'| = 1\ , \tag{10.2.15}$$

$$\sum_{n=0}^{\infty} 2^{n\alpha} h_{y,n} \equiv \infty\ , \tag{10.2.16}$$

and that if for some $s \in \mathbf{N}$

$$\sum_{n=0}^{\infty} 2^{ns} h_{y,n}(y) < \infty\ , \quad \textit{then} \quad \sum_{n=0}^{\infty} D^\beta f_{y,n}(y) = 0 \tag{10.2.17}$$

for all β, $|\beta| \leq s$.

Suppose furthermore that there are functions h_n *satisfying*

$$h_n = \sum_{k\in\mathbf{Z}^N} t_{nk}\chi_{nk}\ , \quad t_{nk} \geq 0\ , \tag{10.2.18}$$

$$\frac{t_{nk}}{t_{nk'}} \geq 2^{-N-2} \quad \textit{if} \quad |k-k'| = 1\ , \tag{10.2.19}$$

$$\left\| \{2^{n\alpha} h_n\}_0^\infty \right\|_{L^p(l^2)} < \infty\ , \tag{10.2.20}$$

such that for every $y \in \mathbf{R}^N$ *there is a nonnegative integer* $n(y)$ *with*

$$h_{y,n} \leq h_n \quad \textit{if } n \geq n(y)\ . \tag{10.2.21}$$

Then, for any $\varepsilon > 0$ *and compact* $K \subset \mathbf{R}^N$, *there is a set* $\Gamma = \{\gamma_1, \ldots, \gamma_u\} \subset \mathbf{N} \times \mathbf{Z}^N$, *points* $y(\gamma) \in \mathbf{R}^N$, *and functions* $\varphi_\gamma \in C_0^\infty(\mathbf{R}^N)$, $\gamma \in \Gamma$, *such that*

$$\sum_{\gamma\in\Gamma} \varphi_\gamma(x) = 1 \quad \textit{on } K\ ,$$

and

$$\left\| \textstyle\sum_{\gamma\in\Gamma} \varphi_\gamma f_{y(\gamma)} \right\|_{\alpha,p} \leq \varepsilon\ .$$

Proof. For $y \in \mathbf{R}^N$, and $n \in \mathbf{N}$ we define $k(n,y) \in \mathbf{Z}^N$ by the requirement $y \in Q_{n,k(n,y)}$. Let $n_0 \in \mathbf{N}$. Applying Lemma 10.1.9 to the functions f_y, $f_{y,n}$ and $h_{y,n}$, we find integers $n_1(y)$ and $n_2(y)$ such that $n_1(y) \geq \max(n_0, n(y))$, and

$$\sum_{n=n_1(y)}^{n_2(y)} \frac{h_{y,n}}{d_{y,n}} \geq 1\ ,$$

where

$$d_{y,n} = \sup_{|\beta|\leq S}\ \sup_{x\in 2Q_{n,k(n,y)}} 2^{-n|\beta|} |D^\beta \Phi_{y,n}(x)|\ ,$$

and

$$\Phi_{y,n} = \sum_{i=0}^{n} f_{y,i} \; .$$

Set $V_y = \operatorname{int} \frac{4}{3} Q_{\gamma(y)}$, where $\gamma(y) = \big(n_2(y), k(n_2(y), y)\big)$. Then $\bigcup_{y \in \mathbf{R}^N} V_y = \mathbf{R}^N$, and V_y is open for each y. Consequently there are points $z_1, z_2, \ldots, z_r$, such that $\bigcup_{i=1}^{r} V_{z_i} \supset K$. Set

$$\Gamma = \bigcup_{i=1}^{r} \bigcup_{n=n_1(z_i)}^{n_2(z_i)} \{(n, k(n, z_i))\} \; ,$$

and for $(n, k) \in \Gamma$

$$a_{nk} = \max_{z_i \in Q_{nk}} \frac{t_{z_i,nk}}{d_{z_i,n}} \; . \tag{10.2.22}$$

Let $y(n, k) = z_i$, where z_i is any point in the set $\{z_1, \ldots, z_r\}$ where the maximum in (10.2.22) is attained. It follows from the construction that on K

$$\sum_{(n,k) \in \Gamma} a_{nk} \chi(\tfrac{4}{3} Q_{nk}) \geq 1 \; .$$

We apply Lemma 10.2.6 to the numbers $\{a_{nk}\}_{(n,k)\in\Gamma}$ to find functions $\{\omega_{nk}\}_{(n,k)\in\Gamma}$. Set

$$\begin{aligned} f_{nk} &= f_{y(n,k)} \; , \\ f'_{nk} &= \Phi_{y(n,k),n} \; , \\ f''_{nk} &= f_{y(n,k)} - \Phi_{y(n,k),n} = \sum_{l=n+1} f_{y(n,k),l} \; , \end{aligned}$$

for $(n, k) \in \Gamma$. It follows from Lemma 10.2.7 that

$$\left\| \textstyle\sum_{(n,k)\in\Gamma} \omega_{nk} f_{y(n,k)} \right\|_{\alpha,p} \leq A \, \|\{2^{n\alpha} h_n\}_{n_0}^{\infty}\|_{L^p(l^2)} \; .$$

The lemma follows from (10.2.20) if n_0 is chosen large enough.

Proof of Theorem 10.2.5. It is sufficient to construct functions f_y, $f_{y,n}$, $h_{y,n}$, and h_n, such that the conditions of Lemma10.2.8 are satisfied, and such that the functions f_y, $y \in \mathbf{R}^N$, can be represented as

$$f_y = \sum_{j=0}^{\infty} \alpha_{y,j} f_j \; ,$$

$\alpha_{y,0} = 1$, and for each y only finitely many $\alpha_{y,j}$ are different from 0.

In fact, by Lemma 10.2.8, applied to the set $K = \operatorname{supp} f_0$, there is a finite set $\Gamma \subset \mathbf{N} \times \mathbf{Z}^N$, and there are functions $\varphi_\gamma \in C_0^\infty$, and points $y(\gamma) \in \mathbf{R}^{\mathbf{N}}$ with $\gamma \in \Gamma$, such that

$$\left\| f_0 + \sum_{\gamma\in\Gamma} \varphi_\gamma \sum_{j=1}^\infty \alpha_{y(\gamma),j} f_j \right\|_{\alpha,p} = \left\| \sum_{\gamma\in\Gamma} \varphi_\gamma f_{y(\gamma)} \right\|_{\alpha,p} < \varepsilon \ .$$

Before we can construct these functions we need to carry out some preparatory work.

By Corollary 10.1.6, applied to the functions f_j, $j = 0, 1, \dots$, we can find functions f_{jn} and h_{jn} satisfying the following conditions:

$$\begin{aligned}
&f_j = \sum_{n=0}^\infty f_{jn}\,, \quad f_{jn} \in C^\infty\,, \quad j \in \mathbf{N}\ ;\\
&h_{jn} = \sum_{k\in\mathbf{Z}^N} t_{jnk}\chi_{nk}\,, \quad t_{jnk} \geq 0\ ;\\
&|D^\beta f_{jn}| \leq 2^{n|\beta|} h_{jn} \quad \text{for } |\beta| \leq S, \text{ (recall that } S > \alpha)\ ;\\
&\left\| \{2^{n\alpha} h_{jn}\}_{n\geq 0} \right\|_{L^p(l^2)} < \infty\ ;\\
&\sum_{n=0}^\infty 2^{n\alpha} h_{jn}(x) = \infty \quad \text{for all } x \in \mathbf{R}^N \text{ and } j \in \mathbf{N}\ .
\end{aligned} \tag{10.2.23}$$

If, moreover, for some $s \in \mathbf{N}$, $s < \alpha$,

$$\sum_{n=0}^\infty 2^{ns} h_{jn}(y) < \infty\ ,$$

then

$$\sum_{n=0}^\infty D^\beta f_{jn}(y) = D^\beta f_j(y) \quad \text{for } |\beta| \leq s\ .$$

(Recall that as agreed after Theorem 10.1.4 we are assuming that y is a Lebesgue point for $D^\beta f_j$ and that its value $D^\beta f_j(y)$ is defined in the sense of Lebesgue.) In addition, f_j has a differential $P_y^s f$ of order s such that

$$P_y^s f = \sum_{|\beta|\leq s} D^\beta f_j(y) \frac{(x-y)^\beta}{\beta!}\ .$$

We define sets E_s, $s \in \mathbf{N}$, $s < \alpha$, by saying that E_s^c is the set of all $y \in \mathbf{R}^N$ such that

$$\sum_{n=0}^\infty 2^{ns} h_{jn}(y) < \infty \quad \text{for all } j \in \mathbf{N}\ ,$$

and

$$P_y^s f_0 \in \mathcal{L}(\{P_y^s f_j\}_{j\geq 1})\ .$$

It follows from the conditions of the theorem and from Proposition 4.7.2 that $C_{\alpha-s,p}(E_s) = 0$. By Proposition 4.7.2 there are now functions $\tilde{h}_{sn}$ such that

$$\tilde{h}_{sn} = \sum_{k \in \mathbf{Z}^{\mathbf{N}}} \tilde{t}_{snk} \chi_{nk} , \quad \tilde{t}_{snk} \geq 0 ,$$
$$\left\| \{2^{n\alpha} \tilde{h}_{sn}\}_{n \geq 0} \right\|_{L^p(l^2)} < \infty , \tag{10.2.24}$$
$$\sum_{n=0}^{\infty} 2^{ns} \tilde{h}_{sn}(x) = \infty \quad \text{for all } x \in E_s .$$

Arguing as in the proof of Corollary 10.1.7, we can assume that

$$\tilde{h}_{sn}(x + k2^{-n}) \leq 2^{N+2} \tilde{h}_{sn}(x) \quad \text{for } k \in \mathbf{Z}^N, \quad |k| = 1 .$$

Now all preparations are finished for constructing the functions f_y, $f_{y,n}$, $h_{y,n}$, and the numbers $\alpha_{y,j}$. Let $y \in \mathbf{R}^N$. Set

$$s(y) = -1 \quad \text{if } y \in E_0 ;$$
$$s(y) = \max\{ s \in \mathbf{N} : y \in E_s^c \} \quad \text{if } y \notin E_0 .$$

We define $\alpha_{y,j}$ in the following way:

$$\alpha_{y,0} = 1 ;$$
$$\alpha_{y,j} = 0 \quad \text{for } j \geq 1 \text{ if } s(y) = -1 ;$$
$$\alpha_{y,j} = -\Delta_j \quad \text{for } j \geq 1 \text{ if } s(y) \geq 0 ,$$

where the numbers Δ_j satisfy $P_y^{s(y)} f_0 = \sum_{j \geq 1} \Delta_j P_y^{s(y)} f_j$, and only finitely many of the Δ_j are non-zero. (We observe that the existence of such numbers Δ_j, $j \geq 1$, follows from the fact that $P_y^{s(y)} f_0 \in \mathcal{L}(\{P_y^{s(y)} f_j\}_{j \geq 0})$, which is a consequence of the condition $y \in E_{s(y)}^c$ and the definition of the sets $E_{s(y)}$.)

Set

$$f_y = \sum_{j=0}^{\infty} \alpha_{yj} f_j ;$$
$$f_{yn} = \sum_{j=0}^{\infty} \alpha_{yj} f_{jn} ;$$
$$h_{yn} = \sum_{j=0}^{\infty} |\alpha_{yj}| h'_{jn} + \sum_{0 \leq s < \alpha} \tilde{h}'_{sn} .$$

The conditions (10.2.12) – (10.2.17) of the lemma follow at once from the definitions of the functions f_y, $f_{y,n}$, $h_{y,n}$, the inequalities $h_{y,n} \geq \sum_{0 \leq s < \alpha} \tilde{h}_{sn}$, and the corresponding properties of the functions f_j, f_{jn}, h_{jn}, and $\tilde{h}_{sn}$.

In order to finish the proof it remains to construct functions h_n, $n \in \mathbf{N}$, that satisfy the conditions (10.2.18) – (10.2.21) of the lemma.

First of all we notice that in order to find $n(y)$ such that (10.2.21) is satisfied, it is enough to show the inequalities

$$h_n \geq \tilde{h}_{sn} , \quad 0 \leq s < \alpha , \quad n \in \mathbf{N} ,$$

and the following claim:

For any $u \in \mathbf{N}$ and any $j \in \mathbf{N}$ there is an integer $n(j, u)$, such that

$$h_n \geq 2^u h_{jn}\,, \qquad \text{if} \quad n \geq n(j, u)\ . \tag{10.2.25}$$

We define integers $n'(j, u)$ so that

$$\left\| \{2^{n\alpha} h_{jn}\}_{n \geq n'(j,u)} \right\|_{L^p(l^2)} \leq 2^{-2u}\ , \tag{10.2.26}$$

observing that the existence of such numbers follows from (10.2.23), and set

$$h_n = \sum_{0 \leq s < \alpha} \tilde{h}_{sn} + \sum_{j=0}^{\infty} \sum_{u=j}^{\infty} 2^u \chi_{[n'(j,u),\infty)}(n) h_{jn}\ ,$$

where $\chi_{[n'(j,u),\infty)}(n) = 1$ if $n'(j, u) \leq n$, and $\chi_{[n'(j,u),\infty)}(n) = 0$ otherwise. We check that h_n satisfies the required conditions. The existence of $n(j, u)$ satisfying (10.2.25) follows immediately from the construction. The inequality (10.2.20) is a consequence of (10.2.24) and the inequalities

$$\begin{aligned} &\left\| \left\{ 2^{n\alpha} \sum\nolimits_{j=0}^{\infty} \sum\nolimits_{u=j}^{\infty} 2^u \chi_{[n'(j,u),\infty)}(n) h_{jn} \right\}_{n \geq 0} \right\|_{L^p(l^2)} \\ &\qquad \leq \sum_{j=0}^{\infty} \sum_{u=j}^{\infty} 2^u \| \{2^{n\alpha} h_{jn}\}_{n \geq n'(j,u)} \|_{L^p(l^2)} \\ &\qquad \leq \sum_{j=0}^{\infty} \left(\sum_{u=j}^{\infty} 2^u 2^{-2u} \right) \leq \sum_{j=0}^{\infty} 2^{1-j} \leq 4\ , \end{aligned}$$

which follow from (10.2.26).

Condition (10.2.19) is a trivial consequence of the corresponding properties of the functions h_{jn} and $\tilde{h}_{sn}$.

10.3 Further Results

10.3.1. Unlike the spaces $W^{m,p}(\mathbf{R}^N)$, $p > 1$, the spaces $W^{m,1}(\mathbf{R}^N)$ are not Lizorkin–Triebel spaces, and the method of proof of Theorem 10.1.1 cannot be extended to cover this case. The result remains true, however, and Netrusov [346] has even proved a stronger result. (A partial result was proved independently by Anders Carlsson [95].)

In order to state Netrusov's theorem we need the fact (see Section 6.5.1) that if $f \in W^{m,1}(\mathbf{R}^N)$, then quasiall points in the sense of Hausdorff measure Λ_{N-m} are Lebesgue points. Here we interpret Λ_{N-m} as the counting measure if $m \geq N$. We say that $\mathrm{Tr}_E\, f = 0$ if $D^\beta f(x) = 0$ for $\Lambda_{N-|\beta|}$-q.e. x in E for all multiindices β with $|\beta| < m$. The result is the following:

Let $L \in \mathbf{N}$, and $E \subset \mathbf{R}^N$. Then there exist functions η_j, $j = 1, 2, \ldots$, on $\mathbf{R}^N$, such that $0 \leq \eta_j \leq 1$, $\eta_j(x) = 0$ on a neighborhood of E, and such that for any function $f \in W^{m,1}(\mathbf{R}^N)$, $m = 1, 2, \ldots, L$, satisfying $\mathrm{Tr}_E\, f = 0$:

(a) $\|\eta_j f\|_{W^{m,1}} \le A\|f\|_{W^{m,1}}$, where A is a constant independent of f;
(b) $\lim_{j\to\infty} \|f - \eta_j f\|_{W^{m,1}} = 0$.

The proof depends on the following covering lemma, which is of independent interest. We denote by $\widetilde{\Lambda}_{N-m}$ the "dyadic Hausdorff measure", defined by means of coverings by dyadic cubes instead of balls, cf. Section 7.6.9.

Let $\{Q_\gamma\}$, $\gamma \in \mathbf{N}\times\mathbf{Z}^N$ be the dyadic cubes defined in the beginning of the chapter, let $E \subset Q_{00}$, and let $0 \le m \le N$. Then, for any λ, $0 < \lambda < 1$, there is a $\Gamma \subset \mathbf{N}\times\mathbf{Z}^N$ such that

(a) for all $\gamma \in \Gamma$

$$\widetilde{\Lambda}_{N-m}(Q_\gamma \cap E) \ge \lambda \widetilde{\Lambda}_{N-m}(Q_\gamma)\ ;$$

(b) for all $\gamma' \in \mathbf{N}\times\mathbf{Z}^N$

$$\sum_{\{\gamma\in\Gamma : Q_\gamma \subset Q_{\gamma'}\}} \widetilde{\Lambda}_{N-m}(Q_\gamma) \le \widetilde{\Lambda}_{N-m}(Q_{\gamma'})\ ;$$

(c)

$$\Lambda_{N-m}\big(E \setminus \textstyle\bigcup_{\gamma\in\Gamma} Q_\gamma\big) = 0\ .$$

It is remarkable that the functions η_j can be chosen independently of f for a given set E. Whether this is true in the general case considered in Theorem 10.1.1 is an open problem (see Hedberg [217]). In special cases it is possible, and the proof of Theorem 9.5.1 can, in fact, be modified to show this in the case of $W^{m,p}$, $1 < p < \infty$, for any compact set K that is $(1, p)$-thick at each of its points. See Hedberg [212], Theorem 3.1. On the other hand the problem is unsolved for general compact sets K with $C_{m,p}(K) = 0$.

10.4 Notes

The main results of this chapter, Theorems 10.1.1 and 10.2.3, are due to Yu. V. Netrusov. They are contained in two more general theorems, formulated for $p > 0$ and for a class of spaces containing the Besov and Lizorkin–Triebel spaces, which were announced by Netrusov in [343]. The proofs will appear in [347].

For the earlier history of Theorem 10.1.1 we refer to the notes to Chapter 9. The theorem is well known (for very general spaces) in the case when the set E is a smooth hypersurface. See J. L. Lions and E. Magenes [270, 271, 272], and H. Triebel [403, 404, 405]. The problem whether the theorem is true for arbitrary closed sets for the general spaces $L^{\alpha,p}$ and $B_\alpha^{p,q}$ obtained by interpolation between Sobolev spaces, was posed by Hedberg in 1978 in the first edition of the "Problem Book" (see [215]). Beyond some results for rather regular sets (see e.g. G. Forsling [152], and J. Marschall [289]), the only significant earlier progress on this problem known to the authors is due to J. Mateu and J. Verdera [292], and J. Orobitg [354], who by methods entirely different from those exposed here proved a result corresponding to Theorem 10.1.1 for, respectively, the spaces $F_2^{1,2}(\mathbf{R}^2)$, and $F_\alpha^{1,2}(\mathbf{R}^N)$, $0 < \alpha < 2$. These results are also contained in the theorem of Netrusov. A result for $p < 1$ was proved by J. Mateu and J. Orobitg [291]. See also the survey paper by Verdera [414].

The proofs of Theorems 10.1.1 and 10.2.3 are not previously published (even in preprint form). They were presented in lectures given by Netrusov in Linköping in March and November 1992, and they are included here with Netrusov's permission and active collaboration.

A sharper form of Theorem 10.1.4 is proved in Netrusov [341, 342]. (The statements (b) and (d) in this theorem are well known, and due to T. Bagby and W. P. Ziemer, [46]. See also the notes to Chapter 6.)

Lemma 10.1.8(a) is due to N. H. Abel [1], and is of some importance in the history of mathematics. Abel's note was written in order to refute a claim that a monotone positive series $\sum a_n$ converges if and only if $\lim na_n = 0$. The response by the unfortunate L. Olivier is amusing reading, and shows something of the background against which Abel's result should be seen. J. Hadamard [188] wrote about this theorem in 1902: "On sait comment Abel a fait entrer l'étude de la convergence des séries dans une voie nouvelle en montrant l'impossibilité d'obtenir, par une règle unique, une condition nécessaire et suffisante de convergence." Abel based his elegant proof on the inequality $\log(1+x) < x$ for $x > 0$, which he proved in the same paper, using it to obtain the estimate $\sum_{m=1}^{n} a_m/A_{m-1} > \log A_n - \log a_0$. See also the book by E. M. Polishchuk and T. O. Shaposhnikova [358] (or T. O. Shaposhnikova and V. G. Maz′ya [377]). Both parts (a) and (b) of the lemma appear as exercises (with hints) in W. Rudin [366], Chapter 3.

Theorem 10.2.2 is Theorem 1 in H. Whitney [431]. The proof was simplified by B. Malgrange [284], Chapter II, Theorem 1.3. Theorem 10.2.1 is proved in [431], Theorem 3, but is a special case of an older theorem which is true for continuous functions on a locally compact Hausdorff space. See e.g. L. H. Loomis [279], 19G, p. 57. Ideals in Sobolev algebras and other related algebras have been studied by L. G. Hanin [189] (and earlier papers), and E. M. Dyn′kin and L. G. Hanin [132].

11. Rational and Harmonic Approximation

We conclude the book by applying some of the results of earlier chapters to certain approximation problems in L^p-norm for analytic and harmonic functions. By duality these problems can be reformulated as "stability problems" in Sobolev spaces, which can be given complete solutions in terms of capacities. The main results depend on Theorem 9.1.3 (or Theorem 10.1.1), but mostly only on the easy case when the space is $W^{1,p}$. The more complicated case of spaces involving higher derivatives is used only in the last section, Section 11.5, but even this section can very well be read before studying the proof of the theorem.

In this chapter we prefer to change notation, and write q instead of p' for the exponent conjugate to p, i.e., $p^{-1}+q^{-1}=1$ for $1 \le p, q \le \infty$.

11.1 Approximation and Stability

Let $E \subset \mathbf{C}$ be a measurable set and let $1 \le q \le \infty$. We denote by $L^q_a(E)$ the subspace of $L^q(E)$ consisting of the functions holomorphic in E^0, and by $R^q(E)$ the closure in $L^q(E)$ of the set of functions with holomorphic extensions to neighborhoods of E. Clearly

$$R^q(E) \subset L^q_a(E) \subset L^q(E) \ ,$$

and we are interested in characterizing the sets E such that $R^q(E) = L^q_a(E)$, or, if $E^0 = \emptyset$, such that $R^q(E) = L^q(E)$.

If E is compact then $R^q(E)$ is the closure of the restrictions to E of the rational functions with poles off E. In fact, by Runge's theorem (see e.g. Rudin [367], Theorem 13.6) any function holomorphic on a neighborhood of a compact set E can be uniformly approximated on E by rational functions with poles on E^c. This is the most important situation, but it is not much harder to include the general case. If E is open, then by the definition $R^q(E) = L^q_a(E)$.

We can consider the functions on E to be extended to all of $\mathbf{C}$. In other words, we can view $L^q_a(E)$ as the subspace of $L^q(\mathbf{C})$ consisting of functions holomorphic in E^0, and $R^q(E)$ as the closure in $L^q(\mathbf{C})$ of the set of functions whose restrictions to E have holomorphic extensions to neighborhoods of E. Seen this way, $L^q_a(E) = L^q_a(E^0)$.

Before investigating the above approximation problem we want to determine the open G for which $L_a^q(G)$ is trivial in the sense that it contains only the zero function.

Proposition 11.1.1. *Let $G \subset \mathbf{C}$ be open, and $1 \leq q < \infty$. Then $L_a^q(G)$ is trivial in exactly the following cases:*

(a) $q = 1$, *and G^c consists of at most two points;*
(b) $1 < q < 2$, *and G^c consists of at most one point;*
(c) $2 \leq q < \infty$, *and $C_{1,p}(G^c) = 0$.*

Proof. Cases (a) and (b) are obvious. Let $2 \leq q < \infty$ and suppose that $C_{1,p}(G^c) > 0$. Then there is a positive measure μ with compact support contained in G^c such that $G_1 * \mu \in L^q(\mathbf{C})$, and consequently $I_1 * \mu \in L^q_{\text{loc}}(\mathbf{C})$. Then $f(z) = \int (\zeta - z)^{-1}\, d\mu(\zeta)$ is holomorphic in G, $f \in L^q_{\text{loc}}(\mathbf{C})$, and $|f(z)| = O(|z|^{-1})$, as $|z| \to \infty$. If $q > 2$ it follows that $f \in L_a^q(G)$, and f is not identically zero since $\lim_{|\zeta| \to \infty} z f(z) = \mu(G^c)$.

If $q = 2$ we modify the construction by choosing disjoint compact subsets K_1 and K_2 of G^c with $C_{1,2}(K_i) > 0$, $i = 1, 2$, and $\mu = \nu_1 - \nu_2$ with non-zero $\nu_i \in \mathcal{M}^+(K_i)$. If we choose ν_i so that $\nu_1(K_1) = \nu_2(K_2)$ and $I_1 * \nu_i \in L^2_{\text{loc}}(\mathbf{C})$, it follows that $|f(z)| = O(|z|^{-2})$, as $|z| \to \infty$, and thus $f \in L_a^2(G)$. In order to prove that f does not vanish identically it is enough to choose a simple closed contour separating K_1 from K_2, and integrate using Cauchy's theorem.

The converse follows from the Liouville theorem, and Theorem 2.7.4. In fact, there it is proved that K is removable for holomorphic functions in L^q if $N_{1,p}(K) = 0$, and $N_{1,p}(K) = C_{1,p}(K)$.

We treat the approximation problem formulated above by duality, and we start by identifying the annihilators of $L_a^q(E)$ and $R^q(E)$, where we consider the latter as subspaces of $L^q(\mathbf{C})$. Consequently, the annihilators are subspaces of $L^p(\mathbf{C})$, which we denote by $L_a^q(E)^\perp$ and $R^q(E)^\perp$. Note that by the discussion above,

$$L_a^q(E)^\perp = L_a^q(E^0)^\perp\ . \tag{11.1.1}$$

The elements of $W^{1,p}$ and other function spaces will now be complex valued.

As usual, the operator $\frac{1}{2}(\partial_1 + i\partial_2)$ is denoted $\bar\partial$, and consequently the Cauchy–Riemann equation is written $\bar\partial f = 0$.

Lemma 11.1.2. *Let $G \subset \mathbf{C}$ be open and bounded, let $1 < p, q < \infty$, and suppose that $g \in L^p(\mathbf{C})$. Then g belongs to the annihilator of $L_a^q(G) \subset L^q(\mathbf{C})$ if and only if (in the sense of distributions) $g = \bar\partial\varphi$ for some $\varphi \in W_0^{1,p}(G)$.*

Proof. By H. Weyl's lemma (see e.g. L. Schwartz [373], Ch. VI, §10) a function $f \in L^q(\mathbf{C})$ is holomorphic in G if and only if $\bar\partial f = 0$ on G in the sense of distributions, i.e., if and only if

$$\langle f, \bar\partial\varphi\rangle = \int_G f\, \bar\partial\varphi\, dm = 0$$

for all $\varphi \in C_0^\infty(G)$. Thus, by the Hahn–Banach theorem (see e.g. Rudin [368], Theorem 4.7), a function $g \in L^p(\mathbf{C})$ annihilates $L_a^q(G)$ if and only if g belongs to the closure in $L^p(\mathbf{C})$ of the set of functions $\{\bar{\partial}\varphi : \varphi \in C_0^\infty(G)\}$.

Suppose that $g \in L^p(\mathbf{C})$, and that there is a sequence $\{\varphi_n\}_1^\infty$ of functions in $C_0^\infty(G)$ such that $\lim_{n\to\infty} \|g - \bar{\partial}\varphi_n\|_p = 0$. Then, by the classical Pompeiu formula, φ_n can be represented as

$$\varphi_n(z) = -\frac{1}{\pi}\int_G \frac{\bar{\partial}\varphi_n(\zeta)}{\zeta - z}\, dm(\zeta) \ . \tag{11.1.2}$$

(This formula is just another way of writing (1.2.4). See also e.g. L. Hörmander [229], Theorem 1.2.1, p. 3.) By the L^p-boundedness of the Riesz transform (Theorem 1.1.4) it follows that the sequence $\{\nabla\varphi_n\}_1^\infty$ is Cauchy in $L^p(\mathbf{C})$. If G is bounded it follows from the Poincaré inequality that $\{\varphi_n\}_1^\infty$ is Cauchy in $L^p(\mathbf{C})$ and converges to a function $\varphi \in W_0^{1,p}(G)$ such that $\bar{\partial}\varphi = g$.

Conversely, if $g = \bar{\partial}\varphi$ for $\varphi \in W_0^{1,p}(G)$, then clearly there are $\varphi_n \in C_0^\infty(G)$ such that $\lim_{n\to\infty} \|g - \bar{\partial}\varphi_n\|_p = 0$.

We want to extend the above lemma to unbounded open sets. We denote

$$\dot{W}_0^{1,p}(G) = \overline{C_0^\infty(G)} \ , \tag{11.1.3}$$

where the closure is taken in the norm $\|\nabla\varphi\|_{L^p(G)}$. Now there is a problem, in that this space is not always a space of functions. For example, if $G = \mathbf{C}$ and $p \geq 2$, one can easily construct a sequence $\{\varphi_n\}_1^\infty$ in $C_0^\infty(\mathbf{C})$ such that $\lim_{n\to\infty} \|\nabla\varphi_n\|_p = 0$, but $\{\varphi_n\}_1^\infty$ does not converge, even in the sense of distributions. See J. Deny and J. L. Lions [122], Ch. I, Remarque 1b, p. 319.

Lemma 11.1.3. *Let $G \subset \mathbf{C}$ be open. Then the elements in $\dot{W}_0^{1,p}(G)$ are functions in $L^1_{\text{loc}}(\mathbf{C})$, if*

(a) $1 < p < 2$ *and G is arbitrary*;
(b) $p = 2$ *and* $C_{1,2}(G^c) > 0$;
(c) $p > 2$ *and* $G^c \neq \emptyset$.

Remark. The condition $C_{1,2}(G^c) > 0$ is also necessary. See Deny and Lions [122], Ch. II, Th. 2.1, p. 350.

Proof. It is enough to prove that for any ball B there is a constant A such that $\|\varphi\|_{L^1(B)} \leq A\, \|\nabla\varphi\|_{L^p(\mathbf{C})}$ for all $\varphi \in C_0^\infty(G)$.

If $1 < p < 2$ this is an immediate consequence of the Sobolev inequality, $\|\varphi\|_{L^{p^*}(\mathbf{C})} \leq A\, \|\nabla\varphi\|_{L^p(\mathbf{C})}$, where $p^* = 2p/(2-p)$.

If $p > 2$ we choose a point $a \in G^c$, and observe that if $\varphi \in C_0^\infty(G)$, then by (11.1.2)

$$\varphi(z) = \varphi(z) - \varphi(a) = -\frac{1}{\pi}\int_G \left(\frac{1}{\zeta - z} - \frac{1}{\zeta - a}\right)\bar{\partial}\varphi(\zeta)\, dm(\zeta)$$
$$= \int_G K(\zeta, z)\bar{\partial}\varphi(\zeta)\, dm(\zeta) \ .$$

Now, for any z the kernel $K(\zeta, z) = O(|\zeta|^{-2})$ as $|\zeta| \to \infty$. It follows that $K(\cdot, z) \in L^q(\mathbf{C})$ for $1 < q < 2$, and then Hölder's inequality easily gives that $\sup_{z \in B} |\varphi(z)| \leq A \|\nabla\varphi\|_{L^p(\mathbf{C})}$. (In fact, it is not hard to prove that there is A independent of z such that $|\varphi(z)| \leq A |z|^{1-(2/p)} \|\nabla\varphi\|_{L^p(\mathbf{C})}$.)

It remains to treat the case $p = 2$. The proof is a modification of the last argument. If $C_{1,2}(G^c) > 0$, it follows from Theorem 2.5.1 that we can find a probability measure μ with compact support on G^c such that $G_1 * \mu \in L^2(\mathbf{C})$, and thus $I_1 * \mu \in L^2_{\text{loc}}(\mathbf{C})$. Then for $\varphi \in C_0^\infty(G)$

$$\varphi(z) = \varphi(z) - \int \varphi(a)\, d\mu(a) = -\frac{1}{\pi} \int_G \left(\frac{1}{\zeta - z} - \int \frac{d\mu(a)}{\zeta - a} \right) \bar{\partial}\varphi(\zeta)\, dm(\zeta)$$

$$= \int_G K(\zeta, z)\, \bar{\partial}\varphi(\zeta)\, dm(\zeta) \ .$$

Now fix z and $\delta > 0$. It is easily seen that $K(\zeta, z) = O(|\zeta|^{-2})$, as $|\zeta| \to \infty$, and then it follows from the choice of μ that $K(\cdot, z) \in L^2\big(G \setminus B(z, \delta)\big)$. For $\zeta \in B(z, \delta)$ we have $|K(\zeta, z)| \leq |I_1(z)| + |I_1 * \mu(\zeta)|$. It follows that

$$|\varphi(z)| \leq G_1 * |\bar{\partial}\varphi|(z) + A\, \|\bar{\partial}\varphi\|_2 \ ,$$

where A depends on δ and μ, and consequently, by (1.1.6) and (1.2.13)

$$\|\varphi\|_{L^2(B)} \leq (1 + A|B|^{1/2}) \|\bar{\partial}\varphi\|_2 \ .$$

Remark. The reader who is familiar with classical potential theory will notice that in the last proof (for $p = 2$) we could have chosen the measure μ produced by sweeping the Dirac measure δ_z to ∂G. This is possible since $C_{1,2}(\partial G) = C_{1,2}(G^c) > 0$. In this way one obtains the representation formula

$$\varphi(z) = -\frac{1}{\pi} \int_G \partial \Gamma_G(\zeta, z)\, \bar{\partial}\varphi(\zeta)\, dm(\zeta) \ , \quad \varphi \in C_0^\infty(G) \ , \qquad (11.1.4)$$

where $\Gamma_G(\cdot, z)$ is the Green's function for G with pole at z. See Deny and Lions [122]. The formula is easily seen to be valid $(1, 2)$-q.e. for all $\varphi \in \dot{W}_0^{1,2}(G)$.

The following lemma now follows immediately.

Lemma 11.1.4. *Let* $1 < p, q < \infty$, *let* $G \subset \mathbf{C}$ *be open and such that* $L_a^q(G)$ *contains non-zero functions. Suppose that* $g \in L^p(\mathbf{C})$. *Then* g *belongs to the annihilator of* $L_a^q(G) \subset L^q(\mathbf{C})$ *if and only if (in the sense of distributions)* $g = \bar{\partial}\varphi$ *for some* $\varphi \in \dot{W}_0^{1,p}(G)$.

Note that the functions φ in the lemma are determined uniquely by g. In fact, the only function $\varphi \in \dot{W}_0^{1,p}(G)$ such that $\bar{\partial}\varphi = 0$ is the zero function. This follows easily from (11.1.2) and the extensions of this formula used in the proof of Lemma 11.1.3.

Functions φ in $\dot{W}_0^{1,p}(G)$ clearly belong to $W^{1,p}(\mathbf{C})$ locally, in the sense that $\eta\varphi \in W^{1,p}(\mathbf{C})$ for any $\eta \in C_0^\infty(\mathbf{C})$. Thus, they can be assumed to be continuous if $p > 2$, and $(1, p)$-quasicontinuous ($(1, p)$-q.c.) if $1 < p \leq 2$.

We now extend the definition of $\dot{W}_0^{1,p}(G)$ to arbitrary sets.

Definition 11.1.5. (a) Let $2 < p < \infty$ and suppose that $\mathbf{C} \setminus E \neq \{\emptyset\}$. Then

$$\dot{W}_0^{1,p}(E) = \{\varphi : \nabla\varphi \in L^p(\mathbf{C}),\ \varphi \in C(\mathbf{C}),\ \varphi(z) = 0 \text{ on } E^c\}\ .$$

(b) Let $1 < p \leq 2$, and suppose that $C_{1,2}(E^c) > 0$ if $p = 2$. Then

$$\dot{W}_0^{1,p}(E) = \{\varphi : \nabla\varphi \in L^p(\mathbf{C}),\ \varphi \text{ is } (1,p)\text{-q.c.},\ \varphi(z) = 0\ (1,p)\text{-q.e. on } E^c\}\ .$$

It follows immediately from Theorem 9.1.3 (the easy case $m = 1$; see Section 9.2) that for open sets this definition agrees with (11.1.3).

We can now easily determine the annihilator of $R^q(E)$.

Lemma 11.1.6. *Let $E \subset \mathbf{C}$ be a Borel set, let $1 < p, q < \infty$, and suppose that $g \in L^p(\mathbf{C})$. Suppose that $R^q(E) \neq \{0\}$. Then g belongs to the annihilator of $R^q(E) \subset L^q(\mathbf{C})$ if and only if $g = \bar{\partial}\varphi$, where $\varphi \in \dot{W}_0^{1,p}(E)$.*

Proof. Consider $L_a^q(G)$ for any open G as the subspace of $L^q(\mathbf{C})$ consisting of the functions holomorphic in G. The annihilator of $R^q(E)$ is

$$R^q(E)^\perp = \bigcap_{G \supset E} L_a^q(G)^\perp\ ,$$

where the intersection is taken over all open sets containing E. Let $g \in R^q(E)^\perp$. It follows from Lemma 11.1.4 that $g = \bar{\partial}\varphi$, where $\varphi \in \dot{W}_0^{1,p}(G)$ for all open G containing E. If E is compact it follows immediately that $\operatorname{supp}\varphi \subset E$.

If $2 < p < \infty$ functions $\varphi \in \dot{W}_0^{1,p}(G)$ are continuous and vanish on G^c, and (a) follows for general sets E. For $1 < p \leq 2$ and general E we have $\varphi(x) = 0$ $(1,p)$-q.e. on every closed subset of E^c. But E is Borel, and hence by the capacitability theorem (Theorem 2.3.11) $\varphi(x) = 0$ $(1,p)$-q.e. on E^c.

Conversely, if $g = \bar{\partial}\varphi$, where $\varphi(z) = 0$ $(1,p)$-q.e. on E^c, then if E is compact, clearly $\varphi \in W_0^{1,p}(G)$ for all open G containing E. In the general case $\varphi(z) = 0$ $(1,p)$-q.e. (everywhere if $2 < p < \infty$) on G^c for every open G containing E. It follows from Lemma 11.1.4 that $g \in \bigcap_{G \supset E} L_a^q(G)^\perp = R^q(E)^\perp$.

Lemmas 11.1.4 and 11.1.6 motivate the following definition.

Definition 11.1.7. A set $E \subset \mathbf{C}$ is called $(1,p)$-*stable* if $\dot{W}_0^{1,p}(E^0) = \dot{W}_0^{1,p}(E)$.

It follows that all planar sets are $(1,p)$-stable for $2 < p < \infty$. In fact, if φ is continuous and $\varphi(z) = 0$ on E^c, then $\varphi(z) = 0$ on ∂E.

Theorem 11.1.8. *Let $1 < q < \infty$ and let $G \subset \mathbf{C}$ be an open set such that $L_a^q(G) \neq \{0\}$.*

If $1 < q < 2$, then rational functions of the type $r(z) = \sum_n a_n(z - z_n)^{-1}$, $z_n \in \partial G$, are dense in $L_a^q(G)$.

If $2 \leq q < \infty$, then functions in $L_a^q(G)$ of the type

$$r(z) = \int_{\partial G} \frac{d\nu(\zeta)}{z - \zeta}\ , \quad \nu \in \mathcal{M}(\partial G)\ ,$$

are dense in $L_a^q(G)$.

Remark. The theorem is true for $q = 1$ also. See Section 11.6.1.

Proof. Let $1 < q < \infty$, let $g \in L^p(G)$, and suppose that $\int_G g\,r\,dm = 0$ for all functions r of the type described in the theorem.

In the case $2 < q < \infty$ we denote $\tilde{g}(z) = \int_G g(\zeta)(\zeta - z)^{-1}\,dm(\zeta)$, and then (in the sense of distributions) $\bar{\partial}\tilde{g} = -\pi g$. If we set $r(z) = \int_{\partial G}(\zeta - z)^{-1}d\nu(\zeta)$ for an arbitrary $\nu \in \mathcal{M}^+(G^c) \cap L^{-1,q}$, we have $r \in L^q_a(G)$, and then by the assumption $\int_G g\,r\,dm = 0$. Fubini's theorem gives $\int_G g\,r\,dm = \int_{\partial G} \tilde{g}\,d\nu$, since $\tilde{g}$ clearly is $(1, p)$-quasicontinuous and ν-integrable. But this is only possible if $\tilde{g}(z) = 0$ $(1, p)$-q.e. on ∂G. It follows that $\tilde{g}|_G \in \dot{W}^{1,p}_0(G)$, and thus $g \in L^q_a(G)^\perp$ by Lemma 11.1.4.

For $1 < q \le 2$ we let $K(\zeta, z)$ be the function constructed in the proof of Lemma 11.1.3. We set $\tilde{g}(z) = \int_G g(\zeta)K(\zeta, z)\,dm(\zeta)$, and then we still have $\bar{\partial}\tilde{g} = -\pi g$.

In the case $1 < q < 2$ it follows by choosing $r(\zeta) = K(\zeta, z)$ that $\tilde{g}(z) = 0$ for all $z \in \partial G$. Moreover $\tilde{g}$ is continuous in the whole plane, and hence the restriction $\tilde{g}|_G$ belongs to $\dot{W}^{1,p}_0(G)$, and thus $g \in L^q_a(G)^\perp$.

In the case $q = 2$, $K(\zeta, z)$ depends on a measure $\mu \in \mathcal{M}^+(G^c) \cap L^{-1,2}$ as in the proof of Lemma 11.1.3. If we set $r(\zeta) = \int_{G^c} K(\zeta, z)\,d\nu(z)$ for an arbitrary compactly supported $\nu \in \mathcal{M}^+(G^c) \cap L^{-1,2}$, we have $r \in L^2_a(G)$ if $\nu(G^c) = \mu(G^c)$, and thus $\int_G g\,r\,dm = 0$. It follows again that $\int_G \tilde{g}\,d\nu = 0$, and thus $\tilde{g}(z) = 0$ $(1, p)$-q.e. on ∂G. The claim follows as before.

Remark. If G is bounded, we can simplify the proof by using the Cauchy kernel $(\zeta - z)^{-1}$ for all q instead of $K(\zeta, z)$, which was introduced only in order to insure rapid enough decay at infinity.

We can now formulate the main result of this section.

Theorem 11.1.9. (a) *Let $1 < q < 2$. Then $R^q(E) = L^q_a(E)$ for all measurable $E \subset \mathbf{C}$.*

(b) *Let $2 \le q < \infty$, and let $E \subset \mathbf{C}$ be a Borel set. Then $R^q(E) = L^q_a(E)$ if and only if E is $(1, p)$-stable.*

Proof. This is an immediate consequence of the definition of $(1, p)$-stability, put together with Lemmas 11.1.4 and 11.1.6, and the fact that, as remarked in (11.1.1), $L^q_a(E)^\perp = L^q_a(E^0)^\perp$.

It is easy to see that $(1, p)$-stability is a local property of sets, in the sense that if E is covered by a union of open sets $\{G\}$ and each $E \cap G$ is $(1, p)$-stable, then E is $(1, p)$-stable.

In the following sections we shall give a number of equivalent characterizations of $(1, p)$-stable sets. See in particular Theorem 11.4.1.

Clearly, in Definition 11.1.7 it is of no importance if the elements in the function spaces are real or complex valued, since the real and imaginary parts can be approximated separately. Thus, the L^q-approximation problem by analytic functions is equivalent to a problem in real analysis, which can just as well be formulated for sets in $\mathbf{R}^N$. In the next section we shall treat this question.

We want to make some related remarks, concerning the set function that is known as *analytic capacity*.

If $K \subset \mathbf{C}$ is compact, and $f \in L_a^q(K^c)$, then f has a Laurent expansion, $f(z) = \sum_{n=1}^{\infty} a_n z^{-n}$. We denote the coefficient a_1 by $f'(\infty)$, and then

$$f'(\infty) = \lim_{|z|\to\infty} zf(z) = \frac{1}{2\pi i}\int_c f(z)\,dz \ ,$$

where c is any simple closed contour surrounding K. For $2 < q \le \infty$ we define

$$\gamma_q(K) = \sup\{\,|f'(\infty)| : f \in L_a^q(K^c),\ \|f\|_{L^q(K^c)} \le 1\,\}\ .$$

For $q = \infty$ this is the *analytic capacity* of K, a quantity that plays an important role in the theory of *uniform* rational approximation; see the notes at the end of the chapter.

For $q = 2$ the definition does not give a non-zero quantity, and we have to modify it. If Ω is a bounded domain containing K, we define

$$\gamma_2(K;\Omega) = \sup\left\{\frac{1}{2\pi}\left|\int_c f(z)\,dz\right| : f \in L_a^2(\Omega\setminus K),\ \|f\|_{L^2(\Omega\setminus K)} \le 1\right\}\ ,$$

where c is a contour as above, contained in Ω.

The following proposition, which has no counterpart for $\gamma = \infty$, shows that there is no need for an analytic q-capacity for $q < \infty$. Cf. Theorem 2.7.2.

Proposition 11.1.10. *Let $K \subset \mathbf{C}$ be compact, and let Ω be a bounded, simply connected domain containing K. If $2 < q < \infty$ then*

$$\gamma_q(K) = \frac{1}{\pi}\inf\{\,\|\bar\partial\omega\|_p : \omega \in C_0^\infty(\mathbf{C}),\ \omega|_K = 1\,\}\ ; \tag{11.1.5}$$

if $q = 2$, then

$$\gamma_2(K;\Omega) = \frac{1}{\pi}\inf\{\,\|\bar\partial\omega\|_2 : \omega \in C_0^\infty(\Omega),\ \omega|_K = 1\,\}\ ; \tag{11.1.6}$$

and there are constants A_1 and A_2 such that

$$A_1 C_{1,p}(K)^{1/p} \le \gamma_q(K) \le A_2 C_{1,p}(K)^{1/p}\ , \quad 2 < q < \infty\ , \tag{11.1.7}$$

and (with constants depending on Ω)

$$A_1 C_{1,2}(K)^{1/2} \le \gamma_2(K;\Omega) \le A_2 C_{1,2}(K)^{1/2}\ . \tag{11.1.8}$$

Proof. Let $\omega \in C_0^\infty$ be such that $\omega(z) = 1$ on a neighborhood of K. Then, if c is a suitable contour in $\{\,z : \omega(z) = 1\,\}\setminus K$, by Green's formula,

$$\begin{aligned}\frac{1}{2\pi i}\int_c f(z)\,dz &= \frac{1}{2\pi i}\int_c f(z)\omega(z)\,dz = -\frac{1}{2\pi i}\int_{\mathbf{C}} \bar\partial\big(f(z)\omega(z)\big)\,d\bar z\wedge dz \\ &= -\frac{1}{\pi}\int_{\mathbf{C}} f(z)\bar\partial\omega(z)\,dm(z)\ .\end{aligned} \tag{11.1.9}$$

The inequalities

$$\gamma_q(K) \le \frac{1}{\pi} \inf\{ \|\bar\partial\omega\|_p : \omega \in C_0^\infty(\mathbf{C}),\ \omega|_K = 1\} \ , \tag{11.1.10}$$

and

$$\gamma_2(K;\Omega) \le \frac{1}{\pi} \inf\{ \|\bar\partial\omega\|_2 : \omega \in C_0^\infty(\Omega),\ \omega|_K = 1\} \tag{11.1.11}$$

follow immediately from the Hölder inequality.

The equality (11.1.6) now follows by choosing functions ω_n in (11.1.9) that converge to the harmonic measure for K with respect to Ω, i.e., the harmonic function ω in $\Omega \setminus K$ that solves the Dirichlet problem with boundary data 1 on K, and 0 on $\partial\Omega$. The function $f = -\partial\omega/\|\partial\omega\|_{L^2(\Omega\setminus K)}$ is holomorphic, and gives equality in (11.1.11) with $f'(\infty) = \gamma_2(K;\Omega)$.

For $2 < q < \infty$ we use the fact that by (11.1.9) the mapping $f \mapsto f'(\infty)$ is a linear functional on $L_a^q(K^c)$ represented by the function $-\frac{1}{\pi}\bar\partial\omega \in L^p(K^c)$. By a standard application of the Hahn–Banach theorem (see e.g. W. Rudin [368], Exercise 4.19, p. 113),

$$\begin{aligned}\gamma_q(K) &= \tfrac{1}{\pi} \sup\{ |\langle f, \bar\partial\omega\rangle| : f \in L_a^q(K^c),\ \|f\|_{L^q(K^c)} \le 1\} \\ &= \tfrac{1}{\pi} \inf\{ \|\bar\partial\omega + g\|_{L^p(K^c)} : g \in L_a^q(K^c)^\perp \} \ .\end{aligned}$$

Lemma 11.1.4 now easily gives (11.1.5). The estimates (11.1.7) and (11.1.8) follow from the L^p-boundedness of the Riesz transform (Theorem 1.1.4), and the equivalence of $C_{1,p}$ and $\dot C_{1,p}$; see Section 5.1.

Remark. In Section 7.6.2 we defined a condenser capacity, $C_{1,2}(K;\Omega)$. It is clear from the above proof that $\gamma_2(K;\Omega) = \frac{1}{2}C_{1,2}(K;\Omega)^{1/2}$. Moreover, the extremal ω is the capacitary potential for K with respect to the Green function $\Gamma_\Omega(\cdot, z)$, cf. (11.1.4) above.

11.2 Approximation by Harmonic Functions in Gradient Norm

The stability of sets in $\mathbf{R}^N$ is closely connected to the Dirichlet problem. In fact, one can prove that a compact $K \subset \mathbf{R}^N$ is $(1,2)$-stable in the sense of Definition 11.1.7 if and only if it is stable for the classical Dirichlet problem under perturbations of the domain in a sense defined by M. V. Keldysh and M. A. Lavrent′ev in the 1930's. We do not intend to go into this interesting circle of ideas here, but some references are given in the notes at the end of the chapter. Here we wish to relate $(1,p)$-stability to approximation by harmonic functions in the $W^{1,q}$-norm. In order not to complicate the picture we now limit ourselves to bounded sets.

Let $G \subset \mathbf{R}^N$, $N \ge 2$, be open and bounded. We denote by $D^q(G)$ the subspace of $W^{1,q}(\mathbf{R}^N)$ consisting of functions harmonic in G. If E is bounded we define

$$D^q(E) = \overline{\bigcup_{G\supset E} D^q(G)} \ ,$$

with the union taken over all open bounded G containing E, and the closure taken in $W^{1,q}(\mathbf{R}^N)$, and as in Definition 11.1.5, $W_0^{1,p}(E) = \dot{W}_0^{1,p}(E)$ consists of those $(1, p)$-quasicontinuous $\varphi \in W^{1,p}(\mathbf{R}^N)$ for which $\varphi(x) = 0$ $(1, p)$-q.e. on E^c.

It should be noticed that in general functions in $W^{1,q}(G)$ cannot be extended to $W^{1,q}(\mathbf{R}^N)$, and thus $D^q(G)$ is different from the subspace of $W^{1,q}(G)$ consisting of functions harmonic in G.

In order to compare $D^q(E^0)$ and $D^q(E)$, we want to identify their annihilators. We can assume that all functions involved vanish outside of some large ball B, and this allows us to use the norm on $\dot{W}_0^{1,q}(B)$, i.e. $\|\nabla u\|_{L^q(B)}$, as a norm on $W^{1,q}$.

Suppose that G is open, $G \subset \frac{1}{2}B$, and let T be a distribution in the dual of $\dot{W}_0^{1,q}(B)$. Suppose that $\langle T, u\rangle = 0$ for all $u \in D^q(G) \subset \dot{W}_0^{1,q}(B)$. Then clearly $\operatorname{supp} T \subset \overline{G}$, and T is bounded on $W^{1,q}(\mathbf{R}^N)$, i.e., $T \in W^{-1,p}(\mathbf{R}^N) = L^{-1,p}(\mathbf{R}^N)$. Let $\varphi = I_2 * T$, so that $\varphi \in W^{1,p}(B)$, and $T = -\Delta\varphi$. It follows that for any test function $u \in C_0^\infty(B)$ we have

$$\begin{aligned}\langle T, u\rangle &= -\langle T, I_2 * \Delta u\rangle = -\langle I_2 * T, \Delta u\rangle \\ &= -\int_B \varphi\, \Delta u\, dm = \int_B \nabla\varphi \cdot \nabla u\, dm\ ,\end{aligned}$$

and by continuity this equality extends to all $u \in \dot{W}_0^{1,q}(B)$.

We denote by $D^q(G)^\perp$ the subspace of $W^{1,p}(B)$ consisting of functions φ such that $\int_B \nabla\varphi \cdot \nabla u\, dm = 0$ for all $u \in D^q(G) \subset \dot{W}_0^{1,q}(B)$.

If $\varphi \in C_0^\infty(G)$ it follows that $\varphi \in D^q(G)^\perp$, since $\int_B \nabla\varphi \cdot \nabla u\, dm = \int_G \varphi\, \Delta u\, dm = 0$ if $u \in D^q(G)$, and it follows by continuity that $\dot{W}_0^{1,p}(G) \subset D^q(G)^\perp$. On the other hand, if $u \in \dot{W}_0^{1,p}(B)$ and $\int_G \Delta\varphi\, u\, dm = 0$ for all $\varphi \in C_0^\infty(G)$, then by the Weyl lemma u has to be harmonic in G. By the Hahn–Banach theorem, $D^q(G)^\perp \subset \dot{W}_0^{1,p}(G)$.

We have proved the first part of the following lemma. The second part follows as in the proof of Lemma 11.1.6.

Lemma 11.2.1. *Let* $1 < q < \infty$. *If* $G \subset \mathbf{R}^N$ *is bounded and open, then* $D^q(G)^\perp = W_0^{1,p}(G)$.

If $E \subset \mathbf{R}^N$ *is a bounded Borel set, then* $D^q(E)^\perp = W_0^{1,p}(E)$.

We extend the definition of stable sets (see Definitions 11.1.7 and 11.1.5).

Definition 11.2.2. A bounded set $E \subset \mathbf{R}^N$ is called $(1, p)$-*stable* if

$$W_0^{1,p}(E^0) = W_0^{1,p}(E)\ .$$

It follows that all bounded $E \subset \mathbf{R}^N$ are $(1, p)$-stable for $N < p < \infty$. In fact, if φ is continuous and $\varphi(z) = 0$ on E^c, then $\varphi(z) = 0$ on ∂E.

The following is an immediate consequence.

Theorem 11.2.3. *If* $1 < q < N/(N-1)$, *then* $D^q(E^0) = D^q(E)$ *for all bounded* $E \subset \mathbf{R}^N$.

If $N/(N-1) \le q < \infty$, *then* $D^q(E^0) = D^q(E)$ *for a bounded Borel set* $E \subset \mathbf{R}^N$ *if and only if* E *is* $(1, p)$-*stable.*

One can also prove a theorem analogous to Theorem 11.1.8.

Theorem 11.2.4. *Let $1 < q < \infty$ and let $G \subset \mathbf{R}^N$ be a bounded open set.*

If $1 < q < N/(N-1)$, then finite sums of the type $r(x) = \sum_n a_n I_2(x - x_n)$, $x_n \in \partial G$, are dense in $D^q(G)$.

If $N/(N-1) \le q < \infty$, then potentials in $D^q(G)$ of the type $I_2 * \nu$, $\nu \in \mathcal{M}(\partial G)$, are dense in $D^q(G)$. (If $N = 2$ the kernel $I_2(x)$ should be replaced by the logarithmic kernel $\log(1/|x|)$.)

We omit the proof. If $q > N/(N-1)$ the proof again extends without change to unbounded sets.

11.3 Stability of Sets Without Interior

Before we characterize stable sets in general we want to consider the stability of nowhere dense sets. If E is a set without interior points, then clearly E is $(1, p)$-stable if and only if $W_0^{1,p}(E) = \{0\}$. Sets with the latter property are also called *sets of uniqueness* for $W^{1,p}$. We will study sets of uniqueness in the more general context of the spaces $L^{\alpha,p}$, where α can be any positive real number.

Definition 11.3.1. A set $E \subset \mathbf{R}^N$ is a set of uniqueness for $L^{\alpha,p}(\mathbf{R}^N)$ if the only element $f \in L^{\alpha,p}(\mathbf{R}^N)$ such that $f(x) = 0$ (α, p)-q.e. on E^c is the zero element.

Clearly no set with interior can be a set of uniqueness for $L^{\alpha,p}$. If $\alpha p > N$, all functions in $L^{\alpha,p}(\mathbf{R}^N)$ are continuous, so in this case a set is a set of uniqueness for $L^{\alpha,p}$ if and only if it has no interior.

Theorem 11.3.2. *Let $\alpha > 0$, $1 < p \le N/\alpha$, and let $E \subset \mathbf{R}^N$ be a Borel set without interior points. Then the following are equivalent:*

(a) *E is a set of uniqueness for $L^{\alpha,p}$, i.e., E is (α, p)-stable;*
(b) *$C_{\alpha,p}(G \setminus E) = C_{\alpha,p}(G)$ for all open G;*
(c) *$C_{\alpha,p}(B(x,\delta) \setminus E) = C_{\alpha,p}(B(x,\delta))$ for all open balls $B(x,\delta)$;*
(d) *For almost all x (with respect to Lebesgue measure)*

$$\limsup_{\delta\to 0} \frac{C_{\alpha,p}(B(x,\delta) \setminus E)}{\delta^N} > 0 \ . \tag{11.3.1}$$

Remark 1. The equivalence of (c) and (d) is surprising, since (d) appears to be a much weaker condition (cf. Propositions 5.1.2 – 5.1.4). This is an instance of the so called *instability of capacity*. See Section 11.6.2 below.

Remark 2. By means of the above theorem it is easy to construct compact, nowhere dense sets of non-uniqueness for $L^{\alpha,p}(\mathbf{R}^N)$ for any $\alpha > 0$ and $1 < p \le N/\alpha$. One can define a set $K = \overline{B} \setminus (\bigcup_i B_i)$, where $\overline{B}$ is a closed ball, and $\{B_i\}_1^\infty$ is a dense collection of disjoint open balls contained in B whose radii are so small that $\sum C_{\alpha,p}(B_i) < C_{\alpha,p}(B)$. Then (b) in the theorem is not satisfied by K.

Proof. We first prove that (d) implies (a) and (b). Suppose that (11.3.1) holds a.e. Let $u = G_\alpha * f$, $f \in L^p$. Then $u(x_0)$ is well defined (α, p)-q.e., and more precisely, at all x_0 such that

$$\int_{\mathbf{R}^N} G_\alpha(x_0 - y)\,|f(y)|\,dy < \infty \ . \tag{11.3.2}$$

Choose x_0 so that (11.3.1) and (11.3.2) hold at x_0, and so that

$$\lim_{\delta\to 0} \frac{1}{\delta^N} \int_{|x-x_0|<\delta} |f(x) - f(x_0)|^p\,dx = 0 \ . \tag{11.3.3}$$

This condition is satisfied for a.e. x_0 by a simple modification of the Lebesgue differentiation theorem (see e.g. Stein [389], Section I.5.7).

Without loss of generality we set $x_0 = 0$. For any $\delta > 0$ we can find a probability measure ν_δ with support in $B(0,\delta) \setminus E$ such that

$$\|G_\alpha * \nu_\delta\|_q \le 2C_{\alpha,p}(B(0,\delta) \setminus E)^{-1/p} \ . \tag{11.3.4}$$

(See Corollary 2.5.2.) We shall show that for a suitable sequence $\{\delta_i\}_1^\infty$

$$\lim_{\delta_i\to 0} \int_{\mathbf{R}^N} u\,d\nu_{\delta_i} = u(0) \ . \tag{11.3.5}$$

We have

$$\int_{\mathbf{R}^N} u\,d\nu_\delta = \int_{\mathbf{R}^N} (G_\alpha * f)\,d\nu_\delta = \int_{\mathbf{R}^N} (G_\alpha * \nu_\delta)\,f\,dx \ .$$

Let $\varepsilon > 0$ be arbitrary. By (11.3.2) we can choose $\rho > 0$ so that

$$\int_{|x|<\rho} G_\alpha(x)\,|f(x)|\,dx < \varepsilon. \tag{11.3.6}$$

Let δ be arbitrary, $0 < \delta < \frac{1}{2}\rho$. Then

$$\begin{aligned}
\left|\int_{\mathbf{R}^N} (G_\alpha * \nu_\delta) f\,dx - u(0)\right| &= \left|\int_{\mathbf{R}^N} \big((G_\alpha * \nu_\delta) - G_\alpha\big) f\,dx\right| \\
&\le \int_{|x|<\rho} G_\alpha|f|\,dx + \left|\int_{|x|<2\delta} (G_\alpha * \nu_\delta) f\,dx\right| \\
&\quad + \int_{2\delta\le|x|<\rho} (G_\alpha * \nu_\delta)|f|\,dx + \int_{|x|\ge\rho} \big|(G_\alpha * \nu_\delta) - G_\alpha\big||f|\,dx \\
&= I_1(\delta) + |I_2(\delta)| + I_3(\delta) + I_4(\delta) \ .
\end{aligned}$$

Here $I_1(\delta) < \varepsilon$ by (11.3.6), and $\lim_{\delta\to 0} I_4(\delta) = 0$, since $\lim_{\delta\to 0}(G_\alpha * \nu_\delta)(x) = G_\alpha(x)$, uniformly for $|x| \ge \rho$. The second integral is written

$$\begin{aligned}
I_2(\delta) &= \int_{|x|<2\delta} (G_\alpha * \nu_\delta)(f - f(0))\,dx + f(0)\int_{|x|<2\delta} (G_\alpha * \nu_\delta)\,dx \\
&= I_2'(\delta) + I_2''(\delta) \ ,
\end{aligned}$$

where, using (11.3.4),

$$I_2'(\delta) \le \|G_\alpha * \nu_\delta\|_q \Big(\int_{|x|<2\delta} |f - f(0)|^p \, dx\Big)^{1/p}$$
$$\le 2\Big(\frac{\delta^N}{C_{\alpha,p}(B(0,\delta)\setminus E)}\Big)^{1/p} \Big(\frac{1}{\delta^N}\int_{|x|<\delta} |f - f(0)|^p \, dx\Big)^{1/p} .$$

Thus, by the assumption (11.3.1), and by (11.3.3), there is a sequence $\{\delta_i\}_1^\infty$ such that $\lim_{\delta_i \to 0} I_2'(\delta_i) = 0$. Moreover, $\lim_{\delta \to 0} I_2''(\delta) = 0$, since

$$|I_2''(\delta)| = |f(0)| \int d\nu_\delta(y) \int_{|x|<2\delta} G_\alpha(x-y)\, dx \le |f(0)| \int_{|x|<3\delta} G_\alpha(x)\, dx \ .$$

Finally,

$$I_3(\delta) = \int d\nu_\delta(y) \int_{2\delta \le |x| < \rho} G_\alpha(x-y)|f(x)|\, dx$$
$$\le \int_{|x|<\rho} G_\alpha(\tfrac{1}{2}x)|f(x)|\, dx \le A \int_{|x|<\rho} G_\alpha(x)|f(x)|\, dx < A\,\varepsilon$$

by (1.2.14) and (11.3.6). Since $\varepsilon > 0$ is arbitrary, (11.3.5) follows.

Now suppose that $u \in L_0^{\alpha,p}(K)$ for some compact set $K \subset E$. Then $u(x) = 0$ on E^c, and it follows from (11.3.5) that $u(x) = 0$ a.e., and consequently everywhere. This proves that $L_0^{\alpha,p}(E) = \{0\}$, i.e., (d) implies (a).

Similarly, suppose that $f \in L^p$, $f \ge 0$, and that $u = G_\alpha * f \ge 1$ on $G \setminus E$ for some open G. (If there is no such f, then $C_{\alpha,p}(G \setminus E) = \infty$, and there is nothing to prove.) Then (11.3.5) implies that $u(x) \ge 1$ a.e. on G, and thus $u(x) \ge 1$ everywhere on G (see Proposition 2.6.7). It follows from the definition of capacity that $C_{\alpha,p}(G \setminus E) \ge C_{\alpha,p}(G)$, and thus (d) implies (b).

That (b) implies (c) and (d) is trivial, so in order to finish the proof it is enough to prove that (a) implies (b). Suppose that there is an open G such that $C_{\alpha,p}(G \setminus E) < C_{\alpha,p}(G)$. Let $u = G_\alpha * f$ be the capacitary function for $G \setminus E$, so that $u(x) \ge 1$ (α, p)-q.e. on $G \setminus E$, and $\|f\|_p^p = C_{\alpha,p}(G \setminus E)$. Then $u(x) < 1$ on a subset of G of positive Lebesgue measure, since otherwise $C_{\alpha,p}(G \setminus E) = C_{\alpha,p}(G)$. By Corollary 3.3.4 there is a constant A and a function $v = \Phi \circ u$ such that $\|v\|_{\alpha,p} < A\,\|f\|_p$, and $v(x) = 1$ (α, p)-q.e. on $G \setminus E$. $\Phi(t)$ can be chosen so that $\Phi(t) < 1$ for $t < 1$, and then $v(x) < 1$ on a subset of G of positive measure. Hence, for a suitable smooth function φ with support in B, the function $\varphi(1 - v)$ belongs to $L_0^{\alpha,p}(E)$ without being identically zero. This finishes the proof of the theorem.

11.4 Stability of Sets with Interior

We now give several equivalent conditions for a set $E \subset \mathbf{R}^N$ to be $(1, p)$-stable, and thus to allow holomorphic or harmonic approximation in the sense of Theorems 11.1.9 and 11.2.3. We recall that the set of points where a set S is (α, p)-thin is denoted $e_{\alpha,p}(S)$; see Definition 6.3.7.

Theorem 11.4.1. *Let $E \subset \mathbf{R}^N$ be an arbitrary set, and let $1 < p \leq N$. Then the following are equivalent:*

(a) *E is $(1, p)$-stable;*
(b) *$C_{1,p}(G \setminus E) = C_{1,p}(G \setminus E^0)$ for all open G;*
(c) *There is $\eta > 0$ such that $C_{1,p}(G \setminus E) \geq \eta\, C_{1,p}(G \setminus E^0)$ for all open G;*
(d) *For $(1, p)$-q.e. $x \in \partial E$*

$$\liminf_{n\to\infty} \frac{C_{1,p}(B_n(x) \setminus E)}{C_{1,p}(B_n(x) \setminus E^0)} > 0\ ;$$

(e) *$C_{1,p}(\partial E \cap e_{1,p}(E^c)) = 0$;*
(f) *$e_{1,p}(E^c) = e_{1,p}((E^0)^c)$;*
(g) *$C_{1,p}\big(e_{1,p}(E^c) \setminus e_{1,p}((E^0)^c)\big) = 0$.*

Proof. We first prove that (c) implies (a). Let φ be a $(1, p)$-quasicontinuous function such that $\varphi(x) = 0$ $(1, p)$-q.e. on E^c. We recall (see Proposition 6.4.9) that a set U is $(1, p)$-quasiopen if for every $\varepsilon > 0$ there is an open H containing U such that $C_{1,p}(H \setminus U) < \varepsilon$. In particular, the set $\{x : \varphi(x) \neq 0\}$ is quasiopen (see Proposition 6.4.10). Thus, if (c) is assumed, then (a) will follow if we can show that (c) extends to all quasiopen sets U. In fact, we can choose $U = \{x : \varphi(x) \neq 0\}$, and then (c) gives $C_{1,p}(U \setminus E^0) \leq \eta^{-1} C_{1,p}(U \setminus E) = 0$.

To see that (c) extends in this way is easy. Let $\varepsilon > 0$, and let U be contained in an open H with $C_{1,p}(H \setminus U) < \varepsilon$. Then, if (c) is assumed,

$$\begin{aligned} C_{1,p}(U \setminus E^0) &\leq C_{1,p}(H \setminus E^0) \leq \eta^{-1} C_{1,p}(H \setminus E) \\ &\leq \eta^{-1}\big(C_{1,p}(U \setminus E) + C_{1,p}(H \setminus U)\big)\ . \end{aligned}$$

But ε is arbitrary, and this proves the claim.

Now assume that (b) is not satisfied, i.e., there is an open G such that $C_{1,p}(G \setminus E) < C_{1,p}(G \setminus E^0)$. Let $u = G_1 * f$ be the capacitary function for $G \setminus E$, so that $u(x) \geq 1$ $(1, p)$-q.e. on $G \setminus E$, and $\|f\|_p^p = C_{1,p}(G \setminus E)$. Then there is a compact $K \subset G \cap \partial E$ such that $u(x) < 1$ on K, and $C_{1,p}(K) > 0$, since otherwise $C_{1,p}(G \setminus E) = C_{1,p}(G \setminus E^0)$. By Theorem 3.3.1 the space $W^{1,p}$ is closed under truncation, so if we set $v(x) = \min\{u(x), 1\}$, then $v \in W^{1,p}$, and $v(x) = 1$ $(1, p)$-q.e. on $G \setminus E$. Let ω be any function in $C_0^\infty(G)$ which is 1 on K. Then the function $\omega(1 - v)$ belongs to $W_0^{1,p}(E)$ without belonging to $W_0^{1,p}(E^0)$. It follows that (a) implies (b), and we have proved the equivalence of (a), (b), and (c).

The equivalence of (b) – (g) is obtained by choosing $S = E^c$ and $T = (E^0)^c$ in the following more general theorem.

Theorem 11.4.2. *Let $1 < p < \infty$, and $0 < \alpha \leq N/p$. Let S and T be arbitrary subsets of $\mathbf{R}^N$ with $S \subset T$. Then the following statements are equivalent:*

(a) *$C_{\alpha,p}(G \cap S) = C_{\alpha,p}(G \cap T)$ for all open G;*
(b) *There is $\eta > 0$ such that $C_{\alpha,p}(G \cap S) \geq \eta\, C_{\alpha,p}(G \cap T)$ for all open G;*

(c) *For* (α, p)*-q.e.* $x \in T \setminus S$

$$\liminf_{n\to\infty} \frac{C_{\alpha,p}(B_n(x) \cap S)}{C_{\alpha,p}(B_n(x) \cap T)} > 0 \ ;$$

(d) $C_{\alpha,p}(T \cap e_{\alpha,p}(S)) = 0$;
(e) $e_{\alpha,p}(S) = e_{\alpha,p}(T)$;
(f) $C_{\alpha,p}\big(e_{\alpha,p}(S) \setminus e_{\alpha,p}(T)\big) = 0$.

Proof. We first observe that (d) implies (a). In fact, let $f \in L^p$, $f \geq 0$, be such that $u(x) = G_\alpha * f(x) \geq 1$ on $G \cap S$. By Theorem 6.4.5 u is (α, p)-finely continuous (α, p)-quasieverywhere. (Alternatively we could use Theorem 6.4.5, which says that every quasicontinuous function is also finely continuous quasieverywhere.) Thus, if $u(x) \geq 1$ q.e. on $G \cap S$, then $u(x) \geq 1$ (α, p)-q.e. on $G \cap T \setminus e_{\alpha,p}(S)$. But then (d) implies that $u(x) \geq 1$ (α, p)-q.e. on $G \cap T$, and (a) follows from the definition of capacity.

We now invoke the definition of thinness, Definition 6.3.7. It is clear from (6.3.2) that (a) implies (e), and similarly (c) implies (f). The Kellogg lemma, Corollary 6.3.17, gives $C_{\alpha,p}\big(T \cap e_{\alpha,p}(T)\big) = 0$. Thus each of (e) and (f) implies (d). That (a) implies (b), and (b) implies (c) is trivial. The theorem is proved.

11.5 Approximation by Harmonic Functions and Higher Order Stability

The L^q-approximation problems considered in Section 11.1 can be extended to approximation by harmonic functions, or by solutions of more general elliptic equations. Then new difficulties appear, because if the equation is of order α, then the dual problem takes place in $W^{\alpha,p}$, and for $\alpha \geq 2$ these spaces are not closed under truncation, which makes them much more complicated to deal with than $W^{1,p}$.

We first extend Definitions 11.1.5 and 11.1.7. We recall the definition of the trace operator given in (9.1.1):

For $f \in L^{\alpha,p}$ and $E \subset \mathbf{R}^N$, $\mathrm{Tr}_E f = 0$ means that for all multiindices with $0 \leq |\beta| < \alpha$, $D^\beta f(x) = 0$ on E if $(\alpha - |\beta|)p > N$, and $D^\beta f(x) = 0$ $(\alpha - |\beta|, p)$-q.e. on E if $(\alpha - |\beta|)p \leq N$.

Definition 11.5.1. Let $E \subset \mathbf{R}^N$, let $\alpha > 0$ and $1 < p < \infty$. Then

$$L_0^{\alpha,p}(E) = \{\varphi : \varphi \in L^{\alpha,p}(\mathbf{R}^N),\ \mathrm{Tr}_{E^c}\, \varphi = 0\} \ .$$

We recall that by Theorem 9.1.3 or Theorem 10.1.1

$$L_0^{\alpha,p}(E) = \overline{\bigcup_{K\subset E} L_0^{\alpha,p}(K)} \ , \tag{11.5.1}$$

where the union is taken over all compact subsets of E, and the closure is taken in $L^{\alpha,p}$. Consequently, for open sets Definition 11.5.1 agrees with the usual definition, i.e.

$$L_0^{\alpha,p}(G) = \overline{C_0^\infty(G)} \quad \text{for } G \text{ open} . \tag{11.5.2}$$

Definition 11.5.2. Let $E \subset \mathbf{R}^N$, $\alpha > 0$, and $1 < p < \infty$. Then E is (α, p)-*stable* if

$$L_0^{\alpha,p}(E^0) = L_0^{\alpha,p}(E) .$$

Now let $\mathcal{L}$ be an elliptic partial differential operator of order α with constant coefficients, let $E \subset \mathbf{R}^N$ be measurable, and let $1 \le q \le \infty$. In analogy with Section 11.1 we denote by $L^q_{\mathcal{L}}(E)$ the subspace of $L^q(E)$ consisting of the functions u satisfying the equation $\mathcal{L}u = 0$ in the interior of E. Similarly, $R^q_{\mathcal{L}}(E)$ denotes the closure in $L^q(E)$ of the functions u having extensions that satisfy $\mathcal{L}u = 0$ on some neighborhood of E.

In order to concentrate on the essential difficulties, we consider bounded sets E and assume that $\alpha < N$. Then we can easily prove the following theorem.

Theorem 11.5.3. *Let $E \subset \mathbf{R}^N$ be a bounded Borel set, let $\mathcal{L}$ be an elliptic partial differential operator of order $\alpha < N$ with constant coefficients, and let $1 < q < \infty$. Then $R^q_{\mathcal{L}}(E) = L^q_{\mathcal{L}}(E)$ if and only if E is (α, p)-stable.*

Remark. If $q < N/(N-1)$, i.e., $p > N$, then all sets are (α, p)-stable for integer α by Theorem 9.1.3, and $R^q_{\mathcal{L}}(E) = L^q_{\mathcal{L}}(E)$ for all measurable sets E.

Proof. Let $g \in L^p(\mathbf{R}^N)$. We first prove as in Lemma 11.1.2 that g belongs to the annihilator of $L^q_{\mathcal{L}}(G)$ for a bounded open G if and only if $g = \mathcal{L}^*\varphi$ for some $\varphi \in W_0^{m,p}(G)$ with $\mathcal{L}^*$ denoting the adjoint operator. But it is well known that a function $f \in L^q(\mathbf{R}^N)$ is a solution to $\mathcal{L}f = 0$ in G in the classical sense if and only if it is a solution in the sense of distributions, i.e.,

$$\langle f, \mathcal{L}^*\varphi \rangle = \int_G f\, \mathcal{L}^*\varphi\, dm = 0 \quad \text{for all } \varphi \in C_0^\infty(G) .$$

(If $\mathcal{L}$ is the Laplacian this is the Weyl lemma. (See e.g. L. Schwartz [373], Ch. VI, §10.) In general it is a consequence of the hypoellipticity of elliptic operators with constant coefficients; see e.g. L. Hörmander [230], Theorem 4.4.1, and Theorem 7.1.22.) By the Hahn–Banach theorem the annihilator of $L^q_{\mathcal{L}}(G)$ is the closure in $L^p(\mathbf{R}^N)$ of the set of functions $\{ \mathcal{L}^*\varphi : \varphi \in C_0^\infty(G) \}$. It is a consequence of the ellipticity of $\mathcal{L}$ and of the theory of singular integrals that this closure is the set $\{ \mathcal{L}^*\varphi : \varphi \in W_0^{m,p}(G) \}$. In fact, φ can be written as $\varphi = \mathcal{E} * \mathcal{L}^*\varphi$, where $\mathcal{E}$ is the fundamental solution of $\mathcal{L}^*$, and convolution with $\mathcal{L}^*$ is a continuous operation from L^p to $W^{m,p}$. (If $\mathcal{L}$ is a power of the Laplacian, this is just the fact that the operators $\partial_i \partial_j \Delta^{-1}$ are bounded on L^p by the boundedness of the Riesz transform; see Theorem 1.1.4.) It follows also that the functions φ in $W_0^{m,p}(G)$ are uniquely determined by $g = \mathcal{L}^*\varphi$ in $L^q_{\mathcal{L}}(G)^\perp$.

If E is an arbitrary set it follows as in Lemma 11.1.6 that

$$R^q_{\mathcal{L}}(E)^{\perp} = \bigcap_{G \supset E} L^q_{\mathcal{L}}(G)^{\perp} ,$$

the intersection being taken over all open G containing E. Thus, if $g \in R^q_{\mathcal{L}}(E)^{\perp}$, then $g = \mathcal{L}^*\varphi$, where $\varphi \in \bigcap_{G \supset E} W_0^{m,p}(G)$. It follows that $\mathrm{Tr}_F\, \varphi = 0$ for all closed $F \subset E^c$, and by the capacitability of E^c, $\mathrm{Tr}_{E^c}\, \varphi = 0$, i.e., $\varphi \in W_0^{m,p}(E)$.

Conversely, if $g = \mathcal{L}^*\varphi$, where $\varphi \in W_0^{m,p}(E)$, then it is easily seen as in Lemma 11.1.6 that $g \in R^q_{\mathcal{L}}(E)^{\perp}$.

The theorem follows in the same way as Theorem 11.1.9 from the fact that $L^q_{\mathcal{L}}(E)^{\perp} = L^q_{\mathcal{L}}(E^0)^{\perp}$.

The (α, p)-stable sets do not have such a simple characterization for general α as in the case $\alpha = 1$, except for nowhere dense sets, for which a complete solution of the problem was given in Section 11.3.

We first observe that Theorem 11.4.2 immediately gives a number of equivalent *sufficient* conditions for E to be (α, p)-stable, for example the following one.

Theorem 11.5.4. *Let $\alpha > 0$, and $1 < p < \infty$. A set $E \subset \mathbf{R}^N$ is (α, p)-stable if E^c is $(\alpha - k, p)$-thick $(\alpha - k, p)$-q.e. on ∂E, i.e., if*

$$C_{\alpha-k,p}(\partial E \cap e_{\alpha-k,p}(E^c)) = 0 \quad \textit{for all } k \in \mathbf{N},\ k < \alpha\ .$$

Remark 1. If $(\alpha - k)p > N$, then $\partial E \cap e_{\alpha-k,p}(E^c) = \emptyset$, and the condition is automatically fulfilled.

Remark 2. This theorem was actually proved (for integer α and closed sets E) in the course of proving Theorem 9.1.3.

Proof. Let $\varphi \in L_0^{\alpha,p}(E)$, and suppose that E satisfies the condition of the theorem. If β is a multiindex with $(\alpha - |\beta|)p > N$, then φ is continuous and vanishes on ∂E. If $\alpha - |\beta| \le N/p$, then $D^\beta\varphi$ is $(\alpha - |\beta|, p)$-finely continuous quasieverywhere, and the condition implies that $D^\beta\varphi$ vanishes q.e. on ∂E.

Conversely, it is easy to see that a *necessary* condition for E to be (α, p)-stable is that $C_{\alpha,p}(\partial E \cap e_{\alpha,p}(E^c)) = 0$, or equivalently, that E satisfies $C_{\alpha,p}(G \setminus E) = C_{\alpha,p}(G \setminus E^0)$ for all open G.

In fact, if the last condition is not satisfied, then the last part of the proof of Theorem 11.3.2 shows that there is a nonzero function $\varphi \in L_0^{\alpha,p}(E) \setminus L_0^{\alpha,p}(E^0)$.

There is, however, no reason to believe that the further conditions in Theorem 11.5.4 are also necessary. The proof of Theorem 11.3.2 breaks down, because if $\alpha > 1$, and $C_{\alpha-1,p}(G \setminus E) < C_{\alpha-1,p}(G \setminus E^0)$, then one can also construct a function $\varphi \neq 0$ belonging to $L_0^{\alpha-1,p}(E) \setminus L_0^{\alpha-1,p}(E^0)$, but this function is in general not the derivative of a function in $L_0^{\alpha,p}(E) \setminus L_0^{\alpha,p}(E^0)$.

Moreover, if E has no interior, then by Theorem 11.3.2 it is both necessary and sufficient for E to be (α, p)-stable that $C_{\alpha,p}(G \setminus E) = C_{\alpha,p}(G)$ for all open G. In particular, if $\alpha p > N$, then this condition is always satisfied, and all nowhere dense sets are stable. But by removing an infinite number of disjoint suitably small balls from the unit ball B it is easy to construct a nowhere dense compact E, such that $C_{\alpha-k,p}(B \setminus E) < C_{\alpha-k,p}(B)$ for all $k < \alpha$ such that $(\alpha - k)p \leq N$. It follows that the conditions in Theorem 11.5.4 are not necessary in the case of sets without interior.

We know by Theorem 10.1.1 that all sets are (α, p)-stable as soon as $p > N$, whereas, as remarked after Definition 11.3.1, all nowhere dense sets are (α, p)-stable for $\alpha p > N$. The following theorem shows that the presence of an interior really complicates the situation, and that the stable sets with interior cannot be characterized in such simple terms as when there is no interior.

Theorem 11.5.5. *Let $1 < p \leq N$ and let $\alpha \geq 1$ be an integer. Then there exists a compact $K \in \mathbf{R}^N$ such that K is not (α, p)-stable.*

Proof. It is enough to construct a set K and a function $\varphi \in W^{\alpha,p}(\mathbf{R}^N)$ such that $\operatorname{supp}\varphi \subset K$, and $\partial_N^{\alpha-1}\varphi(x) \neq 0$ on a subset of ∂K with positive $(1, p)$-capacity. We carry out the construction only in the borderline case $p = N$, which is the most delicate one.

Let B_0 be the unit ball in $\mathbf{R}^N$ and denote by D the $(N-1)$-dimensional ball $D = \{x \in \mathbf{R}^N : x_N = 0, |x| \leq \frac{1}{2}\}$. We shall choose a sequence of suitable disjoint balls $\{B_k\}_1^\infty$, where $B_k = B(x_k, r_k)$ with $x_k \in D$, and set $K = \overline{B}_0 \setminus (\bigcup_1^\infty B_k)$.

Let $R_k > r_k$, and let $\eta_k \in C^\infty(0, \infty)$ be such that $0 \leq \eta_k \leq 1$, $\eta_k(r) = 1$ for $0 \leq r \leq r_k$, $\eta_k(r) = 0$ for $r \geq R_k$, and $|\eta_k^{(j)}(r)| \leq Ar^{-j}(\log R_k/r_k)^{-1}$ for $1 \leq j \leq \alpha$. Such a function can be constructed by modifying the function $(\log R_k/r_k)^{-1}\log^+ R_k/r$. Set $\psi_k(x) = \eta_k(|x - x_k|)$, and choose a function $\varphi_0 \in C_0^\infty(B_0)$ such that $\varphi_0(x) = x_N^{\alpha-1}$ in a neighborhood of D.

It is easily verified that $\sum_{|\sigma|=\alpha} \int |D^\sigma(\varphi_0\psi_k)|^N\,dx \leq A\,(\log R_k/r_k)^{1-N}$, if R_k is small enough. Now choose the balls $\{B(x_k, R_k)\}_1^\infty$ so that they have pairwise disjoint closures and are dense in D, and so that $\sum_{k=1}^\infty R_k^{N-1} < 2^{1-N}$. Finally, choose $\{r_k\}_1^\infty$ so that $\sum_{k=1}^\infty (\log R_k/r_k)^{1-N} < \infty$, and set $\varphi = \varphi_0(1 - \sum_{k=1}^\infty \psi_k)$. Clearly $\varphi \in W^{\alpha,N}$, and $\operatorname{supp}\varphi \subset K$. But every $x \in D$ that is not contained in one of the balls $B(x_k, R_k)$ is a boundary point of K. On the line perpendicular to D through such a point x we have $\varphi = \varphi_0$, and thus $\partial_N^{\alpha-1}\varphi(x) = (\alpha-1)!$. Since the set of such points has positive $(N-1)$-dimensional measure, φ has the desired properties.

An easy modification gives the following example.

Theorem 11.5.6. *Let $1 < p \leq N - 1$, and let $\alpha \geq 1$ be an integer. Then there is a compact set $K \subset \mathbf{R}^N$ with connected complement such that K is not (α, p)-stable.*

Proof. Let $K_0 \subset \mathbf{R}^{N-1}$ be the set constructed in the proof of Theorem 11.5.5, and set $K = K_0 \times [0, 1]$. Let $\varphi \in W^{\alpha,N-1}(\mathbf{R}^{N-1})$ be the function constructed in Theorem 11.5.5, let $\psi \in C_0^\infty(0, 1)$, and set $\Phi(x) = \varphi(x')\psi(x'')$ for $x' \in \mathbf{R}^{N-1}$, $x'' \in \mathbf{R}$. Then Φ has the desired properties.

Theorem 11.5.5, while showing that the situation is complicated, also gives a clue as to how one could find a characterization of the stable sets. It is clear that one has to find a way of measuring sets that takes into account their behavior in different directions.

More precisely, we make the following definition of a "polynomial capacity". It is analogous to the definition of $N_{\alpha,p}$ given in Definition 2.7.1. The space of polynomials of degree at most s is denoted $\mathfrak{P}_s$.

Definition 11.5.7. Let $E \subset \mathbf{R}^N$, $1 < p < \infty$, α be a positive integer, and let $\sigma \in \mathbf{N}^N$ be a multiindex with $|\sigma| < \alpha$. Let Ω be open with $\overline{E} \subset \Omega$. Then

$$N_{\alpha,p;\sigma}(E;\,\Omega) = \inf \|x^\sigma - f - \pi\|^p_{W^{\alpha,p}(\Omega)} \;,$$

the infimum being taken over all $f \in W^{\alpha,p}(\Omega)$ with $\mathrm{Tr}_E\, f = 0$, and over all $\pi \in \mathfrak{P}_{|\sigma|}$ such that $\pi(x) = \sum_{\gamma \neq \sigma} a_\gamma x^\gamma$.

Proposition 11.5.8. *Under the assumptions in Definition* 11.5.7

$$N_{\alpha,p;\sigma}(E;\,\Omega) = \sup_T |\langle T, x^\sigma\rangle|^p \;,$$

where the supremum is taken over all $T \in W^{-\alpha,q}(\Omega)$ *such that* $\operatorname{supp} T \subset E$, $\|T\|_{W^{-\alpha,q}(\Omega)} = 1$, *and* $\langle T, x^\gamma\rangle = 0$ *for all multiindices* $\gamma \neq \sigma$ *with* $|\gamma| \le |\sigma|$.

Remark. If Corollary 9.1.7 is taken into account, it follows that it is enough to consider distributions T of the form $T = D^{\beta_1}\mu_1 + \cdots + D^{\beta_n}\mu_n$, where the μ_i are measures such that $\operatorname{supp}\mu_i \subset E$ and $D^{\beta_i}\mu_i \in L^{-\alpha,q}$, $|\beta_i| < \alpha$.

Proof. This is an immediate consequence of the Hahn–Banach theorem.

Proposition 11.5.9. *Let B be a ball, and let $E \subset \frac{1}{2}B$. Under the assumptions in Definition* 11.5.7 *there is a constant A such that*

$$N_{\alpha,p;\sigma}(E;\,B) \ge A\, N_{\alpha-|\sigma|,p}(E;\,B) \ge A\, C_{\alpha-|\sigma|,p}(E) \;.$$

Proof. For f and π as in Definition 11.5.7 it is obvious that

$$\|x^\sigma - f - \pi\|_{W^{\alpha,p}(B)} \ge \sigma!\, \|1 - D^\sigma f\|_{W^{\alpha-|\sigma|,p}(B)} \ge \sigma!\, \inf_g \|1 - g\|_{W^{\alpha-|\sigma|,p}(\Omega)} \;,$$

where the infimum is taken over all $g \in W^{\alpha-|\sigma|,p}(B)$ such that $\mathrm{Tr}_E\, g = 0$. But for $E \subset \frac{1}{2}B$, the last quantity in the chain is comparable to

$$\inf\{\, \|g\|_{W^{\alpha-|\sigma|,p}(2B)} : g \in W_0^{\alpha-|\sigma|,p}(2B),\ g|_E = 1 \,\} \ge N_{\alpha,p}(E) \ge C_{\alpha,p}(E) \;,$$

where the last two inequalities are immediate consequences of the definitions.

The complication added by introducing these new capacities is in the nature of things, and (by Theorem 11.5.3) the following theorem gives a solution of the problem of harmonic L^q-approximation which is as satisfactory as Theorem 11.4.1.

Theorem 11.5.10. *Let $E \subset \mathbf{R}^N$ be a Borel set, let $1 < p < \infty$, and α a positive integer. Then the following conditions are equivalent*:

(a) *E is (α, p)-stable.*
(b) $N_{\alpha,p;\sigma}(G\backslash E;\Omega) = N_{\alpha,p;\sigma}(G\backslash E^0;\Omega)$ *for all open G and Ω such that $\overline{G} \subset \Omega$, and all $|\sigma| < \alpha$.*
(c) *There is $\eta > 0$ such that $N_{\alpha,p;\sigma}(B \setminus E; 2B) \geq \eta\, N_{\alpha,p;\sigma}(B \setminus E^0; 2B)$ for all open balls B and all $|\sigma| < \alpha$.*
(d) *For all σ such that $0 < (\alpha - |\sigma|)p \leq N$*

$$\sum_{n=0}^{\infty}\Big(2^{n(N-(\alpha-|\sigma|)p)} N_{\alpha,p;\sigma}\big(B_n(x) \setminus E^0\,;\, B_{n-1}(x)\big)\Big)^{q-1} = \infty \qquad (11.5.3)$$

for $(\alpha - |\sigma|, p)$-q.e. $x \in \partial E$.

Proof. We first prove that (d) implies (a). It is enough to prove that $f \in W_0^{\alpha,p}(E)$ implies $\mathrm{Tr}_{\partial E} f = 0$, i.e., $D^\sigma f(x) = 0$ for $(\alpha - |\sigma|), p)$-q.e. $x \in \partial E$.

The idea of the proof is to use Proposition 6.3.12 (of N. G. Meyers), i.e. the fact that for any $s > 0$, and for any finite $\mu \in \mathcal{M}^+$

$$W^{\mu}_{s,p}(x) = \sum_{n=0}^{\infty}\big(2^{n(N-sp)}\mu(B_n(x))\big)^{p-1} < \infty \qquad (11.5.4)$$

for (s, p)-q.e. $x \in \mathbf{R}^N$. Cf. Remark 2 following Theorem 9.5.1.

Let $f \in W^{\alpha,p}(\mathbf{R}^N)$. For any ball $B(\delta)$ with radius δ, and $s = 0, 1, \ldots, \alpha - 1$, there is a polynomial $\pi \in \mathfrak{P}_s$ such that

$$\|\nabla^k(f-\pi)\|_{L^p(B(\delta))} \leq A\,\delta^{s+1-k}\|\nabla^{s+1} f\|_{L^p(B(\delta))}\ , \qquad k = 0, 1, \ldots, s\ ,$$

(see Corollary 8.1.4), and thus

$$\inf_{\pi\in\mathfrak{P}_s} \|f-\pi\|_{W^{\alpha,p}(B(\delta))} \leq A \sum_{k=s+1}^{\alpha} \|\nabla^k f\|_{L^p(B(\delta))}\ ,$$

with A independent of δ for $\delta \leq 1$. Thus, by applying (11.5.4) to the measure $\mu = \big(\sum_{k=s+1}^{\alpha} |\nabla^k f|^p\big)m$, we see that for $(\alpha - s, p)$-q.e. x there are polynomials $\pi_n \in \mathfrak{P}_s$, $s = 1, 2, \ldots, \alpha - 1$, such that

$$\sum_{n=0}^{\infty}\Big(2^{n(N-(\alpha-s)p)}\|f-\pi_n\|^p_{W^{\alpha,p}(B_n(x))}\Big)^{q-1} < \infty\ . \qquad (11.5.5)$$

On the other hand, by Definition 11.5.7, for $f \in W_0^{\alpha,p}(E)$, for any polynomial $\pi(y) = \sum_{|\gamma|\leq s} a_\gamma y^\gamma$, and any σ with $|\sigma| = s$, we have

$$\|f-\pi\|^p_{W^{\alpha,p}(B_n(x))} \geq |a_\sigma|\, N_{\alpha,p;\sigma}\big(B_{n+1}(x) \setminus E;\, B_n(x)\big)\ .$$

Denote $\pi_n(y) = \sum_{|\gamma|\leq s} a_{n,\gamma} y^\gamma$. Here the coefficients $a_{n,\gamma}$ depend on x. As $n \to \infty$, π_n converges for $(\alpha - s, p)$-q.e. x to π, the differential at x of f, in the sense that $\lim_{n\to\infty} a_{n,\sigma} = a_\sigma$ exists for $0 \leq |\sigma| \leq s$, and $a_\sigma = D^\sigma f(x)/\sigma!$. This follows from the construction of the polynomials π_n as averages (see (8.1.2)), and Propositions 6.1.2 and 6.1.3.

Let E_s be the subset of ∂E where both the latter fact, (11.5.3) with $|\sigma| = s$, and (11.5.5) hold, and let $x \in E_s$. Then, for each σ with $|\sigma| = s$,

$$\|f - \pi_n\|^p_{W^{\alpha,p}(B_n(x))} \geq |a_{n,\sigma}|\, N_{\alpha,p;\sigma}\big(B_{n+1}(x) \setminus E\ ;\ B_n(x)\big)\ ,$$

and

$$\sum_{n=0}^{\infty}\Big(2^{n(N-(\alpha-s)p)} N_{\alpha,p;\sigma}\big(B_{n+1}(x) \setminus E^0\ ;\ B_n(x)\big)\Big)^{q-1} = \infty\ .$$

It follows that $D^\sigma f(x) = \lim_{n\to\infty} a_{n\sigma} = 0$ for $|\sigma| = s$. Once we know this, we can apply (11.5.3) and (11.5.5) for $|\sigma| = s-1, s-2, \ldots, 0$, whence $D^\sigma f(x) = 0$ for $|\sigma| \leq s$ for all $x \in E_s$, i.e. $(\alpha - s, p)$-q.e. on ∂E. This is true for $s = 0, 1, \ldots, \alpha - 1$, i.e., $\mathrm{Tr}\,|_{\partial E} f = 0$. This proves that (d) implies (a).

In order to prove that (a) implies (b) we again apply Definition 11.5.7. The extremal function $f + \pi$ exists by uniform convexity, and (a) implies that the extremal for $N_{\alpha,p;\sigma}(G \setminus E; \Omega)$ is also extremal for $N_{\alpha,p;\sigma}(G \setminus E^0; \Omega)$. This proves the claim.

Finally, we prove that (c) implies (d). But this follows immediately from Proposition 11.5.9, and the Kellogg property, Corollary 6.3.17.

11.6 Further Results

11.6.1. L. Bers [52] (see also [53]) proved that for any open $G \subset \mathbf{C}$ the rational functions with poles on ∂G are dense in $L^1_a(G)$. If $g \in L^\infty$ and $g \in L^1_a(G)^\perp$, then as in Lemma 11.1.3, $g = \bar\partial\varphi$, where $\varphi(z) = 0$ on G^c. It is still true that $\int_G fg\,dm$ can be approximated by $\int_G f\bar\partial\varphi_n\,dm$, where $\operatorname{supp}\varphi_n \subset G$, but the simple proof by truncation in Section 9.2 cannot be used when $p = \infty$. Bers used the estimates $|\varphi(z) - \varphi(z+h)| = O(|h|\log(1/|h|)$, as $|h| \to 0$, and $|\varphi(z)| = O(|z|\log|z|))$, as $|z| \to \infty$, and an "Ahlfors mollifier" ω_n to construct approximating functions $\varphi_n = \omega_n\varphi$. This work was prior to the results of V. P. Havin and others exposed in this chapter. See also the notes to Chapter 9.

11.6.2. As is well known, Lebesgue measure has the following instability property: If $E \subset \mathbf{R}^N$ is measurable, then $\lim_{\delta\to 0}|E \cap B(x,\delta)|/|B(x,\delta)|$ exists for a.e. $x \in \mathbf{R}^N$ and equals either 0 or 1. A. G. Vitushkin (see [415, 416]) discovered in connection with his investigation of uniform rational approximation that analytic capacity γ_∞ has a stronger instability property: If $E \subset \mathbf{C}$ is arbitrary, then for a.e. $z \in \mathbf{C}$ either $\lim_{\delta\to 0}\delta^{-2}\gamma_\infty(E \cap B(z,\delta)) = 0$, or $\lim_{\delta\to 0}\delta^{-1}\gamma_\infty(E \cap B(z,\delta)) = 1$. A. A. Gonchar [185, 186], and Yu. A. Lysenko and B. M. Pisarevskiĭ [283] proved a similar instability result for classical harmonic capacity $\dot{C}_{1,2}$.

Extending this result and Theorem 11.3.2, C. Fernström [147] proved that if $E \subset \mathbf{R}^N$ is Borel, then for a.e. $x \in \mathbf{R}^N$ either

$$\lim_{\delta\to 0}\delta^{-N} C_{\alpha,p}(E \cap B(x,\delta)) = 0\ ,$$

or

$$\lim_{\delta \to 0} C_{\alpha,p}(B(x,\delta))^{-1} C_{\alpha,p}(E \cap B(x,\delta)) = 1 \ .$$

In [148] Fernström proved instability results for Hausdorff content. See also J. Mateu and J. Orobitg [291], P. Mattila and J. Orobitg [293], and H. Federer [141], Section 3.3.22, p. 309.

Problems somewhat related to the instability of capacity have been studied in connection with the Dirichlet problem in domains with many holes. See e.g. the book by E. Ya. Khruslov and V. A. Marchenko [246], and D. Cioranescu and F. Murat [106]. A recent paper which contains many references to the literature is G. Dal Maso and A. Garroni [112].

11.6.3. The proofs given here for approximation in L^q are functional analytic and non-constructive. On the contrary, the proofs of the main theorems on uniform rational approximation are constructive and depend on an approximation scheme of A. G. Vitushkin (see Vitushkin [416] and T. W. Gamelin [175], Chapter VIII), and no functional analytic proofs are available. Per Lindberg [267, 268] was able to modify Vitushkin's scheme in a crucial way and proved constructively that Theorem 11.4.1(b) or (c) (for open balls) are necessary and sufficient conditions for $R^q(E) = L^q_a(E)$. Lindberg's method was extended by T. Bagby [44] to prove part of Theorem 11.5.10. See also the book by N. N. Tarkhanov [396].

11.6.4. If $K \subset \mathbf{C}$ is compact, a point $z \in K$ is said to be a *bounded point evaluation* for $R^q(K)$ if the map $z \mapsto f(z)$, defined for a dense subset of $R^q(K)$, extends as a bounded linear functional on $R^q(K)$. For $q < 2$ this is only possible for interior points of K, but for $q \geq 2$ there may be bounded point evaluations on the boundary. In fact, z is a bounded point evaluation for $R^q(K)$, $2 \leq p < \infty$, if and only if $\sum_{n=0}^{\infty} 2^{np} C_{1,p}(B_n(z) \setminus K) < \infty$. See Hedberg [209] and C. Fernström [146].

It is natural to believe that for a set K with no interior the absence of bounded point evaluations for $R^q(K)$ would be sufficient for $R^q(K) = L^q(K)$. However, using the above criterion, Fernström [146] gave a counterexample to this conjecture.

Fernström and J. C. Polking [149] extended these results to solutions of elliptic equations of any order in N dimensions, and gave a complete analysis of the relationship between the existence of bounded point evaluations and the density in $L^q(K)$ of solutions of elliptic equations.

T. Kolsrud [256, 258] gave a similar characterization of bounded point evaluations for harmonic functions in $W^{1,2}$.

11.7 Notes

11.1. As early as 1922, T. Carleman [91] (Hilfssatz I) proved, essentially, that if $G \subset \mathbf{C}$ is a domain bounded by a Jordan curve, then polynomials are dense in $L^q(G)$ for $q > 0$. His proof depended on properties of conformal mapping discovered by E. Lindelöf. (See also J. L. Walsh [425], Chapter 2, Theorem 14.) There are many extensions of this result; see e.g. the surveys by S. N. Mergelyan [316], and M. S. Mel′nikov and S. O. Sinanyan [315].

The problem of characterizing sets that allow *rational* approximation in L^q attracted attention only much later, and after important progress had been done on the corresponding problem for uniform approximation. Some early work is due to S. O. Sinanyan [378] (and other papers; see [315]), who adapted the constructive method for uniform approximation of Mergelyan. As discussed in Section 11.6.1 above, the problem was solved for $q = 1$ by L. Bers [52]. The breakthrough came, however, with V. P. Havin's paper [200], where it was observed that the L^2-problem could be solved in terms of classical logarithmic capacity, or equivalently, in terms of the fine topology of H. Cartan. Havin also pointed out the need for a study of the fine continuity properties of functions in $W^{1,p}$ in order to solve the problem in general. This work was followed by the papers by V. P. Havin and V. G. Maz′ya [201], T. Bagby [42, 43], and L. I. Hedberg [206, 207]. Proposition 11.1.1 is found in L. Carleson [92], Theorem VI.1, and Hedberg [211], Theorem 2. Lemma 11.1.2 was proved in [201], and Proposition 11.1.10 in [206].

See also [315], and N. N. Tarkhanov [396, 397] for more detailed historic surveys.

Lemma 11.1.3, which is due to J. Deny and J. L. Lions [122], Ch. II, Th. 2.1, p. 350, is contained in a more general result of A. A. Khvoles and V. G. Maz′ya [247] (see also Maz′ya [308], Section 11.2, p. 425, and Section 11.8, p. 451), a theorem which goes back to Hörmander and Lions [231]; see also [122], p. 368.

The problem of *uniform* approximation of analytic functions on compact sets, and the related theory of function algebras, have attracted great attention. This is not the place to go into these problems; we refer the interested reader to the book by T. W. Gamelin [175]. It is interesting to observe that the necessary and sufficient conditions for uniform approximation of A. G. Vitushkin [416], expressed in terms of analytic capacity, are similar in form to those in Theorem 11.4.1.

If conditions are sought on a set K for *polynomials* to be dense in $L^q_a(K)$, or in $L^q_a(w, K)$ where w is a positive weight function, or a measure, interesting problems of a different nature appear. The papers by J. E. Brennan [78, 79], and J. E. Thomson [398] are good sources of information on this subject, in addition to [316] and [315] .

Approximation problems similar to those treated in this chapter have been studied in many norms other than the L^p and uniform norms. We refer to the surveys by N. N. Tarkhanov [397], and J. Verdera [414], both of which contain extensive bibliographies.

11.2. The $(1, 2)$-stability of sets is closely related to the stability of the Dirichlet problem, and to the problem of uniform approximation by harmonic functions. In fact, it follows from a classical theorem of M. V. Keldysh [242] that the $(1, 2)$-stable compact sets $K \subset \mathbf{R}^N$ are precisely those compact sets which have the property that every function in $C(K)$, harmonic in K^0, can be uniformly approximated on K by functions harmonic on neighborhoods of K. This theorem, and its relation to Havin's theorem on L^2-approximation by holomorphic functions, and to potential theoretic sweeping (balayage), is discussed in detail in the expository paper Hedberg [218].

11.3. The equivalence of (b), (c) and (d) in Theorem 11.3.2 is due to A. A. Gonchar [185, 186], and Yu. A. Lysenko and B. M. Pisarevskiĭ [283] in the case of classical capacity $\dot{C}_{1,2}$. (See also Section 11.6.2 above.) The theorem was proved in the same way as here for more general radial kernels in Hedberg [207]. Earlier versions are found in Hedberg [206] and J. C. Polking [359]. The existence of nowhere dense sets of uniqueness for $L^{\alpha,p}$, $\alpha p = N$, was proved (differently) by Polking [360].

The uniqueness property of a set for $L^{\frac{1}{2},2}(\mathbf{R})$ or for $L^{\frac{1}{2},2}$ on the circle (when $C_{1/2,2}$ is essentially classical logarithmic capacity) is equivalent to the removability of the set for a class of analytic functions with finite Dirichlet integral, $\int |f'(z)|^2\, dx\, dy$. In this case condition (c) in the theorem is due to L. V. Ahlfors and A. Beurling [28], Theorem 14. Extensions of this result are found in L. Carleson [92], Theorem VI.2, Hedberg [205], Theorem 1 (due to Carleson), and Hedberg [211].

11.4. Theorems 11.4.1 and 11.4.2 are essentially due to T. Bagby [43], and Hedberg [207]. The results could be completed after Wolff had proved that the Kellogg property extends to $(1, p)$-capacities for all $p > 1$ (Corollary 6.3.17); see Hedberg and Wolff [219]. See also Hedberg [218] for the case $q = 2$.

In the terminology of G. Choquet [105], a set S is called *$C_{\alpha,p}$-representative* for a set T if $S \subset T$ and (a) in Theorem 11.4.2 is satisfied. Choquet proved, among other things, a theorem ("des cages de Faraday grillagées") which is essentially the equivalence of (a) and (c) in Theorem 11.4.2 for the Newtonian capacity $C_{1,2}$.

11.5. Stability of sets in the sense discussed here was investigated by I. Babuška [41]; see also B.-W. Schulze and G. Wildenhain [371], Ch. IX, §5.2, and G. Wildenhain [435]. Sufficient conditions were given by Polking [359] and Hedberg [210, 212]. Theorems 11.5.5 and 11.5.6 were proved in Hedberg [212] in response to a question by A. A. Gonchar. È. M. Saak [369] has given a necessary and sufficient condition for $(\alpha, 2)$-stability expressed in the capacity $N_{\alpha,2}$. Theorem 11.5.10 is due to T. Bagby [44] and Yu. V. Netrusov [343]. Bagby proved the equivalence of the approximation property for solutions of elliptic equations with conditions (b) and (c). His proof is constructive, further developing P. Lindberg's [268] adaptation of Vitushkin's [416] technique (see Section 11.6.3), and does not depend on Theorem 9.1.3. In [45] Bagby also pointed out that (d) is a consequence of (c) and asked if (d) implies the approximation property. Netrusov solved this problem, and at the same time obtained the much more general result announced in [343]. The proof given here is his, and is previously unpublished. Capacities closely related to $N_{\alpha,p;\sigma}$ were defined by V. G. Maz'ya [305]; see also his book [308], Section 10.3.2, and K. Nyström [351].

References

1. Abel, N. H.: Note sur le mémoire de Mr. L. Olivier No. 4 du second tome de ce journal, ayant pour titre "remarques sur les séries infinies et leur convergence". Suivie d'une remarque de Mr. L. Olivier sur le même objet, *J. reine angew. Math.* **3** (1828) 79–82. Reprinted in *Œuvres complètes de Niels Henrik Abel, t. I*, 399–402, Christiania, 1881.
2. Adams, D. R.: Traces of potentials arising from translation invariant operators, *Ann. Scuola Norm. Sup. Pisa Cl. Sci.* **25** (1971) 203–217.
3. Adams, D. R.: Maximal operators and capacity, *Proc. Amer. Math. Soc.* **34** (1972) 152–156.
4. Adams, D. R.: Traces of potentials. II, *Indiana Univ. Math. J.* **22** (1973) 907–918.
5. Adams, D. R.: A note on Riesz potentials, *Duke Math. J.* **42** (1975) 765–778.
6. Adams, D. R.: On the existence of capacitary strong type estimates in $\mathbf{R}^N$, *Ark. mat.* **14** (1976) 125–140.
7. Adams, D. R.: Sets and functions of finite L^p-capacity, *Indiana Univ. Math. J.* **27** (1978) 611–627.
8. Adams, D. R.: Quasi-additivity and sets of finite L^p-capacity, *Pacific J. Math.* **79** (1978) 283–291.
9. Adams, D. R.: Capacity and the obstacle problem, *Appl. Math. Optim.* **8** (1981) 39–57.
10. Adams, D. R.: *Lectures on L^p-Potential Theory*, Department of Mathematics, University of Umeå, 1981.
11. Adams, D. R.: The exceptional sets associated with the Besov spaces, in *Linear and Complex Analysis Problem Book. 199 Research Problems* (V. P. Havin, S. V. Hruščëv, N. K. Nikol'skii, eds.), Lecture Notes in Math. **1043**, 515–518, Springer, Berlin Heidelberg, 1984. Also in *Linear and Complex Analysis Problem Book 3, Part II* (V. P. Havin, N. K. Nikolski, eds.), Lecture Notes in Math. **1574**, 169–172, Springer, Berlin Heidelberg, 1994.
12. Adams, D. R.: Weighted nonlinear potential theory, *Trans. Amer. Math. Soc.* **297** (1986) 73–94.
13. Adams, D. R.: A note on the Choquet integrals with respect to Hausdorff capacity, in *Function Spaces and Applications*, Proc., Lund 1986 (M. Cwikel, J. Peetre, Y. Sagher, H. Wallin, eds.), Lecture Notes in Math. **1302**, 115–124, Springer, Berlin Heidelberg, 1988.
14. Adams, D. R.: A sharp inequality of J. Moser for higher order derivatives, *Ann. of Math.* **128** (1988) 385–398.
15. Adams, D. R.: The classification problem for the capacities associated with the Besov and Triebel–Lizorkin spaces, in *Approximation and Function Spaces, Banach Center Publications* **22**, 9–24, PWN Polish Scientific Publishers, Warsaw, 1989.
16. Adams, D. R.: Choquet integrals in potential theory, *Publ. Math.* **42** (1998) 3–66.
17. Adams, D. R. and Frazier, M.: BMO and smooth truncation in Sobolev spaces, *Studia Math.* **89** (1988) 241–260.

18. Adams, D. R. and Frazier, M.: Composition operators on potential spaces, *Proc. Amer. Math. Soc.* **114** (1992) 155–165.
19. Adams, D. R. and Heard, A.: The necessity of the Wiener test for some semi-linear elliptic equations, *Indiana Univ. Math. J.* **41** (1992) 109–124.
20. Adams, D. R. and Hedberg, L. I.: Inclusion relations among fine topologies in non-linear potential theory, *Indiana Univ. Math. J.* **33** (1984) 117–126.
21. Adams, D. R. and Lewis, J. L.: Fine and quasi connectedness in nonlinear potential theory, *Ann. Inst. Fourier (Grenoble)* **35**:1 (1985) 57–73.
22. Adams, D. R. and Meyers, N. G.: Thinness and Wiener criteria for non-linear potentials, *Indiana Univ. Math. J.* **22** (1972) 169–197.
23. Adams, D. R. and Meyers, N. G.: Bessel potentials. Inclusion relations among classes of exceptional sets, *Indiana Univ. Math. J.* **22** (1973) 873–905.
24. Adams, D. R. and Pierre, M.: Capacitary strong type estimates in semilinear problems, *Ann. Inst. Fourier (Grenoble)* **41** (1991) 117–135.
25. Adams, D. R. and Polking, J. C.: The equivalence of two definitions of capacity, *Proc. Amer. Math. Soc.* **37** (1973) 529–534.
26. Adams, Robert A.: *Sobolev Spaces*, Academic Press, New York, 1975.
27. Ahlfors, L. V.: Finitely generated Kleinian groups, *Amer. J. Math.* **86** (1964) 413–429. Reprinted in *Lars Valerian Ahlfors: Collected Papers, Vol. 2*, 273–290, Birkhäuser, Boston, 1982.
28. Ahlfors, L. V. and Beurling, A.: Conformal invariants and function-theoretic null-sets, *Acta Math.* **83** (1950) 101–129. Reprinted in *Lars Valerian Ahlfors: Collected Papers, Vol. 1*, 406–434, Birkhäuser, Boston, 1982, and in *The Collected Works of Arne Beurling, Vol. 1, Complex Analysis*, 171–199, Birkhäuser, Boston, 1989.
29. Aikawa, H.: Tangential boundary behavior of Green potentials and contractive properties of L^p-capacities, *Tokyo Math. J.* **9** (1986) 223–245.
30. Aikawa, H.: Comparison of L^p-capacity and Hausdorff measure, *Complex Variables* **15** (1990) 223–232.
31. Aikawa, H.: Quasiadditivity of Riesz capacity, *Math. Scand.* **69** (1991) 15–30.
32. Aikawa, H.: Quasiadditivity of capacity and minimal thinness, *Ann. Acad. Sci. Fenn. Ser. A. I. Math.* **18** (1993) 65–75.
33. Aikawa, H. and Borichev, A. A.: Quasiadditivity and measure property of capacity and the tangential boundary behavior of harmonic functions, *Trans. Amer. Math. Soc.*, to appear.
34. Aïssaoui, N. and Benkirane, A.: Capacités dans les espaces d'Orlicz, *Ann. Sci. Math. Québec* **18** (1994) 1–23.
35. Anger, B.: Representations of capacities, *Math. Ann.* **229** (1977) 245–258.
36. Aronszajn, N.: Les noyaux pseudo-reproduisants, *C. R. Acad. Sci. Paris* **226** (1948) 456–458. Noyaux pseudo-reproduisants et complétion des classes hilbertiennes, *ibid.*, 537–539. Complétion fonctionelle de certaines classes hilbertiennes, *ibid.*, 617–619. Propriétés de certaines classes hilbertiennes complétées, *ibid.*, 700–702.
37. Aronszajn, N.: Potentiels besseliens, *Ann. Inst. Fourier (Grenoble)* **15** (1965) 43–58.
38. Aronszajn, N., Mulla, F., and Szeptycky, P.: On spaces of potentials connected with L^p-classes, *Ann. Inst. Fourier (Grenoble)* **13**:2 (1963) 211–306.
39. Aronszajn, N. and Smith, K. T.: Functional spaces and functional completion, *Ann. Inst. Fourier (Grenoble)* **6** (1956) 125–185 (report, Lawrence, Kansas, 1954).
40. Aronszajn, N. and Smith, K. T.: Theory of Bessel potentials, I, *Ann. Inst. Fourier (Grenoble)* **11** (1961) 385–475.
41. Babuška, I.: Устойчивость областей определения по отношению к основным задачам теории дифференциальных уравнений с частными производными, главным образом в связи с теорией упругости (Stability of the domain with respect to the fundamental problems in the theory of partial differential equations, mainly in connection with the theory of elasticity), I, II, *Czechoslovak Math. J.* **11** (1961) 76–105, and 165–203.

42. Bagby, T.: L_p approximation by analytic functions, *J. Approx. Th.* **5** (1972) 401–404.
43. Bagby, T.: Quasi topologies and rational approximation, *J. Funct. Anal.* **10** (1972) 259–268.
44. Bagby, T.: Approximation in the mean by solutions of elliptic equations, *Trans. Amer. Math. Soc.* **281** (1984) 761–784.
45. Bagby, T.: Approximation in the mean by harmonic functions, in *Linear and Complex Analysis Problem Book. 199 Research Problems* (V. P. Havin, S. V. Hruščёv, N. K. Nikol′skii, eds.), Lecture Notes in Math. **1043**, 466–470, Springer, Berlin Heidelberg, 1984. Also in *Linear and Complex Analysis Problem Book 3, Part II* (V. P. Havin, N. K. Nikolski, eds.), Lecture Notes in Math. **1574**, 117–120, Springer, Berlin Heidelberg, 1994.
46. Bagby, T. and Ziemer, W. P.: Pointwise differentiability and absolute continuity, *Trans. Amer. Math. Soc.* **191** (1974) 129–148.
47. Baras, P. and Pierre, M.: Singularités éliminables pour des équations semi-linéaires, *Ann. Inst. Fourier* (*Grenoble*) **34**:1 (1984) 185–206.
48. Bauman, P.: A Wiener test for nondivergence structure, second-order elliptic equations, *Indiana Univ. Math. J.* **34** (1985) 825–844.
49. Beckner, W.: Sharp Sobolev inequalities on the sphere and the Moser–Trudinger inequality, *Ann. of Math.* **138** (1993) 213–242.
50. Benkirane, A. and Gossez, J.-P.: An approximation theorem in higher order Orlicz–Sobolev spaces and applications, *Studia Math.* **92** (1989) 231–255.
51. Berger, M. and Schechter, M.: Embedding theorems and quasilinear elliptic boundary value problems for unbounded domains, *Trans. Amer. Math. Soc.* **172** (1972) 261–278.
52. Bers L.: An approximation theorem, *J. Analyse Math.* **14** (1965) 1–4.
53. Bers L.: L_1-approximation of analytic functions, *J. Indian Math. Soc.* **34** (1970) 193–201.
54. Besicovitch, A. and Taylor, S. J.: On complementary intervals of a closed set of zero Lebesgue measure, *J. London Math. Soc.* **29** (1954) 449–459.
55. Besov, O. V. (Бесов, О. В.): О некотором семействе функциональных пространств. Теоремы вложения и продолжения (On a family of function spaces. Imbedding and extension theorems), *Dokl. Akad. Nauk SSSR* **126** (1959) 1163–1165.
56. Besov, O. V. (Бесов, О. В.): Исследование одного семейства функциональных пространств в связи с теоремами вложения и продолжения (Investigation of a family of function spaces in connection with imbedding and extension theorems), *Trudy Mat. Inst. Steklov.* **60** (1961) 42–81.
57. Beurling, A.: *Études sur un problème de majoration*, Thesis, Uppsala University, 1933. Reprinted in *The Collected Works of Arne Beurling, Vol. 1, Complex Analysis*, 1–107, Birkhäuser, Boston, 1989.
58. Beurling, A.: Ensembles exceptionnels, *Acta Math.* **72** (1940) 1–13. Reprinted in *The Collected Works of Arne Beurling, Vol. 1, Complex Analysis*, 121–133, Birkhäuser, Boston, 1989.
59. Beurling, A.: Sur les spectres des fonctions, in *Analyse harmonique*, Colloque, Nancy 1947, 9–29, Centre National de la Recherche Scientifique, Paris, 1949. Reprinted in *The Collected Works of Arne Beurling, Vol. 2, Harmonic Analysis*, 125–145, Birkhäuser, Boston, 1989.
60. Beurling, A.: On the spectral synthesis of bounded functions, *Acta Math.* **81** (1949) 225–238. Reprinted in *The Collected Works of Arne Beurling, Vol. 2, Harmonic Analysis*, 93–106, Birkhäuser, Boston, 1989.
61. Beurling, A. and Deny, J.: Espaces de Dirichlet. I. Le cas élémentaire, *Acta Math.* **99** (1958) 203–224. Reprinted in *The Collected Works of Arne Beurling, Vol. 2, Harmonic Analysis*, 179–200, Birkhäuser, Boston, 1989.

62. Beurling, A. and Deny, J.: Dirichlet spaces, *Proc. Nat. Acad. Sci. USA* **45** (1959) 208–215. Reprinted in *The Collected Works of Arne Beurling, Vol. 2, Harmonic Analysis*, 201–208, Birkhäuser, Boston, 1989.
63. Biroli, M.: Wiener estimates for solutions of elliptic equations in nondivergence form, in *Potential Theory*, Proc., Prague 1987 (J. Král, I. Netuka, J. Lukeš, J. Veselý, eds.), 47–52, Plenum Press, New York, 1988.
64. Boccardo, L. and Murat, F.: Remarques sur l'homogénéisation de certains problèmes quasi-linéaires, *Portugal. Math.* **41** (1982) 535–562.
65. Bochner, S.: *Vorlesungen über Fouriersche Integrale*, Leipzig, 1932. Reprinted by Chelsea, New York, 1948.
66. Bourdaud, G.: Fonctions qui opèrent sur les espaces de Sobolev, *Sém. Anal. Harm. Orsay* (1980–81) 6–17.
67. Bourdaud, G.: Le calcul fonctionnel dans les espaces de Sobolev, *Invent. Math.* **104** (1991) 435-446.
68. Bourdaud, G.: La trivialité du calcul fonctionnel dans l'espace $H^{3/2}(\mathbf{R}^4)$, *C. R. Acad. Sci. Paris Sér. I Math.* **314** (1992) 187–190.
69. Bourdaud, G.: Fonctions qui opèrent sur les espaces de Besov et de Triebel, *Ann. Inst. H. Poincaré. Anal. Non Linéaire* **10** (1993) 413–422.
70. Bourdaud, G. and Kateb, M. E. D.: Calcul fonctionnel dans l'espace de Sobolev fractionnaire, *Math. Z.* **210** (1992) 607–613.
71. Bourdaud, G. and Meyer, Y.: Fonctions qui opèrent sur les espaces de Sobolev, *J. Funct. Anal.* **97** (1991) 351–360.
72. Branson, Th. P., Chang, S.-Y. A., and Yang, P. C.: Estimates and extremals for zeta function determinants on four-manifolds, *Comm. Math. Phys.* **149** (1992) 241–262.
73. Brelot, M.: Points irréguliers et transformations continues en théorie du potentiel, *J. Math. pures appl.* **19** (1940) 319–337.
74. Brelot, M.: Sur les ensembles effilés, *Bull. Sci. Math.* **68** (1944) 12–36.
75. Brelot, M.: Sur l'allure des fonctions harmoniques et sousharmoniques à la frontière, *Math. Nachr.* **4** (1950–51) 298–307.
76. Brelot, M.: Existence theorem for n-capacities, *Ann. Inst. Fourier* (*Grenoble*) **5** (1953–54) 297–304.
77. Brelot, M.: *On Topologies and Boundaries in Potential Theory*, Lecture Notes in Math. **175**, Springer, Berlin, 1971.
78. Brennan, J. E.: Weighted polynomial approximation, quasianalyticity and analytic continuation, *J. reine angew. Math.* 357 (1985) 23–50.
79. Brennan, J. E.: Weighted polynomial approximation and quasianalyticity for general sets, *Algebra i Analiz* **6**:4 (1994) 69–89.
80. Brezis, H. and Browder, F. E.: Sur une propriété des espaces de Sobolev, *C. R. Acad. Sci. Paris* **287** (1978) 113–115.
81. Brezis, H. and Browder, F. E.: A property of Sobolev spaces, *Comm. Partial Differential Equations* **9** (1979) 1077–1083.
82. Brezis, H. and Browder, F. E.: Some properties of higher order Sobolev spaces, *J. Math. pures appl.* **61** (1982) 245–259.
83. Brezis, H. and Browder, F. E.: Initial–boundary value problems for strongly nonlinear parabolic operators, *to appear*.
84. Browder, F. E.: Degree theory for nonlinear mappings, in *Nonlinear Functional Analysis and Its Applications*, Proc., Berkeley 1983 (F. E. Browder, ed.), Proc. Sympos. Pure Math. **45**:1, 203–226, Amer. Math. Soc., Providence, Rhode Island, 1986.
85. Browder, F. E.: Strongly nonlinear parabolic equations of higher order, *Atti Accad. Naz. Lincei* **77** (1986) 159–172.
86. Burenkov, V. I. (Буренков, В. И.): О приближении функций из пространства $W_p^r(\Omega)$ финитными функциями для произвольного открытого множества Ω, *Trudy Mat. Inst. Steklov.* **131** (1974) 51–63. English translation:

On the approximation of functions in the space $W_p^r(\Omega)$ by functions with compact support for an arbitrary open set Ω, *Proc. Steklov Inst. Math.* **131** (1974) 51–36.

87. Caffarelli, L. and Kinderlehrer, D.: Potential methods in variational inequalities, *J. Analyse Math.* **37** (1980) 285–295.
88. Calderón, A. P.: Lebesgue spaces of differentiable functions and distributions, in *Partial Differential Equations*, Proc. Sympos. Pure Math. **4**, 33–49, Amer. Math. Soc., Providence, Rhode Island, 1961.
89. Calderón, A. P. and Zygmund, A.: On the existence of certain singular integrals, *Acta Math.* **88** (1952) 85–139.
90. Calderón, C. P., Fabes, E. B., and Riviere, N. M.: Maximal smoothing operators, *Indiana Univ. Math. J.* **23** (1974) 889–898.
91. Carleman, T.: Über die Approximation analytischer Funktionen durch lineare Aggregate von vorgegebenen Potenzen, *Ark. mat. astr. fys.* **17**:9 (1922) 1–30. Reprinted in *Edition complète des articles de Torsten Carleman*, 215–244, Institut Mittag-Leffler, Djursholm, 1960.
92. Carleson, L.: *Selected Problems on Exceptional Sets*, Van Nostrand, Princeton, New Jersey, 1967 (mimeographed, Uppsala University, 1961). Russian translation (with notes by V. P. Havin): *Избранные проблемы теории исключительных множеств*, Mir, Moscow, 1971.
93. Carleson, L. and Chang, S.-Y. A.: On the existence of an extremal for an inequality of J. Moser, *Bull. Sci. Math.* **110** (1986) 113–127.
94. Carlsson, Anders, Inequalities of Poincaré-Wirtinger type, *Linköping Studies in Science and Technology. Thesis* **232**, Linköping, 1990.
95. Carlsson, Anders, A note on spectral synthesis, *Report LiTH-MAT-R-93-03*, Linköping University, 1993.
96. Carlsson, Anders and Maz'ya, V. G.: On approximation in weighted Sobolev spaces and self-adjointness, *Math. Scand.* **74** (1994) 111–124.
97. Cartan, H.: Théorie générale du balayage en potentiel newtonien, *Ann. Univ. Grenoble, Math. Phys.* **22** (1946) 221–280.
98. Chandrasekharan, K.: *Classical Fourier Transforms*, Springer, Berlin Heidelberg, 1989.
99. Chang, S.-Y. A.: Extremal functions in a sharp form of Sobolev inequality, in *Proceedings of the International Congress of Mathematicians, August 3–11, 1986, Vol. 1*, Berkeley, California, (A. M. Gleason, ed.), 715–723, Amer. Math. Soc., Providence, Rhode Island, 1987.
100. Chang, S.-Y. A. and Marshall, D. E.: On a sharp inequality concerning the Dirichlet integral, *Amer. J. Math.* **107** (1985) 1015–1033.
101. Chang, S.-Y. A. and Yang, P. C.: Spectral invariants of conformal metrics, in *Harmonic Analysis*, Proc., Sendai 1990 (S. Igari, ed.), 51–60, Springer, Tokyo, 1991.
102. Choquet, G.: Theory of capacities, *Ann. Inst. Fourier* (*Grenoble*) **5** (1953–54) 131–295.
103. Choquet, G.: Forme abstraite du théorème de capacitabilité, *Ann. Inst. Fourier* (*Grenoble*) **9** (1959) 83–89.
104. Choquet, G.: Sur les points d'effilément d'un ensemble. Application à l'étude de la capacité, *Ann. Inst. Fourier* (*Grenoble*) **9** (1959) 91–101.
105. Choquet, G.: Convergence vague et suites de potentiels newtoniens, *Bull. Sci. Mat.* **99** (1975) 157–164.
106. Cioranescu, D. and Murat, F.: Un terme étrange venu d'ailleurs, in *Nonlinear Partial Differential Equations and Their Applications: Collège de France Seminar. Volume II* (H. Brezis and J. L. Lions, eds.), 98–138, and *ibid., Volume III*, 154–178, Pitman, Boston–London, 1982, 1983.
107. Clarkson, J. A.: Uniformly convex spaces, *Trans. Amer. Math. Soc.* **40** (1936) 396–414.

108. Coifman, R. and Fefferman, C.: Weighted norm inequalities for maximal functions and singular integrals, *Studia Math.* **5** (1974) 241–250.
109. Cohn, W. S. and Verbitsky, I. E.: Non-linear potential theory on the ball, with applications to exceptional and boundary interpolation sets, *Michigan Math. J.*, **42** (1995) 79–97.
110. Dahlberg, B. E. J.: A note on Sobolev spaces, in *Harmonic Analysis in Euclidean Spaces*, Proc., Williamstown 1978 (S. Wainger and G. Weiss, eds.), Proc. Sympos. Pure Math. **35**:1, 183–185, Amer. Math. Soc., Providence, Rhode Island, 1979.
111. Dahlberg, B. E. J.: Regularity properties of Riesz potentials, *Indiana Univ. Math. J.* **28** (1979) 257–268.
112. Dal Maso, G. and Garroni, A.: New results on the asymptotic behaviour of Dirichlet problems in perforated domains, *Math. Models Methods Appl. Sci.* **4** (1994) 373–407.
113. Dal Maso, G. and Mosco, U.: Wiener criteria and energy decay for relaxed Dirichlet problems, *Arch. Rational Mech. Anal.* **95** (1986) 345–387.
114. Dal Maso, G., Mosco, U., and Vivaldi, M. A.: A pointwise regularity theory for the two-obstacle problem, *Acta Math.* **163** (1989) 57–107.
115. Daubechies, I.: *Ten Lectures on Wavelets*, CBMS-NSF Regional Conf. Ser. in Appl. Math. **61**, SIAM, Philadelphia, Pennsylvania, 1992.
116. Davie A. M. and Øksendal, B.: Analytic capacity and differentiability properties of finely harmonic functions, *Acta Math.* **149** (1982) 127–152.
117. Dellacherie, C.: *Ensembles analytiques, capacités, mesures de Hausdorff*, Lecture Notes in Math. **295**, Springer, Berlin Heidelberg, 1972.
118. Dellacherie, C. and Meyer, P.-A.: *Probabilités et potentiel*, Hermann, Paris, 1975. English translation: *Probabilities and Potential*, North-Holland Mathematics Studies **29**, North-Holland, Amsterdam, 1978.
119. Deny, J.: Les potentiels d'énergie finie, *Acta Math.* **82** (1950) 107–183.
120. Deny, J.: Sur la convergence de certaines intégrales de la théorie du potentiel, *Arch. der Math.* **5** (1954) 367–370.
121. Deny, J.: Méthodes hilbertiennes en théorie du potentiel, in *Potential Theory*, Centro Internazionale Matematico Estivo (C.I.M.E.), I. Ciclo, Stresa 1969 (M. Brelot, coordinatore), 123–201, Edizioni Cremonese, Rome, 1970.
122. Deny, J. and Lions, J. L.: Les espaces du type de Beppo Levi, *Ann. Inst. Fourier (Grenoble)* **5** (1953–54) 305–370.
123. Diestel, J.: *Geometry of Banach Spaces: Selected Topics*, Springer, Berlin Heidelberg, 1975.
124. Doob, J. L.: Applications to analysis of a topological definition of smallness of a set, *Bull. Amer. Math. Soc.* **72** (1966) 579–600.
125. Doob, J. L.: *Classical Potential Theory and Its Probabilistic Counterpart*, Springer, New York, 1984.
126. Du Plessis, N.: *An Introduction to Potential Theory*, Hafner Publishing Co., Darien, Connecticut, 1970.
127. Dynkin, E. B.: A probabilistic approach to one class of nonlinear differential equations, *Probab. Theor. Relat. Fields* **89** (1991) 89–115.
128. Dynkin, E. B.: Superprocesses and partial differential equations, *Ann. Probab.* **21** (1993) 1185–1262.
129. Dynkin, E. B.: *An Introduction to Branching Measure-Valued Processes*, CRM Monograph Series, Vol. 6, Amer. Math. Soc., Providence, Rhode Island, 1994.
130. Dynkin, E. B. and Kuznetsov, S. E.: Superdiffusions and removable singularities for quasilinear partial differential equations, *Comm. Pure Appl. Math.* **49** (1996) 125–176.
131. Dynkin, E. B. and Kuznetsov, S. E.: Solutions of $Lu = u^\alpha$ dominated by L-harmonic functions, *J. Anal. Math.* **68** (1996) 15–37.
132. Dyn′kin, E. M. and Hanin, L. G.: Spectral synthesis of ideals in Zygmund algebras: the asymptotic Cauchy problem approach, *Michigan Math. J.* **43** (1996) 539–557.

133. Evans, G. C.: Application of Poincaré's sweeping-out process, *Proc. Nat. Acad. Sci. USA* **37** (1935) 226–253.
134. Evans, G. C.: On potentials of positive mass. Part I, *Trans. Amer. Math. Soc.* **37** (1935) 226–253.
135. Evans, G. C.: On potentials of positive mass. Part II, *Trans. Amer. Math. Soc.* **38** (1935) 201–236.
136. Evans, L. C. and Gariepy, R. F.: Wiener's criterion for the heat equation, *Arch. Rational Mech. Anal.* **78** (1982) 293–314.
137. Evans, L. C. and Gariepy, R. F.: *Measure Theory and Fine Properties of Functions*, CRC Press, Boca Raton, Florida, 1992.
138. Fabes, E. B., Jerison, D. S., and Kenig, C. E.: The Wiener test for degenerate elliptic equations, *Ann. Inst. Fourier* (*Grenoble*) **32** (1982) 151–182.
139. Falconer, K. J.: *The Geometry of Fractal Sets*, Cambridge University Press, Cambridge, 1985.
140. Fan, K.: Minimax theorems, *Proc. Nat. Acad. Sci. USA* **39** (1953) 42–47.
141. Federer, H.: *Geometric Measure Theory*, Springer, Berlin Heidelberg New York, 1969.
142. Federer, H. and Ziemer, W. P.: The Lebesgue set of a function whose distribution derivatives are p-th power summable, *Indiana Univ. Math. J.* **22** (1972) 139–158.
143. Fefferman, C.: The uncertainty principle, *Bull. Amer. Math. Soc.* **9** (1983) 129–206.
144. Fefferman, C. and Stein, E. M.: Some maximal inequalities, *Amer. J. Math.* **93** (1971) 107–115.
145. Fefferman, R.: A theory of entropy in Fourier analysis, *Adv. Math.* **30** (1978) 171–201.
146. Fernström, C.: Bounded point evaluations and approximation in L^p by analytic functions, in *Spaces of Analytic Functions*, Proc., Kristiansand, Norway 1975 (O. B. Bekken, B. K. Øksendal, A. Stray, eds.), Lecture Notes in Math. **512**, 65–68, Springer, Berlin Heidelberg,1976.
147. Fernström, C.: On the instability of capacity, *Ark. mat.* **15** (1977) 241–252.
148. Fernström, C.: On the instability of Hausdorff content, *Math. Scand.* **60** (1987) 19–30.
149. Fernström, C. and Polking, J. C.: Bounded point evaluations and approximation in L^p by solutions of elliptic partial differential equations, *J. Funct. Anal.* **28** (1978) 1–20.
150. Folland, G. B.: *Introduction to Partial Differential Equations*, Princeton University Press, Princeton, New Jersey, 1976.
151. Fontana, L.: Sharp borderline Sobolev inequalities on compact Riemannian manifolds, *Comment. Math. Helv.* **68** (1993) 415–454.
152. Forsling, G.: Approximation by smooth functions in some function spaces, *Linköping Studies in Science and Technology. Thesis* **213**, Linköping, 1990.
153. Frazier, M. and Jawerth, B.: Decomposition of Besov spaces, *Indiana Univ. Math. J.* **34** (1985) 777–799.
154. Frazier, M. and Jawerth, B.: The φ-transform and applications to distribution spaces, in *Function Spaces and Applications*, Proc., Lund 1986 (M. Cwikel, J. Peetre, Y. Sagher, H. Wallin, eds.), Lecture Notes in Math. **1302**, 223–246, Springer, Berlin Heidelberg, 1988.
155. Frazier, M. and Jawerth, B.: A discrete transform and decompositions of distribution spaces, *J. Funct. Anal.* **93** (1990) 34–170.
156. Frazier, M., Jawerth, B., and Weiss, G.: *Littlewood–Paley Theory and the Study of Function Spaces*, CBMS Regional Conf. Ser. in Math. **79**, Amer. Math. Soc., Providence, Rhode Island, 1991.
157. Friedrichs, K.: Die Randwert- und Eigenwertprobleme aus der Theorie der elastischen Platten, *Math. Ann.* **98** (1928) 205–247.
158. Frostman, O.: Potentiel d'équilibre et capacité des ensembles avec quelques applications à la théorie des fonctions, *Medd. Lunds Univ. Mat. Sem.* **3** (1935) 1–118.

159. Frostman, O.: Les points irréguliers dans la théorie du potentiel et le critère de Wiener, *Kungl. Fysiografiska Sällsk. i Lund förh.* **9**:2 (1939) 1–10.
160. Fuglede, B.: Extremal length and functional completion, *Acta Math.* **98** (1957) 171–219.
161. Fuglede, B.: On the theory of potentials in locally compact spaces, *Acta Math.* **103** (1960) 139–215.
162. Fuglede, B.: On generalized potentials of functions in the Lebesgue classes, *Math. Scand.* **8** (1960) 287–304.
163. Fuglede, B.: Le théorème du minimax et la théorie fine du potentiel, *Ann. Inst. Fourier (Grenoble)* **15**:1 (1965) 65–88.
164. Fuglede, B.: Quasi topology and fine topology, *Séminaire de Théorie du Potentiel (Brelot–Choquet–Deny)* **10** (1965-1966) no.12.
165. Fuglede, B.: Applications du théorème minimax à l'étude de diverses capacités, *C. R. Acad. Sci. Paris Sér. A-B* **266** (1968) 921–923.
166. Fuglede, B.: The quasi topology associated with a countably subadditive set function, *Ann. Inst. Fourier (Grenoble)* **21**:1 (1971) 123–169.
167. Fuglede, B.: Connexion en topologie fine et balayage des mesures, *Ann. Inst. Fourier (Grenoble)* **21**:3 (1971) 227–244.
168. Fuglede, B.: *Finely Harmonic Functions*, Lecture Notes in Math. **289**, Springer, Berlin Heidelberg, 1972.
169. Fuglede, B.: Fonctions BLD et fonctions finement surharmoniques, in *Séminaire de Théorie du Potentiel, Paris, No. 6*, Proc. (F. Hirsch, G. Mokobodzki, eds.), Lecture Notes in Math. **906**, 126–157, Springer, Berlin Heidelberg, 1982.
170. Fuglede, B.: Fine potential theory, in *Potential Theory—Surveys and Problems*, Proc., Prague 1987 (J. Král, J. Lukeš, I. Netuka, J. Veselý, eds.), Lecture Notes in Math. **1344**, 81–97, Springer, Berlin Heidelberg, 1988.
171. Fuglede, B.: A simple proof that certain capacities decrease under contraction, *Hiroshima Math. J.* **19** (1989) 567–573.
172. Fukushima, M.: Two topics related to Dirichlet forms: quasi everywhere convergences and additive functionals, in *Dirichlet Forms*, Proc., Varenna 1992 (G. Dell'Antonio, U. Mosco, eds.), Lecture Notes in Math. **1563**, 21–53, Springer, Berlin Heidelberg, 1992.
173. Fukushima, M. and Kaneko, H.: On (r, p)-capacities for general Markovian semigroups, in *Infinite Dimensional Analysis and Stochastic Processes*, Proc., Bielefeld 1983 (S. Albeverio, ed.), Pitman, Boston–London, 1985.
174. Gagliardo, E.: Ulteriori proprietà, di alcune classi di funzioni in più variabili, *Ricerche Mat.* **8** (1959) 24–51.
175. Gamelin, T. W.: *Uniform Algebras*, Prentice Hall, Englewood Cliffs, New Jersey, 1969.
176. García-Cuerva, J. and Rubio de Francia, J. L.: *Weighted Norm Inequalities and Related Topics*, North-Holland Mathematics Studies **116**, North-Holland, Amsterdam, 1985.
177. Gariepy, R. and Ziemer, W. P.: A regularity condition at the boundary for solutions of quasilinear elliptic equations, *Arch. Rational Mech. Anal.* **67** (1977) 25–39.
178. Garofalo, N. and Lanconelli, E.: Wiener's criterion for parabolic equations with variable coefficients and its consequences, *Trans. Amer. Math. Soc.* **308** (1988) 811–836.
179. Garofalo, N. and Segala, F.: Estimates of the fundamental solution and Wiener's criterion for the heat equation on the Heisenberg group, *Indiana Univ. Math. J.* **39** (1990) 1155–1196.
180. Garsia, A. M.: *Letter to J. Moser*, March 29, 1972.
181. Gauss, C. F.: Allgemeine Lehrsätze in Beziehung auf die im verkehrten Verhältnisse des Quadrats der Entfernung wirkenden Anziehungs- und Abstossungs-Kräfte, *Resultate aus den Beobachtungen des magnetischen Vereins im Jahre 1839*, Leipzig,

1840. Reprinted in *Carl Friedrich Gauss Werke* **5**, 197–242, Königl. Gesellschaft der Wissenschaften, Göttingen, 1877.
182. Giaquinta, M.: *Multiple Integrals in the Calculus of Variations and Nonlinear Elliptic Systems*, Princeton University Press, Princeton, New Jersey, 1983.
183. Gilbarg, D. and Trudinger, N. S.: *Elliptic Partial Differential Equations of Second Order*, Second edition, Springer, Berlin Heidelberg, 1983.
184. Gol′dshteĭn, V. M. (Гольдштейн, В. М.) and Reshetnyak, Yu. G. (Решетняк, Ю. Г.): *Введение в теорию функций с обобщенными производными и квазиконформные отображения*, Nauka, Moscow, 1983. Revised English translation: *Quasiconformal Mappings and Sobolev Spaces*, Kluwer Acad. Publ., Dordrecht–Boston–London, 1990.
185. Gonchar, A. A. (Гончар, А. А.): О равномерном приближении непрерывных функций гармоническими (On the uniform approximation of continuous functions by harmonic functions), *Izv. Akad. Nauk SSSR Ser. Mat.* **27** (1963) 1239–1250.
186. Gonchar, A. A. (Гончар, А. А.): О свойстве неустойчивости гармонической емкости, *Dokl. Akad. Nauk SSSR* **165** (1965) 479–481. English translation: On the property of instability of harmonic capacity, *Soviet Math.* **6** (1965) 1458–1460.
187. Grun-Rehomme, M.: Caractérisation du sous-différentiel d'intégrandes convexes dans les espaces de Sobolev, *J. Math. pures appl.* **56** (1977) 149–156.
188. Hadamard, J.: Deux théorèmes d'Abel sur la convergence des séries, *Acta Math.* **26** (1902) 177–183. Reprinted in *Œuvres de Jacques Hadamard, t. 1*, 327–333, Centre National de la Recherche Scientifique, Paris, 1968.
189. Hanin, L. G.: Closed ideals in algebras of smooth functions, *Dissertationes Math. (Rozprawy Mat.)* **371** (1998).
190. Hanner, O.: On the uniform convexity of L^p and l^p, *Ark. mat.* **3** (1955) 239–244.
191. Hansson, K.: On a maximal imbedding theorem of Sobolev type and spectra of Schrödinger operators, *Linköping Studies in Science and Technology. Dissertations* **26**, Linköping, 1978.
192. Hansson, K.: Imbedding theorems of Sobolev type in potential theory, *Math. Scand.* **45** (1979) 77–102.
193. Hansson, K.: Continuity and compactness of certain convolution operators, *Inst. Mittag-Leffler Report* **9** (1982).
194. Hardy, G. H. and Littlewood, J. E.: Some properties of fractional integrals, I, *Math. Z.* **27** (1928) 565–606.
195. Hardy, G. H. and Littlewood, J. E.: A maximal theorem with function-theoretic applications, *Acta Math.* **54** (1930) 81–116.
196. Harvey, R. and Polking, J. C.: Removable singularities of solutions of linear partial differential equations, *Acta Math.* **125** (1970) 39–56.
197. Harvey, R. and Polking, J. C.: A notion of capacity that characterizes removable singularities, *Trans. Amer. Math. Soc.* **169** (1972) 183–195.
198. Hatano, K.: Bessel capacity of symmetric generalized Cantor sets, *Hiroshima Math. J.* **17** (1987) 149–156.
199. Hatano, K.: A note on (α, p)-thinness of symmetric generalized Cantor sets, *Mem. Fac. Ed. Shimane Univ. Natur. Sci.* **22** (1988) 1–4.
200. Havin, V. P. (Khavin, V. P.) (Хавин, В. П.): Аппроксимация аналитическими функциями в среднем, *Dokl. Akad. Nauk SSSR* **178** (1968) 1025–1028. English translation: Approximation in the mean by analytic functions, *Soviet Math. Dokl.* **9** (1968) 245–248.
201. Havin, V. P. (Khavin, V. P.) (Хавин, В. П.) and Maz′ya, V. G. (Мазья, В. Г.): Об аппроксимации в среднем аналитическими функциями, *Vestnik Leningrad. Univ. Mat. Mekh. Astronom.* **23**:13 (1968) 62–74. English translation: On approximation in the mean by analytic functions, *Vestnik Leningrad Univ. Math.* **1** (1974) 231–245.

202. Havin, V. P. (Khavin, V. P.) (Хавин, В. П.) and Maz′ya, V. G. (Мазья, В. Г.): Нелинейный аналог ньютоновского потенциала и метрические свойства (p, l)-емкости, *Dokl. Akad. Nauk SSSR* **194** (1970) 770–773. English translation: A nonlinear analogue of the Newtonian potential and metric properties of the (p, l)-capacity, *Soviet Math. Dokl.* **11** (1970) 1294–1298.
203. Havin, V. P. (Khavin, V. P.) (Хавин, В. П.) and Maz′ya, V. G. (Мазья, В. Г.): Нелинейная теория потенциала, *Uspekhi Mat. Nauk* **27**:6 (1972) 67–138. English translation: Non-linear potential theory, *Russian Math. Surveys* **27**:6 (1972) 71–148.
204. Havin, V. P. (Khavin, V. P.) (Хавин, В. П.) and Maz′ya, V. G. (Мазья, В. Г.): Приложения (p, l)-емкости к нескольким задачам теории исключительных множеств, *Mat. Sb.* **90(132)** (1973) 558–591. English translation: Use of (p, l)-capacity in problems of the theory of exceptional sets, *Math. USSR-Sb.* **19** (1973) 547–580.
205. Hedberg, L. I.: The Stone–Weierstrass theorem in certain Banach algebras of Fourier type, *Ark. mat.* **6** (1965) 77–102.
206. Hedberg, L. I.: Approximation in the mean by analytic functions, *Trans. Amer. Math. Soc.* **163** (1972) 157–171.
207. Hedberg, L. I.: Non-linear potentials and approximation in the mean by analytic functions, *Math. Z.* **129** (1972) 299–319.
208. Hedberg, L. I.: On certain convolution inequalities, *Proc. Amer. Math. Soc.* **36** (1972) 505–510.
209. Hedberg, L. I.: Bounded point evaluations and capacity, *J. Funct. Anal.* **10** (1972) 269–280.
210. Hedberg, L. I.: Approximation in the mean by solutions of elliptic equations, *Duke Math. J.* **40** (1973) 9–16.
211. Hedberg, L. I.: Removable singularities and condenser capacities, *Ark. mat.* **12** (1974) 181–201.
212. Hedberg, L. I.: Two approximation problems in function spaces, *Ark. mat.* **16** (1978) 51–81.
213. Hedberg, L. I.: Spectral synthesis in Sobolev spaces, and uniqueness of solutions of the Dirichlet problem, *Acta Math.* **147** (1981) 237–264.
214. Hedberg, L. I.: On the Dirichlet problem for higher-order equations, in *Conference on Harmonic Analysis in Honor of Antoni Zygmund*, Proc., Chicago 1981 (W. Beckner, A. P. Calderón, R. Fefferman, P. W. Jones, eds.), 620–633, Wadsworth, Belmont, California, 1983.
215. Hedberg, L. I.: Spectral synthesis in Sobolev spaces, in *Linear and Complex Analysis Problem Book. 199 Research Problems* (V. P. Havin, S. V. Hruščëv, N. K. Nikol'skii, eds.), Lecture Notes in Math. **1043**, 435–437, Springer, Berlin Heidelberg, 1984. Also in *Linear and Complex Analysis Problem Book 3, Part II* (V. P. Havin, N. K. Nikolski, eds.), Lecture Notes in Math. **1574**, 174–176, Springer, Berlin Heidelberg, 1994.
216. Hedberg, L. I.: Nonlinear potential theory and Sobolev spaces, in *Nonlinear Analysis, Function Spaces and Applications, Vol. 3*, Proc. Spring School, Litomyšl 1986 (M. Krbec, A. Kufner, J. Rákosník, eds.), 5–30, Teubner, Leipzig, 1987.
217. Hedberg, L. I.: Nonlinear potential theory, in *Potential Theory*, Proc. Int. Conf. Potential Theory, Nagoya 1990 (M. Kishi, ed.), 43–54, de Gruyter, Berlin New York, 1991.
218. Hedberg, L. I.: Approximation by harmonic functions, and stability of the Dirichlet problem, *Expo. Math.* **11** (1993) 193–259.
219. Hedberg, L. I. and Wolff, Th. H.: Thin sets in nonlinear potential theory, *Ann. Inst. Fourier* (*Grenoble*) **33**:4 (1983) 161–187.
220. Heinonen, J., Kilpeläinen, T., and Malý, J.: Connectedness in fine topologies, *Ann. Acad. Sci. Fenn. Ser. A. I. Math.* **15** (1990) 107–123.

221. Heinonen, J., Kilpeläinen, T., and Martio, O.: *Nonlinear Potential Theory of Degenerate Elliptic Equations*, Oxford University Press, Oxford, 1993.
222. Helms, L. L.: *Introduction to Potential Theory*, Wiley–Interscience, New York, 1969.
223. Hempel, J. A., Morris, G. R., and Trudinger, N. S.: On the sharpness of a limiting case of the Sobolev imbedding theorem, *Bull. Austr. Math. Soc.* **3** (1970) 369–373.
224. Hestenes, M. R.: Extension of the range of a differentiable function, *Duke Math. J.* **8** (1941) 183–192.
225. Hewitt, E. and Stromberg, K.: *Real and Abstract Analysis*, Springer, New York, 1965.
226. Hirschman, I. I., Jr.: A convexity theorem for certain groups of transformations, *J. Analyse Math.* **2** (1953) 209–218.
227. Hörmander, L.: Estimates for translation invariant operators in L^p spaces, *Acta Math.* **104** (1960) 93–140.
228. Hörmander, L.: *Linear Partial Differential Operators*, Springer, Berlin Heidelberg, 1963.
229. Hörmander, L.: *An Introduction to Complex Analysis in Several Variables*, Van Nostrand, Princeton, New Jersey, 1966.
230. Hörmander, L.: *The Analysis of Linear Partial Differential Operators I*, Springer, Berlin Heidelberg, 1983.
231. Hörmander, L. and Lions, J. L.: Sur la completion par rapport à une integrale de Dirichlet, *Math. Scand.* **4** (1956) 259–270.
232. Janson, S.: On functions with derivatives in H^1, in *Harmonic Analysis and Partial Differential Equations*, Proc., El Escorial 1987 (J. García-Cuerva, ed.), Lecture Notes in Math. **1384**, 193–201, Springer, Berlin Heidelberg, 1989.
233. Jawerth, B.: Some observations on Besov and Lizorkin–Triebel spaces, *Math. Scand.* **40** (1977) 94–104.
234. Jawerth, B., Pérez, C., and Welland, G.: The positive cone in Triebel–Lizorkin spaces and the relation among potential and maximal operators, in *Harmonic Analysis and Partial Differential Equations*, Proc. Conf. Boca Raton 1988 (M. Milman and T. Schonbek, eds.), Contemporary Mathematics **107**, 71–91, Amer. Math. Soc., Providence, Rhode Island, 1989.
235. John, F.: *Plane Waves and Spherical Means Applied to Partial Differential Equations*, Interscience, New York, 1955.
236. John, F.: *Partial Differential Equations*, Springer, Berlin Heidelberg, 1971.
237. Jones, P. W.: Quasiconformal mappings and extendability of functions in Sobolev spaces, *Acta Math.* **147** (1981) 71–88.
238. Jonsson, A. and Wallin, H.: *Function Spaces on Subsets of* $\mathbf{R}^N$, Harwood Acad. Publ., New York, 1984.
239. Kahane, J.-P.: Dimension capacitaire et dimension de Hausdorff, in *Théorie du Potentiel*, Proc., Orsay 1983 (G. Mokobodzki, D. Pinchon, eds.), Lecture Notes in Math. **1096**, 393–400, Springer, Berlin Heidelberg, 1984.
240. Kahane, J.-P.: Ensembles aléatoires et dimensions, in *Recent Progress in Fourier Analysis*, Proc., El Escorial 1983 (I. Peral, J.-L. Rubio de Francia, eds.), 65–121, North-Holland, Amsterdam, 1985.
241. Katznelson, Y.: *An Introduction to Harmonic Analysis*, Wiley, New York, 1968.
242. Keldysh, M. V. (Келдыш, М. В.): О разрешимости и устойчивости задачи Дирихле, *Uspekhi Mat. Nauk* **8** (1941) 171–231. Reprinted in *М. В. Келдыш, Избранные труды. Математика* (*M. V. Keldysh, Selected Works. Mathematics*), 237–296, Nauka, Moscow, 1985. English translation: On the solvability and stability of the Dirichlet problem, *Amer. Math. Soc. Translations* (2) **51** (1966) 1–73.
243. Kellogg, O. D.: Unicité des fonctions harmoniques, *C. R. Acad. Sci. Paris* **187** (1928) 526–527.
244. Kellogg, O. D.: *Foundations of Potential Theory*, J. Springer, Berlin, 1929. Reprinted, Dover, New York, 1954.

245. Kerman, R and Sawyer, E. T.: The trace inequality and eigenvalue estimates for Schrodinger operators, *Ann. Inst. Fourier* (*Grenoble*) **36**:4 (1986) 207–228.
246. Khruslov E. Ya. (Хруслов Е. Я.) and Marchenko, V. A. (Марченко, В. А.): *Краевые задачи в областях с мелкозернистой границей* (*Boundary value problems in domains with fine-grained border*), Naukova Dumka, Kiev, 1974.
247. Khvoles, A. A. (Хволес, А. А.) and Maz'ya, V. G. (Мазья, В. Г.): О вложении пространства $\mathring{L}_p^l(\Omega)$ в пространство обобщенных функций (On imbedding the space $\mathring{L}_p^l(\Omega)$ into the space of generalized functions), *Trudy Tbiliss. Mat. Inst. Razmadze Akad. Nauk Gruzin. SSR* **66** (1981) 70–83.
248. Kilpeläinen, T.: Weighted Sobolev spaces and capacity, *Ann. Acad. Sci. Fenn. Ser. A. I. Math.* **19** (1994) 95–113.
249. Kilpeläinen, T. and Malý, J.: Degenerate elliptic equations with measure data and nonlinear potentials, *Ann. Scuola Norm. Sup. Pisa Cl. Sci.* (4) **19** (1992) 591–613.
250. Kilpeläinen, T. and Malý, J.: The Wiener test and potential estimates for quasilinear elliptic equations, *Acta Math.* **172**, 137–161, 1994.
251. Kinderlehrer, D. and Stampacchia, G.: *An Introduction to Variational Inequalities and Their Applications*, Academic Press, New York, 1980.
252. Kirszbraun, M. D.: Über die zusammenziehende und Lipschitzsche Transformationen, *Fund. Math.* **22** (1934) 77–108.
253. Kishi, M.: An existence theorem in potential theory, *Nagoya Math. J.* **27** (1966) 133–137.
254. Kneser, H.: Sur un théorème fondamental de la théorie des jeux, *C. R. Acad. Sci. Paris* **234** (1952) 2418–2420.
255. Kolsrud, T.: Bounded point evaluations and balayage, *Ark. mat.* **20** (1982) 137–146.
256. Kolsrud, T.: A uniqueness theorem for higher order elliptic partial differential equations, *Math. Scand.* **51** (1982) 323–332.
257. Kolsrud, T.: Condenser capacities and removable sets in $W^{1,p}$, *Ann. Acad. Sci. Fenn. Ser. A. I. Math.* **8** (1983) 343–348.
258. Kolsrud, T.: Fine potential theory in Dirichlet spaces, *Osaka J. Math.* **23** (1986) 337–361.
259. Köthe, G, *Topological Vector Spaces, I*, Springer, Berlin, 1969.
260. Lanconelli, E.: Sul problema di Dirichlet per l'equatione del calore, *Ann. Math. Pura Appl.* **97** (1973) 83–114.
261. Landes, R.: A note on strongly nonlinear parabolic equations of higher order, *Differential and Integral Equations* **3** (1990) 851–862.
262. Landis, E. M. (Ландис, Е. М.): Необходимые и достаточные условия регулярности граничной точки для задачи Дирихле для уравнения теплопроводности, *Dokl. Akad. Nauk AN SSSR* **185** (1969) 517–520. English translation: Necessary and sufficient conditions for regularity of a boundary point in the Dirichlet problem for the heat-conduction equation, *Soviet Math. Dokl.* **10** (1969) 380–384.
263. Landis, E. M. (Ландис, Е. М.): *Уравнения второго порядка эллиптического и параболического типов* (*Second order equations of elliptic and parabolic types*), Nauka, Moscow, 1971.
264. Landis, E. M.: Application of the potential theory to the study of qualitative properties of solutions of the elliptic and parabolic equations, in *Potential Theory—Surveys and Problems*, Proc., Prague 1987 (J. Král, J Lukeš, I. Netuka, J. Veselý, eds.), Lecture Notes in Math. **1344**, 133–153, Springer, Berlin Heidelberg, 1988.
265. Landis, E. M. (Ландис, Е. М.): О регулярности граничной точки для уравнения теплопроводности (On the regularity of a boundary point for the heat equation), in *Qualitative Theory of Boundary Value Problems of Mathematical Physics*, Collection dedicated to F. G. Maksudov (V. K. Kalantarov et al., eds.), 66–96, Ehlm, Baku, 1991.

266. Landkof, N. S. (Ландкоф, Н. С.): *Основы современной теории потенциала*, Nauka, Moscow, 1966. English translation: *Foundations of Modern Potential Theory*, Springer, Berlin, 1972.
267. Lindberg, P.: L^p-approximation by analytic functions in an open region, *Dissertation*, Uppsala, 1977.
268. Lindberg, P.: A constructive method for L^p-approximation by analytic functions, *Ark. mat.* **20** (1982) 61–68.
269. Lindqvist, P. and Martio, O.: Two theorems of N. Wiener for solutions of quasilinear elliptic equations, *Acta Math.* **155** (1985) 153–171.
270. Lions, J. L. and Magenes, E.: Problèmes aux limites non homogènes IV, *Ann. Scuola Norm. Sup. Pisa Cl. Sci.* (3) **15** (1961) 311–326.
271. Lions, J. L. and Magenes, E.: Problemi ai limiti non omogenei V, *Ann. Scuola Norm. Sup. Pisa Cl. Sci.* (3) **16** (1962) 1–44.
272. Lions, J. L. and Magenes, E.: *Problèmes aux limites non homogènes et applications, I*, Dunod, Paris, 1968.
273. Littman, W.: Polar sets and removable singularities of partial differential equations, *Ark. mat.* **7** (1967) 1–9.
274. Littman, W.: A connection between α-capacity and $m-p$ polarity, *Bull. Amer. Math. Soc.* **73** (1967) 862–866.
275. Littman, W., Stampacchia, G., and Weinberger, H.: Regular points for elliptic equations with discontinuous coefficients, *Ann. Scuola Norm. Sup. Pisa Cl. Sci.* (3) **17** (1963) 45–79.
276. Lizorkin, P. I. (Лизоркин, П. И.): Операторы, связанные с дробным дифференцированием, и классы дифференцируемых функций, *Trudy Mat. Inst. Steklov.* **117** (1973) 212–243. English translation: Operators connected with fractional differentiation, and classes of differentiable functions, *Proc. Steklov Inst. Math.* **117** (1972) 251–286.
277. Lizorkin, P. I.: Свойства функций из пространств $\Lambda^r_{p,\theta}$, *Trudy Mat. Inst. Steklov.* **131** (1974) 158–181. English translation: Properties of functions in the spaces $\Lambda^r_{p,\theta}$, *Proc. Steklov Inst. Math.* **131** (1974) 165–188.
278. Loewner, C.: On the conformal capacity in space, *J. Math. Mech.* **8** (1959) 411–414.
279. Loomis, L. H.: *An Introduction to Abstract Harmonic Analysis*, Van Nostrand, New York, 1953.
280. Lukeš, J. and Malý, J.: Thinness, Lebesgue density and fine topologies (An interplay between real analysis and potential theory), in *Summer School in Potential Theory, Joensuu 1990* (I. Laine, ed.), 35–70, University of Joensuu, Publications in Sciences, **26**, Joensuu, 1992.
281. Lusin, N. (Luzin, N. N.) (Лузин, Н. Н.): *Leçons sur les ensembles analytiques et leurs applications*, Gauthier-Villars, Paris, 1930.
282. Lyons, T.: Finely holomorphic functions, *J. Funct. Anal.* **37** (1980) 1–18.
283. Lysenko, Yu. A. (Лысенко, Ю. А.) and Pisarevskiĭ, B. M. (Писаревский, Б. М.): Неустойчивость гармонической емкости и аппроксимация непрерывных функций гармоническими функциями, *Mat. Sb.* **76** (**118**) (1968) 52–71. English translation: The instability of harmonic capacity and the approximation of continuous functions by harmonic functions, *Math. USSR Sb.* **5** (1968) 53–72.
284. Malgrange, B.: *Ideals of Differentiable Functions*, Tata Institute of Fundamental Research, Bombay, and Oxford University Press, London, 1966.
285. Malý, J.: Hölder type quasicontinuity, *Potential Analysis* **2** (1993) 249–254.
286. Malliavin, P.: Impossibilité de la synthèse spectrale sur les groupes abéliens non compacts, *Publ. Math. Inst. Hautes Etudes Sci. Paris* **1959**, 61–68.
287. Malliavin, P.: Infinite dimensional analysis, *Bull. Sci. Math.* (2) **117** (1993) 63–90.
288. Marcinkiewicz, J.: Sur les multiplicateurs des séries de Fourier, *Studia Math.* **8** (1939) 78–91.

289. Marschall, J.: The trace of Sobolev–Slobodeckij spaces on Lipschitz domains, *Manuscripta Math.* **58** (1987) 47–65.
290. Marshall, D. E.: A new proof of a sharp inequality concerning the Dirichlet integral, *Ark. mat.* **27** (1989) 131–137.
291. Mateu, J. and Orobitg, J.: Lipschitz approximation by harmonic functions and some applications to spectral synthesis, *Indiana Univ. Math. J.* **39** (1990) 703–736.
292. Mateu, J. and Verdera, J.: BMO harmonic approximation in the plane and spectral synthesis for Hardy–Sobolev spaces, *Rev. Mat. Iberoamericana* **4** (1988) 291–318.
293. Mattila, P. and Orobitg, J.: On some properties of Hausdorff content related to instability, *Ann. Acad. Sci. Fenn. Ser. A. I. Math.* **19** (1994) 393–398.
294. Maz'ya, V. G. (Мазья, В. Г.): p-проводимость и теоремы вложения некоторых функциональных пространств в пространство C, *Dokl. Akad. Nauk SSSR* **140** (1961) 299–302. English translation: The p-conductivity and theorems on imbedding certain function spaces into a C-space, *Soviet Math.* **2** (1961) 1200–1203.
295. Maz'ya, V. G. (Мазья, В. Г.): Об отрицательном спектре многомерного оператора Шредингера, *Dokl. Akad. Nauk SSSR* **144** (1962) 721–722. English translation: The negative spectrum of the n-dimensional Schrödinger operator, *Soviet Math.* **3** (1962) 808–810.
296. Maz'ya, V. G. (Мазья, В. Г.): О задаче Дирихле для эллиптических уравнений произвольного порядка в неограниченных областях, *Dokl. Akad. Nauk SSSR* **150** (1963) 1221–1224. English translation: The Dirichlet problem for elliptic equations of arbitrary order in unbounded regions, *Soviet Math.* **4** (1963) 860–863.
297. Maz'ya, V. G. (Мазья, В. Г.): О регулярности на границе решений эллиптических уравнений и конформного отображения, *Dokl. Akad. Nauk SSSR* **152** (1963) 1297–1300. English translation: Regularity at the boundary of solutions of elliptic equations and conformal mapping, *Soviet Math.* **4** (1963) 1547–1551.
298. Maz'ya, V. G. (Мазья, В. Г.): К теории многомерного оператора Шредингера (On the theory of the n-dimensional Schrödinger operator), *Izv. Akad. Nauk SSSR, Ser. Mat.* **28** (1964) 1145–1172.
299. Maz'ya, V. G. (Мазья, В. Г.): Полигармоническая емкость в теории первой краевой задачи (Polyharmonic capacity in the theory of the first boundary value problem), *Sibirsk. Mat. Zh.* **6** (1965) 127–148.
300. Maz'ya, V. G. (Мазья, В. Г.): О поведении вблизи границы решения задачи Дирихле для эллиптического уравнения второго порядка в дивергентной форме, *Mat. Zametki* **2** (1967) 209–220. English translation: Behavior, near the boundary, of solutions of the Dirichlet problem for a second order elliptic equation in divergent form, *Math. Notes* **2** (1967) 610–617.
301. Maz'ya, V. G. (Мазья, В. Г.): Классы множеств и мер, связанные с теоремами вложения (Classes of sets and measures connected with imbedding theorems), in *Теоремы вложения и их приложения* (*Imbedding Theorems and Their Applications*), Proc., Baku 1966 (L. D. Kudryavtsev (Л. Д. Кудрявцев), ed.), 142–159, Nauka, Moscow, 1970.
302. Maz'ya, V. G. (Мазья, В. Г.): О непрерывности в граничной точке решений квазилинейных эллиптических уравнений, *Vestnik Leningrad. Univ. Mat. Mekh. Astronom.* **25**:13 (1970) 42–55. Correction, *ibid.* **27**:1 (1972) 160. English translation: On the continuity at a boundary point of solutions of quasilinear equations, *Vestnik Leningrad Univ. Math.* **3** (1976) 225–242.
303. Maz'ya, V. G. (Мазья, В. Г.): О некоторых интегральных неравенствах для функций многих переменных, *Problemy Matematicheskogo Analiza, Leningrad. Univ.* **3** (1972) 33–68. English translation: On certain integral inequalities for functions of many variables, *J. Soviet Math.* **1** (1973) 205–234.
304. Maz'ya, V. G. (Мазья, В. Г.): Об устранимых особенностях ограниченных решений квазилинейных эллиптических уравнений любого порядка,

Zap. Nauchn. Sem. Leningrad. Otdel. Mat. Inst. Steklov. (*LOMI*) **26** (1972) 116–130. English translation: Removable singularities of bounded solutions of quasilinear elliptic equations of any order, *J. Soviet Math.* **3** (1975) 480–492.

305. Maz′ya, V. G. (Мазья, В. Г.): О (p, l)-емкости, теоремах вложения и спектре самосопряженного эллиптического оператора, *Izv. Akad. Nauk SSSR, Ser. Mat.* **37** (1973) 356–385. English translation: On (p, l)-capacity, imbedding theorems and the spectrum of a selfadjoint elliptic operator, *Math. USSR-Izv.* **7** (1973) 357–387.
306. Maz′ya, V. G. (Мазья, В. Г.): Behaviour of solutions to the Dirichlet problem for the biharmonic operator at a boundary point, in *Equadiff IV*, Proc., Prague, 1977 (J. Fábera, Ed.) *Lecture Notes in Math.* **703**, 250–262, Springer, Berlin Heidelberg, 1979.
307. Maz′ya, V. G. (Мазья, В. Г.): О суммируемости по произвольной мере функций из пространств С. Л. Соболева–Л. Н. Слободецкого (On summability with respect to an arbitrary measure of functions in Sobolev–Slobodeckiĭ spaces), *Zap. Nauchn. Sem. Leningrad. Otdel. Mat. Inst. Steklov.* (*LOMI*) **92** (1979) 192–202.
308. Maz′ya (Maz'ja), V. G.: *Sobolev Spaces*, Springer, Berlin New York, 1985. Russian edition, *Пространства С. Л. Соболева*, Izd. Leningrad. Univ., Leningrad, 1985.
309. Maz′ya, V. G.: Unsolved problems connected with the Wiener criterion, in *The Legacy of Norbert Wiener: A Centennial Symposium* (Proc., Cambridge, Massachusetts, 1994), *Proc. Symp. Pure Math.* **60**, 199–208, Amer. Math. Soc., Providence, Rhode Island, 1997.
310. Maz′ya, V. G. (Мазья, В. Г.) and Donchev, T. (Дончев, Т.): О регулярности по Винеру граничной точки для полигармонического оператора, *C. R. Acad. Bulgare Sci.* **36** (1983) 177–179. English translation: On Wiener regularity at a boundary point for a polyharmonic operator, *Amer. Math. Soc. Transl.* (2) **137** (1987) 53–55.
311. Maz′ya, V. G. and Netrusov, Yu. V.: Some counterexamples for the theory of Sobolev spaces on bad domains, *Potential Anal.* **4** (1995) 47–65.
312. Maz′ya, V. G. and Preobrazhenskiĭ, S. P.: Estimates for capacities and traces of potentials, *Internat. J. Math. Math. Sci.* **7** (1984) 41–63. (Also preprint: Мазья, В. Г. and Преображенский, С. П., Об оценках (l, p)-емкостей и следах потенциалов, *Wissenschaftlige Informationen*, Technische Hochschule, Karl-Marx-Stadt, Sekt. Math., No. 28 (1981).)
313. Maz′ya, V. G. and Shaposhnikova, T. O.: *The Theory of Multipliers in Spaces of Differentiable Functions*, Pitman, Boston–London, 1985. Russian edition (with additions), *Мультипликаторы в пространствах дифференцируемых функций*, Izd. Leningrad. Univ., Leningrad, 1986.
314. Maz′ya, V. G. and Verbitsky, I. E.: Capacitary inequalities for fractional integrals with applications to partial differential equations and Sobolev multipliers, *Ark. mat.* **33** (1995) 81–115.
315. Mel′nikov, M. S. (Мельников, М. С.) and Sinanyan, S. O. (Синанян, С. О.): Вопросы теории приближений функций одного комплексного переменного, in *Sovremennye problemy matematiki* (R. V. Gamkrelidze, ed.) **4**, 143–250, Itogi Nauki i Tekhniki, VINITI, Moscow, 1975. English translation: Problems in the theory of approximation of functions of one complex variable, *J. Soviet Math.* **5** (1976) 688–752.
316. Mergelyan, S. N. (Мергелян, С. Н.): О полноте систем аналитических функций, *Uspekhi Mat. Nauk* **8**:4 (1953) 3–63. English translation: On the completeness of systems of analytic functions, *Amer. Math. Soc. Translations* (2) **19** (1962) 109–166.
317. Meyer, Y.: *Ondelettes et opérateurs I, Ondelettes*, Hermann, Paris, 1990. English translation: *Wavelets and Operators*, Cambridge University Press, Cambridge, 1992.

318. Meyers, N. G.: A theory of capacities for potentials of functions in Lebesgue classes, *Math. Scand.* **26** (1970) 255–292.
319. Meyers, N. G.: Continuity of Bessel potentials, *Israel J. Math.* **11** (1972) 271–283.
320. Meyers, N. G.: Taylor expansion of Bessel potentials, *Indiana Univ. Math. J.* **23** (1974) 1043–1049.
321. Meyers, N. G.: Continuity properties of potentials, *Duke Math. J.* **42** (1975) 157–166.
322. Meyers, N. G.: Integral inequalities of Poincaré and Wirtinger types, *Arch. Rational Mech. Anal.* **68** (1978) 113–120.
323. Meyers, N. G. and Serrin, J.: $H = W$, *Proc. Nat. Acad. Sci. USA* **51** (1964) 1055–1056.
324. Michael, J. H. and Ziemer, W. P.: A Lusin type approximation of Sobolev functions by smooth functions, in *Classical Real Analysis*, Proc., Madison, Wisconsin, 1982 (D. Waterman, ed.), Contemporary Mathematics **42**, 135–167, Amer. Math. Soc., Providence, Rhode Island, 1985.
325. Mikhlin, S. G. (Михлин, С. Г.): О мультипликаторах интегралов Фурье (On multipliers of Fourier integrals), *Dokl. Akad. Nauk SSSR* **109** (1956) 701–703.
326. Mikhlin, S. G. (Михлин, С. Г.): Интегралы Фурье и кратные сингулярные интегралы (Fourier integrals and multiple singular integrals), *Vestnik Leningrad. Univ. Ser. Mat. Mech. Astr.* **12**:7 (1957) 143–155.
327. Mikhlin, S. G. (Михлин, С. Г.): *Многомерные сингулярные интегралы и интегральные уравнения* (Singular integrals in several dimensions and integral equations), Fizmatgiz, Moscow, 1962.
328. Miller, K.: Exceptional boundary points for the nondivergence equation which are regular for the Laplace equation and vice-versa, *Ann. Scuola Norm. Sup. Pisa Cl. Sci.* (3) **22** (1968) 315–330.
329. Minty, G. J.: On the extension of Lipschitz, Lipschitz–Hölder continuous, and monotone functions, *Bull. Amer. Math. Soc.* **76** (1970) 334–339.
330. Mizuta, Y.: Fine differentiability of Riesz potentials, *Hiroshima Math. J.* **8** (1978) 505–514.
331. Molchanov, A. M. (Молчанов, А. М.): Об условиях дискретности спектра самосопряженных дифференциальных уравнений второго порядка (On conditions for discreteness of the spectrum of selfadjoint differential equations of the second order), *Trudy Moskov. Mat. Obshch.* **2** (1953) 169–200.
332. Morawetz, C.: L^p-inequalities, *Bull. Amer. Math. Soc.* **75** (1969) 1299–1302.
333. Morrey, C. B., Jr.: *Multiple Integrals in the Calculus of Variations*, Springer, Berlin Heidelberg New York, 1966.
334. Moser, J.: A sharp form of an inequality by N. Trudinger, *Indiana Univ. Math. J.* **20** (1971) 1077–1092.
335. Muckenhoupt, B. and Wheeden, R. L.: Weighted norm inequalities for fractional integrals, *Trans. Amer. Math. Soc.* **192** (1974) 261–274.
336. Nagel, A., Rudin, W., and Shapiro, Joel H.: Tangential boundary behavior of functions in Dirichlet type spaces, *Ann. of Math.* **116** (1982) 331–360.
337. Netrusov, Yu. V. (Нетрусов, Ю. В.): Теоремы вложения пространств Бесова в идеальные пространства, *Zap. Nauchn. Sem. Leningrad. Otdel. Mat. Inst. Steklov.* (*LOMI*) **159** (1987) 69–82. English translation: Imbedding theorems of Besov spaces in Banach lattices, *J. Soviet Math.* **47**:6 (1989) 2871–2881.
338. Netrusov, Yu. V. (Нетрусов, Ю. В.): Теоремы вложения пространств Лизоркина–Трибеля, *Zap. Nauchn. Sem. Leningrad. Otdel. Mat. Inst. Steklov.* (*LOMI*) **159** (1987) 103–112. English translation: Embedding theorems for Lizorkin–Triebel spaces, *J. Soviet Math.* **47**:6 (1989) 2896–2903.
339. Netrusov, Yu. V. (Нетрусов, Ю. В.): Множества особенностей функций из пространств Бесова (Exceptional sets of functions from Besov spaces), in *Vsesoyuznaya shkola po teorii funkciĭ* (*12–22 oktyabrya 1987*). *Tezisy dokladov*, Izdatel′stvo Erevanskogo universiteta, Erevan, 1987.

340. Netrusov, Yu. V. (Нетрусов, Ю. В.): Теоремы вложения для следов пространств Бесова и Лизоркин–Триебеля, *Dokl. Akad. Nauk SSSR* **298** (1988) 1326–1330. English translation: Imbedding theorems for traces of Besov spaces and Lizorkin–Triebel spaces, *Soviet Math. Dokl.* **37** (1988) 270–273.
341. Netrusov, Yu. V. (Нетрусов, Ю. В.): Множества особенностей функций из пространств типа Бесова и Лизоркина–Трибеля, *Trudy Matem. Inst. Steklov.* **187** (1989) 162–177. English translation: Sets of singularities of functions in spaces of Besov and Lizorkin–Triebel type, *Proc. Steklov Inst. Math.* **187** (1990) 185–203.
342. Netrusov, Yu. V. (Нетрусов, Ю. В.): Метрические оценки емкостей множеств в пространствах Бесова, *Trudy Matem. Inst. Steklov.* **190** (1989) 159–185. English translation: Metric estimates of the capacities of sets in Besov spaces, *Proc. Steklov Inst. Math.* **190** (1992) 167–192.
343. Netrusov, Yu. V. (Нетрусов, Ю. В.): Спектральный синтез в пространствах гладких функций, *Ross. Akad. Nauk Dokl.* **325** (1992) 923–925. English translation: Spectral synthesis in spaces of smooth functions, *Russian Acad. Sci. Dokl. Math.* **46** (1993) 135–137.
344. Netrusov, Yu. V. (Нетрусов, Ю. В.): Оценки емкостей, связанных с пространствами Бесова, *Zap. Nauchn. Sem. S.-Peterburg. Otdel. Mat. Inst. Steklov.* (*POMI*) **201** (1992) 124–156. English translation: Estimates of capacities associated with Besov spaces, *J. Math. Sci.* **78** (1996) 199–217.
345. Netrusov, Yu. V. (Нетрусов, Ю. В.): Нелинейная аппроксимация функций из пространств Бесова–Лоренца в равномерной метрике, *Zap. Nauchn. Sem. S.-Peterburg. Otdel. Mat. Inst. Steklov.* (*POMI*) **204** (1993) 61–81. English translation: Uniform-metric nonlinear approximation of functions from Besov–Lorentz spaces, *J. Math. Sci.* **79** (1996) 1308–1319.
346. Netrusov, Yu. V. (Нетрусов, Ю. В.): Спектральный синтез в пространстве Соболева, порожденном интегральной метрикой (Spectral synthesis in the Sobolev space associated with integral metric), *Zap. Nauchn. Sem. S.-Peterburg. Otdel. Mat. Inst. Steklov.* (*POMI*) **217** (1994) 92–111. English translation: *J. Math. Sci.* **85** (1997) 1814–1826.
347. Netrusov, Yu. V. (Нетрусов, Ю. В.): In preparation.
348. von Neumann, J.: Zur Theorie der Gesellschaftsspiele, *Math. Ann.* **100** (1928) 295–320.
349. Nikol′skiĭ, S. M. (Никольский, С. М.): *Приближение функций многих переменных и теоремы вложения*, Nauka, Moscow, 1969. English translation: *Approximation of Functions of Several Variables and Imbedding Theorems*, Springer, Berlin Heidelberg, 1975.
350. Nirenberg, L.: On elliptic partial differential equations, *Ann. Scuola Norm. Sup. Pisa Cl. Sci.* (3) **13** (1959) 115–162.
351. Nyström, K.: An estimate of a polynomial capacity, *Potential Anal.* **9** (1998) 217–227.
352. Ohtsuka, M.: Capacité d'ensembles de Cantor généralisés, *Nagoya Math. J.* **11** (1957) 151–160.
353. O'Neil, R.: Convolution operators and $L(p, q)$ spaces, *Duke Math. J.* **30** (1963) 129–142.
354. Orobitg, J.: On spectral synthesis in some Hardy–Sobolev spaces, *Proc. Roy. Irish Acad. Sect. A* **92** (1992) 205–223.
355. Peetre, J.: On spaces of Triebel–Lizorkin type, *Ark. mat.* **13** (1975) 123–130. Correction, *ibid.* **14** (1976) 299.
356. Peetre, J.: *New Thoughts on Besov Spaces*, Mathematics Department, Duke University, Durham, North Carolina, 1976.
357. Pérez, C.: Two weighted inequalities for potential and fractional type maximal inequalities, *Indiana Univ. Math. J.* **43** (1994) 663–683.

358. Polishchuk, E. M. (Полищук, Е. М.) and Shaposhnikova, T. O. (Шапошникова, Т. О.): *Жак Адамар* (*Jacques Hadamard*), Nauka, Leningrad, 1990.
359. Polking, J. C.: Approximation in L^p by solutions of elliptic partial differential equations, *Amer. J. Math.* **94** (1972) 1231–1244.
360. Polking, J. C.: A Leibniz formula for some differentiation operators of fractional order, *Indiana Univ. Math. J.* **21** (1972) 1019–1029.
361. Polking, J. C.: A survey of removable singularities, in *Seminar on Nonlinear Partial Differential Equations*, Proc. (S. S. Chern, ed.), Math. Sci. Research Institute Pub. **2**, Springer, New York, 1984.
362. Reshetnyak, Yu. G. (Решетняк, Ю. Г.): О понятии емкости в теории функций с обобщенными производными, *Sibirsk. Mat. Zh.* **10** (1969) 1109–1138. English translation: On the concept of capacity in the theory of functions with generalized derivatives, *Siberian Math. J.* **10** (1969) 818–842.
363. Reshetnyak, Yu. G.: О граничном поведении функций с обобщенными производными, *Sibirsk. Mat. Zh.* **13** (1972) 411–419. English translation: On the boundary behavior of functions with generalized derivatives, *Siberian Math. J.* **13** (1972) 285–290.
364. Rickman, S.: *Quasiregular Mappings*, Springer, Berlin Heidelberg, 1993.
365. Rogers, C. A.: *Hausdorff Measures*, Cambridge University Press, Cambridge, 1970.
366. Rudin, W.: *Principles of Mathematical Analysis*, Third edition, McGraw-Hill, New York, 1976.
367. Rudin, W.: *Real and Complex Analysis*, Third edition, McGraw-Hill, New York, 1986.
368. Rudin, W.: *Functional Analysis*, Second edition, McGraw-Hill, New York, 1991.
369. Saak, È. M. (Саак, Э. М.): Емкостный критерий для области с устойчивой задачей Дирихле для эллиптических уравнений высших порядков, *Mat. Sb.* **100**:2 (1976) 201–209. English translation: A capacity criterion for a domain with stable Dirichlet problem for higher order elliptic equations, *Math. USSR Sbornik* **29**:2 (1976) 177–185.
370. Sawyer, E. T.: A characterization of two weight norm inequalities for fractional and Poisson integrals, *Trans. Amer. Math. Soc.* **308** (1988) 533–545.
371. Schulze, B.-W. and Wildenhain, G.: *Methoden der Potentialtheorie für elliptische Differentialgleichungen beliebiger Ordnung*, Akademie-Verlag, Berlin, 1977.
372. Schwartz, L.: Sur une propriété de synthèse spectrale dans les groupes noncompacts, *C. R. Acad. Sci. Paris* **227** (1948) 424–426.
373. Schwartz, L.: *Théorie des distributions*, 2nd ed., Hermann, Paris, 1966.
374. Serrin, J. Local behavior of solutions of quasi-linear equations, *Acta Math.* **111** (1964) 247–302.
375. Serrin, J. Removable singularities of solutions of elliptic equations, *Arch. Rational Mech. Anal.* **17** (1964) 65–78.
376. Shapiro, H. S.: *Smoothing and Approximation of Functions*, Van Nostrand, New York, 1969.
377. Shaposhnikova, T. O. and Maz'ya, V. G.: *Jacques Hadamard, A Universal Mathematician*, Amer. Math. Soc., Providence, R.I.; London Math. Soc., London, 1998.
378. Sinanyan, S. O. (Синанян, С. О.): Аппроксимация аналитическими функциями и полиномами в среднем по площади, *Mat. Sb.* **69** (1966) 546–578. English translation: Approximation by polynomials and analytic functions in the areal mean, *Amer. Math. Soc. Translations* (2) **74** (1968) 91–124.
379. Sion, M.: On capacitability and measurability, *Ann. Inst. Fourier* (*Grenoble*) **13** (1963) 88–99.
380. Sjödin, T.: Bessel potentials and extension of continuous functions on compact sets, *Ark. mat.* **13** (1975) 263–271.

381. Sjödin, T.: Non-linear potential theory in Lebesgue spaces with mixed norm, in *Potential Theory*, Proc., Prague 1987 (J. Král, I. Netuka, J. Lukeš, J. Veselý, eds.), 325–331, Plenum Press, New York, 1988.
382. Skrypnik, I. V.: *Nonlinear Elliptic Boundary Value Problems*, Teubner, Leipzig, 1986.
383. Sobolev, S. L. (Соболев, С. Л.): Об одной краевой задаче для полигармонических уравнений, *Mat. Sb.* (*N. S.*) **2(44)** (1937) 465–499. English translation: On a boundary value problem for polyharmonic equations, *Amer. Math. Soc. Translations* (2) **33** (1963) 1–40.
384. Sobolev, S. L.: Об одной теореме функционального анализа, *Mat. Sbornik* (*N. S.*) **4(46)** (1938) 471–497. English translation: On a theorem in functional analysis, *Amer. Math. Soc. Translations* (2) **34** (1963) 39–68.
385. Sobolev, S. L.: *Некоторые применения функционального анализа в математической физике*, Izd. LGU, Leningrad, 1950. Third, revised edition, Nauka, Moscow, 1988. English translation: *Applications of Functional Analysis in Mathematical Physics*, Amer. Math. Soc., Providence, Rhode Island, 1963. Third edition, *Some Applications of Functional Analysis in Mathematical Physics*, Amer. Math. Soc., Providence, Rhode Island, 1991.
386. Souslin, M. (Suslin, M. Ya.) (Суслин, М. Я.): Sur une définition des ensembles mesurables *B* sans nombres transfinis, *C. R. Acad. Sci. Paris* **164** (1917) 88–91.
387. Stampacchia, G.: The spaces $L^{p,\lambda}$ and $N^{p,\lambda}$ and interpolation, *Ann. Scuola Norm. Sup. Pisa Cl. Sci.* **19** (1965) 443–462.
388. Stein, E. M.: The characterization of functions arising as potentials I, II, *Bull. Amer. Math. Soc.* **67** (1961) 102–104, and *ibid.* **68** (1962) 577–582.
389. Stein, E. M.: *Singular Integrals and Differentiability of Functions*, Princeton University Press, Princeton, New Jersey, 1970.
390. Stein, E. M.: *Harmonic Analysis: Real-Variable Methods, Orthogonality, and Oscillatory Integrals*, Princeton University Press, Princeton, New Jersey, 1993.
391. Stein, E. M. and Weiss, G.: *Introduction to Fourier Analysis on Euclidean Spaces*, Princeton University Press, Princeton, New Jersey, 1971.
392. Strichartz, R. S.: Multipliers on fractional Sobolev spaces, *J. Math. Mech.* **16** (1967) 1031–1060.
393. Strichartz, R. S.: A note on Trudinger's extension of Sobolev's inequalities, *Indiana Univ. Math. J.* **21** (1972) 841–842.
394. Strichartz, R. S.: H^p Sobolev spaces, *Colloq. Math.* **60/61** (1990) 129–139.
395. Taylor, S. J.: On the connexion between Hausdorff measure and generalized capacity, *Proc. Cambridge Philos. Soc.* **57** (1961) 524–531.
396. Tarkhanov, N. N. (Тарханов, Н. Н.): *Ряд Лорана для решений эллиптических систем* (The Laurent series for solutions of elliptic systems), Nauka, Novosibirsk, 1991.
397. Tarkhanov, N. N. (Тарханов, Н. Н.): Аппроксимация на компактах решениями систем с сюрьективным символом, *Uspekhi Mat. Nauk* **48**:5 (1993) 107–146. English translation: Approximation on compact sets by solutions of systems with surjective symbol, *Russian Math. Surveys* **48**:5 (1994) 103–145.
398. Thomson, J. E.: Approximation in the mean by polynomials, *Ann. of Math.* **133** (1991) 477-507.
399. Titchmarsh, E. C.: *Introduction to the Theory of Fourier Integrals*, Second edition, Oxford University Press, Oxford, 1948.
400. Torchinsky, A.: *Real-Variable Methods in Harmonic Analysis*, Academic Press, Orlando, Florida, 1986.
401. Treves, F.: *Basic Linear Partial Differential Equations*, Academic Press, New York, 1975.
402. Triebel, H.: Spaces of distributions of Besov type on Euclidean n-space. Duality, interpolation, *Ark. mat.* **11** (1973) 13–64.

403. Triebel, H.: Boundary values for Sobolev-spaces with weights. Density of $\mathcal{D}(\Omega)$ in $W^s_{p,\gamma_0,\dots,\gamma_r}(\Omega)$ and in $H^s_{p,\gamma_0,\dots,\gamma_r}(\Omega)$ for $s > 0$ and $r = [s - \frac{1}{p}]^-$, *Ann. Scuola Norm. Sup. Pisa Cl. Sci.* (3) **27** (1973) 73–96.

404. Triebel, H.: *Interpolation Theory, Function Spaces, Differential Operators*, VEB Deutscher Verlag der Wissenschaften, Berlin, 1973.

405. Triebel, H.: *Theory of Function Spaces*, Akademische Verlagsgesellschaft Geest & Portig K.-G., Leipzig, 1983, and Birkhäuser Verlag, Basel, 1983.

406. Triebel, H.: *Theory of Function Spaces II*, Birkhäuser Verlag, Basel, 1992.

407. Trudinger, N. S.: On imbeddings into Orlicz spaces and some applications, *J. Math. Mech.* **17** (1967) 473–483.

408. Turesson, B. O.: Nonlinear potential theory and weighted Sobolev spaces, *Linköping Studies in Science and Technology. Dissertations* **387**, Linköping, 1995.

409. Ugaheri, T.: On the general potential and capacity, *Japanese J. Math.* **20** (1950) 37–43.

410. Väisälä, J.: *Lectures on n-Dimensional Quasiconformal Mappings*, Lecture Notes in Math. **229**, Springer, Berlin Heidelberg, 1971.

411. Vasilesco, F.: Sur la continuité du potentiel à travers des masses, *C. R. Acad. Sci. Paris* **200** (1935) 1173–1174.

412. Verbitsky, I. E.: Weighted norm inequalities for maximal operators and Pisier's theorem on factorization through $L^{p\infty}$, *Integral Equations Operator Theory* **15** (1992) 124–153.

413. Verbitsky, I. E. and Wheeden, R. L.: Weighted trace inequalities for fractional integrals and applications to semilinear equations, *J. Funct. Anal.* **129** (1995) 221–241.

414. Verdera, J.: Removability, capacity and approximation, in *Complex Potential Theory* (*Montreal 1993*), *NATO Adv. Sci. Inst. Ser. C Math. Phys. Sci.* **439**, 419–473, Kluwer Acad. Publ., Dordrecht, 1994.

415. Vitushkin, A. G. (Витушкин, А. Г.): Условия на множество, необходимые и достаточные для возможности равномерного приближения аналитическими (или рациональными) функциями всякой непрерывной на этом множестве функции (Necessary and sufficient conditions a set should satisfy in order that any function continuous on it can be approximated uniformly by analytic or rational functions), *Dokl. Akad. Nauk SSSR* **128** (1959) 17–20.

416. Vitushkin, A. G. (Витушкин, А. Г.): Аналитическая емкость множеств в задачах теории приближений, *Uspekhi Mat. Nauk* **22**:6 (1967) 141–199. English translation: Analytic capacity of sets in problems of approximation theory, *Russian Math. Surveys* **22**:6 (1967) 139–200.

417. Vodop'yanov, S. K. (Водопьянов, С. К.): Теория потенциала на однородных группах, *Mat. Sb.* **180**:1 (1989) 51–77. English translation: Potential theory on homogeneous groups, *Math. USSR Sb.* **66** (1990) 59–81.

418. Vodop'yanov, S. K. (Водопьянов, С. К.): L_p-теория потенциала и квазиконформные отображения на однородных группах (L_p-potential theory and quasiconformal mappings on homogeneous groups), in *Современные проблемы геометрии и анализа* (*Contemporary problems of geometry and analysis*), Trudy Instituta Matematiki **14**, 45–89, Nauka, Novosibirsk, 1989.

419. Vodop'yanov, S. K. (Водопьянов, С. К.): L_p-теория потенциала для обобщенных ядер и ее приложения (L_p-potential theory for generalized kernels and applications), *Preprint* **6** (1990), Math. Inst. Sib. Branch AN SSSR, Novosibirsk.

420. Vodop'yanov, S. K. (Водопьянов, С. К.): Весовая L_p-теория потенциала на однородных группах, *Sibirsk. Mat. Zh.* **33**:2 (1992) 29–48. English translation: Weighted L_p potential theory on homogeneous groups, *Siberian Math. J.* **33** (1992) 201–218.

421. Wallin, H.: Continuous functions and potential theory, *Ark. mat.* **5** (1963) 55–84.

422. Wallin, H.: A connection between α-capacity and L^p-classes of differentiable functions, *Ark. mat.* **5** (1964) 331–341.
423. Wallin, H.: On the existence of boundary values of a class of Beppo Levi functions, *Trans. Amer. Math. Soc.* **120** (1965) 510–525.
424. Wallin, H.: Metrical characterization of conformal capacity zero, *J. Math. Anal. Appl.* **58** (1977) 298–311.
425. Walsh, J. L.: *Interpolation and Approximation by Rational Functions in the Complex Domain*, Second edition, Amer. Math. Soc., Providence, Rhode Island, 1960.
426. Watson, G. N.: *A Treatise on the Theory of Bessel Functions*, Second edition, Cambridge University Press, Cambridge, 1944.
427. Webb, J. R. L.: Boundary value problems for strongly nonlinear elliptic equations, *J. London Math. Soc.* **21** (1980) 123–132.
428. Wermer, J.: *Potential Theory*, Lecture Notes in Math. **408**, Springer, Berlin, 1974.
429. Wheeden, R. L.: A characterization of some weighted norm inequalities for the fractional maximal function, *Studia Math.* **107** (1993) 257–272.
430. Whitney, H.: Analytic extensions of differentiable functions defined in closed sets, *Trans. Amer. Math. Soc.* **36** (1934) 63–89.
431. Whitney, H.: On ideals of differentiable functions, *Amer. J. Math.* **70** (1948) 635–658.
432. Wiener, N.: Certain notions in potential theory, *J. Math. and Phys.* **3** (1924) 24–51. Reprinted in *Norbert Wiener: Collected Works with Commentaries, Vol. 1*, 364–391, MIT Press, Cambridge, Massachusetts, 1976.
433. Wiener, N.: The Dirichlet problem, *J. Math. and Phys.* **3** (1924) 127–146. Reprinted in *Norbert Wiener: Collected Works with Commentaries, Vol. 1*, 394–413, MIT Press, Cambridge, Massachusetts, 1976.
434. Wiener, N.: The ergodic theorem, *Duke Math. J.* **5** (1939) 1–18. Reprinted in *Norbert Wiener: Collected Works with Commentaries, Vol. 1*, 672–689, MIT Press, Cambridge, Massachusetts, 1976.
435. Wildenhain, G.: Potential theory methods for higher order elliptic equations, in *Potential Theory—Surveys and Problems*, Proc., Prague 1987 (J. Král, J Lukeš, I. Netuka, J. Veselý, eds.), Lecture Notes in Math. **1344**, 181–195, Springer, Berlin Heidelberg, 1988.
436. Yudovich, V. I. (Юдович, В. И.): О некоторых оценках, связанных с интегральными операторами и решениями эллиптических уравнений, *Dokl. Akad. Nauk SSSR* **138** (1961) 805–808. English translation: Some estimates connected with integral operators and with solutions of elliptic equations, *Soviet Math.* **2** (1961) 746–749.
437. Ziemer, W. P.: Uniform differentiability of Sobolev functions, *Indiana Univ. Math. J.* **37** (1988) 789–799.
438. Ziemer, W. P.: *Weakly Differentiable Functions*, Springer, New York, 1989.
439. Zygmund, A.: On a theorem of Marcinkiewicz concerning interpolation of operators, *J. Math. pures appl.* **35** (1956) 223–248.
440. Zygmund, A.: *Trigonometric Series*, Second edition, Cambridge University Press, Cambridge, 1959.

Index

List of Symbols

Symbols listed in order of appearance

Chapter 1

A various unspecified constants, 1

$\mathbf{N}$ natural numbers, 1

$\mathbf{Z}$ integers, 1

$\mathbf{R}$ real numbers, 1

$\mathbf{R}^N$ Euclidean N-space, 1

$x \cdot y = x_1 y_1 + \cdots + x_N y_N$ Euclidean scalar product, 1

$|x| = (x \cdot x)^{1/2}$ Euclidean norm, 1

$\overline{E}$ closure of E, 1

E^c, $\mathbf{R}^N \setminus E$ complement of E, 1

E^0, $\operatorname{int} E$ interior of E, 1

$|E|$, dm, dx Lebesgue measure, 1

χ_E, $\chi(E)$ characteristic function, 1

$B(x_0, r)$ open ball, 1

$B_n(x_0)$, B_n dyadic ball, 1

f_+, f_- positive and negative parts of f, 1

$(\partial/\partial x_j) f = \partial_j f$ partial derivative, 2

$\nabla f = (\partial_1 f, \ldots, \partial_N f)$ gradient, 2

$\Delta f = (\partial_1^2 + \cdots + \partial_N^2) f$ Laplacian, 2

$\sigma = (\sigma_1, \ldots, \sigma_N)$ multiindex, 2

$|\sigma| = \sigma_1 + \cdots + \sigma_N$ length of multiindex, 2

$D^\sigma f$ partial derivative, 2

$x^\sigma = x_1^{\sigma_1} \cdots x_N^{\sigma_N}$ 2

$\sigma! = \sigma_1! \cdots \sigma_N!$ 2

$P_x^m f(x+y)$ Taylor polynomial, 2

$\nabla^m f$ higher gradient, 2

$|\nabla^m f|$ 2

$L^p(E)$, L^p, $L^p(\nu)$ Lebesgue spaces, 2

$\|f\|_{L^p(E)}$, $\|f\|_p$, L^p-norms 2

$L_+^p(E)$ positive cone in $L^p(E)$, 2

$L^{p'}(E)$, $p' = p/(p-1)$ dual space of $L^p(E)$, 2

$C(E)$ space of continuous functions, 2

$C_0(\Omega)$ compactly supported functions, 2

$C^m(\Omega)$ space of m times differentiable functions, 2

$C_0^m(\Omega)$ compactly supported functions in $C^m(\Omega)$, 2

$\mathcal{M}(\Omega)$ dual of $C_0(\Omega)$, Radon measures, 2

$\mathcal{M}^+(\Omega)$ positive measures, 2

Mf, $M_0 f$ Hardy–Littlewood maximal function, 3

$M_\alpha f$ fractional maximal function, 3

$M_{\alpha,\delta} f$ inhomogeneous maximal function, 3

$L^p(l^q)$ 3

$f * g$ convolution, 4

$C^\infty(\Omega)$ 4

$C_0^\infty(\Omega)$ compactly supported test functions, 4

$\mathcal{S}$ Schwartz class of test functions, 4

$\mathcal{D}'$ space of distributions, 4

$\mathcal{S}'$ space of temperate (tempered) distributions, 4

$\langle \cdot, \cdot \rangle$ pairing, 4

$\mathcal{F}$ Fourier transformation, 5

$\mathcal{F}f$, $\widehat{f}$ Fourier transform, 5

$\mathcal{R}_j$ Riesz transform, 5

R_j singular Riesz kernel, 5

$W^{\alpha,p}(\Omega)$, $W^{\alpha,p}$ Sobolev space, 6, 7

$W_0^{\alpha,p}(\Omega)$ Sobolev space, 7

$\dot{W}_0^{\alpha,p}(\Omega)$ homogeneous Sobolev space, 8

ω_{N-1} area of $(N-1)$-dimensional unit sphere, 8

I_2 Newton kernel, 8

$\mathcal{I}_\alpha$ Riesz potential operator, 9

I_α Riesz kernel, 9

γ_α constant in I_α, 9

G_α Bessel kernel, 9

$\mathcal{G}_\alpha$ Bessel potential operator, 10

J_ν Bessel function, 11

K_ν modified Bessel function, Macdonald function, 11

$L^{\alpha,p}$ Bessel potential space, 13

$\dot{L}^{\alpha,p}$ Riesz potential space, 13

p^* Sobolev exponent, 14

Chapter 2

$C(\cdot)$ capacity, 17

$C'_{\alpha,p}$, $\alpha \in \mathbf{N}$ capacity, 19

Δ_p p-Laplace operator, p-Laplacian, 20

ω_K subset of $\mathcal{S}$, 20

$C_{\alpha,p}$, $\alpha > 0$, (α, p)-capacity, Bessel capacity, 20, 38

$\overline{\omega}_K$ closure of ω_K, 20

F^K capacitary potential, 21, 22

f^K capacitary function, 21, 22

μ^K capacitary measure, 21

$V^\mu_{\alpha,p}$ nonlinear potential, 21

M measure space, 24

$\mathcal{G}f$, $\check{\mathcal{G}}\mu$ potentials, 24

$\mathcal{E}_g$ mutual energy, 24

Ω_E subset of $L^p(\nu)$, 25

$C_{g,p}$ capacity, 25

$\overline{\Omega}_E$ closure of Ω_E, 27

$V^\mu_{g,p}$ nonlinear potential, 36

$\dot{C}_{\alpha,p}$, $\alpha > 0$ Riesz capacity, 38

$\dot{V}^\mu_{\alpha,p}$ Riesz nonlinear potential, 38

$\dot{W}^\mu_{\alpha,p}$ Wolff potential, 45

$W^\mu_{\alpha,p}$ Wolff potential, 45

$\widetilde{\omega}_K$ subset of $\mathcal{S}$, 45

$N_{\alpha,p}$ distribution capacity, 45

$\mathcal{L}$ partial differential operator, 47

$\mathcal{L}^*$ adjoint operator, 47

Chapter 3

p^* Sobolev exponent, 56

f^* non-increasing rearrangement, 60

f^{**} average of f^*, 60

$[\alpha]$ integral part of α, 68

$\mathcal{D}^\alpha$ operator, 68

$\mathcal{S}^\alpha$ operator, 72

$I_{\alpha,\delta}$ modified Riesz kernel, 74

$L^p(l^q)$ 78

$\mathcal{L}^{p,\lambda}$ Morrey–Campanato space, 79

BMO functions of bounded mean oscillation, 79

Chapter 4

$B^{p,q}_\alpha$ Besov space, 86

$l^q(L^p)$ 86, 104, 117

$\mathfrak{P}$ space of polynomials, 86

$\dot{B}^{p,q}_\alpha$ homogeneous Besov space, 86

$F^{p,q}_\alpha$ Lizorkin–Triebel space, 92

$\dot{F}^{p,q}_\alpha$ homogeneous Lizorkin–Triebel space, 92

$\{r_n(t)\}_0^\infty$ Rademacher functions, 92

$B^{p,q}_{\alpha;\lambda}$ weighted Besov space, 97

u^{**} Peetre maximal function, 98

$\{\eta_n^{**}u\}_0^\infty$ Peetre maximal function, 98

$L^p(l^q)$ 104, 117

$h_\alpha(x, y, n)$ kernel, 104

$\mathcal{H}_\alpha f$, $\check{\mathcal{H}}_\alpha\mu$ potentials, 104

$C(\cdot\,; B^{p,q}_\alpha)$ Besov capacity, 105

$C(\cdot\,; F^{p,q}_\alpha)$ Lizorkin–Triebel capacity, 105

$\mathcal{E}_{h_\alpha}$ mutual energy, 105

$\mathcal{V}^\mu_{\alpha,p}$ Besov nonlinear potential, 106

$\mathcal{V}^\mu_{\alpha,p,q}$ Besov or Lizorkin–Triebel nonlinear potential, 107, 108

$W^\mu_{\alpha,p}$ Wolff potential, 109, 110

$\dot{W}^\mu_{\alpha,p}$ homogeneous Wolff potential, 109, 110

Q_{nk} dyadic cubes, 111

χ_{nk} characteristic function for Q_{nk}, 111

$l(Q)$ side of cube, 111, 246

λQ concentric cube, 111

$\widetilde{Q}_{nk} = 3Q_{nk}$ expanded cube, 111

$\widetilde{\chi}_{nk}$ characteristic function for $\widetilde{Q}_{nk}$, 111
a_{nk} atoms, 111
$s_n(x) = \sum_{k\in\mathbf{Z}^N} s_{nk}\chi_{nk}(x)$ 111
$\mathfrak{h}_\alpha(x, n, k)$ kernel, 117
$\mathfrak{H}_\alpha s$, $\check{\mathfrak{H}}_\alpha \mu$ atomic potentials, 117
$\mu_{nk} = \mu(Q_{nk})$ 117
$\mathcal{E}_{\mathfrak{h}_\alpha}$ mutual energy, 117
$b_\alpha^{p,q}$ 117
$f_\alpha^{p,q}$ 118
$C(\cdot\,; b_\alpha^{p,q})$ Besov capacity, 118
$C(\cdot\,; f_\alpha^{p,q})$ Lizorkin–Triebel capacity, 118
$\mathfrak{V}^\mu_{\alpha,p}$ atomic nonlinear potential, 119
$\mathfrak{V}^\mu_{\alpha,p,q}$ atomic nonlinear potential, 120
$\widetilde{\mu}_{nk} = \mu(\widetilde{Q}_{nk})$ 121
$\mathcal{E}(u, F)$ local degree of approximation, 122
$\mathcal{E}^\alpha$ operator, 122

Chapter 5

B_r ball, 129
$\Lambda_h^{(\rho)}$, $\Lambda_h^{(\infty)}$ Hausdorff content (Hausdorff capacity), 132
Λ_h, Λ_α Hausdorff measure, 132
$\mathcal{Q}$, $\mathcal{Q}_n$ meshes of cubes, 136
$\mathcal{L} = \{l_k\}_0^\infty$ 142
$E_{\mathcal{L}}$ Cantor set, 143

Chapter 6

$f|_E$ trace, 158
$\widetilde{f}$ quasicontinuous representative, 158
$b(E)$ set where E is thick, 165, 176
$e(E)$ set where E is thin, 165, 176
$b_{\alpha,p}(E)$ set where E is (α, p)-thick, 166
$e_{\alpha,p}(E)$ set where E is (α, p)-thin, 166
$\mathcal{C}_{\alpha,p}$ (α, p)-capacity 167
$\widetilde{E}$ fine closure of E, 176
BV functions of bounded variation, 181
BV^α 181

Chapter 7

μ_E restriction of μ, 191
$\Theta_q(\mu)$ 192
$\Gamma_{\alpha,p}(\cdot)$ norm, 199
$\mathbf{K}_u$ convex set in L^p, 199
$L^p(C_{\alpha,p})$ space of quasicontinuous functions, 200
$C_{1,p}(K; \Omega)$ condenser capacity, 209
$C_{2,p}^+(K; \Omega)$ condenser capacity, 209
$\widetilde{C}_{m,p}(F; \Omega)$ condenser capacity, 209
$\widetilde{W}^{m,p}(\Omega)$ Sobolev space, 209
$\kappa(\mu; t)$ 211
$\widetilde{\Lambda}_\alpha^{(\infty)}$ dyadic Hausdorff capacity, 212
$L^1(\Lambda_{N-\alpha}^{(\infty)})$ space of quasicontinuous functions, 212

Chapter 8

$f|_E$, $f|_E = 0$ trace, 219
$\widetilde{f}$ quasicontinuous representative, 219
$C_{k,p;\delta}$ modified capacity, 223
$c_{k,p}$ relative capacity, 223
E_m extension operator, 227
$\mathfrak{P}_m$ space of polynomials of degree $\leq m$, 228

Chapter 9

$f|_E$, $f|_E = 0$ trace, 234
$\widetilde{f}$ quasicontinuous representative, 234
Tr_E trace operator, 234
$\mathrm{Ker}\,\mathrm{Tr}_E$ kernel of Tr_E, 234
$[f]_n$ average over dyadic ball, 245
$\mathcal{Q}_n$ meshes of cubes, 246
Q_{ni} cubes in $\mathcal{Q}_n$, 246
$l(Q)$ side of cube, 246
rQ concentric cube, 246
$\widetilde{Q}_{ni} = 7Q_{ni}$ 246
$Q(f)$ average over Q, 247
λ_{ni} 247
$\rho_n(x)$ 248
G_n 248, 252

$\rho(x)$ 248
$\Gamma(x)$ 249
$\overline{\lambda}_n, \underline{\lambda}_n$ 249
λ^*_{ni} 251
$\rho^*_n(x)$ 251
$\rho'_n(x)$ 252
G_n, G'_n 252
$\Gamma_0(x)$ 252
$U_\lambda, U'_\lambda, U''_\lambda$ 263, 264, 272

Chapter 10

$L_0^{\alpha,p}(E)$ closure of functions with compact support, 281
Q_{nk} dyadic cube, 282
$\widetilde{Q}_{nk} = 3Q_{nk}$ 282
$\chi_{nk}, \widetilde{\chi}_{nk}$ characteristic functions, 282
$\mathfrak{M}_E$ ideal, 293
$\mathfrak{P}, \mathfrak{P}_m$ spaces of polynomials, 293
$\mathfrak{I}_m$ polynomial ideal, 293
$\mathfrak{P}/\mathfrak{I}_m$ quotient ring, 293
$\mathfrak{E}_y$ ideal, 294
$\mathcal{L}(V)$ linear hull, 294

Chapter 11

$q = p/(p-1)$ 305
$L_a^q(E)$ space of analytic functions, 305
$R^q(E)$ closure of rational functions, 305
$L_a^q(E)^\perp$ 306
$R^q(E)^\perp$ 306
$\bar{\partial}$ Cauchy–Riemann operator, 306
$\dot{W}_0^{1,p}(E)$ 309, 313
$f'(\infty)$ 311
$\gamma_q(K)$ analytic capacity, 311
$D^q(G)$, $D^q(E)$ spaces of harmonic functions, 312
$D^q(G)^\perp$ 313
Tr_E trace operator, 318
$L_0^{\alpha,p}(E)$ 318
$\mathcal{L}$ partial differential operator, 319
$L^q_{\mathcal{L}}(E)$ space of solutions to $\mathcal{L}u = 0$, 319
$R^q_{\mathcal{L}}(E)$ space of solutions to $\mathcal{L}u = 0$, 319
$R^q_{\mathcal{L}}(E)^\perp, L^q_{\mathcal{L}}(G)^\perp$ 320
$\mathfrak{P}_s$ space of polynomials of degree $\leq s$, 322
$N_{\alpha,p;\sigma}(E;\Omega)$ polynomial capacity, 322

Grundlehren der mathematischen Wissenschaften

A Series of Comprehensive Studies in Mathematics

A Selection

219. Duvaut/Lions: Inequalities in Mechanics and Physics
220. Kirillov: Elements of the Theory of Representations
221. Mumford: Algebraic Geometry I: Complex Projective Varieties
222. Lang: Introduction to Modular Forms
223. Bergh/Löfström: Interpolation Spaces. An Introduction
224. Gilbarg/Trudinger: Elliptic Partial Differential Equations of Second Order
225. Schütte: Proof Theory
226. Karoubi: K-Theory. An Introduction
227. Grauert/Remmert: Theorie der Steinschen Räume
228. Segal/Kunze: Integrals and Operators
229. Hasse: Number Theory
230. Klingenberg: Lectures on Closed Geodesics
231. Lang: Elliptic Curves. Diophantine Analysis
232. Gihman/Skorohod: The Theory of Stochastic Processes III
233. Stroock/Varadhan: Multidimensional Diffusion Processes
234. Aigner: Combinatorial Theory
235. Dynkin/Yushkevich: Controlled Markov Processes
236. Grauert/Remmert: Theory of Stein Spaces
237. Köthe: Topological Vector Spaces II
238. Graham/McGehee: Essays in Commutative Harmonic Analysis
239. Elliott: Probabilistic Number Theory I
240. Elliott: Probabilistic Number Theory II
241. Rudin: Function Theory in the Unit Ball of C^n
242. Huppert/Blackburn: Finite Groups II
243. Huppert/Blackburn: Finite Groups III
244. Kubert/Lang: Modular Units
245. Cornfeld/Fomin/Sinai: Ergodic Theory
246. Naimark/Stern: Theory of Group Representations
247. Suzuki: Group Theory I
248. Suzuki: Group Theory II
249. Chung: Lectures from Markov Processes to Brownian Motion
250. Arnold: Geometrical Methods in the Theory of Ordinary Differential Equations
251. Chow/Hale: Methods of Bifurcation Theory
252. Aubin: Nonlinear Analysis on Manifolds. Monge-Ampère Equations
253. Dwork: Lectures on ρ-adic Differential Equations
254. Freitag: Siegelsche Modulfunktionen
255. Lang: Complex Multiplication
256. Hörmander: The Analysis of Linear Partial Differential Operators I
257. Hörmander: The Analysis of Linear Partial Differential Operators II
258. Smoller: Shock Waves and Reaction-Diffusion Equations
259. Duren: Univalent Functions
260. Freidlin/Wentzell: Random Perturbations of Dynamical Systems
261. Bosch/Güntzer/Remmert: Non Archimedian Analysis – A System Approach to Rigid Analytic Geometry
262. Doob: Classical Potential Theory and Its Probabilistic Counterpart
263. Krasnosel'skiĭ/Zabreĭko: Geometrical Methods of Nonlinear Analysis
264. Aubin/Cellina: Differential Inclusions
265. Grauert/Remmert: Coherent Analytic Sheaves
266. de Rham: Differentiable Manifolds
267. Arbarello/Cornalba/Griffiths/Harris: Geometry of Algebraic Curves, Vol. I
268. Arbarello/Cornalba/Griffiths/Harris: Geometry of Algebraic Curves, Vol. II

269. Schapira: Microdifferential Systems in the Complex Domain
270. Scharlau: Quadratic and Hermitian Forms
271. Ellis: Entropy, Large Deviations, and Statistical Mechanics
272. Elliott: Arithmetic Functions and Integer Products
273. Nikol'skiĭ: Treatise on the Shift Operator
274. Hörmander: The Analysis of Linear Partial Differential Operators III
275. Hörmander: The Analysis of Linear Partial Differential Operators IV
276. Liggett: Interacting Particle Systems
277. Fulton/Lang: Riemann-Roch Algebra
278. Barr/Wells: Toposes, Triples and Theories
279. Bishop/Bridges: Constructive Analysis
280. Neukirch: Class Field Theory
281. Chandrasekharan: Elliptic Functions
282. Lelong/Gruman: Entire Functions of Several Complex Variables
283. Kodaira: Complex Manifolds and Deformation of Complex Structures
284. Finn: Equilibrium Capillary Surfaces
285. Burago/Zalgaller: Geometric Inequalities
286. Andrianaov: Quadratic Forms and Hecke Operators
287. Maskit: Kleinian Groups
288. Jacod/Shiryaev: Limit Theorems for Stochastic Processes
289. Manin: Gauge Field Theory and Complex Geometry
290. Conway/Sloane: Sphere Packings, Lattices and Groups
291. Hahn/O'Meara: The Classical Groups and K-Theory
292. Kashiwara/Schapira: Sheaves on Manifolds
293. Revuz/Yor: Continuous Martingales and Brownian Motion
294. Knus: Quadratic and Hermitian Forms over Rings
295. Dierkes/Hildebrandt/Küster/Wohlrab: Minimal Surfaces I
296. Dierkes/Hildebrandt/Küster/Wohlrab: Minimal Surfaces II
297. Pastur/Figotin: Spectra of Random and Almost-Periodic Operators
298. Berline/Getzler/Vergne: Heat Kernels and Dirac Operators
299. Pommerenke: Boundary Behaviour of Conformal Maps
300. Orlik/Terao: Arrangements of Hyperplanes
301. Loday: Cyclic Homology
302. Lange/Birkenhake: Complex Abelian Varieties
303. DeVore/Lorentz: Constructive Approximation
304. Lorentz/v. Golitschek/Makovoz: Construcitve Approximation. Advanced Problems
305. Hiriart-Urruty/Lemaréchal: Convex Analysis and Minimization Algorithms I. Fundamentals
306. Hiriart-Urruty/Lemaréchal: Convex Analysis and Minimization Algorithms II. Advanced Theory and Bundle Methods
307. Schwarz: Quantum Field Theory and Topology
308. Schwarz: Topology for Physicists
309. Adem/Milgram: Cohomology of Finite Groups
310. Giaquinta/Hildebrandt: Calculus of Variations I: The Lagrangian Formalism
311. Giaquinta/Hildebrandt: Calculus of Variations II: The Hamiltonian Formalism
312. Chung/Zhao: From Brownian Motion to Schrödinger's Equation
313. Malliavin: Stochastic Analysis
314. Adams/Hedberg: Function Spaces and Potential Theory
315. Bürgisser/Clausen/Shokrollahi: Algebraic Complexity Theory
316. Saff/Totik: Logarithmic Potentials with External Fields
317. Rockafellar/Wets: Variational Analysis
318. Kobayashi: Hyperbolic Complex Spaces
319. Bridson/Haefliger: Metric Spaces of Non-Positive Curvature
320. Kipnis/Landim: Scaling Limits of Interacting Particle Systems
321. Grimmett: Percolation
322. Neukirch: Algebraic Number Theory
323. Neukirch/Schmidt/Wingberg: Cohomology of Number Fields
324. Liggett: Stochastic Interacting Systems: Contact, Voter and Exclusion Processes

Printing: Mercedes-Druck, Berlin
Binding: Stürtz AG, Würzburg